Probability and its Applications

A Series of the Applied Probability Trust

Springer
New York
Berlin
Heidelberg
Barcelona
Hong Kong
London
Milan
Paris
Singapore
Tokyo

Probability and its Applications

Anderson: Continuous-Time Markov Chains.
Azencott/Dacunha-Castelle: Series of Irregular Observations.
Bass: Diffusions and Elliptic Operators.
Bass: Probabilistic Techniques in Analysis.
Choi: ARMA Model Identification.
de la Peña/Giné: Decoupling: From Dependence to Independence.
Galambos/Simonelli: Bonferroni-type Inequalities with Applications.
Gani (Editor): The Craft of Probabilistic Modelling.
Grandell: Aspects of Risk Theory.
Gut: Stopped Random Walks.
Guyon: Random Fields on a Network.
Kallenberg: Foundations of Modern Probability.
Last/Brandt: Marked Point Processes on the Real Line.
Leadbetter/Lindgren/Rootzén: Extremes and Related Properties of Random Sequences and Processes.
Nualart: The Malliavin Calculus and Related Topics.
Rachev/Rüschendorf: Mass Transportation Problems. Volume I: Theory.
Rachev/Rüschendorf: Mass Transportation Problems. Volume II: Applications.
Resnick: Extreme Values, Regular Variation and Point Processes.
Shedler: Regeneration and Networks of Queues.
Todorovic: An Introduction to Stochastic Processes and Their Applications.

Víctor H. de la Peña   Evarist Giné

Decoupling

From Dependence to Independence

Randomly Stopped Processes
U-Statistics and Processes
Martingales and Beyond

Springer

Víctor H. de la Peña
Department of Statistics
Columbia University
New York, NY 10027
USA

Evarist Giné
Department of Mathematics
University of Connecticut, Storrs
Storrs, CT 06269
USA

Mathematics Subject Classifications (1991): 60E15, 60G42, 60G50, 60G40, 60F05, 60F17, 60J30, 60F15, 62E20

Library of Congress Cataloging-in-Publication Data
de la Peña, Víctor.
Decoupling : from dependence to independence / Víctor de la Peña, Evarist Giné.
p. cm. — (Probability and its applications)
Includes bibliographical references and index.
ISBN 0-387-98616-2 (hc : alk. paper)
1. Decoupling (Mathematics) I. Giné, Evarist, 1944– . II. Title. III. Series: Springer series in statistics. Probability and its applications.
QA295.D355 1999
515′.243—dc21 98-30322

Printed on acid-free paper.

Production managed by A. Orrantia; manufacturing supervised by Jeffrey Taub.
Typeset by TeX Consultants, Three Rivers, CA.
Printed and bound by Maple-Vail Book Manufacturing Group, York, PA.
Printed in the United States of America.

9 8 7 6 5 4 3 2 1

ISBN 0-387-98616-2 Springer-Verlag New York Berlin Heidelberg SPIN 10689929

A mis padres, Ernesto y Carmen
To my Teacher, Michael Klass
(V.H. de la P.)

A la meva muller, Rosalind; filles, Núria i Roser;
i pares, Joan i Maria
(E.G.)

Preface

This book presents the theory and several applications of the decoupling principle, which provides a general approach for handling complex problems involving dependent variables. Its main tools consist of inequalities used for breaking (decoupling) the dependence structure in a broad class of problems by introducing enough independence so that they can be analyzed by means of standard tools from the theory of independent random variables.

Since decoupling reduces problems on dependent variables to problems on related (conditionally) independent variables, we begin with the presentation of a series of results on sums of independent random variables and (infinite-dimensional) vectors, which will be useful for analyzing the decoupled problems and which at the same time are tools in developing the decoupling inequalities. These include several recent definitive results, such as an extension of Lévy's maximal inequalities to independent and identically distributed but not necessarily symmetric random vectors, the Khinchin–Kahane inequality (Khinchin for random vectors) with best constants, and sharp decompositions of the L_p norm of a sum of independent random variables into functions that depend on their marginals only. A consequence of the latter consists of the first decoupling result we present, namely, comparing the L_p norms of sums of arbitrary positive random variables or of martingale differences with the L_p norms of sums of independent random variables with the same (one-dimensional) marginal distributions. With a few subjects, such as Hoffmann-Jørgensen's inequality, we compromise between sharpness and expediency and take a middle, practical road.

Concerning decoupling itself, we choose to introduce it by developing in great detail two of the main areas where it has been most successfully applied. These are 1) randomly stopped sums of independent random variables and processes with independent increments and 2) U-statistics and U-processes. There are two main reasons for starting with these "particular cases" rather than with the general theory: on the one hand, these examples motivate very clearly the general theory and, on the other hand, the general theory does not and cannot imply the strongest results in these two important areas.

The effect of decoupling on randomly stopped sums of independent random variables (and processes with independent increments) consists in creating independence between the stopping time and the variables so that the stopped sum can be treated conditionally as a sum of a fixed number of independent random variables. This is done for Banach space valued variables (sums) and processes. These results constitute striking generalizations of Wald's equation. A special case of the result on stopped processes consists of an extension of the Burkholder–Gundy inequality for randomly stopped Brownian motion to Banach valued processes with independent increments. An advantage of having these results in Banach spaces is that they apply also to some real valued processes that do not have independent increments, such as Bessel processes, by realizing them as functionals of Banach valued independent increments processes. Another advantage is that the constants involved in the approximations are independent of dimension. The decoupling results are then applied to the study of the first passage time for the absolute value (or the norm, in the Banach case) of a process with independent increments by developing a natural, yet surprising connection with boundary crossing by nonrandom functions, with applications that highlight the relevance of working in a Banach setting.

An important area of applications of the decoupling principle is the theory of U-statistics. Such statistics arise in the definition of unbiased estimators, including the sample variance of a sequence of i.i.d. random variables, and as higher-order terms in von Mises expansions of smooth statistical functionals. A U-statistic is the average value over the sample $X_1, \ldots, X_n$ of a function of several variables h; so, it involves a multiple sum of $m!\binom{n}{m}$ terms $h(X_{i_1}, \ldots, X_{i_m})$. Decoupling reduces the U-statistic to an average of terms of the form $h(X^1_{i_1}, \ldots, X^m_{i_m})$, where each entry of h (say, the jth entry) is filled up with the terms of a different, independent copy of the original sequence of random variables (say, $\{X^j_i : i = 1, \ldots, n\}$). This produces enough independence so that this average, conditioned on all but one of the independent sequences, becomes a sum of independent random variables. Decoupling is very different from Hoeffding's decompositions and constitutes an additional extremely powerful tool for analyzing U-statistics and processes. For instance, if the kernel has conditional mean zero, then decoupling allows for symmetrization and randomization, which are tools used to freeze the variables involved and/or reduce the problems to ones involving a weighted sum of Bernoulli random variables. Decoupling has played a central role in recent advances in the asymptotic theory of U-statistics, and has produced, among others, optimal results on the central limit theorem and very sharp results on the law of the iterated logarithm and on exponential inequalities. It has had even a more pivotal role in the development of the theory of U-processes, the analogue of empirical processes for U-statistics. We present a rather complete account of the asymptotic theory of U-statistics and U-processes, as well as a few statistical applications of the latter, e.g., to multidimensional M-estimators (including analogues of the sample median in several dimensions), and to the analysis of truncated data. As part of this study, we give a unified account of the construction of the chaos decomposition of the L_2 space of a Gaussian process and the proof of the central limit theorem for degenerate U-statistics.

In fact, we consider decoupling of a generalized form of U-statistics, with the kernel h varying with the multiindex $(i_1, \ldots, i_m)$. In this generality, the results presented apply as well to multilinear forms in independent random variables, which constitute one of the first objects to which decoupling was applied, motivated by multiple stochastic integration. A generalization of another historically important decoupling result comparing tail probabilities of Gaussian polynomials also follows in a straightforward way by combining decoupling of U-statistics with the central limit theorem.

The latter part of the book is devoted to the general theory of decoupling. More specifically, consider an arbitrary sequence $\{d_i\}$ of real random variables and let $\mathcal{F}_i$ be an increasing sequence of σ-fields to which it is adapted (we can take $\mathcal{F}_i$ to be the σ-field generated by $d_1, \ldots, d_i$). Let $\mathcal{G}$ be another σ-field contained in $\mathcal{F}_\infty$. Then, a sequence of random variables $\{e_i\}$ is a decoupled version of $\{d_i\}$ (with respect to $\{\mathcal{F}_i\}$ and $\mathcal{G}$), if i) $\mathcal{L}(e_i|\mathcal{F}_{i-1}) = \mathcal{L}(d_i|\mathcal{F}_{i-1})$, ii) the sequence $\{e_i\}$ is conditionally independent given $\mathcal{G}$, and iii) $\mathcal{L}(e_i|\mathcal{F}_{i-1}) = \mathcal{L}(e_i|\mathcal{G})$. Condition i) indicates proximity between the two sequences and allows for comparison of some of their characteristics such as moments of sums or maxima, etc. Sequences related by condition i) are said to be tangent to each other. Condition ii) expresses the fact that the sequence $\{e_i\}$ is more independent than the sequence $\{d_i\}$, and iii) allows for transfer to the original sequence $\{d_i\}$ of properties of the sequence $\{e_i\}$ related to its conditional independence given $\mathcal{G}$. Conditions ii) and iii) together are known as the C.I. (conditional independence) condition. Then, a general decoupling result is simply an inequality relating $\mathbb{E}\Phi(\sum d_i)$ and $\mathbb{E}\Phi(\sum e_i)$, where $\Phi(x)$ could be $|x|^p$, $\exp|x|^\alpha$, or $I_{x>t}$, in order to compare moments, exponential moments, or even the distributions of the sums; also, sums can be replaced by other functionals, such as maxima. It is important to stress the fact that decoupled sequences always exist and therefore, decoupling inequalities in this general context have a broad appeal as they always apply (the drawback is that they are not always useful). Moment inequalities hold in great generality and exponential moment inequalities hold also quite generally, but tail probability inequalities, which do hold in the case of Banach space valued U-statistics, do not hold in general. There are other types of inequalities such as, e.g., comparison of weak moments.

There is a very close connection between decoupling inequalities in this general setting and martingale inequalities. For instance, it can be shown that the square function martingale inequality of Burkholder–Davis–Gundy is equivalent to a decoupling inequality for martingales. There is also interplay between martingale and decoupling inequalities: for example, the Burkholder–Rosenthal inequality for martingale differences implies a decoupling inequality, and a sharp decoupling inequality paired with a sharp version of Rosenthal's inequality for sums of independent variables implies the Burkholder–Rosenthal inequality with best constants. Also, it is possible to give unified proofs of martingale inequalities and decoupling inequalities for conditionally symmetric sequences. These relationships are explored in depth in the chapters on decoupling of tangent sequences.

We also present the principle of conditioning, which is a general method for obtaining limit (in distribution and almost everywhere) results for sums of dependent variables based on analogous results for sums of independent variables. As applications we give a proof of the Brown–Eagleson central limit theorem for martingales by applying the principle of conditioning along with the Lindeberg–Feller central limit theorem for sums of independent random variables. As another consequence of this result we provide a proof of the central limit theorem for a sequence of (arbitrarily dependent) two by two tables. This result is relevant in the theory of biostatistics and provides a situation in which martingale methods do not seem to apply but the decoupling approach succeeds. Other applications of the general theory of decoupling that we present in detail include a general method for extending exponential inequalities for sums of independent variables to the ratio of a martingale over its conditional variance, an extension of Wald's equation to U-statistics, estimation of moments of randomly stopped U-statistics and an extension to U-statistics of Anscombe's theorem, convergence of moments included.

The decoupling approach to handling problems with dependent random variables can be traced back to a result of Burkholder and McConnell included in Burkholder (1983) which represents a step in extending the theory of martingales to Banach spaces. Therefore, it can be said that decoupling (for tangent sequences) was born as a natural continuation to the martingale approach in order to handle problems that traditionally could not be handled by means of martingale tools. A typical inequality for martingales due to Burkholder and Gundy compares a martingale to its square function, which is the square root of the sum of squares of its martingale difference sequence. This in effect transforms a problem involving martingales into one involving sums of non-negative variables and provides sufficient advantage in developing solutions to the problem in case. The idea of replacing the square function of a martingale by a decoupled (conditionally independent) version of the martingale was proposed in order to avoid problems with the definition of a square function in a Banach space. The first general decoupling inequality for tangent sequences was obtained by Zinn (1985) and extended by Hitczenko (1988).

A turning point in the theory of decoupling for tangent sequences has been Kwapień and Woyczynski (1991) (available as a preprint in 1986). It is shown in this paper that one can always obtain a decoupled tangent sequence to any adapted sequence, hence making general decoupling inequalities widely applicable. This work also develops the theory of semimartingale integrals via decoupling inequalities. Decoupling has been quite effective in treating problems involving U-statistics, multilinear forms and randomly stopped sums of independent random variables. Decoupling of multilinear forms was first considered by McConnell and Taqqu (1986) with a view towards the development of a theory of multiple stochastic integration. Their article, which precedes Zinn's and Kwapień and Woyczynski's, provided great impetus to the theory. Concerning randomly stopped sums, Klass (1988, 1990) obtained definitive decoupling results for variables in general Banach spaces. Kwapień and Woyczynski (1992) contains the development of the theory up to that point, including several L_p and tail probability inequalities, and uses decoupling to develop a general theory of stochastic integration. Building upon Kwapień's (1987) extension of McConnell and Taqqu's result, de la Peña (1992) further extended decoupling to a general class of random variables that contain both multilinear forms and U-statistics with values in general Banach spaces, hence providing a springboard to a wealth of results, initiated by Arcones and Giné (1993), on the general theory of U-statistics and more generally U-processes, the latter introduced by Nolan and Pollard (1987). Kwapień (1987) and Kwapień and Woyczynski (1992) proved the first tail probability decoupling inequalities for quadratic forms and multilinear forms of independent random variables. Giné and Zinn (1994) obtained a decoupling and symmetrization inequality for U-statistics. The definitive decoupling result along this line of work is de la Peña and Montgomery-Smith (1994), which provides a tail probability decoupling inequality for generalized U-statistics. Concerning the general theory of decoupling, recent developments include work of de la Peña (1994), with the first general exponential decoupling inequality, and of Hitczenko (1994), who extended this result by providing L_p inequalities with constants independent of p. A more detailed account of the history of the developments of decoupling can be found at the end of each chapter.

This book is addressed to researchers in Probability and Statistics and to advanced graduate students. Thus, the exposition is at the level of a second graduate course. For instance, we do not include a proof of Doob's maximal inequality, but we do include one for the Burkholder–Davis–Gundy inequality. This text contains as well a self-contained section on weak convergence of processes, sufficient for the study of U-processes. Except for relying on material from standard first year graduate courses, we only occasionally refer the reader to material not presented in this book. We have successfully incorporated some of the material from this book in our first year graduate probability courses, including Levy's and Hoffmann-Jørgensen's inequalities, the development of Wald's equations, exponential inequalities, some of the decoupling inequalities for U-statistics, which require only basic facts about conditional expectation and conditional Jensen's inequality, and several applications to the asymptotic theory of U-statistics.

The content of the book divides naturally into four parts. 1) Chapter 1, on sums of independent random variables; 2) Chapter 2, on randomly stopped sums and processes; 3) Chapters 3–4–5 on U-statistics and U-processes; and 4) Chapters 6–7–8 on the general theory of decoupling, with applications. The last three parts can be read independently of each other. On the other hand, the material in Chapter 1 is used in each of the other parts, occassionally in Chapters 2, 6, 7, and 8, but more often in Chapters 3, 4, and 5.

In the process of writing this book we benefited from past and present discussions with several people, including D. Alemayehu, M. Arcones, D. Burkholder, R. Dudley, N. Eisenbaum, J. Engel, P. Hitczenko, Z. Jin, M. Klass, S. Kou, S. Kwapień, R. Latała, S. Montgomery-Smith, D. Pollard, J. Szulga, J. Wellner, M. Yang, C. H. Zhang, and J. Zinn.

We would like to thank Doodmatie Kalicharan for excellent typing of part of this book.

New York
Storrs
April 1998

Víctor Hugo de la Peña Díaz Infante
Evarist Giné i Masdéu

Contents

1

Sums of Independent Random Variables

The theory of decoupling aims at reducing the level of dependence in certain problems by means of inequalities that compare the original sequence to one involving independent random variables. It is therefore important to have information on results dealing with functionals of independent random variables.

In this chapter we collect several types of results on sums of independent random variables that will be used throughout. We consider aspects of estimation of tail probabilities and moments that are relevant to the theory of decoupling and develop them to the extent needed, and, in a few instances, a little more.

We begin with the classical Lévy maximal inequalities, bounding the tail probabilities of the maximum of the norm of a sum of independent symmetric random vectors by the tail probabilities of the norm of the last sum, that is, the reflection principle for symmetric random walk extended to random variables taking values in a Banach space. Then, we also present analogous maximal inequalities for sums of arbitrary independent identically distributed random vectors. The proofs in the Banach space case are not more difficult than for the real case.

A way to prove integrability for (classes of) random variables is to obtain bounds for tail probabilities in terms of the squares of these same probabilities at lower levels. This is illustrated by the Hoffmann-Jørgensen type inequalities that we present in Section 2, which bound the pth moment of a sum of independent centered random vectors by a constant times the same moment of their maximum plus the pth power of a quantile. They are important as a means of upgrading stochastic boundedness (or weak convergence) of sequences of variables to uniform boundedness (or to convergence) of their moments. This very useful type of inequality originates with Kolmogorov's converse to his maximal inequality. As an application, we present a kind of reversed Jensen's inequality for exponentials. Other instances of the use of these inequalities can be found in subsequent chapters.

Next, we come to estimation of moments, starting with Khinchin's inequalities (Section 3). In their crudest form these inequalities assert that on the span of a Rademacher sequence all the L_p norms are equivalent. The inequality comparing the L_1 and L_2 norms is proved here for Rademacher linear combinations with coefficients in a Banach space (Khinchin–Kahane inequality) and with best constant. Extension to all moments and to Rademacher chaos is done in Chapter 3. This will be a basic ingredient in the asymptotic theory of U-statistics.

Finally, we consider the question of finding two-sided bounds for the L_p norm of a sum of independent random variables in terms of quantities that involve only one-dimensional integrals with respect to the probability laws of the individual summands. This is the subject of the last two sections, where several approaches are developed. Hoffmann-Jørgensen's inequality together with a quite precise estimate of the moments of the maximum of independent variables is used in one of the approaches (which carries to infinite dimensions), whereas the L function bounds, which constitute the most precise approach, do imply Rosenthal's and Hoffmann-Jørgensen's inequalities in $\mathbb{R}$ with essentially best constants. The K function approach, which was chronologically the first, is also briefly discussed. We present three applications of these inequalities. One, already mentioned, is Rosenthal's and Hoffmann-Jørgensen's inequalities for real variables with constants of the best order. Another, computes, up to multiplicative constants independent of p, the L_p norm of linear combinations of Rademacher variables. The third compares moments of sums of arbitrary positive random variables and martingales to moments of sums of independent positive and/or centered variables with the same individual distributions; these inequalities constitute an example of decoupling inequalities.

The developments just described require certain lemmas on truncation, randomization, etc., that are elementary but quite useful.

1.1 Lévy-Type Maximal Inequalities

This section is devoted to the extension of the classical Lévy inequalities for sums of independent symmetric random vectors to sums of not necessarily symmetric, but i.i.d., random vectors, possibly with different constants. These inequalities hold in great generality but, in order to avoid measurability considerations we assume the variables take values in a *separable* Banach space B.

For completeness sake, we begin with Lévy's inequalities, that are used all over this text.

THEOREM 1.1.1. *Let X_i, $1 \le i \le n$, be independent symmetric B-valued random variables. Then, for every $t > 0$,*

$$\Pr\Big\{\max_{1\le k\le n} \Big\|\sum_{i=1}^{k} X_i\Big\| > t\Big\} \le 2\Pr\Big\{\Big\|\sum_{i=1}^{n} X_i\Big\| > t\Big\} \tag{1.1.1}$$

and

$$\Pr\Big\{\max_{1\le k\le n} \|X_k\| > t\Big\} \le 2\Pr\Big\{\Big\|\sum_{i=1}^{n} X_i\Big\| > t\Big\}. \tag{1.1.2}$$

In particular,

$$\mathbb{E}\Big(\max_{1\le k\le n} \Big\|\sum_{i=1}^{k} X_i\Big\|\Big)^p \le 2\mathbb{E}\Big\|\sum_{i=1}^{n} X_i\Big\|^p, \qquad \mathbb{E}\Big(\max_{1\le k\le n} \|X_k\|\Big)^p \le 2\mathbb{E}\Big\|\sum_{i=1}^{n} X_i\Big\|^p$$

for all $p > 0$.

PROOF. We set $S_k = \sum_{i=1}^k X_i$, $k = 1, \dots, n$. The sets

$$A_k := \big\{\|S_i\| \le t \text{ for } 1 \le i \le k-1,\ \|S_k\| > t\big\}$$

are disjoint and

$$\Big\{\max_{1\le k\le n} \|S_k\| > t\Big\} = \cup_{k=1}^n A_k.$$

(A_k is the event "the random walk S_i exceeds the level t for the first time at time k.") For each $k \le n$ we define

$$S_n^{(k)} := S_k - X_{k+1} - \cdots - X_n$$

and note that, by symmetry and independence, the joint probability law of the n variables $(X_1, \dots, X_n)$ is the same as that of $(X_1, \dots, X_k, -X_{k+1}, \dots, -X_n)$, so that S_n and $S_n^{(k)}$ both have the same law. On the one hand, we obviously have that

$$\Pr\big[A_k \cap \{\|S_n\| > t\}\big] = \Pr\big[A_k \cap \{\|S_n^{(k)}\| > t\}\big],$$

and on the other hand,

$$A_k = \big[A_k \cap \{\|S_n\| > t\}\big] \cup \big[A_k \cap \{\|S_n^{(k)}\| > t\}\big]$$

since otherwise there would exist $\omega \in A_k$ such that

$$2\|S_k(\omega)\| = \|S_n(\omega) + S_n^{(k)}(\omega)\| \le 2t,$$

a contradiction with the definition of A_k. The last two identities imply that

$$\Pr A_k \le 2\Pr\big[A_k \cap \{\|S_n\| > t\}\big],\ k = 1, \dots, n,$$

and therefore,

$$\Pr\Big\{\max_{1\le k\le n}\|S_k\|>t\Big\}=\sum_{k=1}^{n}\Pr A_k\le 2\sum_{k=1}^{n}\Pr\big[A_k\cap\{\|S_n\|>t\}\big]$$
$$\le 2\Pr\big\{\|S_n\|>t\big\},$$

which gives inequality (1.1.1). The second inequality is proved in the same way if we redefine A_k as

$$A_k := \{\|X_i\|\le t \text{ for } 1\le i\le k-1,\ \|X_k\|>t\}$$

and $S_n^{(k)}$ as

$$S_n^{(k)} := -X_1-\cdots-X_{k-1}+X_k-X_{k+1}-\cdots-X_n.$$

The statements about expected values follow from (1.1.1) and (1.1.2) by integation by parts ($\int|\xi|^p dP = p\int t^{p-1}\Pr\{|\xi|>t\}dt$). □

If the random vectors are not symmetric, we have the following weaker inequality:

PROPOSITION 1.1.2. *Let X_i, $i\le n$, be independent B-valued random variables. Then, for all $t\ge 0$,*

$$\Pr\Big\{\max_{1\le k\le n}\Big\|\sum_{i=1}^{k}X_i\Big\|>t\Big\}\le 3\max_{k\le n}\Pr\Big\{\Big\|\sum_{i=1}^{k}X_i\Big\|>\frac{t}{3}\Big\}. \tag{1.1.3}$$

PROOF. Almost as in the previous proof, we define, for all $u, v\ge 0$ and $1\le k\le n$, $A_k=\big\{\|S_i\|\le u+v$ for $i<k$, and $\|S_k\|>u+v\big\}$. The sets A_k are disjoint and their union is $\big\{\max_{1\le k\le n}\|S_k\|>u+v\big\}$. Therefore,

$$\Pr\{\|S_n\|>u\}\ge\Pr\Big\{\|S_n\|>u,\ \max_{1\le k\le n}\|S_k\|>u+v\Big\}$$
$$\ge\sum_{k=1}^{n}\Pr\big\{A_k\cap\{\|S_n-S_k\|\le v\}\big\}$$
$$=\sum_{k=1}^{n}\Pr\{A_k\}\Pr\{\|S_n-S_k\|\le v\}$$
$$\ge\Big[1-\max_{k\le n}\Pr\{\|S_n-S_k\|>v\}\Big]\Pr\Big\{\max_{1\le k\le n}\|S_k\|>u+v\Big\}.$$

This is the well-known Lévy–Ottaviani inequality. Taking $u=t/3$ and $v=2t/3$ in this inequality gives

$$\Pr\Big\{\max_{1\le k\le n}\|S_k\|>t\Big\}\le\frac{\Pr\{\|S_n\|>t/3\}}{1-\max_{k\le n}\Pr\{\|S_n-S_k\|>2t/3\}}$$
$$\le\frac{\max_{k\le n}\Pr\{\|S_k\|>t/3\}}{1-2\max_{k\le n}\Pr\{\|S_k\|>t/3\}}.$$

This proves inequality (1.1.3) if $\max_{k\le n}\Pr\{\|S_k\|>t/3\}<1/3$. But (1.1.3) is trivially satisfied otherwise. □

The above two inequalities are classical. Next we will extend Lévy's inequality to sequences of random vectors which are not necessarily symmetric, but which are i.i.d. The crucial point for these extension consists of the following theorem.

THEOREM 1.1.3. *If X_i, $i \in \mathbb{N}$, are independent identically distributed B-valued random variables, then, for $1 \le j \le k < \infty$,*

$$\Pr\Big\{\Big\|\sum_{i=1}^{j} X_i\Big\| > t\Big\} \le 3\Pr\Big\{\Big\|\sum_{i=1}^{k} X_i\Big\| > t/10\Big\}. \tag{1.1.4}$$

For $k = 2$ this theorem has a surprisingly simple proof: Let X, Y, Z, be i.i.d. Then,

$$\begin{aligned}\Pr\big\{\|X\| > t\big\} &= \Pr\big\{\|(X+Y)+(X+Z)-(Y+Z)\| > 2t\big\}\\ &\le 3\Pr\big\{\|X+Y\| > 2t/3\big\}.\end{aligned} \tag{1.1.4$'$}$$

The general case is more delicate. Its proof rests on the lemma that follows. First, an auxiliary definition: we say that x is a *t-concentration point* for the random vector X if $\Pr\{\|X - x\| > t\} < 1/3$.

LEMMA 1.1.4. *Let X_i, $i \in \mathbb{N}$, be i.i.d. random vectors. If $S_j = \sum_{i=1}^{j} X_i$ has a t-concentration point s_j for $1 \le j \le k$, then*

$$\|ks_j - js_k\| \le 3(k+j)t. \tag{1.1.5}$$

PROOF. First we observe that for X and Y arbitrary, if x is a t-concentration point for X, y is a t-concentration point for Y and z is a t-concentration point for $X + Y$ then

$$\|x + y - z\| \le 3t. \tag{1.1.6}$$

To see this just note

$$\begin{aligned}\Pr\{\|x+y-z\| > 3t\} &= \Pr\{\|X - x + Y - y - (X + Y - z)\| > 3t\}\\ &\le \Pr\{\|X-x\| > t\} + \Pr\{\|Y-y\| > t\} + \Pr\{\|X+Y-z\| > t\} < 1,\end{aligned}$$

so that $\Pr\{\|x+y-z\| \le 3t\} > 0$ and therefore (1.1.6) holds since x, y, and z are nonrandom. To prove the lemma we now proceed by induction. The lemma obviously holds for $j = k$, and (1.1.6) gives it for $k = 2$. Hence, it suffices to show that if the lemma holds for $1 \le j < r$ for all $r < k$, then it also holds for $1 \le j < k$. Now,

$$js_k - ks_j = js_k - (k-j)s_j - js_j = (js_{k-j} - (k-j)s_j) + j(s_k - s_j - s_{k-j}).$$

Hence, applying (1.1.6) and the induction hypothesis, we obtain

$$\begin{aligned}\|js_k - ks_j\| &\le \|js_{k-j} - (k-j)s_j\| + j\|s_k - s_{k-j} - s_j\|\\ &\le 3(k-j+j)t + 3jt = 3(k+j)t.\end{aligned}$$

□

PROOF OF THEOREM 1.1.3. We distinguish three cases. Suppose first that $\Pr\{\|S_{k-j}\| > 9t/10\} \le 1/3$. Then, independence of S_j and $S_k - S_j$, together with the fact that S_{k-j} and $S_k - S_j$ have the same distribution, give

$$\Pr\big\{\|S_j\| > t\big\} \le \Pr\big\{\|S_j\| > t,\ \|S_k - S_j\| \le 9t/10\big\} + \frac{1}{3}\Pr\big\{\|S_j\| > t\big\},$$

and therefore

$$\Pr\big\{\|S_j\| > t\big\} \le \frac{3}{2}\Pr\big\{\|S_j\| > t,\ \|S_k - S_j\| \le 9t/10\big\} \le \frac{3}{2}\Pr\big\{\|S_k\| > t/10\big\}.$$

Next we asume that there exists some $1 \le i \le k$ such that S_i does not have any $(t/10)$-concentration points. Then

$$\Pr\Big\{\|S_i + X_{i+1} + \cdots + X_k\| > \frac{t}{10}\,\Big|\, X_{i+1},\ldots,X_k\Big\} \ge \frac{1}{3},$$

and therefore

$$\Pr\big\{\|S_k\| > t/10\big\} \ge \frac{1}{3} \ge \frac{1}{3}\Pr\{\|S_j\| > t\}$$

for all $1 \le j \le k$.

Suppose, finally, that $\Pr\big\{\|S_{k-j}\| > 9t/10\big\} > 1/3$ and that S_i has a $(t/10)$-concentration point s_i for all $1 \le i \le k$. Then

$$\{\|S_{k-j}\| > 9t/10\} \cap \{\|S_{k-j} - s_{k-j}\| \le t/10\} \ne \emptyset$$

and therefore, $\|s_{k-j}\| \ge 4t/5$. Hence, by Lemma 1.1.4,

$$\|s_k\| \ge \frac{k}{k-j}\|s_{k-j}\| - 3\frac{2k-j}{k-j}\frac{t}{10} \ge \frac{4kt}{5(k-j)} - \frac{6kt}{10(k-j)} > \frac{t}{5}.$$

This gives

$$\Pr\big\{\|S_k\| \ge t/10\big\} \ge \Pr\big\{\|S_k - s_k\| \le t/10\big\} \ge \frac{2}{3} \ge \frac{2}{3}\Pr\{\|S_j\| > t\}. \qquad \square$$

Combining Theorem 1.1.3 and Proposition 1.1.2 we readily obtain an analog of Lévy's inequality for sums of i.i.d random variables.

THEOREM 1.1.5. *If X_i, $i \in \mathbb{N}$, are independent identically distributed B-valued random variables, then, for $1 \le k \le n < \infty$ and all $t > 0$,*

$$\Pr\Big\{\max_{1\le k\le n}\Big\|\sum_{i=1}^{k} X_i\Big\| > t\Big\} \le 9\Pr\Big\{\Big\|\sum_{i=1}^{n} X_i\Big\| > \frac{t}{30}\Big\}. \tag{1.1.7}$$

Theorems 1.1.3 and 1.1.5, which are quite recent (see the notes at the end of this chapter), constitute an important addition to the theory of sums of i.i.d. variables. They have consequences for decoupling, to be seen in Chapter 3. To immediately illustrate their usefulness, we end this section with an application to a *contraction principle* and to *randomization* for sums of i.i.d. random vectors. For this result (and only for it) we assume that the Banach space B where the variables lie is over the complex numbers.

COROLLARY 1.1.6. *There are universal constants $0 < c_1, c_2 < \infty$ such that if X_i, $i \in \mathbb{N}$, are either i.i.d. or independent and symmetric, and if α_i, $1 \le i \le n$, are any complex numbers such that $|\alpha_i| \le 1$, then, for all $n \in \mathbb{N}$ and $t \ge 0$,*

$$\Pr\Big\{\Big\|\sum_{i=1}^{n} \alpha_i X_i\Big\| > t\Big\} \le c_1 \Pr\Big\{\Big\|\sum_{i=1}^{n} X_i\Big\| > c_2 t\Big\}. \tag{1.1.8}$$

PROOF. Let k, $1 \le k \le n$, be fixed. The constants α_i, $1 \le i \le k$, can be assumed to be real and in decreasing order $-1 \le \alpha_k \le \cdots \le \alpha_1 \le 1$. So, we can write $\alpha_j = -1 + \sum_{i=j}^{k} \sigma_i$ for all $j \le k$, where $\sigma_i \ge 0$ for all $i \le k$ and $\sum_{i=1}^{k} \sigma_i \le 2$. Then,

$$\begin{aligned}
\Big\|\sum_{j=1}^{k} \alpha_j X_j\Big\| &= \Big\|\sum_{j=1}^{k}\Big(\sum_{i=j}^{k} \sigma_i - 1\Big)X_j\Big\| \\
&= \Big\|\sum_{j=1}^{k} \sigma_j \sum_{i=1}^{j} X_i - \sum_{i=1}^{k} X_i\Big\| \\
&\le \Big(\sum_{j=1}^{k} \sigma_j\Big) \max_{1\le j\le k}\Big\|\sum_{i=1}^{j} X_i\Big\| + \Big\|\sum_{i=1}^{k} X_i\Big\| \\
&\le 3 \max_{1\le j\le k}\Big\|\sum_{i=1}^{j} X_i\Big\|.
\end{aligned}$$

Now the result follows from Theorem 1.1.5 in the i.i.d. case and from Theorem 1.1.1 in the symmetric case. □

REMARK 1.1.7. *Randomization.* It is clear (Fubini's theorem) that the constants α_i in the previous Corollary can be replaced by random variables θ_i uniformly bounded by 1 and independent from the sequence $\{X_j\}$.

REMARK 1.1.8. *Measurability.* The previous propositions can also be proved in the following more general context: The variables X_i are the coordinates of a product probability space $(S^{\mathbb{N}}, \mathcal{S}^{\mathbb{N}}, P^{\mathbb{N}})$, the norm $\|\cdot\|$ is the sup over a not necessarily countable family $\mathcal{F}$ of measurable functions on S and Pr is replaced by outer probability $\Pr^*$. The measurable outer envelope $\|\cdot\|_{\mathcal{F}}^*$ of the sup norm over $\mathcal{F}$ works essentially as a norm and $\Pr^*\{\|\cdot\|_{\mathcal{F}} > t\} = \Pr\{\|\cdot\|_{\mathcal{F}}^* > t\}$ for all $t \ge 0$, and this is essentially all that is needed to show that the above proofs work with only formal changes in the setting just decribed. A good available reference for calculus with outer probabilities is van der Vaart and Wellner (1996). We skip the details, but anticipate that this more general context is the natural one for U-processes, to be discussed in Chapter 4.

1.2 Hoffmann-Jørgensen Type Inequalities

The classical Kolmogorov converse maximal inequality for sums of centered uniformly bounded random variables X_i bounds the second moment of S_n in terms of the common bound of the variables and a quantile of $\max_{1\le k\le n}|S_k|$ (or a quantile of $|S_n|$ in the symmetric or in the i.i.d. cases by the Lévy-type inequalities in Section 1.1). This inequality is basic in the proof of the three-series theorem, to show that convergence of a series of independent variables implies convergence of the series of truncated variances. The converse Kolmogorov inequality basically extends to unbounded variables, with the place of the uniform bound taken in this case by the second moment of the maximum of the variables. Also, the variables can be Banach valued and the second moment can be replaced by the pth moment for any p. The resulting inequalities go by the name of Hoffmann-Jørgensen or Hoffmann-Jørgensen type inequalities. In this section we present versions of these inequalities which allow comparison of different moments; although the dependence on p of the multiplicative constants involved is very important, we try to reach a balance between difficulty of proofs and usefulness (for us here) and as a result the constants we obtain miss optimality by a factor of $\log p$ as $p \to \infty$ with q fixed (e.g., in Theorem 1.2.5). These inequalities are extremely useful for extracting uniform integrability from weak convergence of sums of independent random variables (somewhat as in the three-series theorem, but without truncation); however, this most important application will not be pursued here. We will develop instead two interesting (but somewhat more special) consequences that will be put to use in some of the next chapters. One is an exponential inequality that can be interpreted as a sort of reverse Jensen's inequality for the exponential of a sum of independent variables. The other one (postponed to Section 1.4) consists of a sort of moment comparison, which, in the case of Hilbert space valued random variables, yields equivalent upper and lower estimates of moments of sums of independent random vectors in terms of integrals of functions of the individual summands only, a subject to be considered in detail in the last two sections of this chapter.

We emphasize that the main ingredient for the inequalities in this section is a bound for the tail probabilities of $\max_{1\le k\le n}|S_k|$ in terms of the square of the tail probabilities of the same variable at a smaller level and the tail probabilities of $\max_{1\le k\le n}|X_k|$.

Besides the notation

$$S_k = \sum_{i=1}^{k} X_i, \qquad k \in \mathbb{N},$$

already used in the previous section, it will be convenient to employ the notations

$$X_n^* := \max_{1\le k\le n}\|X_k\|, \qquad S_n^* := \max_{1\le k\le n}\|S_k\|, \quad \text{and} \quad S_{n*}^* := \max_{1\le j<k\le n}\|S_k - S_j\|$$

at least in proofs, if not in statements. We should also mention that we will use, here and elsewhere, the notation

$$\|X\|_p := \left(\mathbb{E}\|X\|^p\right)^{1/p}$$

for all $p > 0$, although this is a (pseudo) norm only for $p \ge 1$.

Here is the basic tail inequality.

PROPOSITION 1.2.1. *Let X_i, $i \le n$, be independent random vectors with values in a separable Banach space B and set $S_k = \sum_{i=1}^k X_i$. Then, for every $s, t, u > 0$,*

$$\Pr\Big\{\max_{1\le k\le n}\|S_k\| > s+t+u\Big\} \le \Pr\Big\{\max_{1\le k\le n}\|X_k\| > u\Big\} + \Pr\Big\{\max_{1\le k\le n}\|S_k\| > t\Big\}\Pr\Big\{\max_{1\le j<k\le n}\|S_k - S_j\| > s\Big\}. \tag{1.2.1}$$

PROOF. We set $A_j = \{\|S_i\| \le t \text{ for } 1 \le i \le j-1, \|S_j\| > t\}$ and note $\{S_n^* > t\} = \cup_{j=1}^n A_j$, as in the proof of Lévy's inequality. On A_j, $\|S_k\| \le t$ for all $k < j$ and $\|S_k\| \le \|S_{j-1}\| + \|X_j\| + \|S_k - S_j\|$ if $k \ge j$. Therefore, on A_j,

$$S_n^* \le t + X_n^* + \max_{k:j<k\le n}\|S_k - S_j\|.$$

The sets A_j are disjoint and, for each j, A_j depends only on $X_1, \dots, X_j$ and is therefore independent of $\max_{k:j<k\le n}\|S_k - S_j\|$. This gives

$$\Pr\big\{A_j, S_n^* > s+t+u\big\} \le \Pr\big\{A_j, X_n^* > u\big\} + \Pr\{A_j\}\Pr\Big\{\max_{k:j<k\le n}\|S_k - S_j\| > s\Big\}.$$

Summing up over j yields (1.2.1). □

Inequality (1.2.1) is most effective in the following three instances:

1) If the variables X_i are *real non-negative*, or are *non-negative functions* in a Banach space of functions whose norm satifies $\|f\| \le \|g\|$ whenever $0 \le f \le g$, then $\|S_n\| = S_n^* \ge \max_{1\le j,k\le n}\|S_k - S_j\|$ and, in this case, inequality (1.2.1) just becomes

$$\Pr\big\{\|S_n\| > s+t+u\big\} \le \Pr\big\{X_n^* > u\big\} + \Pr\big\{\|S_n\| > t\big\}\Pr\big\{\|S_n\| > s\big\}. \tag{1.2.2}$$

2) If the variables X_i are *symmetric*, then, because of Lévy's inequality (Theorem 1.1.1), we may as well just consider $\|S_n\|$ instead of S_n^*. Then, the main estimate in the proof above becomes

$$\Pr\big\{A_j, \|S_n\| > s+t+u\big\} \le \Pr\big\{A_j, X_n^* > u\big\} + \Pr\{A_j\}\Pr\big\{\|S_n - S_j\| > s\big\},$$

and the last probability is dominated by $2\Pr\{\|S_n\| > s\}$ (by applying Lévy's inequality to the partial sums $X_n + \cdots + X_{j+1}$, $j = n-1, \dots, 0$). The resulting inequality,

$$\Pr\{\|S_n\| > s+t+u\} \le \Pr\big\{X_n^* > u\big\} + 4\Pr\big\{\|S_n\| > t\big\}\Pr\big\{\|S_n\| > s\big\}, \tag{1.2.3}$$

is Hoffmann-Jørgensen's original inequality when $s = t$.

3) Finally, if the variables X_i are i.i.d., then we can apply Theorem 1.1.5 (Lévy's inequality for i.i.d. vectors) instead of Theorem 1.1.1, but otherwise proceed as in the symmetric case, and obtain

$$\Pr\{\|S_n\| > 30s + 30t + u\} \le \Pr\{X_n^* > u\} + 81 \Pr\{\|S_n\| > t\} \Pr\{\|S_n\| > s\}. \tag{1.2.4}$$

As a corollary to Proposition 1.2.1, we obtain an upper estimate of the pth moment of S_n^* in terms of the pth moment of X_n^* and a quantile of $\max_{1\le j<k\le n}\|S_k - S_j\|$.

PROPOSITION 1.2.2. *Let X_i, $i \le n$, be independent B-valued random variables, B a separable Banach space, and set $S_k = \sum_{i=1}^k X_i$, $k = 1, \dots, n$. Then, for all $t, p > 0$,*

$$\mathbb{E}\Big(\max_{1\le k\le n}\|S_k\|\Big)^p \le \left[\frac{t^{p/(p+1)} + \big(\big\|\max_{1\le k\le n}\|X_k\|\big\|_p\big)^{p/(p+1)}}{1 - \big(\Pr\{\max_{1\le j<k\le n}\|S_k - S_j\| > t\}\big)^{1/(p+1)}}\right]^{p+1}. \tag{1.2.5}$$

PROOF. If we take $s = \alpha v$, $t = \beta v$, and $u = \gamma v$ in inequality (1.2.1), multiply by pv^{p-1} and integrate with respect to dv, we obtain

$$\begin{aligned}\Big(\frac{1}{\alpha+\beta+\gamma}\Big)^p \mathbb{E}(S_n^*)^p \le{}& \Big(\frac{1}{\gamma}\Big)^p \mathbb{E}(X_n^*)^p \\ &+ \int_0^\infty pv^{p-1}\Pr\{S_n^* > \beta v\}\Pr\{S_{n*}^* > \alpha v\}dv.\end{aligned}$$

Letting $v' = \beta v$ and changing variables, the above turns into

$$\begin{aligned}&\Big(\frac{1}{\alpha+\beta+\gamma}\Big)^p \mathbb{E}(\|S_n\|^*)^p \\ &\quad\le \Big(\frac{1}{\gamma}\Big)^p \mathbb{E}(X_n^*)^p + \Big(\frac{1}{\beta}\Big)^p \int_0^\infty pv^{p-1}\Pr\{S_n^* > v\}\Pr\{S_{n*}^* > \alpha v/\beta\}dv \\ &\quad\le \Big(\frac{1}{\gamma}\Big)^p \mathbb{E}(X_n^*)^p + \Big(\frac{1}{\beta}\Big)^p \int_{\frac{\beta t}{\alpha}}^\infty pv^{p-1}\Pr\{S_n^* > v\}\Pr\{S_{n*}^* > \alpha v/\beta\}dv \\ &\qquad + \Big(\frac{1}{\beta}\Big)^p \int_0^{\frac{\beta t}{\alpha}} pv^{p-1}dv \\ &\quad\le \Big(\frac{1}{\gamma}\Big)^p \mathbb{E}(X_n^*)^p + \Big(\frac{1}{\beta}\Big)^p \mathbb{E}(S_n^*)^p \Pr\{S_{n*}^* > t\} + \Big(\frac{1}{\alpha}\Big)^p t^p.\end{aligned}$$

Using the fact that, for any $a, b, c \ge 0$,

$$\min_{x+y+z=1, x,y,z\ge 0}(x^{-p}a + y^{-p}b + z^{-p}c) = \big(a^{\frac{1}{p+1}} + b^{\frac{1}{p+1}} + c^{\frac{1}{p+1}}\big)^{p+1},$$

minimizing the right-hand side of the above inequality gives

$$\mathbb{E}(S_n^*)^p \le \Big\{\big(\mathbb{E}(X_n^*)^p\big)^{\frac{1}{p+1}} + \big[\mathbb{E}(S_n^*)^p\Pr\{S_{n*}^* > t\}\big]^{\frac{1}{p+1}} + t^{\frac{p}{p+1}}\Big\}^{p+1},$$

which is just inequality (1.2.5). □

Inequality (1.2.5) can be put in a nicer way in each of the three instances mentioned above, namely, for symmetric vectors, for i.i.d. vectors and for non-negative variables and functions. For expediency, we will only do this in the case of symmetric random variables. First, the by now classical Hoffmann-Jørgensen inequality bounding moments of a sum by the corresponding moments of the maximum plus a quantile of the sum:

THEOREM 1.2.3. *For each $p > 0$ there exists a constant K_p such that, if X_i, $i \leq n < \infty$, are independent symmetric B-valued random variables, B a separable Banach space, and if $t_0 \geq 0$ is defined as*

$$t_0 = \inf\{t > 0 : \Pr\{\|S_n\| > t\} \leq 1/8\},$$

then

$$\Big\|\sum_{i=1}^{n} X_i\Big\|_p \leq K_p\big[t_0 + \big\|\max_{1\leq k\leq n}\|X_k\|\big\|_p\big]. \tag{1.2.6}$$

The constant may be taken to be $K_p = 2 \cdot 16^{1/p}(p+1)^{(p+1)/p}$, which is of the order of p as $p \to \infty$.

PROOF. If at the beginning of the previous proof we replace inequality (1.2.1) by inequality (1.2.3), we obtain

$$\mathbb{E}\|S_n\|^p \leq \left[\frac{4^{1/(p+1)}t^{p/(p+1)} + \big(\|X_n^*\|_p\big)^{p/(p+1)}}{1 - \big(4\Pr\{\|S_n\| > t\}\big)^{1/(p+1)}}\right]^{p+1}. \tag{1.2.5s}$$

(Note that similar simplified expressions are possible for positive and for i.i.d. variables.) Now, since $1 - x^\alpha \geq \alpha(1-x)$ for $0 \leq x \leq 1$ and $0 \leq \alpha \leq 1$ (e.g., by the mean value theorem) and since, by convexity, $(a+b)^{(p+1)/p} \leq 2^{1/p}\big(a^{(p+1)/p} + b^{(p+1)/p}\big)$, inequality (1.2.5s) yields

$$\|S_n\|_p \leq \left[\frac{p+1}{1 - 4\Pr\{\|S_n\| > t\}}\right]^{\frac{p+1}{p}} 2^{1/p}\big[4^{1/p}t + \|X_n^*\|_p\big].$$

Hence,

$$\|S_n\|_p \leq 2^{(p+2)/p}(p+1)^{(p+1)/p}\big[4^{1/p}t_0 + \|X_n^*\|_p\big],$$

proving the theorem. □

Note that Chebyshev's inequality and Lévy's inequality imply the converse of inequality (1.2.6). A similar comment applies regarding Theorem 1.2.5 below, as well as to the variations of these theorems for centered and for positive variables.

REMARK 1.2.4. It is clear from the above proof that Theorem 1.2.3 is also true for $\{X_i\}$ i.i.d. B-valued and for $\{X_i\}$ independent non-negative in $\mathbb{R}$ or non-negative functions in, e.g., L_p or $C(S)$, if $1/8$ is replaced by a suitable constant in the definition of t_0. Here and elsewhere $C(S)$ denotes the Banach space of bounded continuous functions on a metric space S, equipped with the sup norm. The constant K_p in these cases is of the same order of magnitude as in the case of symmetric variables.

If in (1.2.5s) we replace t by $2 \cdot 4^{1/q}\|S_n\|_q$, $0 < q \le p$, Chebyshev's inequality bounds the denominator from above by $1 - (1/2)^{q/(p+1)}$, so that

$$\mathbb{E}\|S_n\|^p \le \left[\frac{4^{1/(p+1)}(2 \cdot 4^{1/q} \cdot \|S_n\|_q)^{p/(p+1)} + \left(\|X_n^*\|_p\right)^{p/(p+1)}}{1 - (1/2)^{q/(p+1)}}\right]^{p+1}.$$

Then, proceeding as in the previous proof but taking $\alpha = q/(p+1)$, we obtain

$$\|S_n\|_p \le \left[\frac{2(p+1)}{q}\right]^{\frac{p+1}{p}} 2^{1/p}\left[4^{1/p+1/q} \cdot 2\|S_n\|_q + \|X_n^*\|_p\right],$$

and it is easy to check that, for all $c > 0$,

$$\sup_{p,q:c<q\le p} \frac{q}{p}\left[\frac{2(p+1)}{q}\right]^{\frac{p+1}{p}} 2^{1+1/p}4^{1/p+1/q} < \infty.$$

This gives the following result on comparison of moments:

THEOREM 1.2.5. *For every $c > 0$ there exists a constant $K_c < \infty$ such that, if X_i, $i \le n < \infty$, are independent symmetric B-valued random variables, B a separable Banach space, and $c \le q < p$,*

$$\Big\|\sum_{i=1}^n X_i\Big\|_p \le K_c\frac{p}{q}\Big[\Big\|\sum_{i=1}^n X_i\Big\|_q + \Big\|\max_{1\le k\le n}\|X_k\|\Big\|_p\Big]. \tag{1.2.7}$$

Just like with Theorem 1.2.3, Theorem 1.2.5 is true for $\{X_i\}$ i.i.d. B-valued, and for $\{X_i\}$ independent non-negative random variables or functions. Moreover, Theorem 1.2.5 extends to independent *centered* not necessarily symmetric random variables via a very simple and useful randomization/symmetrization lemma involving *Rademacher* variables. A Rademacher random variable ε is a symmetric variable on the two points set $\{-1, 1\}$ (a "random sign"). Usually, $\{\varepsilon_i\}$ will denote a sequence of independent Rademacher variables independent from any other variables appearing in formulas with them.

LEMMA 1.2.6. *Let X_i, $i \le n$, be independent centered B-valued random variables with norms in L_p for some $p \ge 1$ and let ε_i, $i \le n$, be independent Rademacher random variables independent of the variables X_i. Then,*

$$2^{-p}\mathbb{E}\Big\|\sum_{i=1}^n \varepsilon_i X_i\Big\|^p \le \mathbb{E}\Big\|\sum_{i=1}^n X_i\Big\|^p \le 2^p\mathbb{E}\Big\|\sum_{i=1}^n \varepsilon_i X_i\Big\|^p, \tag{1.2.8}$$

and

$$\mathbb{E}\max_{k\le n}\Big\|\sum_{i=1}^k X_i\Big\|^p \le 2^{p+1}\mathbb{E}\Big\|\sum_{i=1}^n \varepsilon_i X_i\Big\|^p, \tag{1.2.9}$$

PROOF. Let $\{X_i'\}$ be a copy of the sequence $\{X_i\}$ independent of $\{X_i\}$. Conditional Jensen's inequality, symmetry, and independence give

$$\begin{aligned}\mathbb{E}\|\sum X_i\|^p &= \mathbb{E}\|\sum X_i - \mathbb{E}\sum X_i'\|^p \\ &= \mathbb{E}\|\mathbb{E}(\sum X_i - \sum X_i' | \sum X_i)\|^p \\ &\leq \mathbb{E}\|\sum(X_i - X_i')\|^p \\ &= \mathbb{E}\|\sum \varepsilon_i(X_i - X_i')\|^p \\ &\leq 2^p \mathbb{E}\|\sum \varepsilon_i X_i\|^p.\end{aligned}$$

And, conversely, letting $\mathbb{E}_\varepsilon$ denote integration with respect to the Rademacher variables only, we have, again by application of the same principles (note that whenever independence is present, there is no loss of generality in assuming a product probability space and then conditional Jensen is nothing but unconditional Jensen plus Fubini),

$$\begin{aligned}\mathbb{E}\|\sum \varepsilon_i X_i\|^p &\leq 2^{p-1}\mathbb{E}_\varepsilon\mathbb{E}\|\sum_{i:\varepsilon_i=1} X_i\|^p + 2^{p-1}\mathbb{E}_\varepsilon\mathbb{E}\|\sum_{i:\varepsilon_i=-1} X_i\|^p \\ &= 2^{p-1}\mathbb{E}_\varepsilon\mathbb{E}\|\sum_{i:\varepsilon_i=1} X_i + \sum_{i:\varepsilon_i=-1}\mathbb{E}X_i\|^p \\ &\quad + 2^{p-1}\mathbb{E}_\varepsilon\mathbb{E}\|\sum_{i:\varepsilon_i=-1} X_i + \sum_{i:\varepsilon_i=1}\mathbb{E}X_i\|^p \\ &\leq 2^p\mathbb{E}\|\sum X_i\|^p.\end{aligned}$$

So, (1.2.8) is proved. (1.2.9) follows from Lévy's inequality once we incorporate the max into the norm. For this, we define the random vectors $\mathbf{X}_1 := (X_1, \overset{n)}{\ldots}, X_1)$, $\mathbf{X}_i := (0, \overset{i-1)}{\ldots}, 0, X_i, \ldots, X_i)$, $1 < i \leq n$, which take values in the Banach space $(B^n, |||\cdot|||)$, with norm $|||(x_1, \ldots, x_n)||| := \max \|x_i\|$, the maximum of the norms of the coordinates. Then,

$$|||\sum_{i=1}^n \mathbf{X}_i||| = \max_{1\leq k\leq n}\|\sum_{i=1}^k X_i\| \quad \text{and} \quad |||\sum_{i=1}^n \varepsilon_i\mathbf{X}_i||| = \max_{1\leq k\leq n}\|\sum_{i=1}^k \varepsilon_i X_i\|.$$

Hence, inequality (1.2.8) applied to these vectors gives

$$\mathbb{E}\max_k\|\sum_{i=1}^k X_i\|^p \leq 2^p\mathbb{E}\max_k\|\sum_{i=1}^k \varepsilon_i X_i\|^p$$

and now (1.2.9) follows from Lévy's inequality (Theorem 1.2.1). □

COROLLARY 1.2.7. *The conclusion of Theorem 1.2.5 with $1 \leq q < p$ holds for B-valued random variables X_i independent and centered. Moreover, the variable $\|\sum_{i=1}^n X_i\|$ at the left side of inequality (1.2.7) can be replaced by the variable $\max_{1\leq k\leq n}\|\sum_{i=1}^k X_i\|$.*

PROOF. It follows immediately by applying Theorem 1.2.5 to the symmetric variables $\varepsilon_i X_i$, using the inequalities in Lemma 1.2.6. □

Before considering an application, we should mention that the constants in these theorems are not best possible. Talagrand (1989) shows that for each fixed q in Theorem 1.2.5, the right order of the multiplicative constant as $p \to \infty$ is not p but $p/\log p$. In Section 1.5 we will prove this result for the particular case $B = \mathbb{R}$.

Theorems 1.2.3 and 1.2.5 are very useful tools for the study of sums of independent random vectors, as mentioned above. We now illustrate their use by obtaining an exponential inequality that can be interpreted as a sort of reversed Jensen's inequality for exponentials of sums of independent random variables.

THEOREM 1.2.8. *Let X_i, $i \le n$, be independent, centered B-valued random variables such that $\max_{1\le k\le n} \|X_k\|_\infty < \infty$. Then, there exist constants $C > 0$ and $\delta > 0$, independent of X_i, B, and n, such that, for every $p \ge 1$ and every*

$$\lambda \le \frac{\delta p}{\|S_n\|_p \vee \left(\max_{1\le k\le n} \|X_k\|_\infty\right)},$$

we have

$$\mathbb{E}\exp\big(\lambda \|S_n\|\big) \le 3\exp\big(C\lambda(\|S_n\|_p + \max_{1\le k\le n} \|X_k\|_\infty)\big). \tag{1.2.10}$$

PROOF. Let r be the largest integer not exceeding p. Expanding the exponential into power series and integrating, we obtain

$$\begin{aligned}
\mathbb{E}\exp\big(\lambda\|S_n\|\big) &= 1 + \sum_{j=1}^{\infty} \frac{\lambda^j}{j!}\|S_n\|_j^j \\
&= 1 + \sum_{j=1}^{r} \frac{\lambda^j}{j!}\|S_n\|_j^j + \sum_{j=r+1}^{\infty} \frac{\lambda^j}{j!}\|S_n\|_j^j \\
&\le 1 + \sum_{j=1}^{r} \frac{\lambda^j}{j!}\|S_n\|_p^j + \sum_{j=r+1}^{\infty} \frac{\lambda^j}{j!}\|S_n\|_j^j \\
&\le \exp\big(\lambda\|S_n\|_p\big) + \sum_{j=r+1}^{\infty} \frac{\lambda^j}{j!}\|S_n\|_j^j .
\end{aligned}$$

So, the proof reduces to estimating the last sum. Corollary 1.2.7 implies that, for $j \ge p$,

$$\|S_n\|_j^j \le \left(\frac{Kj}{p}\right)^j \left(\|S_n\|_p^j + \|X_n^*\|_j^j\right) \le \left(\frac{Kj}{p}\right)^j \left(\|S_n\|_p^j + \|X_n^*\|_\infty^j\right),$$

where $K = 4K_1$, K_1 being the constant from (1.2.7) for $c = 1$ (the factor 4 accounts for the fact that the variables are centered but not necessarily symmetric). Using the inequality $\frac{j^j}{j!} \le e^j$ gives

$$\sum_{j=r+1}^{\infty} \frac{\lambda^j}{j!} \|S_n\|_j^j \le \sum_{j=r+1}^{\infty} e^j K^j \Big(\frac{\lambda}{p}\Big)^j \big(\|S_n\|_p^j + \|X_n^*\|_\infty^j\big).$$

If we now choose λ such that

$$\frac{eK\lambda}{p} \max\big(\|S_n\|_p, \|X_n^*\|_\infty\big) \le \frac{1}{2},$$

then the last sum is the sum of two geometric series each with ratio at most $1/2$. The sum of each of these series does not exceed twice the respective first term, and therefore we have

$$\begin{aligned}\sum_{j=r+1}^{\infty} \frac{\lambda^j}{j!} \|S_n\|_j^j &\le 2e^{r+1} K^{r+1} \Big(\frac{\lambda}{p}\Big)^{r+1} \big(\|S_n\|_p^{r+1} + \|X_n^*\|_\infty^{r+1}\big) \\ &\le 2e^p K^p \Big(\frac{\lambda}{p}\Big)^p \big(\|S_n\|_p^p + \|X_n^*\|_\infty^p\big)\end{aligned}$$

(where the last inequality holds by the definition of λ). Since $x^p \le \exp(px)$ for $x > 0$, it follows that

$$\sum_{j=r+1}^{\infty} \frac{\lambda^j}{j!} \|S_n\|_j^j \le 2\exp\big(eK\lambda(\|S_n\|_p + \|X_n^*\|_\infty)\big),$$

proving the proposition. □

REMARK 1.2.9. *Measurability.* As in Section 1.2, the results in the present section also hold in the more general context described in Remark 1.2.8 above, just for the reasons given there.

1.3 The Khinchin–Kahane Inequalities

A useful technique made possible by decoupling consists in randomization by Rademacher variables (independent ± 1s) in expressions of the form $\sum h(X_i, X_j)$ and more general, followed by conditional use of the excellent integrability properties of Rademacher polynomials. In this section we describe a classical inequality that is at the base of these properties for linear combinations of Rademacher variables, namely, Khinchin's inequality. We will study integrability properties of Rademacher polynomials in Chapter 3, Section 2, and we will defer the proof of at least one result from the present section to that chapter in order to avoid repetition.

A *Rademacher sequence* $\{\varepsilon_i\}_{i=1}^{\infty}$ is a family of independent identically distributed random variables whose common law is

$$\Pr\{\varepsilon_i = 1\} = 1 - \Pr\{\varepsilon_i = -1\} = 1/2.$$

The variables ε_i are referred to as *Rademacher variables*. We have encountered these variables before, in Lemma 1.2.6.

The classical Khinchin inequalities express the fact that *in the linear span of a Rademacher sequence all the L_p norms ($0 < p < \infty$) are equivalent.* These inequalities have been extended and refined in several directions over the years. The first extension we require replaces the real coefficients by Banach space valued coefficients (actually, B needs not be complete, just a normed linear space). The refinement of interest to us concerns the constants $C_{q,2}$ between the L_q and the L_2 norms, $q \geq 2$, in

$$\left\| \sum \varepsilon_i x_i \right\|_q \leq C_{q,2} \left\| \sum \varepsilon_i x_i \right\|_2,$$

where here and elsewhere we set $\|X\|_q := \left(\mathbb{E}\|X\|^q\right)^{1/q}$, as is customary. In order to infer accurate exponential integrability behavior of Rademacher linear combinations from these inequalities, we need to find constants $C_{p,2}$ with optimal dependence on p as $p \to \infty$. In Section 2 of Chapter 3, Theorem 3.2.2, we will prove the following theorem:

THEOREM 1.3.1. *Let F be a normed linear space, let $\{x_i\} \subset F$ be a finite (or countably infinite) subset of F, and let $1 < p < q < \infty$. Then,*

$$\left\| \sum \varepsilon_i x_i \right\|_q \leq \left(\frac{q-1}{p-1}\right)^{1/2} \left\| \sum \varepsilon_i x_i \right\|_p. \tag{1.3.1}$$

The constants in these inequalities may not be the best possible, but they are of the right order as $q \to \infty$ while p remains fixed. This can be seen for x real using the central limit theorem: the CLT for Rademacher finite sums holds with convergence of all moments—precisely, e.g., by the uniform integrability provided by inequality (1.3.1)—and the qth moment of a normal variable grows like $(q/e)^{q/2}$. However, for p close to 1 these constants behave very badly, and (1.3.1) does not even apply if $p = 1$. Our object in this section is to obtain the best possible constants in this range, particularly for $p = 1$ and $q = 2$.

THEOREM 1.3.2. *Let $\{x_i\}_{i=1}^n \subset F$, F a normed linear space, and $n < \infty$, and let $\{\varepsilon_i\}_{i=1}^n$ be a Rademacher sequence. Then,*

$$\left\| \sum \varepsilon_i x_i \right\|_2 \leq \sqrt{2} \left\| \sum \varepsilon_i x_i \right\|_1, \tag{1.3.2}$$

and the constant is best possible.

Before proceeding to the proof of the theorem in full generality, we will give Littlewood's (1930) proof of (1.3.2) in the case of *real* coefficients and with a worse constant ($\sqrt{3}$ instead of $\sqrt{2}$). We do this because the method of proof, besides being extremely simple, is also generally useful as a way of obtaining *converse*

Hölder inequalities between lower moments once one has such inequalities for higher ones.

Here is Littlewood's proof. Let $Z = \sum \varepsilon_i x_i$, x_i real. Then, since the variables ε_i are independent, centered and $\varepsilon_i^2 = 1$, a simple computation gives

$$\mathbb{E}Z^2 = \sum x_i^2 \quad \text{and} \quad \mathbb{E}Z^4 = \sum x_i^4 + 3\sum_{i \neq j} x_i^2 x_j^2 \leq 3\Big(\sum x_i^2\Big)^2 = 3(\mathbb{E}Z^2)^2. \tag{1.3.3}$$

A very natural approach to take, in order to go from here to an inequality between $\mathbb{E}|Z|$ and $\mathbb{E}Z^2$, is to naively relate $\mathbb{E}|Z|$ to both $\mathbb{E}Z^2$ and $\mathbb{E}Z^4$ by Hölder, as follows. For $\alpha < 1$, we can write

$$\mathbb{E}Z^2 = \mathbb{E}\big(|Z|^\alpha |Z|^{2-\alpha}\big) \leq \big(\mathbb{E}|Z|\big)^\alpha \big(\mathbb{E}|Z|^{(2-\alpha)/(1-\alpha)}\big)^{1-\alpha},$$

where we are using Hölder with $1/p = \alpha$ and $1/q = 1 - \alpha$. Then we can solve for α in

$$\frac{2-\alpha}{1-\alpha} = 4,$$

which gives $\alpha = 2/3$, and the above equation becomes

$$\mathbb{E}Z^2 \leq \big(\mathbb{E}|Z|\big)^{2/3}\big(\mathbb{E}|Z|^4\big)^{1/3}. \tag{1.3.4}$$

Hence, by (1.3.3),

$$\mathbb{E}|Z| \geq \big(\mathbb{E}Z^2\big)^{3/2}/\big(\mathbb{E}Z^4\big)^{1/2} \geq \frac{1}{\sqrt{3}}\big(\mathbb{E}Z^2\big)^{1/2},$$

which is inequality (1.3.2) with $\sqrt{2}$ replaced by $\sqrt{3}$.

In a sense, Littlewood's observation on how to get down from higher moments to lower moments in *hypercontractivity or converse Hölder inequalities* has been superseded by Paley–Zygmund's argument, which actually takes these inequalities *down to quantiles* (= 0th moments!). Paley–Zygmund's argument and some of its consequences are presented in Section 3.2; in particular, it will be shown there that inequality (1.3.2) does imply that, for arbitrary non-negative, nondecreasing functions $f : \mathbb{R}_+ \mapsto \mathbb{R}_+$ and for any $c > 2$,

$$\mathbb{E}f\Big(\big|\sum \varepsilon_i x_i\big|\Big) \geq \frac{1}{8} f\Big(\big(\sum x_i^2\big)^{1/2}/(c\sqrt{2})\Big), \tag{1.3.5}$$

a Khinchin inequality for non-negative nondecreasing functions, hence also for *tail probabilities* [take $f(x) = I_{(t,\infty)}(x)$].

The bound 3 for the ratio between the fourth moment and the square of the second moment of Z is best possible because this is also the ratio between the corresponding moments in the case of normal variables. By analogy, we could predict a best bound of $\pi/2$ for the ratio $\mathbb{E}Z^2/(\mathbb{E}|Z|)^2$. However, $\mathbb{E}(\varepsilon_1 + \varepsilon_2)^2 = 2\mathbb{E}\varepsilon_1^2 = 2$, whereas $\mathbb{E}|\varepsilon_1 + \varepsilon_2| = 1$ (the variable $|\varepsilon_1 + \varepsilon_2|$ takes the values 2 and 0 each with probability 1/2), showing that $\pi/2$ is too small and that the best constant is at least 2. Theorem 1.3.2 tells us that 2 is indeed the best constant, and that this is true, not only in $\mathbb{R}$, but in any normed linear space.

Next, we introduce some notation for the proof of Theorem 1.3.2. Letting $\epsilon = (\epsilon_1, \ldots, \epsilon_n)$ and $\eta = (\eta_1, \ldots, \eta_n)$ denote generic points in $\{-1, 1\}^n$, we set $-\epsilon := (-\epsilon_1, \ldots, -\epsilon_n)$ and define the distance

$$d(\epsilon, \eta) = \mathrm{Card}\{i : \epsilon_i \neq \eta_i\}.$$

For $\alpha = (\alpha_1, \ldots, \alpha_n) \in \{0, 1\}^n$ and $\epsilon = (\epsilon_1, \ldots, \epsilon_n) \in \{-1, 1\}^n$, we also define

$$|\alpha| := \sum_{i=1}^{n} |\alpha_i| \quad \text{and} \quad \omega_\alpha(\epsilon) := \prod_{i=1}^{n} \epsilon_i^{\alpha_i}.$$

The uniform probability measure μ on $\{-1, 1\}^n$ is the product of n copies of the uniform probability on $\{-1, 1\}$; hence, since the functions $f_0(x) = 1 = x^0$ and $f_1(x) = x = x^1$ constitute an orthonormal basis for L_2 of the two-point space, it follows that the set of functions $\{\omega_\alpha : \alpha \in \{0, 1\}^n\}$ is an orthonormal basis of $\mathcal{H} := L_2(\{-1, 1\}^n, \mu)$ (the Walsh basis). Given $x_i \in F$, we finally set X to be the function defined on $\{-1, 1\}^n$ by

$$X(\epsilon) := \Big\| \sum_{i=1}^{n} \epsilon_i x_i \Big\|, \qquad \epsilon = (\epsilon_1, \ldots, \epsilon_n) \in \{-1, 1\}^n,$$

and note that

$$\Big\| \sum \varepsilon_i x_i \Big\|_p^p = \sum_{\epsilon \in \{-1,1\}^n} X^p(\epsilon)/2^n = \int X^p d\mu.$$

Proof of Theorem 1.3.2. Let $L : \mathcal{H} \mapsto \mathcal{H}$ denote the linear operator defined by the formula

$$(Lf)(\epsilon) := \sum_{\eta \in \{-1,1\}^n, d(\epsilon,\eta)=1} f(\eta), \qquad f \in \mathcal{H}.$$

L has a simple expression in terms of the basis ω_α: if η with $d(\epsilon, \eta) = 1$ differs from ϵ at the ith coordinate and $\alpha_i = 0$, then $\omega_\alpha(\eta) = \omega_\alpha(\epsilon)$ whereas $\omega_\alpha(\eta) = -\omega_\alpha(\epsilon)$ if $\alpha_i = 1$; therefore,

$$L(\omega_\alpha) = (n - 2|\alpha|)\omega_\alpha, \qquad \alpha \in \{0, 1\}^n.$$

Note also that if

$$X = \sum_{\alpha \in \{0,1\}^n} a_\alpha \omega_\alpha$$

denotes the expansion of the function X in the basis ω_α, then, since $\omega_\alpha(-\epsilon) = -\omega_\alpha(\epsilon)$ if $|\alpha|$ is odd, and $X(\epsilon) = X(-\epsilon)$, it follows that

$$a_\alpha = 0 \quad \text{if } |\alpha| \text{ is odd}.$$

Then, owing to these considerations, if $\langle\cdot,\cdot\rangle$ denotes the inner product of $\mathscr{H}$, we have

$$\begin{aligned}\langle X, L(X)\rangle &= \Big\langle \sum a_\alpha\omega_\alpha, \sum a_\alpha(n-2|\alpha|)\omega_\alpha\Big\rangle = \sum a_\alpha^2(n-2|\alpha|)\\ &\le na_0^2 + (n-4)\sum_{\alpha\neq 0} a_\alpha^2 = 4a_0^2 + (n-4)\sum a_\alpha^2\\ &= 4\Big(\int X d\mu\Big)^2 + (n-4)\int X^2 d\mu\\ &= 4\Big\|\sum \varepsilon_i x_i\Big\|_1^2 + (n-4)\Big\|\sum\varepsilon_i x_i\Big\|_2^2. \end{aligned} \tag{1.3.6}$$

Now, we will bound $\langle X, L(X)\rangle$ from below. For this we note that, by the definition of the distance d and the triangle inequality,

$$\begin{aligned}(n-2)X(\epsilon) &= \Big\|\sum_{i=1}^n [(n-1)\epsilon_i - \epsilon_i]x_i\Big\|\\ &= \Big\|\sum_{i=1}^n\Big(\sum_{\eta\in\{-1,1\}^n, d(\epsilon,\eta)=1}\eta_i\Big)x_i\Big\|\\ &= \Big\|\sum_{\eta\in\{-1,1\}^n, d(\epsilon,\eta)=1}\Big(\sum_{i=1}^n \eta_i x_i\Big)\Big\|\\ &\le \sum_{\eta\in\{-1,1\}^n, d(\epsilon,\eta)=1}\Big\|\sum_{i=1}^n\eta_i x_i\Big\| = \sum_{\eta\in\{-1,1\}^n, d(\epsilon,\eta)=1} X(\eta).\end{aligned}$$

Therefore,

$$\begin{aligned}2^n\langle X, L(X)\rangle &= \sum_{\epsilon\in\{-1,1\}^n} X(\epsilon)(LX)(\epsilon) = \sum_{\epsilon\in\{-1,1\}^n}\Big(\sum_{\eta\in\{-1,1\}^n, d(\epsilon,\eta)=1} X(\eta)\Big)X(\epsilon)\\ &\ge (n-2)\sum_{\epsilon\in\{-1,1\}^n} X^2(\epsilon) = 2^n(n-2)\Big\|\sum\varepsilon_i x_i\Big\|_2^2. \end{aligned}\tag{1.3.7}$$

Combining inequalities (1.3.6) and (1.3.7) we obtain

$$(n-2)\Big\|\sum\varepsilon_i x_i\Big\|_2^2 \le 4\Big\|\sum\varepsilon_i x_i\Big\|_1^2 + (n-4)\Big\|\sum\varepsilon_i x_i\Big\|_2^2,$$

and inequality (1.3.2) follows by collecting similar terms.

That (1.3.2) is best possible can be checked on $\varepsilon_1+\varepsilon_2$, as mentioned above. □

Theorem 1.3.2 can be used to improve the constants in Theorem 1.1 for q near 1, although not necessarily in an optimal way; for instance, it trivially gives the following estimate for $p \ge 1$,

$$\Big\|\sum\varepsilon_i x_i\Big\|_2 \le \sqrt{2}\Big\|\sum\varepsilon_i x_i\Big\|_1 \le \sqrt{2}\Big\|\sum\varepsilon_i x_i\Big\|_p, \tag{1.3.8}$$

which is better than (1.3.1) for $1 \le p < 3/2$. More interesting is the following observation that extends the domain of validity of inequality (1.3.1) below the L_1 threshold, with best constants.

COROLLARY 1.3.3. *If $q \le p$, $q \in (0,1]$, $p \in (0,2]$, then*

$$\Big\| \sum \varepsilon_i x_i \Big\|_p \le 2^{1/q-1/p} \Big\| \sum \varepsilon_i x_i \Big\|_q \tag{1.3.9}$$

and the constants are best possible.

PROOF. For any random variable X and numbers $r, p, q > 0$, $\theta \in [0,1]$, satisfying

$$\frac{1}{r} = \frac{\theta}{p} + \frac{1-\theta}{q}$$

[so that the numbers $p/(\theta r)$ and $q/((1-\theta)r)$ are conjugate], Hölder's inequality gives

$$\big(\mathbb{E}|X|^r\big)^{1/r} = \big(\mathbb{E}|X|^{\theta r}|X|^{(1-\theta)r}\big)^{1/r} \le \big(\mathbb{E}|X|^p\big)^{\theta/p}\big(\mathbb{E}|X|^q\big)^{(1-\theta)/q}.$$

This inequality shows that the function $f(s) = s \log \mathbb{E}|X|^{1/s}$, $s > 0$, is convex. Then, in particular, since $1/2 \le 1/p$, $1 \le 1/q$, and $1/p \le 1/q$, we have

$$\frac{f(1) - f(1/2)}{1/2} \le \frac{f(1/q) - f(1/p)}{1/q - 1/p},$$

that is,

$$\left(\frac{\|X\|_1}{\|X\|_2}\right)^2 \le \left(\frac{\|X\|_q}{\|X\|_p}\right)^{1/(1/q-1/p)}.$$

For $X = \|\sum \varepsilon_i x_i\|$, the left-hand side of this inequality is larger than or equal to 1/2 by Theorem 1.3.2, proving inequality (1.3.7). Again, optimality checks on $\varepsilon_1 + \varepsilon_2$. □

1.4 Moment Bounds

Let X_i, $i \in \mathbb{N}$, be independent real random variables and, for each i, let P_i be the probability law of X_i. Then

$$\mathbb{E}\Big|\sum_{i=1}^n X_i\Big|^p = \int \cdots \int |x_1 + \cdots + x_n|^p dP_1(x_1) \cdots dP_n(x_n).$$

What we seek here is to approximate this multiple integral by n one-dimensional integrals with respect to the individual laws P_i. We have already obtained (Section 1.3) a solution to this problem in the important particular case $X_i = x_i \varepsilon_i$, $x_i \in H$, where H is a Hilbert space: since in Hilbert space we have

$$\begin{aligned}\mathbb{E}\Big\|\sum_i x_i\varepsilon_i\Big\|^2 &= \mathbb{E}\langle \sum_i x_i\varepsilon_i, \sum_i x_i\varepsilon_i\rangle \\ &= \sum_{i,j}\langle x_i, x_j\rangle \mathbb{E}\varepsilon_i\varepsilon_j = \sum_i \|x_i\|^2,\end{aligned}$$

the inequalities from the previous section become

$$\Big(\sum_i \|x_i\|^2\Big)^{1/2} \le \Big\|\sum_i x_i\varepsilon_i\Big\|_p \le (p-1)^{1/2}\Big(\sum_i \|x_i\|^2\Big)^{1/2}, \qquad p > 2,$$

and

$$(2^{-1/2} \wedge 2^{1/2-1/p})\Big(\sum_i \|x_i\|^2\Big)^{1/2} \le \Big\|\sum_i x_i\varepsilon_i\Big\|_p \le \Big(\sum_i \|x_i\|^2\Big)^{1/2}, \qquad p < 2.$$

Although these bounds are not very tight, particularly for large p, they have the advantage of having a very simple and convenient form (more complicated, tighter two-sided bounds are available, as we will see in the next section). Likewise, the bounds considered in this section will be simple but not tight for large values of p.

First we consider the easier problem of estimating the maximum of a sequence of independent random variables. Then, we present two related approaches to the estimation of sums. Finally, we show, as an application, how these results yield bounds for moments (and related functionals) of martingales and sums of arbitrary positive variables in terms of moments (and related functionals) of sums of independent centered or positive random variables.

1.4.1 Maxima

Let ξ_i, $i \le n$, be independent non-negative random variables in L_p, $p > 0$, and, for $\lambda > 0$, let

$$\delta_0 = \delta_0(\lambda) := \inf\Big[t > 0 : \sum_i \Pr\{\xi_i > t\} \le \lambda\Big]. \qquad (1.4.1)$$

Then, there are on the average about $\lambda \ne 0$ terms among $\xi_1, \dots, \xi_n$ that are larger than δ_0, and this makes it plausible that, on average,

$$\max \xi_i^p \simeq \lambda^{-1} \sum \xi_i^p I_{\xi_i > \delta_0}.$$

Since the expected value of the sum at the right depends only on the marginal distributions, we will have solved our problem if we can make this argument precise. This is done in the following proposition and its corollary.

PROPOSITION 1.4.1. *Let ξ_i, $i \le n$, be a set of independent non-negative random variables in L_p, $p > 0$, and let λ be a positive real number. Then, setting*

$$\delta_0 := \inf\Big[t > 0 : \sum_i \Pr\{\xi_i > t\} \le \lambda\Big], \qquad (1.4.1')$$

we have

$$\begin{aligned} \frac{\lambda}{1+\lambda}\delta_0^p + \frac{p}{1+\lambda}\sum_i \int_{\delta_0}^{\infty} t^{p-1}\Pr\{\xi_i > t\}dt &\le \mathbb{E}\max_i \xi_i^p \\ &\le \delta_0^p + p\sum_i \int_{\delta_0}^{\infty} t^{p-1}\Pr\{\xi_i > t\}dt. \end{aligned} \qquad (1.4.2)$$

PROOF. The right side of inequality (1.4.2) is trivial by integration by parts (Fubini) and in fact holds for any $\delta > 0$:

$$\begin{aligned}\mathbb{E}\max \xi_i^p &= p\int_0^\infty t^{p-1}\Pr\{\max \xi_i > t\}dt \\ &\le p\int_0^\delta t^{p-1}dt + \int_\delta^\infty t^{p-1}\sum \Pr\{\xi_i > t\}dt.\end{aligned}$$

For the left side we first recall the well-known fact that

$$\begin{aligned}\Pr\{\max_i \xi_i > t\} &\ge 1 - \prod_i \big(1 - \Pr\{\xi_i > t\}\big) \\ &\ge 1 - \exp\Big(-\sum_i \Pr\{\xi_i > t\}\Big) \\ &\ge \frac{\sum_i \Pr\{\xi_i > t\}}{1 + \sum_i \Pr\{\xi_i > t\}}\end{aligned}$$

[which follows from the inequalities $1 - x \le \exp(-x)$ and $1 - \exp(-x) \ge x/(1+x)$, valid for $x \ge 0$]. Then, since $\sum_i \Pr\{\xi_i > t\} \le \lambda$ if and only if $t \ge \delta_0$, monotonicity of the function $x/(1+x)$ gives

$$\Pr\{\max_i \xi_i > t\} \ge \begin{cases} \dfrac{1}{1+\lambda}\sum \Pr\{\xi_i > t\} & \text{if } t \ge \delta_0, \\ \dfrac{\lambda}{1+\lambda} & \text{if } t < \delta_0. \end{cases}$$

Therefore, it follows by integration by parts that

$$\begin{aligned}\mathbb{E}\max_i \xi_i^p &= p\Big(\int_0^{\delta_0} + \int_{\delta_0}^\infty\Big) t^{p-1}\Pr\{\max_i \xi_i > t\}dt \\ &\ge \frac{\lambda}{1+\lambda}\delta_0^p + \frac{p}{1+\lambda}\sum_i \int_{\delta_0}^\infty t^{p-1}\Pr\{\xi_i > t\}dt,\end{aligned}$$

proving (1.4.2). □

Although inequality (1.4.2) solves the problem of the evaluation of the moments of the maximum of a sequence of independent random variables, it may be more appealing to have two-sided bounds in terms of truncated moments. This can be easily done:

COROLLARY 1.4.2. *Under the hypothesis of Proposition 1.4.1, and with the same notation, we have that, for every* $\lambda > 0$,

$$\frac{1}{1+\lambda}\sum_i \mathbb{E}\xi_i^p I_{\xi_i > \delta_0} \le \mathbb{E}\max_i \xi_i^p \le \frac{1}{\lambda \wedge 1}\sum_i \mathbb{E}\xi_i^p I_{\xi_i \ge \delta_0} \tag{1.4.3}$$

and

$$\frac{1 \wedge \lambda}{2(1+\lambda)}\Big[\delta_0^p + \sum_i \mathbb{E}\xi_i^p I_{\xi_i > \delta_0}\Big] \le \mathbb{E}\max_i \xi_i^p \le \delta_0^p + \sum_i \mathbb{E}\xi_i^p I_{\xi_i > \delta_0}. \tag{1.4.4}$$

PROOF. By definition of δ_0,

$$\sum_i \Pr\{\xi_i > \delta_0\} \le \lambda \le \sum_i \Pr\{\xi_i \ge \delta_0\},$$

and this implies, by integration by parts, that

$$\sum_i \mathbb{E}\xi_i^p I_{\xi_i \ge \delta_0} \ge \lambda\delta_0^p + p\sum_i \int_{\delta_0}^{\infty} t^{p-1} \Pr\{\xi_i > t\}dt$$

and

$$\sum_i \mathbb{E}\xi_i^p I_{\xi_i > \delta_0} \le \lambda\delta_0^p + p\sum_i \int_{\delta_0}^{\infty} t^{p-1} \Pr\{\xi_i > t\}dt.$$

These two inequalities together with inequality (1.4.2) give (1.4.3) and the left part of (1.4.4). The right side of (1.4.4) is very simple and works for every $\delta \ge 0$:

$$\mathbb{E}\max_i \xi_i^p \le \mathbb{E}\max_i \xi_i^p I_{\xi_i \le \delta} + \mathbb{E}\max_i \xi_i^p I_{\xi_i > \delta} \le \delta^p + \sum \mathbb{E}\xi_i^p I_{\xi_i > \delta}. \qquad \square$$

Inequality (1.4.3) corresponds to the heuristics indicated at the beginning of this subsection, particularly if $\sum \Pr\{\xi_i = \delta_0\} = 0$.

1.4.2 Estimating L_p-Norms in Hilbert Space

Let X_i be independent real symmetric random variables. As we will see in the first lemma below, for all $\delta > 0$,

$$\mathbb{E}\Big|\sum X_i\Big| \simeq \mathbb{E}\Big|\sum X_i I_{|X_i| \le \delta}\Big| + \mathbb{E}\Big|\sum X_i I_{|X_i| > \delta}\Big|. \tag{1.4.5}$$

If we take $\delta = \delta_0(\lambda)$ as in (1.4.1), then we will not lose much by using the triangle inequality to estimate $\mathbb{E}\big|\sum X_i I_{|X_i|>\delta_0}\big|$ by $\sum \mathbb{E}\big|X_i\big| I_{|X_i|>\delta_0}$, intuitively, because most of the summands $X_i I_{|X_i|>\delta_0}$ are zero, and with complete rigor, because, by Corollary 1.4.2 and Lévy's inequality (1.1.2),

$$\sum \mathbb{E}\big|X_i\big| I_{|X_i|>\delta_0} \le (1+\lambda)\mathbb{E}\max|X_i| I_{|X|_i>\delta_0} \le 2(1+\lambda)\mathbb{E}\Big|\sum X_i I_{|X_i|>\delta_0}\Big|.$$

Hence, with this choice of δ the triangle inequality is accurate for estimating the second term in (1.4.5). As for the first term, it is Hölder plus independence (the variance of a sum of independent variables equals the sum of the variances) that is relatively accurate: not only do we have

$$\mathbb{E}\Big|\sum X_i I_{|X_i|\le\delta_0}\Big| \le \Big(\sum \mathbb{E}X_i^2 I_{|X_i|\le\delta_0}\Big)^{1/2},$$

but, by Hoffmann-Jørgensen's inequality (Theorem 1.2.5) with $p = 2$ and $q = 1$, if $\lambda \le 1/8$, then also

$$\begin{aligned}\mathbb{E}\Big(\sum X_i I_{|X_i|\le\delta_0}\Big)^2 &\le K\Big[\mathbb{E}\max|X_i|^2 I_{|X_i|\le\delta_0} + \Big(\mathbb{E}\Big|\sum X_i I_{|X_i|\le\delta_0}\Big|\Big)^2\Big] \\ &\le K\Big[\delta_0^2 + \Big(\mathbb{E}\Big|\sum X_i I_{|X_i|\le\delta_0}\Big|\Big)^2\Big].\end{aligned}$$

Since δ_0 is smaller than a constant times $\mathbb{E}\left|\sum X_i\right|$ by Proposition 1.4.1 and Lévy, we conclude that

$$\mathbb{E}\Big|\sum X_i\Big| \simeq \Big(\sum \mathbb{E}X_i^2 I_{|X_i|\le\delta_0}\Big)^{1/2} + \sum \mathbb{E}|X_i| I_{|X_i|>\delta_0}.$$

Note that the right-hand side of this inequality only depends on the marginal distributions of the variables X_i, thus solving the problem posed at the beginning of this section. The proof of this inequality involves Hoffmann-Jørgensen, Lévy, the estimates of moments of maxima from the previous subsection, and the equivalence (1.4.5).

The above comments simply describe a pattern of proof of a much more general result, which we present now. We begin by justifying equation (1.4.5). One side of it is trivial by the triangle inequality. The other side consists of a truncation principle for symmetric variables that can be proved in the context of separable Banach spaces B.

LEMMA 1.4.3. *If X_i are independent symmetric B-valued random variables, then, for all sets $A \subset B$ measurable and symmetric about the origin and for all $t > 0$,*

$$\Pr\Big\{\Big\|\sum_i X_i I_{X_i\in A}\Big\| > t\Big\} \le 2\Pr\Big\{\Big\|\sum_i X_i\Big\| > t\Big\}. \tag{1.4.6}$$

In particular,

$$\mathbb{E}\Big\|\sum_i X_i I_{X_i\in A}\Big\|^p \le 2\mathbb{E}\Big\|\sum_i X_i\Big\|^p \tag{1.4.7}$$

for all $p > 0$.

PROOF. By integration by parts, only the first inequality requires proof. Then, convexity of the norm and symmetry of the set A and the independent variables X_i easily justify the following estimate:

$$\begin{aligned}\Pr\Big\{\Big\|\sum_i X_i I_{X_i\in A}\Big\| > t\Big\} &\le \Pr\Big\{\Big\|\sum_i X_i I_{X_i\in A} + \sum_i X_i I_{X_i\notin A}\Big\| > t\Big\}\\ &\quad + \Pr\Big\{\Big\|\sum_i X_i I_{X_i\in A} - \sum_i X_i I_{X_i\notin A}\Big\| > t\Big\}\\ &= 2\Pr\Big\{\Big\|\sum_i X_i\Big\| > t\Big\}.\end{aligned}$$

The first step consists of a typical convexity argument (if $\|\alpha\| > t$, then either $\|\alpha+\beta\| > t$ or $\|\alpha-\beta\| > t$). Here is a detailed justification of the second step: we must show that if X is symmetric and A is a symmetric set, then the probability laws of X and $XI_{X\in A} - XI_{X\notin A}$ coincide. To see this we note first that

$\{X \notin A\} = \{-X \notin A\}$ by symmetry of A and then, that this fact and symmetry of X imply that for every measurable $C \subset B$,

$$\begin{aligned}\Pr\{XI_{X\notin A} \in C, X \notin A\} &= \Pr\{-XI_{-X\notin A} \in C, -X \notin A\}\\ &= \Pr\{-XI_{X\notin A} \in C, X \notin A\}.\end{aligned}$$

These two observations now give that, for all measurable $C \subset B$,

$$\begin{aligned}\Pr\{X \in C\} &= \Pr\{XI_{X\in A} + XI_{X\notin A} \in C\}\\ &= \Pr\{XI_{X\in A} \in C, X \in A\} + \Pr\{XI_{X\notin A} \in C, X \notin A\}\\ &= \Pr\{XI_{X\in A} \in C, X \in A\} + \Pr\{-XI_{X\notin A} \in C, X \notin A\}\\ &= \Pr\{XI_{X\in A} - XI_{X\notin A} \in C\}. \qquad \square\end{aligned}$$

It may be interesting to note that this proof gives a slightly more general inequality than (1.4.6), namely

$$\Pr\Big\{\sum_i X_i I_{X_i\in A} \in K^c\Big\} \le 2\Pr\Big\{\sum_i X_i \in K^c\Big\}, \tag{1.4.6$'$}$$

for all convex symmetric subsets $K \subset B$.

THEOREM 1.4.4. *For every $p, q > 0$ there exist finite positive constants $K_1(p)$ and $K_2(p, q)$ such that, if X_i, $i \le n < \infty$, are independent symmetric B-valued random variables with norm in L_p, B a separable Banach space, and if δ_0 is defined as*

$$\delta_0 = \inf\Big[t : \sum_i \Pr\{\|X_i\| > t\} \le 1/8\Big],$$

then,

$$\begin{aligned}&K_1(p)\tilde{K}_1(q)\Big[\Big\|\sum_i X_i I_{\|X_i\|\le\delta_0}\Big\|_q + \Big\|\max_i \|X_i\|\Big\|_p\Big]\\ &\le \Big\|\sum_i X_i\Big\|_p \le K_2(p,q)\Big[\Big\|\sum_i X_i I_{\|X_i\|\le\delta_0}\Big\|_q + \Big\|\max_i \|X_i\|\Big\|_p\Big]. \qquad (1.4.8)\end{aligned}$$

The constants K_i are bounded on subsets of points (p, q) which are bounded and bounded away from zero, $\limsup_{p\to\infty} K_2(p,q)/p < \infty$ for every $q > 0$ fixed, $\liminf_{p\to 0}(K_1(p))^p > 0$ and $\liminf_{q\to 0}(\tilde{K}_1(q))^q > 0$.

PROOF. For the upper bound we write

$$\|S_n\|_p \le \Big\|\sum_i X_i I_{\|X_i\|\le\delta_0}\Big\|_p + \Big\|\sum_i X_i I_{\|X_i\|>\delta_0}\Big\|_p$$

and consider each summand separately. By Hölder for $q \ge p$ and by Theorem 1.2.5 (Hoffmann-Jørgensen) for $q < p$, we have

$$\Big\|\sum_i X_i I_{\|X_i\|\le\delta_0}\Big\|_p \le \begin{cases}\big\|\sum_i X_i I_{\|X_i\|\le\delta_0}\big\|_q & \text{if } q \ge p\\ K_q\frac{p}{q}\big[\big\|\sum_i X_i I_{\|X_i\|\le\delta_0}\big\|_q + \|X_n^*\|_p\big] & \text{if } q \le p\end{cases}$$

where K_q is bounded for q bounded away from zero. As for the second summand we can apply Theorem 1.2.3 (also H–J) with $t_0 = 0$ because, by definition of δ_0,

$$\Pr\Big\{\Big\|\sum_i X_i I_{\|X_i\|>\delta_0}\Big\| > 0\Big\} \le \sum_i \Pr\big\{\|X_i\| > \delta_0\big\} \le 1/8.$$

This gives

$$\Big\|\sum_i X_i I_{\|X_i\|>\delta_0}\Big\|_p \le K_p \|X_n^*\|_p$$

with $\limsup_{p\to\infty} K_p/p < \infty$, proving the upper bound.

For the lower bound, we first note that, by Lévy's inequality (Theorem 1.1.1),

$$\|X_n^*\|_p \le 2^{1/p}\|S_n\|_p. \tag{1.4.9}$$

As for the truncated moments term, we apply Hölder and Lemma 1.4.3 for $q \le p$ and Theorem 1.2.3 (with $t_0 = 8^{1/p}\|S_n\|_p$) for $q > p$, to obtain

$$\Big\|\sum_i X_i I_{\|X_i\|\le\delta_0}\Big\|_q \le \begin{cases} \big\|\sum_i X_i I_{\|X_i\|\le\delta_0}\big\|_p \le 2^{1/p}\|S_n\|_p & \text{if } q \le p \\ K_q\big[8^{1/p}\big\|\sum_i X_i I_{\|X_i\|\le\delta_0}\big\|_p + \delta_0\big] & \text{if } q > p \end{cases}$$

where K_q is as in Theorem 1.2.3, in particular, it is of the order of $16^{1/q}$ for small q. Now, the lower bound follows since, by Lemma 1.4.3,

$$\Big\|\sum_i X_i I_{\|X_i\|\le\delta_0}\Big\|_p \le 2^{1/p}\|S_n\|_p$$

and, by inequality (1.4.2) followed by (1.4.9),

$$\delta_0 \le 9^{1/p}\|X_n^*\|_p \le 18^{1/p}\|S_n\|_p.$$

The properties of the functions $K_1(p)$, $\tilde{K}_1(q)$ follow from the previous estimates together with the definition of K_p in Theorem 1.2.3. □

REMARK 1.4.5. *(Positive random variables and functions)* Suppose X_i are independent non-negative random variables or non-negative functions in a Banach space of functions whose norm satisfies $\|f\| \le \|g\|$ whenever $0 \le f \le g$, such as the L_p spaces or $C(S)$. Then, Lévy's maximal inequalities trivially hold, not only for probabilities but pointwise, and so does the truncation inequality in Lemma 1.4.3. Since Hoffmann-Jørgensen's inequalities hold as well, it follows from the above proof that *the inequalities (1.4.8) hold unchanged* for these variables.

REMARK 1.4.6. *(Centered variables)* By the randomization Lemma 1.2.6, it is obvious that the inequalities (1.4.8) extend to *centered* not necessarily symmetric random variables provided $p \ge 1$, as follows:

$$\begin{aligned} K_1(p)\tilde{K}_1(q)\Big[\Big\|\sum_i \varepsilon_i X_i I_{\|X_i\|\le\delta_0}\Big\|_q + \big\|\max_i \|X_i\|\big\|_p\Big] &\le \Big\|\sum_i X_i\Big\|_p \\ &\le K_2(p,q)\Big[\Big\|\sum_i \varepsilon_i X_i I_{\|X_i\|\le\delta_0}\Big\|_q + \big\|\max_i \|X_i\|\big\|_p\Big], \end{aligned} \tag{1.4.8$'$}$$

where ε_i are independent Rademacher variables indepedent from the variables X_i, and where the constants differ from those in (1.4.8) by a factor of 1/2 for K_1 and 2 for K_2.

REMARK 1.4.7. *(Maxima of partial sums)* It also goes without saying that, by Lemma 1.2.6 for centered variables, by Lévy's inequality for symmetric, and trivially for non-negative, $\|S_n\|_p$ can be replaced by $\|S_n^*\|_p$ in (1.4.8) and (1.4.8′), with the pertinent changes in constants, just as with Hoffmann-Jørgensen inequalities.

Now, we know from Proposition 1.4.1 and Corollary 1.4.2 that the moments of the max can be bounded by expressions that only contain one-dimensional integrals. So, combining these results with Theorem 1.4.4 *for* $q = 1$, we obtain:

THEOREM 1.4.8. 1) *If X_i are independent non-negative real random variables or functions (in a space of functions whose norm satisfies $\|f\| \le \|g\|$ if $0 \le f \le g$), or if they are independent B-valued random variables, symmetric for $p < 1$, centered for $p \ge 1$, then, given $p > 0$, there exist finite positive constants $K_2(p)$ depending only on p such that*

$$\Big\|\sum_i X_i\Big\|_p \le K_2(p)\Big[\sum_i \mathbb{E}\|X_i\| I_{\|X_i\|\le\delta_0} + \Big(\sum_i \mathbb{E}\|X_i\|^p I_{\|X_i\|\ge\delta_0}\Big)^{1/p}\Big], \quad (1.4.10)$$

where

$$\delta_0 = \inf\Big[t : \sum_i \Pr\{\|X_i\| > t\} \le 1/8\Big].$$

2) *If X_i are independent non-negative real random variables or independent non-negative functions in a L_1 space, then, given $p > 0$ there exist finite positive constants $K_1(p)$ and $K_2(p)$ depending on p only, such that, if δ_0 is as above, then*

$$\begin{aligned} K_1(p)\Big[\delta_0 + \sum_i \mathbb{E}\|X_i\| I_{\|X_i\|\le\delta_0} + \Big(\sum_i \mathbb{E}\|X_i\|^p I_{\|X_i\|>\delta_0}\Big)^{1/p}\Big] & \\ \le \Big\|\sum_i X_i\Big\|_p & \quad (1.4.11) \\ \le K_2(p)\Big[\delta_0 + \sum_i \mathbb{E}\|X_i\| I_{\|X_i\|\le\delta_0} + \Big(\sum_i \mathbb{E}\|X_i\|^p I_{\|X_i\|>\delta_0}\Big)^{1/p}\Big]. & \end{aligned}$$

In both cases the constants K_i are bounded on subsets of points p, which are bounded and bounded away from zero, $\limsup_{p\to\infty} K_2(p)/p < \infty$ and $\liminf_{p\to 0}(K_1(p))^p > 0$.

The same comment applies to symmetric or centered variables taking values in a Hilbert space, *provided we take* $q = 2$ because, in a Hilbert space, if X_i are independent centered random vectors, then

$$\mathbb{E}\Big\|\sum_i X_i\Big\|^2 = \sum_i \mathbb{E}\|X_i\|^2.$$

We then have:

THEOREM 1.4.9. *Given $p > 0$ there exist finite positive constants $K_1(p)$ and $K_2(p)$ such that if H is a separable Hilbert space, X_i are independent H-valued random variables with norm in L_p, centered if $p \geq 1$, symmetric if $p < 1$, and if δ_0 is defined by*

$$\delta_0 = \inf\Big[t : \sum_i \Pr\{\|X_i\| > t\} \leq 1/8\Big],$$

then,

$$\begin{aligned} K_1(p)\Big[\delta_0 + \Big(\sum_i \mathbb{E}\|X_i\|^2 I_{\|X_i\| \leq \delta_0}\Big)^{1/2} + \Big(\sum_i \mathbb{E}\|X_i\|^p I_{\|X_i\| > \delta_0}\Big)^{1/p}\Big] \\ \leq \Big\| \sum_i X_i \Big\|_p \qquad (1.4.12) \\ \leq K_2(p)\Big[\delta_0 + \Big(\sum_i \mathbb{E}\|X_i\|^2 I_{\|X_i\| \leq \delta_0}\Big)^{1/2} + \Big(\sum_i \mathbb{E}\|X_i\|^p I_{\|X_i\| > \delta_0}\Big)^{1/p}\Big] \end{aligned}$$

and

$$\Big\| \sum_i X_i \Big\|_p \leq K_2(p)\Big[\Big(\sum_i \mathbb{E}\|X_i\|^2 I_{\|X_i\| \leq \delta_0}\Big)^{1/2} + \Big(\sum_i \mathbb{E}\|X_i\|^p I_{\|X_i\| \geq \delta_0}\Big)^{1/p}\Big]. \qquad (1.4.12')$$

Moreover, the constants behave as in Theorem 1.4.8.

Inequalities (1.4.10,11) and (1.4.12,12′) simply combine (1.4.8) and (1.4.8′) with (1.4.3) and (1.4.4).

As mentioned at the beginning of Section 2, although we have proved Hoffmann–Jørgensen's inequality with constants of the order of p as $p \to \infty$, these constants are in fact of the order of $p/\log p$ (Talagrand, 1989) and therefore, the order of the constants $K_2(p)$ and $K_2(p, q)$ in all the theorems in this subsection can also be improved to $p/\log p$.

1.4.3 K-Function Bounds

We next obtain a second set of two-sided bounds for the moments of $\sum X_i$, with X_i independent. These two-sided bounds, which are also computable with only one-dimensional integrals with respect to the individual laws of the variables, were chronologically the first to be derived. We consider in full only the case of X_i i.i.d. centered and $p = 1$.

Given a random variable $X \not\equiv 0$ symmetric and such that $E|X| < \infty$, the K function $K(y)$ is implicitly defined as

$$K_X(y)^2 = y\mathbb{E}\big[X^2 \wedge (|X| K_X(y))\big] = y\mathbb{E}X^2 I_{|X| \leq K_X(y)} + yK_X(y)\mathbb{E}|X| I_{|X| > K_X(y)}. \qquad (1.4.13)$$

$K_X(y)$ exists and is unique since the function $\mathbb{E}\big[(X^2/t^2) \wedge (|X|/t)\big]$, $t > 0$, is strictly decreasing and continuous, with values ranging from 0 to ∞. As an immediate consequence of the definition we have that

$$n\mathbb{E}X^2 I_{|X|\le K_X(n)} \le K_X(n)^2 \quad \text{and} \quad n\mathbb{E}|X| I_{|X|>K_X(n)} \le K_X(n), \tag{1.4.14}$$

and therefore

$$\begin{aligned}\mathbb{E}\Big|\sum_{i=1}^n X_i\Big| &\le \mathbb{E}\Big|\sum_{i=1}^n X_i I_{|X_i|\le K_X(n)}\Big| + \mathbb{E}\Big|\sum_{i=1}^n X_i I_{|X_i|>K_X(n)}\Big| \\ &\le \big[n\mathbb{E}X^2 I_{|X|\le K_X(n)}\big]^{1/2} + n\mathbb{E}|X| I_{|X|>K_X(n)} \le 2K_X(n).\end{aligned} \tag{1.4.15}$$

Besides truncation, this upper bound uses only the two most conspicuous and elementary inequalities in probability, namely, the triangle inequality and Hölder's inequality for second moments plus independence, just as the upper bound part of (1.4.6). The variance identity for sums of independent random variables constitutes a simple and efficient way of capturing the cancellation occurring when adding independent and centered random variables, but X is not square integrable and therefore we must truncate if we are to take advantage of this phenomenon. The level of truncation chosen in Theorem 1.4.4, δ_0, is the smallest possible so that the average number of terms $|X_i|$ exceeding the level δ_0 is not larger than a fixed number (1/8 in that case), and this ensures that we can safely use the otherwise crude triangle inequality in what remains after truncation. But then, as shown by Theorems 1.4.8 and 1.4.9, δ_0 is not large enough to bound $\mathbb{E}|\sum X_i|$. Here K, larger than δ_0, will be level of truncation and bound at the same time. To see that K is larger than δ_0, we just note that, by Chebyshev's inequality and the definition of the K function,

$$\Pr\{|X| > K_X(n)\} = \Pr\big\{X^2 \wedge (|X|K_X(n)) > K_X(n)^2\big\} \le \frac{1}{n}. \tag{1.4.16}$$

This inequality already gives a lower bound for $\mathbb{E}|\sum_{i=1}^n X_i|$ in terms of the K-function: by Lemma 1.4.3 and elementary probability,

$$\begin{aligned}\mathbb{E}\Big|\sum_{i=1}^n X_i\Big| &\ge \mathbb{E}\Big|\sum_{i=1}^n X_i\Big| I_{\#\{i:|X_i|>K_X(n)\}=1} \\ &= n\mathbb{E}\Big[|X_1| I_{|X_1|>K_X(n)} \prod_{i=2}^n I_{|X_i|\le K_X(n)}\Big] \\ &\ge n\Big(1-\frac{1}{n}\Big)^{n-1}\mathbb{E}|X| I_{|X|>K_X(n)} \\ &> e^{-1} n\mathbb{E}|X| I_{|X|>K_X(n)}.\end{aligned} \tag{1.4.17}$$

We now proceed to prove that $\mathbb{E}\left|\sum X_i\right|$ is bounded from below by a constant times $K_X(n)$. First we observe that, by definition (1.4.13),

$$\text{either} \quad n\mathbb{E}X^2 I_{|X|\le K_X(n)} \ge \frac{1}{2}K_X(n)^2 \quad \text{or} \quad n\mathbb{E}|X|I_{|X|>K_X(n)} \ge \frac{1}{2}K_X(n),$$

and that if the latter holds, then (1.4.17) becomes

$$\mathbb{E}\left|\sum X_i\right| > \frac{1}{4e}K_X(n). \tag{1.4.18}$$

So, we can assume

$$n\mathbb{E}X^2 I_{|X|\le K_X(n)} \ge \frac{1}{2}K_X(n)^2. \tag{1.4.19}$$

Setting $Y_i := X_i I_{|X|\le K_X(n)}$ then, by symmetry and independence of the variables X_i,

$$\begin{aligned}
\mathbb{E}\left(\sum Y_i\right)^4 &= \sum \mathbb{E}Y_i^4 + 6\sum_{i<j}\mathbb{E}Y_i^2Y_j^2 \\
&\le K_X(n)^2\sum \mathbb{E}Y_i^2 + 3\left[\sum \mathbb{E}Y_i^2\right]^2 \\
&= K_X(n)^2\mathbb{E}\left(\sum Y_i\right)^2 + 3\left[\mathbb{E}\left(\sum Y_i\right)^2\right]^2.
\end{aligned}$$

Just as in Littlewood's proof of Khinchin's inequality, Hölder's inequality with $1/p = 2/3$ and $1/q = 1/3$ gives [see (1.3.4)]

$$\mathbb{E}\left(\sum Y_i\right)^2 = \mathbb{E}\left[\left(\sum Y_i\right)^{2/3}\left(\sum Y_i\right)^{4/3}\right] \le \left[\mathbb{E}\left|\sum Y_i\right|\right]^{2/3}\left[\mathbb{E}\left(\sum Y_i\right)^4\right]^{1/3}.$$

Solving for $\mathbb{E}\left|\sum Y_i\right|$ and using the previous bound for the fourth moment, we then obtain

$$\mathbb{E}\left|\sum Y_i\right| \ge \frac{\left[\mathbb{E}\left(\sum Y_i\right)^2\right]^{3/2}}{\left[K_X(n)^2\mathbb{E}\left(\sum Y_i\right)^2 + 3\left[\mathbb{E}\left(\sum Y_i\right)^2\right]^2\right]^{1/2}}.$$

(Note here that the upper bound of the fourth moment in terms of the second allows, via Littlewood's argument, to obtain a lower bound for the first moment in terms of the second, just as in the elementary proof of Khinchin's inequality.) Since the function $y = t^3/(K^2t + 3t^2)$ is nondecreasing for $t \ge 0$, this and assumption (1.4.19) yield

$$\mathbb{E}\left|\sum Y_i\right| \ge \frac{\left[K_X(n)^2/2\right]^{3/2}}{\left[K_X(n)^4/2 + 3K_X(n)^4/4\right]^{1/2}} = \frac{1}{\sqrt{10}}K_X(n).$$

We can now use Lemma 1.4.3 once more and conclude that, under assumption (1.4.19),

$$\mathbb{E}\left|\sum X_i\right| \ge \frac{1}{\sqrt{40}}K_X(n). \tag{1.4.20}$$

Then, it follows from this and (1.4.18) that inequality (1.4.18) holds in all cases, which, in combination with (1.4.15), proves the following:

PROPOSITION 1.4.10. *If X, X_i are i.i.d. symmetric integrable random variables, then, for all $n \in \mathbb{N}$,*

$$\frac{1}{4e} K_X(n) \le \mathbb{E}\Big|\sum_{i=1}^n X_i\Big| \le 2K_X(n), \tag{1.4.21}$$

where $K_X(y)$ is defined by equation (1.4.13).

Extending this theorem to i.i.d. centered random variables and maxima of partial sums by Lemma 1.2.6 is by now routine.

The function $K_X(y)$ can be inverted, and this should facilitate computing with it. Let μ be the probability law of $|X|$. Then, Fubini's theorem justifies the following computation:

$$\begin{aligned}
K_X(y)^2 &= y\mathbb{E}X^2 I_{|X|\le K_X(y)} + yK_X(y)\mathbb{E}|X|I_{|X|>K_X(y)}\\
&= y\int_0^{K_X(y)}\Big(\int_0^x du\Big)x\,d\mu(x) + y\int_{K_X(y)}^\infty\Big(\int_0^{K_X(y)} du\Big)x\,d\mu(x)\\
&= y\int_0^{K_X(y)}\Big(\int_u^\infty x\,d\mu(x)\Big)du\\
&= y\int_0^{K_X(y)} E|X|1_{|X|>u}\,du.
\end{aligned}$$

Hence,

$$y = \frac{K_X(y)^2}{\int_0^{K_X(y)} E|X|1_{|X|>u}\,du}$$

or, $K_X(y)$ is the inverse function of

$$G_X(y) = \frac{y^2}{\int_0^y \mathbb{E}|X|I_{|X|>u}\,du}. \tag{1.4.22}$$

We should finally mention that the following monotonicity properties of the K function, which are easily verified, are useful in applications, as we will see in the next chapter:

$$K_X(y)\nearrow\infty,\quad \frac{K_X(y)^2}{y}\nearrow \mathbb{E}X^2\le\infty,\quad\text{and}\quad \frac{K_X(y)}{y}\searrow 0 \tag{1.4.23}$$

as $y\to\infty$.

Proposition 1.4.10 is a very special case of a result that we now describe in some generality. Instead of moments, one may as well consider integrals of more general functions that do not grow too fast (like, e.g., powers multiplied by logarithms). More concretely, given $\alpha > 0$ fixed, we define $\mathcal{F}_\alpha$ as the class of functions

$$\mathcal{F}_\alpha = \Big\{\Phi : [0,\infty) \mapsto \mathbb{R} : \quad \Phi \text{ is strictly increasing, convex and continuous,} \\ \Phi(0) = 0,\ \Phi(cx) \le c^\alpha \Phi(x) \text{ for all } x > 0 \text{ and } c \ge 2\Big\}. \tag{1.4.24}$$

Given a finite set of independent centered random variables $\mathbf{X} = \{X_1, \dots, X_n\}$, then the K function corresponding to $\mathbf{X}$ and $\Phi \in \mathcal{F}_\alpha$ is defined as $\sum_{i=1}^n E\Phi(|X_i|)$ if this sum is 0 or ∞, and otherwise as the unique postive number y such that

$$\sum_{i=1}^n \mathbb{E}X_i^2 I_{|X_i|\le y} + \frac{y^2}{\Phi(y)} \sum_{i=1}^n \mathbb{E}\Phi(|X_i|) I_{|X_i|>y} = y^2. \tag{1.4.25}$$

This definition is the analog of definition (1.4.13) for $K_X(n)$ and $\Phi(x) = x$. Then, the K function approximation result of Klass (1981) is as follows:

THEOREM 1.4.11. *Let $\mathbf{X} = \{X_1, \dots, X_n\}$ be a finite set of independent, integrable, centered random variables, let $\alpha > 0$ and set $S = \sum X_i$. Then, for all $\Phi \in \mathcal{F}_\alpha$ there exist constants $0 < c_\alpha, C_\alpha < \infty$, depending on α only, such that*

$$c_\alpha \Phi(K_{\mathbf{X},\Phi}) \le E\Phi(|S|) \le C_\alpha \Phi(K_{\mathbf{X},\Phi}). \tag{1.4.26}$$

The constant at the right side of equation (1.4.26) is of the order of α^α, comparable with the constants in Theorems 1.4.8 and 1.4.9.

1.4.4 A General Decoupling for Sums of Arbitrary Positive Random Variables and Martingales

In this subsection we show that, surprisingly, moments of sums of arbitrary non-negative random variables compare to moments of sums of independent random variables with the same individual probability distributions, and that the same is true for martingales. The comparison becomes a trivial identity for $p = 1$ in the case of positive variables and for $p = 2$ in the martingale case. As will become apparent in the proof, what makes these comparisons possible is the availability, from either of the last two subsections, of two-sided bounds for moments of sums of independent random variables that depend only on truncated moments of the *individual* summands.

The results in this section (Theorems 1.4.12 and 1.4.14) may be considered as *decoupling results* for sums of independent non-negative random variables and for martingales in the sense that these sums are compared to related sums of independent random variables.

We should recall that, given an arbitrary sequence of random variables $\{X_i\}_{i=1}^n$, $n \le \infty$, there always exist a sequence $\{Y_i\}_{i=1}^n$ of *independent* random variables with the same individual marginals as the original sequence, that is, such that the probability distribution of Y_i is the same as that of X_i for all i. One such sequence $\{Y_i\}$ simply consists of the set of coordinate functions on the product probability space $\left(\mathbb{R}^n, \mathcal{B}^n, \prod_{i=1}^n \mathcal{L}(X_i)\right)$, where $\mathcal{L}(X)$ denotes the probability law of the random variable X, $\Pr \circ X^{-1}$.

THEOREM 1.4.12. *There exist finite positive constants $c < \infty$, $c_p < \infty$ for $p \geq 1$, such that, if X_i, $i \leq n < \infty$, are non-negative but otherwise arbitrary real random variables in L_p, $p > 0$, and if Y_i are independent random variables satisfying that, for each i, the probability distribution of Y_i equals that of X_i, then*

$$\mathbb{E}\Big(\sum_{i=1}^n X_i\Big)^p \leq c\mathbb{E}\Big(\sum_{i=1}^n Y_i\Big)^p \quad \text{for } 0 < p \leq 1 \tag{1.4.27}$$

and

$$\mathbb{E}\Big(\sum_{i=1}^n Y_i\Big)^p \leq c_p\mathbb{E}\Big(\sum_{i=1}^n X_i\Big)^p \quad \text{for } p \geq 1. \tag{1.4.28}$$

PROOF. If $0 \leq p \leq 1$, then the function $|x|^p$ is concave and $(\sum |x_i|)^p \leq \sum |x_i|^p$. Hence, Theorem 1.4.8 [the left side of inequality (1.4.11)] applied to the variables Y_i gives:

$$\begin{aligned}
\mathbb{E}\big(\sum X_i\big)^p &\leq \mathbb{E}\big(\sum X_i I_{X_i \leq \delta_0}\big)^p + \mathbb{E}\big(\sum X_i I_{X_i > \delta_0}\big)^p \\
&\leq \big(\sum \mathbb{E} X_i I_{X_i \leq \delta_0}\big)^p + \sum \mathbb{E} X_i^p I_{X_i > \delta_0} \\
&= \big(\sum \mathbb{E} Y_i I_{Y_i \leq \delta_0}\big)^p + \sum \mathbb{E} Y_i^p I_{Y_i > \delta_0} \\
&\leq c\mathbb{E}\big(\sum Y_i\big)^p,
\end{aligned}$$

with δ_0 as defined in Theorem 1.4.8 and $c = \sup_{0<p\leq 1}(K_1(p))^{-p} < \infty$, $K_1(p)$ as in (1.4.11) in the same theorem. The second inequality is proved in the same way: if $p \geq 1$, then $|x|^p$ is convex and $\sum |x_i|^p \leq (\sum |x_i|)^p$, so that, by Theorem 1.4.8 [inequality (1.4.10)], we have

$$\begin{aligned}
\frac{1}{2^{p-1}K_2(p)^p}\mathbb{E}\Big(\sum_{i=1}^k Y_i\Big)^p &\leq \big(\sum \mathbb{E} Y_i I_{Y_i \leq \delta_0}\big)^p + \sum \mathbb{E} Y_i^p I_{Y_i \geq \delta_0} \\
&= \big[\mathbb{E}\big(\sum X_i I_{X_i \leq \delta_0}\big)\big]^p + \mathbb{E}\big(\sum X_i^p I_{X_i \geq \delta_0}\big) \\
&\leq \mathbb{E}\big(\sum X_i I_{X_i \leq \delta_0}\big)^p + \mathbb{E}\big(\sum X_i I_{X_i \geq \delta_0}\big)^p \\
&\leq 2\mathbb{E}\big(\sum X_i\big)^p.
\end{aligned}$$

□

It is interesting to realize that inequalities (1.4.27) and (1.4.28) do not reverse in general. Let g_i be i.i.d. $N(0, 1)$ random variables and let τ_i, $i = 0, 1, \ldots$, be i.i.d. with $\Pr\{\tau = 1\} = \alpha = 1 - \Pr\{\tau = 0\}$. We set $X_i = \tau_0 g_i^2$ and $Y_i = \tau_i g_i^2$. Then, letting $a \approx_p b$ denote two-sided inequalities between a and b possibly multiplied by constants that depend only on p, we see that, by Lemma 1.4.13 immediately below this example, and by the properties of normal variables,

$$\mathbb{E}\Big|\sum_{i=1}^{k}\tau_i g_i^2\Big|^p \approx_p \mathbb{E}\Big|\sum_{i=1}^{n}\tau_i g_i\Big|^{2p} = \mathbb{E}\Big[\mathbb{E}\Big(\big(\sum_{i=1}^{n}\tau_i g_i\big)^{2p}\big|\{\tau_i\}\Big)\Big]$$
$$\approx_p \mathbb{E}\Big(\sum_{i=1}^{n}\tau_i^2\Big)^p = \mathbb{E}\Big(\sum_{i=1}^{n}\tau_i\Big)^p.$$

Now we use Theorem 1.4.8 to estimate this last expected value: assuming $n\alpha > 1/8$ it turns out that $\delta_0 = 1$, and therefore

$$\mathbb{E}\Big(\sum_{i=1}^{n}\tau_i\Big)^p \approx_p \big(n\mathbb{E}\tau I_{\tau\le 1}\big)^p + n\mathbb{E}\tau I_{\tau>1} + \delta_0 = (n\alpha)^p + 1 \approx_p (n\alpha)^p.$$

On the other hand, also by the properties of normal variables,

$$\mathbb{E}\Big(\tau_0\sum_{i=1}^{n}g_i^2\Big)^p = \alpha\mathbb{E}\Big(\sum_{i=1}^{n}g_i^2\Big)^p \approx_p \alpha n^p.$$

Then, taking $\alpha = 1/n$ shows that these bounds are of different order of magnitude whenever $p \ne 1$.

Let now $\{X_i\}$ be a martingale difference sequence. Then, by the Burkholder–Davis–Gundy inequalities (Theorem 6.6.2 below, for $\Phi(x) = |x|^p$), we have

$$\mathbb{E}\max_{1\le k\le n}\Big|\sum_{i=1}^{k}X_i\Big|^p \approx_p \mathbb{E}\Big(\sum_{i=1}^{n}X_i^2\Big)^{p/2}, \qquad p \ge 1, \tag{1.4.29}$$

so that estimation of moments of the maximal function of a martingale reduces to estimation of moments of sums of non-negative random variables, and we can apply the last theorem. The following corollary to Khinchin's inequality will help us obtain a neater result; it goes by the name of Marcinkiewicz inequalities, and will be used more than once throughout.

LEMMA 1.4.13. *Given $p \ge 1$, there exist constants c_p, C_p, depending only on p such that, if Y_i, $1 \le i \le n$, are centered, independent and in L_p, then*

$$c_p^{-1}\mathbb{E}\Big(\sum_{i=1}^{n}Y_i^2\Big)^{p/2} \le \mathbb{E}\Big|\sum_{i=1}^{n}Y_i\Big|^p \le C_p\mathbb{E}\Big(\sum_{i=1}^{n}Y_i^2\Big)^{p/2} \tag{1.4.30}$$

and

$$c_p^{-1}\mathbb{E}\Big(\sum_{i=1}^{n}Y_i^2\Big)^{p/2} \le \mathbb{E}\max_{1\le k\le n}\Big|\sum_{i=1}^{k}Y_i\Big|^p \le 2C_p\mathbb{E}\Big(\sum_{i=1}^{n}Y_i^2\Big)^{p/2}. \tag{1.4.31}$$

PROOF. (1.4.30) and (1.4.31) simply combine Khinchin's inequalities in Section 1.3 with Lemma 1.2.6 on symmetrization. □

The reader can easily estimate these constants; for instance, for $p = 1$, $c_p = 1/2^{3/2}$. Note that the Burkholder–Davis–Gundy inequalities are just the extension to martingales of the Marcinkiewicz inequalities.

THEOREM 1.4.14. *There exist finite positive constants $c < \infty$, $c_p < \infty$ for $p \geq 2$ such that, if the variables X_i, $1 \leq i \leq n < \infty$, form a martingale difference sequence and are in L_p for some $p \geq 1$, and if Y_i, $1 \leq i \leq n$, are independent random variables satisfying that, for each i, the probability distribution of Y_i equals that of X_i, then*

$$\mathbb{E} \max_{1 \leq k \leq n} \Big|\sum_{i=1}^{n} X_i\Big|^p \leq c\mathbb{E}\Big|\sum_{i=1}^{n} Y_i\Big|^p \quad \text{for } 1 \leq p \leq 2 \tag{1.4.32}$$

and

$$\mathbb{E} \max_{1 \leq k \leq n} \Big|\sum_{i=1}^{k} Y_i\Big|^p \leq c_p \mathbb{E} \max_{1 \leq k \leq n} \Big|\sum_{i=1}^{k} X_i\Big|^p \quad \text{for } p \geq 2. \tag{1.4.33}$$

PROOF. It follows by combining Theorem 1.4.12 with the Burkholder–Davis–Gundy inequalities (1.4.29) and the Marcinkiewicz inequalities (1.4.30) and (1.4.31). □

Instead of using Theorem 1.4.8 in the proof of the results in this subsection, we could as well have used the K function bounds. And, of course, powers can be replaced by functions Φ satisfying appropriate growth and convexity conditions (in analogy with Theorem 1.4.11).

1.5 Estimates with Sharp Constants for the L_p-Norms of Sums of Independent Random Variables: The L-Function

Let X_i, $i \in \mathbb{N}$, be independent real random variables with probability laws P_i. One may ask whether it is possible to bound the L_p-norm $\| \sum_{i=1}^{n} X_i \|_p$ from above and from below by constants times a quantity $L(\{X_i\}, p)$, computable only with one-dimensional integrals, *in such a way that the multiplicative constants do not depend on n or p, at least for p bounded away from zero.* Moreover, we would like the function L to enjoy certain monotonicity properties that will make it useful when dealing with problems like the one of decoupling the stopping time T in $\mathbb{E}|X_1 + \cdots + X_T|$ (Section 2.1). The answer is positive if the variables are all non-negative or all centered. The centered case reduces to the symmetric case by Lemma 1.2.6.

At the end of the section the L function upper bound is applied to obtain a classical inequality for moments, Rosenthal's inequality, with essentially best constants and, what is almost the same, also to get essentially best constants in Hoffmann-Jørgensen's inequalities for real variables. Moments of linear combinations of Rademacher variables are also considered.

We should recall that, if X is a random variable, we write $\|X\|_p := \left(\mathbb{E}|X|^p\right)^{1/p}$ for all $p > 0$.

We begin with a somewhat loose discussion that will make the L function look more or less natural. For this we need the notion of an Orlicz norm, of which L_p-norms are examples. Given a Young function ψ—that is, a monotone convex function on the non-negative reals, increasing to infinity and such that $\psi(0) = 0$—we denote as $L_\psi(\Omega, \Sigma, P)$ the set of all real random variables ξ on Ω such that $\mathbb{E}\psi(t|\xi|) < \infty$ for some $t > 0$. L_ψ, which is called an Orlicz space, is a normed space, with the norm defined by the equation

$$\|\xi\|_\psi := \inf\{c > 0 : \mathbb{E}\psi(|\xi|/c) \leq 1\}$$

(Krasnoselsky and Rutitsky, 1961). L_p, $p \geq 1$, is the Orlicz space corresponding to the Young function $\psi(x) = x^p$. The collection of random variables with first-order exponential integrability constitute the Orlicz space L_{ψ_1}, with Young function $\psi_1(x) := e^x - 1$. By the properties of the exponential, the ψ_1 norm of a sum of independent non-negative random variables does depend only on one-dimensional integrals and can be written as

$$\|\sum X_i\|_{\psi_1} = \inf\{c > 0 : \sum \log \mathbb{E}\exp(X_i/c) \leq \log 2\}.$$

Can we find a modification of the Young function x^p whose computation involves only one-dimensional integrals? Of course, this is not possible, but it is natural to try with functions that are of the order of x^p for x large (the xs that matter for integrability) but that behave as much as possible as exponentials (given that exponentials have the desired property). One such function is $|1 + x|^p$, which is close to e^{px} for small x and is of the order of x^p for $x > 1$. It is somewhat illustrative to check if this approach works with X, X_i i.i.d. and non-negative: if t_n is defined as the only positive number satisfying the relation

$$n \log \mathbb{E}\left(1 + \frac{X}{t_n}\right) = 1,$$

then, indeed,

$$t_n = \frac{\mathbb{E}X}{e^{1/n} - 1} \simeq n\mathbb{E}X = \mathbb{E}\left(\sum_{i=1}^{n} X_i\right)$$

for large n, with precise two-sided inequalities for every n. This definition, with the required formal changes, will work for any p, and for centered as well as for non-negative random variables.

Let $\{X_i\}$ be a finite set of independent random variables and let $p > 0$. Then the L function is defined as

$$L(\{X_i\}, p) = \inf\left\{t > 0 : \sum \log \mathbb{E}\left|1 + \frac{X_i}{t}\right|^p \leq p\right\}, \tag{1.5.1}$$

a kind of Orlicz norm. If $\mathbb{E}|X|^p < \infty$ and X is not a.s. 0, then, as a function of t, $\mathbb{E}\left|1 + X/t\right|^p$ is continuous at $t \neq 0$ by dominated convergence. Therefore, L satisfies

$$\sum \log \mathbb{E}\left|1 + \frac{X_i}{L\big(\{X_i\}, p\big)}\right|^p = p \tag{1.5.2}$$

if X_i is not 0 a.s. for all i. Moreover, $L(\{X_i\}, p) = 0$ if and only if $X_i = 0$ a.s. for all i.

It is indeed very easy to see that the L function upperbounds the L_p-norm of $\sum X_i$ for independent non-negative variables: since

$$1 + x + y \leq (1 + x)(1 + y), \qquad x, y \geq 0,$$

it follows by independence that

$$\mathbb{E}\left(1 + \frac{\sum X_i}{t}\right)^p \leq \prod_i \mathbb{E}\left(1 + \frac{X_i}{t}\right)^p; \tag{1.5.3}$$

using this and the fact that, by (1.5.2), if $t_0 := L\big(\{X_i\}, p\big)$, then

$$\sum \log \mathbb{E}\left(1 + \frac{X_i}{t_0}\right)^p = p,$$

we obtain

$$\begin{aligned} \frac{1}{t_0^p}\mathbb{E}\Big(\sum X_i\Big)^p &\leq \mathbb{E}\left(1 + \frac{\sum X_i}{t_0}\right)^p \\ &\leq \prod_i \mathbb{E}\left(1 + \frac{X_i}{t_0}\right)^p = e^p \end{aligned} \tag{1.5.4}$$

if the variables X_i are not all 0 a.s. [otherwise, $\mathbb{E}\big(\sum X_i\big)^p = t_0^p = 0$]. What makes this interesting is that there is a similar lower bound. To prove this, we will isolate in a lemma the relevant properties of the functional

$$\phi_p(X) := \mathbb{E}|1 + X|^p. \tag{1.5.5}$$

LEMMA 1.5.1. *Let X_i be independent and non-negative random variables and let $p > 0$. Then,*

a) $\phi_p(\sum X_i) \leq \prod_i \phi_p(X_i)$;

b) $\phi_p\big(2X_1 + \phi_p^{2/p}(X_1)X_2\big) \geq \phi_p(X_1)\phi_p(X_2)$; *and*

c) *if* $\sum_i \log \phi_p(X_i) \leq p$, *then* $\phi_p\big(2e^2 \sum X_i\big) \geq \prod_i \phi_p(X_i)$.

PROOF. a) is just (1.5.3). b) is a consequence of the following two elementary facts: for $t \geq 1$, $x \geq 1$,

$$1 + tx \geq t^{1/2}(1 + x),$$

and, for $x \geq 0$ and $0 \leq y < 1$,

$$1 + 2x + \phi_p^{2/p}(X_1)y \geq 1 + (1+y)x + y = (1+x)(1+y)$$

[note $\phi_p(X_1) \geq 1$]. The first inequality gives

$$\begin{aligned}\big(1 + 2X_1 + \phi_p^{2/p}(X_1)X_2\big)I_{X_2 \geq 1} &\geq \big(1 + \phi_p^{2/p}(X_1)X_2\big)I_{X_2 \geq 1} \\ &\geq \phi_p^{1/p}(X_1)(1 + X_2)I_{X_2 \geq 1},\end{aligned}$$

and, upon raising to the pth power and integrating,

$$\mathbb{E}\big(1 + 2X_1 + \phi_p^{2/p}(X_1)X_2\big)^p I_{X_2 \geq 1} \geq \phi_p(X_1)\mathbb{E}(1 + X_2)^p I_{X_2 \geq 1}.$$

The second inequality yields

$$\mathbb{E}\big(1 + 2X_1 + \phi_p^{2/p}(X_1)X_2\big)^p I_{X_2 < 1} \geq \mathbb{E}(1 + X_1)^p \mathbb{E}(1 + X_2)^p I_{X_2 < 1}.$$

Combining the last two inequalities gives b).

To show that c) holds, setting $Y_k = 2\Big(\prod_{i=1}^k \phi_p^{2/p}(X_i)\Big)\big(\sum_{i=1}^k X_i\big), k = 1, \dots, n,$ it suffices to prove

$$\phi_p(Y_k) \geq \prod_{i=1}^k \phi_p(X_i), \qquad k \leq n,$$

under the assumption that $\sum \log \phi_p(X_i) \leq p$. This is obvious for $k = 1$ since $\phi_p(X_1) \geq 1$ and ϕ_p is monotone increasing. We assume the inequality holds for k and prove it for $k+1$. Since

$$\begin{aligned}Y_{k+1} &= 2\Big[\phi_p^{2/p}(X_{k+1})\prod_{i=1}^k \phi_p^{2/p}(X_i)\Big]\Big[X_{k+1} + \sum_{i=1}^k X_i\Big] \\ &\geq Y_k\phi_p^{2/p}(X_{k+1}) + 2X_{k+1},\end{aligned}$$

monotonicity of ϕ_p, part b) and the induction hypothesis give

$$\begin{aligned}\phi_p(Y_{k+1}) &\geq \phi_p\big(2X_{k+1} + \phi_p^{2/p}(X_{k+1})Y_k\big) \\ &\geq \phi_p(X_{k+1})\phi_p(Y_k) \geq \prod_{i=1}^{k+1}\phi_p(X_i).\end{aligned}$$ □

Now let X_i be an independent non-negative finite sequence and $t_0 := L\big(\{X_i\}, p\big)$, as above. To lowerbound $\| \sum X_i \|_p$ we use property c) in the previous lemma. By (1.5.2) we can apply Lemma 1.5.1 c) to X_i/t_0 and obtain

$$\phi_p\Big(2e^2\frac{\sum X_i}{t_0}\Big) \geq e^p. \tag{1.5.6}$$

Since for $Z \geq 0$,

$$\phi_p(Z) = \mathbb{E}(1 + Z)^p \leq (1 + \|Z\|_p)^p \quad \text{if } p \geq 1$$

and

$$\phi_p(Z) = \mathbb{E}(1+Z)^p \le \mathbb{E}(1+Z^p) = 1 + \|Z\|_p^p \quad \text{if } p < 1,$$

inequality (1.5.6) does become a lower bound for $\|\sum X_i\|_p$ in terms of t_0. This and (1.5.4) give:

THEOREM 1.5.2. *If X_i are independent non-negative random variables and $p > 0$, then*

$$\frac{\left(e^{p\wedge 1}-1\right)^{1/(p\wedge 1)}}{2e^2} L(\{X_i\}, p) \le \left\|\sum X_i\right\|_p \le eL(\{X_i\}, p). \tag{1.5.7}$$

The left side constant is independent of p for $p \ge 1$, and is bounded from below for $p \ge c$, for any $c > 0$.

If the variables X_i are symmetric, then we could use Khinchin's inequality to relate $\|\sum X_i\|_p = \|\sum \varepsilon_i X_i\|_p$ and $\|\sum X_i^2\|_{p/2}^{1/2}$ (see Lemma 1.4.13). For $p \ge 2$,

$$\left\|\sum \varepsilon_i x_i\right\|_p \le (p-1)^{1/2}\left(\sum x_i^2\right)^{1/2}$$

by Theorem 1.1.1. So, if we were to use this inequality in conjunction with Theorem 1.5.2, the majorization constant (at the right side) in the corresponding inequality would tend to infinity as $p \to \infty$. Fortunately, symmetric variables can be treated separately (and analogously) for $p \ge 2$. For $p \le 2$, Theorem 1.5.2 and Khinchin (for $p \le 1$) suffice to obtain acceptable bounds:

COROLLARY 1.5.3. *If X_i are symmetric and independent, then*

$$e^{-1}\left(\frac{e^{p/2}-1}{2}\right)^{1/p} L^{1/2}(\{X_i^2\}, p/2) \le \left\|\sum X_i\right\|_p \le e^{1/2}L^{1/2}(\{X_i^2\}, p/2), \quad p \in (0, 1] \tag{1.5.8}$$

and

$$\begin{aligned}\left[2^{-1/2} \vee (p-1)^{1/2}\right]&\frac{\left(e^{p/2}-1\right)^{1/p}}{\sqrt{2}e} L^{1/2}(\{X_i^2\}, p/2) \\ &\le \left\|\sum X_i\right\|_p \le e^{1/2}L^{1/2}(\{X_i^2\}, p/2), \qquad p \in (1, 2).\end{aligned} \tag{1.5.9}$$

PROOF. For X_i symmetric, if ε_i are independent Rademacher variables independent of the X_i variables, then the sequences $\{X_i\}$ and $\{\varepsilon_i X_i\}$ have the same probability law and therefore

$$\mathbb{E}\left|\sum X_i\right|^p = \mathbb{E}\left|\sum \varepsilon_i X_i\right|^p = \mathbb{E}_X \mathbb{E}_\varepsilon \left|\sum \varepsilon_i X_i\right|^p,$$

where $\mathbb{E}_X$ ($\mathbb{E}_\varepsilon$) denotes expected value with respect to the X (the Rademacher) variables only. Khinchin's inequality with best constants gives, for $0 < p \le 1$,

$$\frac{1}{2^{1-p/2}}\left(\sum X_i^2\right)^{p/2} \le \mathbb{E}_\varepsilon\left|\sum \varepsilon_i X_i\right|^p \le \left(\sum X_i^2\right)^{p/2}$$

(see Corollary 1.3.3). Inequality (1.5.8) then follows from these two observations and Theorem 1.5.2 applied to the sequence $\{X_i^2\}$ for $p/2$. For $1 < p < 2$ Khinchin's inequality [not with best constants: see (1.3.1) and (1.3.8)] is as follows:

$$\left[2^{-1/2} \vee (p-1)^{1/2}\right]^p \left(\sum X_i^2\right)^{p/2} \le \mathbb{E}_\varepsilon \left|\sum \varepsilon_i X_i\right|^p \le \left(\sum X_i^2\right)^{p/2}.$$

Then, (1.5.9) follows similarly from Theorem 1.5.2. □

In order to consider the symmetric case for $p \ge 2$ we re-derive Lemma 1.5.1 in this case.

LEMMA 1.5.4. *Let X_i be independent symmetric random variables and let $p \ge 2$. Then,*

a_s) $\phi_p(\sum X_i) \le \prod_i \phi_p(X_i)$;

b_s) $\phi_p\left(2X_1 + \phi_p^{2/p}(X_1)X_2\right) \ge \phi_p(X_1)\phi_p(X_2)$; *and*

c_s) *if* $\sum \log \phi_p(X_i) \le p$, *then* $\phi_p\left(2e^2 \sum X_i\right) \ge \prod \phi_p(X_i)$.

PROOF. To prove inequality a_s) it suffices to show

$$\phi_p(X_1 + X_2) \le \phi_p(X_1)\phi_p(X_2).$$

Replacing X_i by εX_i and integrating with respect to the Rademacher variables, this follows from the inequality

$$\begin{aligned} &|1+x+y|^p + |1+x-y|^p + |1-x+y|^p + |1-x-y|^p \\ &\quad \le \left(|1+x|^p + |1-x|^p\right)\left(|1+y|^p + |1-y|^p\right) \qquad (1.5.10) \\ &\quad = |1+x+y+xy|^p + |1+x-y-xy|^p + |1-x+y-xy|^p \\ &\qquad + |1-x-y+xy|^p, \end{aligned}$$

that, as we show next, holds for all $x, y \in \mathbb{R}$ and all $p \ge 2$. If we show that, whenever a, b, c, d satisfy $a < b < c < d$ and $b - a = d - c$, the functions of $t \in \mathbb{R}$,

$$g_p(t) = |a+t|^p + |b-t|^p + |c-t|^p + |d+t|^p, \qquad p \ge 2,$$

are nondecreasing for $t \ge 0$, then inequality (1.5.10) obtains by just taking $t = xy$ and choosing a, b, c, d among $1 \pm x \pm y$ depending on the values of x and y. For $p = 2$, $g_2(t) = a^2 + b^2 + c^2 + d^2 + 4t^2$, which is obviously increasing; for $p > 2$, we see that $g_p''(t) \ge 0$ because the function $|x|^p$ is convex and twice differentiable, and that

$$g_p'(0) = p\left(|d|^{p-1} - |c|^{p-1}\right) - p\left(|b|^{p-1} - |a|^{p-1}\right) \ge 0$$

because the function $|x|^{p-1}$ is also convex. Hence $g_p' \ge 0$. This proves inequality (1.5.10) and therefore a_s).

To prove inequality b_s) first we observe that if ε denotes a Rademacher variable, and if $t \geq 1$, $|x| \geq 1$ and $p \geq 1$, then

$$\mathbb{E}|1+\varepsilon t x|^p \geq t^{p/2}\mathbb{E}|1+\varepsilon x|^p. \tag{1.5.11}$$

To see this, we fix $x \geq 1$ (these expected values are symmetric in x), set

$$f(t) = \mathbb{E}|1+\varepsilon t x|^p / t^{p/2},$$

and note that f is nondecreasing on $[1,\infty)$, which is clear because

$$f'(t) = \frac{p}{2t}\frac{(tx-1)(tx+1)^{p-1} + (tx+1)(tx-1)^{p-1}}{(tx+1)^p + (tx-1)^p} \geq 0.$$

Now we can proceed by analogy with the proof of inequality b) in Lemma 1.5.1, using the following observations: a) if X is symmetric, then $\phi_p(X) = \mathbb{E}|1+X|^p \geq 1$ by convexity of the function $|1+x|^p$ and Jensen; b) if X_1 and X_2 are independent and symmetric, and if ε_1, ε_2 are independent Rademacher variables independent of X_1, X_2, then the random vectors (X_1, X_2), $(X_1, \varepsilon_2 X_2)$, $(\varepsilon_1 X_1, X_2)$, and $(\varepsilon_1 X_1, \varepsilon_2 X_2)$ all have the same probability law; and c), by convexity, $\mathbb{E}|a+b\varepsilon|^p \leq \mathbb{E}|a+c\varepsilon|^p$ if $|b| \leq |c|$. For simplicity of notation, we set

$$\varphi_p(x) := |1+x|^p.$$

Then, using Jensen, Fubini, and observation b) for the first inequality that follows, and inequality (1.5.11) for the second, we have

$$\begin{aligned}\mathbb{E}\varphi_p\big(2X_1 + \phi_p^{2/p}(X_1)X_2\big)I_{|X_2|\geq 1} &\geq \mathbb{E}\varphi_p\big(\phi_p^{2/p}(X_1)\varepsilon_2 X_2\big)I_{|X_2|\geq 1}\\ &\geq \phi_p(X_1)\mathbb{E}\varphi_p\big(\varepsilon_2 X_2\big)I_{|X_2|\geq 1}\\ &= \phi_p(X_1)\mathbb{E}\varphi_p\big(X_2\big)I_{|X_2|\geq 1}.\end{aligned}$$

Similarly, observations a), b), and c) justify the following inequalities:

$$\begin{aligned}\mathbb{E}\varphi_p\big(2X_1 + \phi_p^{2/p}(X_1)X_2\big)I_{|X_2|<1} &= \mathbb{E}\mathbb{E}_\varepsilon\varphi_p\big(2X_1 + \phi_p^{2/p}(X_1)\varepsilon_2 X_2\big)I_{|X_2|<1}\\ &\geq \mathbb{E}\mathbb{E}_\varepsilon\varphi_p\big(2X_1 + \varepsilon_2 X_2\big)I_{|X_2|<1}\\ &= \mathbb{E}\varphi_p\big(2\varepsilon_1 X_1 + X_2\big)I_{|X_2|<1}\\ &\geq \mathbb{E}\varphi_p\big((1+X_2)\varepsilon_1 X_1 + X_2\big)I_{|X_2|<1}\\ &= \mathbb{E}\varphi_p\big((1+X_2)X_1 + X_2\big)I_{|X_2|<1}\\ &= \mathbb{E}|1+X_1|^p|1+X_2|^p I_{|X_2|<1}\\ &= \phi_p(X_1)\mathbb{E}\varphi_p\big(X_2\big)I_{|X_2|\geq 1}.\end{aligned}$$

Combining these two strings of inequalities gives b_s).

Now, to prove c_s) from b_s) we can proceed exactly as in the proof of c) in Lemma 1.5.1. □

Lemma 1.5.4 gives, just as in the case of non-negative random variables, the analog of Theorem 1.5.2 for symmetric variables and $p \geq 2$:

THEOREM 1.5.5. *If X_i are independent and symmetric and $p \geq 2$, then*

$$\frac{e-1}{2e^2} L\big(\{X_i\}, p\big) \leq \big\| \sum X_i \big\|_p \leq eL\big(\{X_i\}, p\big). \tag{1.5.12}$$

Corollary 1.5.3 and Theorem 1.5.5 extend to centered random variables by Lemma 1.2.6, as follows:

THEOREM 1.5.6. *The estimates contained in Corollary 1.5.3 for $p \in [1, 2)$ and in Theorem 1.5.5 for $p \geq 2$ extend to independent centered random variables with the following changes: i) $L(\{X_i\}, p)$ should be replaced by $L(\{\varepsilon_i X_i\}, p)$ if $p \geq 2$, and ii) for all $p \geq 1$, the right side constant should be multiplied by 2 and the left side constant by 1/2. Furthermore, at the expense of another factor of $2^{(p+1)/p}$ at the right side, these bounds also hold for $\big\| \max_{1\leq k\leq n} \big|\sum_{i=1}^k X_i\big|\big\|_p$.*

We now take a look at the form of the L function in the important particular case of sequences $\{X_i\}$ of independent *identically distributed* random variables. It has already been mentioned at the beginning of this section that the function $\phi_p(X) = \mathbb{E}|1 + X/t|^p$ is continuous in t if $\mathbb{E}|X|^p < \infty$. Next we observe that $\phi_p(X)$ is strictly decreasing both when X is non-negative and when X is symmetric (and not identically zero). In the non-negative case, by monotonicity of $(1+x)^p$, if it were not strictly monotone, then the random function (of t) $1 + X/t$ would be constant on an interval for almost all values of X, which is not unless $X = 0$ a.s. This argument works for all $p > 0$. In the symmetric case with $p \geq 1$ the function $\mathbb{E}_\varepsilon|1+\varepsilon X/t|^p$ is monotone decreasing in t by convexity; hence, if $\phi_p(X)$ were not strictly monotone, then the function $\mathbb{E}_\varepsilon|1 + \varepsilon X/t|^p$ would also be constant on an interval for almost all values of X. But its derivative is not zero unless $X = 0$ a.s., proving strict monotonicity. A consequence of this is that, for X_i not identically zero, with finite absolute pth moment and either non-negative or symmetric, there is a definition of the L function which is more precise than (1.5.1), namely, *$L\big(\{X_i\}, p\big)$ is the only number t that satisfies the equation*

$$\sum \log \mathbb{E}|1 + X_i/t|^p = p.$$

This definition is made for all $p > 0$ if X is non-negative, and for all $p \geq 1$ if X is symmetric. Now let X, X_i be i.i.d. random variables, all symmetric or all non-negative, and let us relabel the L function as a function of n, as follows:

$$L_{X,p}(n) := L\big(\{X_i\}_{i=1}^n, p\big).$$

Then, $L_{X,p}(n)$ is the only solution of the equation

$$\log \mathbb{E}|1 + X/t|^p = p/n, \tag{1.5.13}$$

and the continuity and strict monotonicity of $\mathbb{E}|1 + X/t|^p$ just discussed imply that the function $L_{X,p}(y)$, obtained by replacing the integer valued variable n by the real positive valued variable y, is strictly monotone increasing in y. Hence, it can be inverted and we have

$$L_{X,p}^{-1}(y) = \frac{p}{\log \mathbb{E}|1 + X/y|^p}. \tag{1.5.14}$$

So, for instance, if a formula (or a bound) for $\mathbb{E}|1+X/y|^p$ is available, one obtains an explicit formula (or a bound) for $L_{X,p}(n)$ by just inverting it. We will only check the usefulness of formula (1.5.14) by proving certain monotonicity properties of $L_{X^2,1/2}(n)$ to be used in Chapter 2. (By Corollary 1.5.3, the square root of this function is equivalent to $\mathbb{E}|\sum X_i|$ if the variables X_i are centered.)

LEMMA 1.5.7. *If $\mathbb{E}|X| < \infty$, then the function $L_{X^2,1/2}(y)/y$ is monotone increasing on $(0,\infty)$ and the function $L_{X^2,1/2}(y)/y^2$ is monotone decreasing.*

PROOF. The first statement is equivalent to the function $y/L^{-1}(y)$ being monotone increasing (we drop the subindices for ease of notation). Now, by (1.5.14),

$$\exp\big[y/L^{-1}(y)\big] = \left[\mathbb{E}\Big(1+\frac{X^2}{y}\Big)^{1/2}\right]^{2y} = \left(1 + \frac{\Big[\mathbb{E}\Big(1+\frac{X^2}{y}\Big)^{1/2} - 1\Big]y}{y}\right)^{2y},$$

and the function

$$\left[\mathbb{E}\Big(1+\frac{X^2}{y}\Big)^{1/2} - 1\right]y = \mathbb{E}\frac{X^2}{(1+X^2/y)^{1/2}+1}$$

is obviously monotone increasing. Hence, since the function $(1+c/y)^y$ is monotone increasing for $c \ge 0$, so is $y/L^{-1}(y)$. The second statement is equivalent to the function $y/(L^{-1}(y))^2$ being monotone decreasing, and it follows, in a similar way, from (decreasing) monotonicity of the functions $-X^2/\big[y^{1/2}+(y+X^2)^{1/2}\big]$ and $(1-c/y)^y$. □

Next we give an important application of the preceding material. It is clear that the bounds in terms of the L function should be useful for obtaining inequalities for moments of sums of independent centered (or non-negative) random variables, particularly when the dependence of the constants on p is an issue. This is the case, for instance, of *Rosenthal's inequalities with constants of the best order in p* as $p \to \infty$, that will now be derived. These inequalities are important because they provide extremely convenient (though not too tight) bounds for the L_p norms of sums of independent random variables for $p \ge 1$ (in the positive case) or $p \ge 2$ (in the case of centered variables). They are related to Hoffmann-Jørgensen's inequalities (particularly, Theorem 1.2.5), and we will also see how the bounds for Rosenthal's inequalities can be applied to improve the constant in Hoffmann-Jørgensen's.

As usual, we begin with non-negative random variables.

Let $p \ge 1$ and let X_i be independent non-negative random variables. Then,

$$\sum X_i^p \le \Big(\sum X_i\Big)^p \quad\text{and}\quad \sum \|X_i\|_1 = \Big\|\sum X_i\Big\|_1 \le \Big\|\sum X_i\Big\|_p.$$

Therefore,

$$\max\Big[\sum \mathbb{E}X_i, \big(\sum \mathbb{E}X_i^p\big)^{1/p}\Big] \le \Big\|\sum X_i\Big\|_p. \tag{1.5.15}$$

For the converse, we estimate the $L(p)$ function of $\{X_i\}$.

LEMMA 1.5.8. *If X_i are independent and non-negative, then the following inequality holds for all $p \ge 1$ and $c > 0$:*

$$L(\{X_i\}, p) \le 2\max\Big[\frac{(1+c)^p}{cp}\big(\sum \mathbb{E}X_i\big), \Big(1+\frac{1}{c}\Big)p^{-1/p}\big(\sum \mathbb{E}X_i^p\big)^{1/p}\Big]. \tag{1.5.16}$$

PROOF. If $0 < c \le x$, then

$$(1+x)^p \le \Big(\frac{x}{c} + x\Big)^p = x^p\Big(1+\frac{1}{c}\Big)^p,$$

whereas for $0 \le x < c$, by convexity,

$$\frac{(1+x)^p - 1}{x} \le \frac{(1+c)^p - 1}{c}$$

so that

$$(1+x)^p \le 1 + (1+c)^p x/c.$$

That is,

$$(1+x)^p \le 1 + (1+c)^p\frac{x}{c} + \Big(1+\frac{1}{c}\Big)^p x^p, \qquad \text{for all } x \ge 0,\ c > 0.$$

Hence,

$$\log \mathbb{E}\Big(1+\frac{X_i}{t}\Big)^p \le \frac{1}{t}\frac{(1+c)^p}{c}\mathbb{E}X_i + \frac{1}{t^p}\Big(1+\frac{1}{c}\Big)^p \mathbb{E}X_i^p.$$

The lemma follows immediately from this inequality and the definition of L. □

THEOREM 1.5.9. *There exist $K < \infty$ such that if X_i are independent and non-negative and $p \ge 1$, then*

$$\begin{aligned}\max\Big[\sum \mathbb{E}X_i, \big(\sum \mathbb{E}X_i^p\big)^{1/p}\Big] &\\ \le \Big\|\sum X_i\Big\|_p &\\ \le K\frac{p}{\log p}\max\Big[\sum \mathbb{E}X_i, \big(\sum \mathbb{E}X_i^p\big)^{1/p}\Big]. &\end{aligned} \tag{1.5.17}$$

PROOF. The left-hand side inequality is just (1.5.15), and the right-hand side follows from Theorem 2.1 (right side inequality) and Lemma 1.5.8 applied with $c = (\log p)/p$. □

Next we consider symmetric random variables (hence, without saying, centered random variables as well: see Lemma 1.2.6). Let X_i be independent symmetric random variables and let $p \geq 2$. By Jensen,

$$\left(\sum \mathbb{E}X_i^2\right)^{1/2} = \left[\mathbb{E}\left(\sum X_i\right)^2\right]^{1/2} \leq \|\sum X_i\|_p,$$

and, since $(\sum a_i^p)^{1/p} \leq (\sum a_i^2)^{1/2}$ for $a_i \geq 0$,

$$\begin{aligned}\sum \mathbb{E}|X_i|^p &\leq \mathbb{E}\left(\sum X_i^2\right)^{p/2} \\ &= \mathbb{E}\left(\mathbb{E}_\varepsilon\left(\sum \varepsilon_i X_i\right)^2\right)^{p/2} \\ &\leq \mathbb{E}\mathbb{E}_\varepsilon\left|\sum \varepsilon_i X_i\right|^p = \mathbb{E}\left|\sum X_i\right|^p.\end{aligned}$$

Hence,

$$\max\left[\left(\sum \mathbb{E}X_i^2\right)^{1/2}, \left(\sum \mathbb{E}|X_i|^p\right)^{1/p}\right] \leq \|\sum X_i\|_p. \tag{1.5.18}$$

For the converse, we again estimate the $L(p)$ function of $\{X_i\}$.

LEMMA 1.5.10. *If X_i are symmetric random variables, then for every $p \geq 3$ and $c \in (0, 1)$,*

$$L(\{X_i\}, p) \leq 2\max\left[\frac{(1+c)^{p/2}}{cp^{1/2}}\left(\sum \mathbb{E}X_i^2\right)^{1/2}, \left(1+\frac{1}{c}\right)p^{-1/p}\left(\sum \mathbb{E}|X_i|^p\right)^{1/p}\right], \tag{1.5.19}$$

and for $2 \leq p \leq 3$,

$$L(\{X_i\}, p) \leq 2\max\left[\left(\sum \mathbb{E}X_i^2\right)^{1/2}, 2p^{-1/p}\left(\sum \mathbb{E}|X_i|^p\right)^{1/p}\right]. \tag{1.5.20}$$

PROOF. As in Lemma 1.5.8, these inequalities are direct consequences of inequalities for the function $|1+x|^p$; actually, in the present case, by randomization, for the function $\mathbb{E}|1+\varepsilon x|^p$. More to the point, inequalities (1.5.19) and (1.5.20) follow directly from the following inequalities:

$$\mathbb{E}|1+\varepsilon x|^p \leq 1 + (1+c)^p x^2/c^2 + (1+1/c)^p|x|^p \tag{1.5.21}$$

valid for all x, $p \geq 3$ and $0 < c < 1$, and

$$\mathbb{E}|1+\varepsilon x|^p \leq 1 + px^2 + 2^p|x|^p, \tag{1.5.22}$$

valid for all x and $2 \leq p \leq 3$ (by symmetry, only non-negative values of x are relevant). To prove them, we consider the functions

$$f_p(x) = 2x^{-2}\mathbb{E}|1+\varepsilon x|^p = \frac{|1+x|^p + |1-x|^p - 2}{x^2}.$$

For $0 < x \leq 1$, we set

$$g_p(x) = x^3 f_p'(x) = px(1+x)^{p-1} - px(1+x)^{p-1} - 2(1+x)^p - 2(1-x)^p + 4,$$

with $g_p(0) = 0$. Then, $g_p'(0) = 0$ and

$$g_p''(x) = p(p-1)(p-2)x\big[(1+x)^{p-3} - (1-x)^{p-3}\big],$$

so that $g_p''(x)$, $0 < x < 1$, is positive for $p > 3$ and negative for $2 < p < 3$. $f_p'(x)$ is accordingly positive or negative. Thus we have that, on $(0, 1]$, $f_p(x)$ is increasing for $p > 3$, decreasing for $2 < p < 3$ and $f_2 \equiv 2$, $f_3 \equiv 6$.

So, if $p \geq 3$, $0 < c < 1$ and $|x| \leq c$, we have [f_p being symmetric and nondecreasing on $(0, 1)$]

$$\mathbb{E}|1 + \varepsilon x|^p = 1 + \frac{1}{2}x^2 f_p(x) \leq 1 + \frac{1}{2}x^2 f_p(c) \leq 1 + (1+c)^p \frac{x^2}{c^2};$$

since, on the other hand, if $|x| > c$, we have

$$\mathbb{E}|1 + \varepsilon x|^p \leq 2(1 + |x|)^p \leq 2\Big(\frac{|x|}{c} + |x|\Big)^p = \Big(1 + \frac{1}{c}\Big)^p |x|^p;$$

this proves inequality (1.5.21).

If $2 \leq p \leq 3$ and $|x| \leq 1$, then, since f_p is nonincreasing on $(0, 1)$ and symmetric, and since $\lim_{x\to 0} f_p(x)$ is the second derivative of $(1+x)^p$ at zero, we have

$$\mathbb{E}|1 + \varepsilon x|^p = 1 + \frac{1}{2}x^2 f_p(x) \leq 1 + \frac{1}{2} f_p(0) x^2 = 1 + \binom{p}{2} x^2 \leq 1 + p x^2,$$

which, together with the fact that $\mathbb{E}|1 + \varepsilon x|^p \leq 2^p |x|^p$ for $|x| \geq 1$, gives inequality (1.5.22). □

THEOREM 1.5.11. *There exist $K < \infty$ such that if X_i are independent and symmetric and $p \geq 2$, then*

$$\begin{aligned} &\max\Big[\Big(\sum \mathbb{E}X_i^2\Big)^{1/2}, \Big(\sum \mathbb{E}|X_i|^p\Big)^{1/p}\Big] \\ &\quad \leq \Big\|\sum X_i\Big\|_p \\ &\quad \leq K \frac{p}{\log p} \max\Big[\Big(\sum \mathbb{E}X_i^2\Big)^{1/2}, \Big(\sum \mathbb{E}|X_i|^p\Big)^{1/p}\Big]. \end{aligned} \tag{1.5.23}$$

PROOF. The left side is just inequality (1.5.19) and the right side follows from Theorem 1.5.5 and Lemma 1.5.10, where we set $c = (\log p)/p$, as before. □

Again, the above inequalities, with a constant $1/2$ at the left side, also hold for independent *centered* not necessarily symmetric random variables, and in fact $\|\sum_{i=1}^n X_i\|_p$ can be replaced by $\big\|\max_{1\leq k\leq n}\big|\sum_{i=1}^k X_i\big|\big\|_p$, as a consequence of Lemma 1.2.6.

The inequalities (1.5.17) and (1.5.23) are known as Rosenthal's inequalities with constants of the best order. To see that the right sides of these inequalities are best possible (up to a constant), we consider the folowing example.

EXAMPLE 1.5.12. Suppose C_p, $p > 1$, satisfies

$$\left\|\sum X_i\right\|_p \leq C_p\Big[\left\|\sum X_i\right\|_1, \left(\sum \mathbb{E}X_i^p\right)^{1/p}\Big]$$

for all X_i independent non-negative and in L_p and all $n < \infty$. For p fixed, let X_i be independent variables such that

$$\Pr\{X_i = 1\} = 1 - \Pr\{X_i = 0\} = \frac{\log p}{p}.$$

Then,

$$\left\|\sum X_i\right\|_p \geq \left(n^p \Pr\{\sum X_i = n\}\right)^{1/p} = n\Big(\frac{\log p}{p}\Big)^{n/p}$$

and $\|\sum X_i\|_1 = n(\log p)/p$, $\left(\sum \mathbb{E}X_i^p\right)^{1/p} = \left[n(\log p)/p\right]^{1/p}$. If we choose n such that $n-1 \leq p/\log p < n$, the L_1 norm of $\sum X_i$ dominates the right side of Rosenthal's inequality and therefore C_p must satisfy

$$\begin{aligned} C_p &\geq \frac{p}{\log p}\Big(\frac{\log p}{p}\Big)^{n/p} \\ &\geq \frac{p}{\log p}\Big(\frac{\log p}{p}\Big)^{1/\log p + 1/p} \\ &\geq \frac{p}{\log p}\,\frac{(\log p)^{1/\log p}}{p^{1/p} p^{1/\log p}}. \end{aligned}$$

So, $C_p(\log p)/p \to 1/e$ as $p \to \infty$, showing that the constant at the right of inequality (1.5.17) is of the right order. In order to show that the same is true for inequality (1.5.23) we could take X_i symmetric such that $\Pr\{X_i = 1\} = \Pr\{X_i = -1\} = \log p/p$ and $\Pr\{X_i = 0\} = 1 - 2\log p/p$, $p > 2$ and reason in a similar way.

Finally, we obtain Hoffmann-Jørgensen's inequality with constants of the best type, as announced in Section 1.2 (we refer here to the inequality in Theorem 1.2.5). This will be a simple consequence of Rosenthal's inequalities and Proposition 1.4.1 on moments of maxima. We only consider the case of centered variables, but it should become clear that the proof works also for non-negative variables.

THEOREM 1.5.13. *For every $q > 0$ there exists $K_q < \infty$ such that, if X_i are independent centered real random variables, then, for all $p > 1$,*

$$\Big\|\max_{1\leq k\leq n}\sum_{i=1}^{n} X_i\Big\|_p \leq K_q\frac{p}{\log p}\Big[\Big\|\sum_{i=1}^{n} X_i\Big\|_q + \Big\|\max_{1\leq k\leq n}|X_i|\Big\|_p\Big]. \qquad (1.5.24)$$

PROOF. By Lemma 1.2.6, it suffices to prove the inequality for $\left\|\sum X_i\right\|_p$, X_i independent and symmetric. If (1.5.24) with this reduction holds for $q = 2$ and all $p \geq 2$, then it holds in general: the inequality for smaller values of q follows from Theorem 1.2.5 (with an irrelevant change in the constants), and Hölder gives it for larger values of q; values of p smaller than 2 are already covered by Theorem 1.2.5.

We now set $\delta_0 = \inf\bigl[t > 0 : \sum \Pr\{|X_i| > t\} \le 1\bigr]$. Then, by inequalities (1.4.3) and (1.4.2) in Corollary 1.4.2 and Proposition 1.4.1, we have

$$\begin{aligned}\sum \mathbb{E}|X_i|^p &= \sum \mathbb{E}|X_i|^p I_{|X_i|>\delta_0} + \sum \mathbb{E}|X_i|^p I_{|X_i|\le\delta_0}\\ &\le 2\mathbb{E}\max|X_i|^p + \delta_0^{p-2}\sum \mathbb{E}X_i^2 I_{|X_i|\le\delta_0}\\ &\le 2\mathbb{E}\max|X_i|^p + 2\mathbb{E}\bigl(\sum X_i\bigr)^2\mathbb{E}\max|X_i|^{p-2}\\ &\le 2\mathbb{E}\max|X_i|^p + 2\mathbb{E}\bigl(\sum X_i\bigr)^2\bigl[\mathbb{E}\max|X_i|^p\bigr]^{(p-2)/p}\\ &\le 4\max\Bigl[\mathbb{E}\max|X_i|^p, \bigl(\mathbb{E}\bigl(\sum X_i\bigr)^2\bigr)^{p/2}\Bigr].\end{aligned}$$

This, together with Rosenthal's inequality (1.5.23), gives

$$\Bigl\|\sum X_i\Bigr\|_p \le K_2\frac{p}{\log p}\Bigl[\Bigl\|\sum X_i\Bigr\|_2 + \bigl\|\max|X_i|\bigr\|_p\Bigr]$$

where $K_2 = 4^{1/2}K$. □

It is easy to see that the symmetric random variables in Example 1.5.12 also show that $p/\log p$ is the right rate in Hoffmann-Jørgensen's inequality (1.5.24).

We began our discussion on moment bounds, at the beginning of Section 1.4, with Khinchin's inequalities. It is clear, e.g., by the central limit theorem, that, as moment estimators, they are not tight for large values of p. It is thus appropriate to conclude this discussion by providing bounds for the L_p-norm of $\sum x_i\varepsilon_i$, $x_i \in \mathbb{R}$, that do not depend on p.

THEOREM 1.5.14. *Let $\{\varepsilon_i\}$ be a Rademacher sequence and let $\{a_i\}$ be a non-decreasing sequence of non-negative real numbers. Then, there exist universal constants c, C such that, for all $p \ge 2$,*

$$c\Bigl(\sum_{1\le i\le p} a_i + \sqrt{p}\bigl(\sum_{i>p} a_i^2\bigr)^{1/2}\Bigr) \le \Bigl\|\sum a_i\varepsilon_i\Bigr\|_p \le C\Bigl(\sum_{1\le i\le p} a_i + \sqrt{p}\bigl(\sum_{i>p} a_i^2\bigr)^{1/2}\Bigr). \tag{1.5.25}$$

PROOF. We sketch the proof only (that is, we leave several elementary but perhaps cumbersome computations to the reader). By Theorem 1.5.5, we only need to compute $L(\{a_i\varepsilon_i\}, p)$. We set

$$\tilde{\phi}_p(x) := \mathbb{E}|1 + x\varepsilon|^p = \frac{1}{2}\Bigl[|1+x|^p + |1-x|^p\Bigr], \quad x \ge 0,$$

where ε is a Rademacher variable. Then, elementary computations give the following bounds:

$$\begin{aligned}&\tilde{\phi}_p(x) \ge 1 + \frac{p(p-1)}{4}x^2 \ge 1 + \frac{p^2x^2}{8},\\ &\max\Bigl[\frac{1}{2}(1+x)^p, 1 + x^p\Bigr] \le \tilde{\phi}_p(x) \le (1+x)^p \le e^{px}\end{aligned}$$

and

$$\tilde{\phi}_p(x) \leq \cosh px \leq 1 + p^2x^2 \quad \text{for } px \leq 1.$$

These inequalities imply

$$\sum \log \mathbb{E}\left|1 + \frac{a_i\varepsilon_i}{t}\right| \approx \sum_{i:pa_i/t\geq 1} \frac{pa_i}{t} + \sum_{i:pa_i/t<1} \frac{p^2a_i^2}{t^2}.$$

It follows from this that the solution to the equation

$$\sum \log \mathbb{E}\left|1 + \frac{a_i\varepsilon_i}{t}\right| = p$$

is of the order $t \approx \sum_{1\leq i\leq p} a_i + \sqrt{p}\left(\sum_{i>p} a_i^2\right)^{1/2}$. That is, $L(\{a_i\varepsilon_i\}, p)$ is of this order, which gives (1.5.25). □

The inequalities in Section 1.4, applied to $\sum a_i\varepsilon_i$ give only bounds that are not significantly different from the bounds obtained using Khinchin's inequalities (they are even worse for large p as the right side constants one obtains are of the order of p instead of $\sqrt{p}$). Theorems 1.5.14 and the previous two applications to the Rosenthal and Hoffmann-Jørgensen inequalities are good illustrations of the applicability of the L-function bounds, in the same way as Theorems 1.4.13 and 1.4.14 are good examples of the usefulness of the moment bounds in Section 1.4.

1.6 References for Chapter 1

The present proof of Lévy classical inequalities in Banach spaces is basically taken from Kahane (1968), whereas most of the rest of the results described in Section 1.4.1, in particular the extension to i.i.d. random vectors and the application to the contraction principle (Kahane, 1968), were obtained by Montgomery-Smith (1994). The simple inequality (1.1.4′) appears in de la Peña and Montgomery-Smith (1993a) and Proposition 1.1.2, partly classical (the Lévy–Ottaviani inequality, Ottaviani, 1939), was proved by Etemadi (1985) with constants 4 and 1/4, and it comes, with constants 3 and 1/3, from Kwapień and Woyczynski (1992).

Propositions 1.2.1, 1.2.2 and Theorem 1.2.3 are modifications by Kwapień and Woyczynski (1992) of—by now classical—results of Hoffmann-Jørgensen (1974), whose version of inequality (1.2.6) had however a worse constant. Theorem 1.2.5 was obtained by Hitczenko (1993), building upon Kwapień and Woyczynski's material. Talagrand (1989) proved a stronger version of Theorem 1.2.8 (see also Ledoux and Talagrand, 1991, Theorem 6.21) but the present statement and its proof are taken from Hitczenko (1993). The elementary but useful Lemma 1.2.6 belongs to the folklore and we do not recall when or from whom we learned it. Theorem 1.2.5 with constants of the order $p/\log p$, which are best for any given q, was discovered by Johnson, Schechtman, and Zinn (1985) in $\mathbb{R}$ and by

Talagrand (1989) in Banach spaces (Kwapień and Szulga, 1991, have a different proof of it).

The Khinchin inequalities for $B = \mathbb{R}$ and different constants than those in Section 1.3 are due to Khinchin (1923), Littlewood (1930), and Paley and Zygmund (1930), and were extended to Banach valued coefficients by Kahane (1964). With best constants in $\mathbb{R}$, Khinchin's inequality comparing first and second moments was obtained by Szarek (1976), whereas Haagerup (1982) obtained it in the case of comparison of any moments with the second, and Latała and Oleszkiewicz (1994) obtained it in Banach spaces, with a very simple proof—simplest even for $\mathbb{R}$. The proof here is a further simplification of their published proof, and comes from an unpublished manuscript by Latała and Oleszkiewicz (1995), generalized by Kwapień, Latała, and Oleszkiewicz (1996).

Klass (1981) obtained the K function bounds for moments of sums of independent random variables (Proposition 1.4.10, Theorem 1.4.11), and these seem to have been the first two-sided bounds in terms of only the one-dimensional marginals. Giné and Zinn (1983) obtained a second set of such bounds valid in infinite dimensions (Theorems 1.4.4, 1.4.8, and 1.4.9, with different constants). Giné and Zinn (1983) also contains Proposition 1.4.1 on estimation of moments of maxima of independent variables. The simple truncation Lemma 1.4.3 is also part of the folklore and in our case it comes from Araujo and Giné (1980).

Theorem 1.4.14 on comparison between martingales and sums of independent variables was first included in de la Peña (1988) and published in de la Peña (1990). Inspired by this, B. Davis obtained the analogous result for sums of positive variables, Theorem 1.4.12, which was then included in de la Peña (1988, 1990). Johnson and Schechtman (1989), generalized these results to some Banach spaces.

The L function estimates, Theorems 1.5.2, 1.5.5, and 1.5.6, were proved by Latała (1997), and solve the problem of finding exact bounds for the L_p-norms of sums of independent variables in terms of only the marginals (the word "exact" meaning in this setting that the multiplicative constants in front of the approximating quantity do not depend on p). Rosenthal's inequality (Theorems 1.4.10, 1.4.12) was obtained by Rosenthal (1970) and with best rates as $p \to \infty$, by Johnson, Schechtman, and Zinn (1985), but the proof here is from Latała (1997); and so is the proof of the exact estimate of the L_p-norms of linear combinations of Rademacher variables for large p given in Theorem 1.5.14, although the result itself was obtained by Hitczenko (1993) (Montgomery-Smith, 1990, has a result in the same spirit for tail probabilities of Rademacher linear combinations with real coefficients, and Dilworth and Montogomery–Smith, 1993, extend the Montgomery-Smith and the Hitczenko results to Rademacher linear combinations with Banach valued coefficients). Best constants in Rosenthal's inequalities (making the bounds attainable) were obtained by Utev (1985) for $p \geq 4$ and by Figiel, Hitczenko, Johnson, Schechtman, and Zinn (1997) for $2 < p \leq 4$.

2

Randomly Stopped Processes With Independent Increments

This chapter is devoted to the study of several general properties of randomly stopped sums of independent random variables and more generally randomly stopped processes with independent increments. Concerning randomly stopped sums of independent random variables, the key result behind the theory of sequential analysis is Wald's first equation. In this chapter we discuss this result and provide (in some sense) the weakest conditions under which it holds. Concerning randomly stopped processes with independent increments, the work of Burkholder and Gundy dealing with inequalities for the L_p-norm of Brownian motion (Corollary 5.3 of this chapter) has been of paramount importance in the development of our understanding of the behavior of such processes. In particular, their inequalities (and Wald's equations) can be used to determine the finiteness of expectations of stopping times associated with the processes in question. The aim of this chapter is to provide a decoupling reinterpretation and extensions of these classical inequalities. We will present an extension of these results to all processes continuous on the right with limits from the left, with independent increments and values in general Banach spaces. Our framework permits an extension in a different direction, dealing with inequalities for more general functions (than powers) like exponentials. As an application we present an example connecting boundary crossing by nonrandom functions to boundary crossing by processes with independent increments.

The chapter is organized as follows: In Section 1 we present Wald's first equation along with a sharp extension of this seminal result. The proof of this extension is based on decoupling concepts as well as a use of the K-function (alternatively the

L-function from Section 1.5). In Section 2 we introduce good lambda inequalities that are used for developing L_p comparisons between correlated processes based on a special type of tail probability comparison, which is frequently manageable. They are used extensively in this chapter as well as in Chapters 6 and 7 in developing general decoupling and martingale inequalities. In Section 3 we present the main result of this chapter, a two-sided decoupling inequality for the L_p-norm of randomly stopped sums of independent random variables and give a proof of the right-hand side of this result. The proof of the left-hand side, being more complicated, is given in Section 4. Continuing in Section 5 we show how to extend the decoupling inequality to the case of continuous time processes with independent increments. The proof of this extension is based on limiting arguments. In the same section we also show how to apply this inequality to obtain sharp conditions for the validity of Wald's equation. These results can be interpreted as extensions of Wald's equations and support the heuristic that stopping times of processes with independent increments can (often) be viewed as random times independent of the associated processes. In Section 6, we develop an extension of the Burkholder–Gundy inequality to general Banach spaces and to the case of exponential bounds. In particular this extension permits the development of similar inequalities for certain processes without an obvious independent increment structure like Bessel processes by viewing them as processes with independent increments with values in an appropriate Banach space. In this section we include an example dealing with first exit times for a vector of random processes with potential applications to problems related to financial markets and related results for Bessel processes. Our final application is presented in Section 7 where we provide a natural extension of the concept of boundary crossing by nonrandom functions to the case of random processes with independent increments.

2.1 Wald's Equations

We begin with Wald's equations, which constitute the cornerstone of the theory of sequential analysis.

THEOREM 2.1.1. *Let $\{X_i\}$ be a sequence of i.i.d. random variables adapted to $\mathcal{F}_i = \sigma(X_1, \ldots, X_i)$, with $\mathbb{E}X_1 = \mu$, $|\mu| < \infty$. Let T be a stopping time adapted to $\sigma\{X_i\}$. Set $S_n = X_1 + \cdots + X_n$. Then*

$$\mathbb{E}S_T = \mu \mathbb{E}T, \qquad \text{whenever } \mathbb{E}T < \infty. \tag{2.1.1}$$

Moreover, if $\mathbb{E}X_1 = 0$ and $\mathbb{E}X_1^2 < \infty$, then

$$\mathbb{E}S_T^2 = \mathbb{E}X_1^2 \mathbb{E}T, \qquad \text{whenever } \mathbb{E}T < \infty. \tag{2.1.2}$$

PROOF. By working on the sequence $\{X_i - \mu\}$ we may assume that $\mathbb{E}X_1 = 0$. Let $X = X^+ - X^-$, where $X^+ = XI_{X\geq 0}$ and $X^- = -XI_{X<0}$. Then $\mathbb{E}X_1 = 0$

implies $\mathbb{E}X_1^+ = \mathbb{E}X_1^- < \infty$. Observe that $I_{T\geq i} = 1 - I_{T\leq i-1}$ is in $\mathcal{F}_{i-1}$. Therefore, X_i and $I_{T\geq i}$ are independent. Using this and letting $T \wedge n = \min\{T, n\}$,

$$\begin{aligned}
\mathbb{E}\sum_{i=1}^{T\wedge n} X_i^+ &= \mathbb{E}\sum_{i=1}^{n} X_i^+ I_{T\geq i} = \sum_{i=1}^{n} \mathbb{E}X_i^+ I_{T\geq i} \\
&= \sum_{i=1}^{n} \mathbb{E}X_i^+ \Pr\{T \geq i\} \qquad \text{(by independence)} \\
&= \mathbb{E}X_1^+ \mathbb{E}\sum_{i=1}^{n} I_{T\geq i} = \mathbb{E}X_1^+ \mathbb{E}(T \wedge n).
\end{aligned}$$

Likewise $\mathbb{E}\sum_{i=1}^{T\wedge n} X_i^- = \mathbb{E}X_1^- \mathbb{E}(T\wedge n)$. Letting $n \to \infty$ and using the monotone convergence theorem gives $\mathbb{E}\sum_{i=1}^{T} X_i^+ = \mathbb{E}TX_1^+$ and $\mathbb{E}\sum_{i=1}^{T} X_i^- = \mathbb{E}TX_1^-$. Therefore,

$$\mathbb{E}S_T = \mathbb{E}\Big(\sum_{i=1}^{T} X_i^+ - \sum_{i=1}^{T} X_i^-\Big) = \mathbb{E}T\mathbb{E}(X_1^+ - X_1^-) = \mathbb{E}T\mathbb{E}X_1 = 0.$$

To get equation (2.1.2) we use the facts that $S_{T\wedge n} = \sum_{i=1}^{n} X_i I_{T\geq i}$ is a martingale with respect to $\{\mathcal{F}_i\}$ and that $I_{T\geq i}$ is independent of X_i, which give

$$\mathbb{E}S_{T\wedge n}^2 = \sum_{i=1}^{n} \mathbb{E}X_i^2 I_{T\geq i} = \sum_{i=1}^{n} \mathbb{E}X_i^2 \Pr\{T \geq i\} = \mathbb{E}X_1^2 \mathbb{E}(T \wedge n).$$

Hence,

$$\lim_{n\to\infty} \mathbb{E}S_{T\wedge n}^2 = \mathbb{E}X_1^2 \mathbb{E}T < \infty, \tag{2.1.3}$$

assuming $\mathbb{E}T < \infty$. Using a similar reasoning we have that

$$\mathbb{E}\big(S_{T\wedge n} - S_{T\wedge m}\big)^2 = \mathbb{E}\Big(\sum_{i=m}^{n} X_i I_{T\geq i}\Big)^2 = \mathbb{E}X_1^2\big(\mathbb{E}T\wedge n - \mathbb{E}T\wedge m\big). \tag{2.1.4}$$

It is easy to see from (2.1.4) that

$$\lim_{m\to\infty} \sup_{n>m} \mathbb{E}\big(S_{T\wedge n} - S_{T\wedge m}\big)^2 = 0. \tag{2.1.5}$$

Equation (2.1.5) along with the Cauchy convergence criterion gives that there exists a random variable (say Y) in L_2 such that $S_{T\wedge n}$ converges in L_2 and in probability to Y. This implies that $\mathbb{E}S_{T\wedge n}^2 \to \mathbb{E}Y^2$. Moreover, convergence in probability to Y implies that there exists an increasing subsequence $m(n) \to \infty$ such that $S_{T\wedge m(n)}$ converges almost surely to Y. Finally $Y = S_T$ since we have also that $\lim_n S_{T\wedge n} = S_T$ almost surely. □

As observed before, $S_T = \sum_{i=1}^{\infty} X_i 1(T \geq i)$ is composed of summands $X_i 1(T \geq i)$ that are products of independent random variables. This motivates the idea of replacing $\{X_i 1(T \geq i)\}$ by $\{\tilde{X}_i 1(T \geq i)\}$, where $\{\tilde{X}_i\}$ is an independent copy of $\{X_i\}$ and independent of T as well. From this replacement we obtain by means of Fubini's theorem the following reformulation of Wald's equations.

THEOREM 2.1.1′. *Let $\{X_i\}$ be a sequence of i.i.d. random variables with $\mathbb{E}X_1 = \mu$, with $|\mu| < \infty$. Let T be a stopping time adapted to $\mathcal{F}_i = \sigma(X_1, \ldots, X_i)$. Let $\{\tilde{X}_i\}$ be an independent copy of $\{X_i\}$, with $\{\tilde{X}_i\}$ independent of T as well. Set $S_n = X_1 + \cdots + X_n$ and $\tilde{S}_n = \tilde{X}_1 + \cdots + \tilde{X}_n$. Then,*

$$\mathbb{E}S_T = \mathbb{E}\tilde{S}_T, \qquad \text{whenever } \mathbb{E}T < \infty. \tag{2.1.1′}$$

Likewise, if $\mathbb{E}X_1 = 0$ and $\mathbb{E}X_1^2 < \infty$, then

$$\mathbb{E}S_T^2 = \mathbb{E}\tilde{S}_T^2 \qquad \text{whenever } \mathbb{E}T < \infty. \tag{2.1.2′}$$

The above results follow directly from Wald's equations in conjunction with Fubini's theorem since, by design, $\tilde{S}_n$ and T are taken to be independent and therefore $\mathbb{E}\tilde{S}_T = \mathbb{E}\big(\mathbb{E}(\sum_{i=1}^T \tilde{X}_i | T)\big) = \mathbb{E}(\sum_{i=1}^T \mathbb{E}\tilde{X}_i) = \mu\mathbb{E}T$ and likewise for $\mathbb{E}\tilde{S}_T^2$.

It is important to observe that the variables S_T and $\tilde{S}_T$ can have drastically different behavior as the following example shows.

EXAMPLE 2.1.2. Let $\{X_i\}$ be i.i.d. with $\Pr\{X_i = 1\} = \Pr\{X_i = -1\} = \frac{1}{2}$. Set $T = \{\inf n : S_n = a \text{ or } S_n = -b\}$ for integers $a, b > 0$. Let $\{\tilde{X}_i\}$ be an independent copy of $\{X_i\}$ (in particular the sequence $\{\tilde{X}_i\}$ is independent of T as well). Then, S_T can take on only two values a or $-b$. In contrast $\tilde{S}_T$ is not restricted to these values.

From Wald's first and second equations (if $\mathbb{E}T < \infty$) since $\mathbb{E}X_1 = 0$ and $\mathbb{E}X_1^2 = 1$, $\mathbb{E}S_T = 0 = a\Pr\{S_T = a\} - b\Pr\{S_T = -b\}$ and $\mathbb{E}S_T^2 = \mathbb{E}T = a^2\Pr\{S_T = a\} + b^2\Pr\{S_T = -b\}$. Solving this system of equations one obtains

$$\mathbb{E}T = ab \quad \text{and} \quad \Pr\{S_T = a\} = \frac{b}{a+b}, \quad \Pr\{S_T = -b\} = \frac{a}{a+b}. \tag{2.1.6}$$

The fact that all this information can be retrieved from the use of Wald's equations motivates our work of extending their applicability to a more general framework, flexible enough to give information on other moments of T, to a wider class of distributions and dependence structures. Before discussing these extensions, we present a result that is helpful in verifying that $\mathbb{E}T < \infty$ for stopping times as in Example 2.1.2, that is, for the first passage time through a nonsymmetric barrier of a sum of i.i.d. random variables.

THEOREM 2.1.3. *Let $\{X_i\}$ be i.i.d. with $\Pr\{X_1 = 0\} < 1$. Define*

$$T_{(a,b)} = \inf\Big\{n \geq 1 : \sum_{i=1}^n X_i \notin (a, b)\Big\}$$

with $T_{(a,b)} = \infty$ *if* $\sum_{i=1}^n X_i \in (a,b)$ *for all* n. *Then, there exists a constant* $0 < \gamma < 1$ *such that*

$$\Pr\{T_{(a,b)} > n\} \leq \gamma^n, \tag{2.1.7}$$

for $n = 1, 2, \ldots$. *Therefore, adding over* n, $\mathbb{E}T_{(a,b)}^k < \infty$, *for* $k = 1, 2, \ldots$, *and* $\mathbb{E}\exp\{\lambda T_{(a,b)}\} < \infty$ *for* $\lambda < \log \frac{1}{\gamma}$.

PROOF. Without loss of generality, we may assume that there exist (ε, δ) with $\delta > 0$, $0 < \varepsilon < 1$ and $\Pr\{X_1 \geq \delta\} \geq \varepsilon > 0$. Take $r > (b-a)/\delta$. Then,

$$\begin{aligned}\Pr\{X_1 + \cdots + X_r > (b-a)\} &\geq \Pr\Big\{X_1 \geq \delta, \ldots, X_r \geq \delta; \quad \sum_{i=1}^r X_i > (b-a)\Big\} \\ &\geq \Pr\{X_1 \geq \delta, \ldots, X_r \geq \delta; \quad r\delta > (b-a)\} \\ &\geq \varepsilon^r.\end{aligned}$$

Therefore,

$$\begin{aligned}\Pr\{T_{(a,b)} > nr\} &= \Pr\Big\{T_{(a,b)} > nr, \quad \sum_{i=(m-1)r+1}^{mr} X_i \leq (b-a), \quad m = 1, \ldots, n\Big\} \\ &\leq \Pr\Big\{\sum_{i=(m-1)r+1}^{mr} X_i \leq (b-a), \quad m = 1, \ldots, n\Big\} \\ &= \prod_{m=1}^n \Pr\Big\{\sum_{i=(m-1)r+1}^{mr} X_i \leq (b-a)\Big\} \\ &\leq (1-\varepsilon^r)^n,\end{aligned}$$

completing the proof of the theorem with $\gamma = (1 - \varepsilon^r)$. □

Using a decoupling approach, one can show that for arbitrary stopping times and arbitrary independent random variables $\{X_i\}$ (not necessarily identically distributed) one has that,

$$0 \leq \mathbb{E}S_T^2 \leq 2\mathbb{E}\tilde{S}_T^2 \tag{2.1.8}$$

and that if the X_is are independent and non-negative then

$$\frac{1}{5}\mathbb{E}\tilde{S}_T^2 \leq \mathbb{E}S_T^2 \leq 2\mathbb{E}\tilde{S}_T^2. \tag{2.1.9}$$

Moreover, for non-negative independent variables and all $p \geq 1$,

$$\mathbb{E}S_T^p \leq 2^{p-1}\mathbb{E}\tilde{S}_T^p. \tag{2.1.10}$$

These inequalities (which are sharp) provide an answer to the problem in sequential analysis of extending Wald's equations to nonzero mean random variables. Observe that because of the independence between $\{\tilde{X}_i\}$ and T,

$$\mathbb{E}\tilde{S}_T^2 = \sum_{n=1}^{\infty} \mathbb{E}\tilde{S}_n^2 \Pr\{T = n\} = \sum_{n=1}^{\infty}\Big((\sum_{i=1}^{n} \mathbb{E}\tilde{X}_i)^2 + \sum_{i=1}^{n} \operatorname{Var}\tilde{X}_i\Big) \Pr\{T = n\}.$$

It is also important to note that the bounds provided for $\mathbb{E}S_T^2$ are set in terms of the marginal distributions of the X_is and T. Specializing to the case of i.i.d. random variables, one obtains $\mathbb{E}\tilde{S}_T^2 = (\mathbb{E}X_1)^2\mathbb{E}T + \mathbb{E}X_1^2\mathbb{E}T^2$ and hence the decoupling bounds provide information on $\mathbb{E}T$ as well as $\mathbb{E}T^2$. In doing the computations above we used the fact that $\tilde{S}_T$ is a sum of conditionally independent random variables. Because of this we say that in $\tilde{S}_T$, the sum and the stopping time are decoupled. In Section 3 we will dwell more on this line of investigation.

Along a different line of research, an alternate set of conditions for Wald's first equation requiring more integrability from X_1 but less of T follows. Let $\{X_i\}$ be i.i.d mean zero. Then,

$$\mathbb{E}S_T = 0 \text{ whenever } \mathbb{E}T^{\frac{1}{\alpha}} < \infty, \tag{2.1.11}$$

and $\mathbb{E}|X_1|^\alpha < \infty$ for $1 < \alpha \leq 2$. We will show how to do this and more. In particular, we will give an answer to the natural question: What is the weakest set of conditions on the stopping time T and the distribution of X_1 under which Wald's first equation holds? The following theorem provides (in some sense) the answer.

THEOREM 2.1.4. *Let $\{X_i\}$ be a sequence of i.i.d. random variables. Assume that $\mathbb{E}X_1 = 0$ (then, formally $\mathbb{E}|X_1| < \infty$). Then,*

$$\mathbb{E}S_T = 0 \tag{2.1.12}$$

whenever

$$\mathbb{E}K_X(T) < \infty,$$

where, $K_X(n)$ is the K-function associated to the random variable X_1 (see (1.4.22)).

It follows from the properties of the K-function that Theorem 2.1.4 implies (2.1.11). To see this use (2.1.14) below, Hölder's inequality, (1.4.30) and the concavity of the function $f(x) = |x|^{\alpha/2}$ to get,

$$K_X(n) \leq 8e\Big(\mathbb{E}|\sum_{i=1}^{n} X_i|^\alpha\Big)^{1/\alpha} \leq 8eC_\alpha\Big(\mathbb{E}(\sum_{i=1}^{n} X_i^2)^{\alpha/2}\Big)^{1/\alpha}$$
$$\leq 8eC_\alpha\Big(\sum_{i=1}^{n} \mathbb{E}|X_i|^\alpha\Big)^{1/\alpha} = 8eC_\alpha n^{1/\alpha}\Big(\mathbb{E}|X_1|^\alpha\Big)^{1/\alpha}$$

Therefore, $\mathbb{E}K_X(T) \leq 4eC_\alpha \mathbb{E}T^{1/\alpha}(\mathbb{E}|X_1|^\alpha)^{1/\alpha} < \infty$ if $\mathbb{E}T^{\frac{1}{\alpha}} < \infty$ and $\mathbb{E}|X_1|^\alpha < \infty$.

A useful equivalent condition to (2.1.12) is

$$\mathbb{E}a_T < \infty, \tag{2.1.13}$$

where $a_n = \mathbb{E}|S_n|$. If the variables are symmetric, this is clear since (1.4.21) gives $\mathbb{E}|S_n| \approx K_X(n)$ and, hence,

$$\mathbb{E}a_T = \sum_{n=1}^{\infty} \mathbb{E}|S_n| \Pr\{T = n\} \approx \sum_{n=1}^{\infty} K_X(n) \Pr\{T = n\} = \mathbb{E}K_X(T).$$

When the variables are just assumed to be i.i.d. mean zero, this approximation is still valid by symmetrization. Let $\{\varepsilon_i\}$ be a sequence of independent Bernoulli random variables with $\Pr\{\varepsilon_i = 1\} = \Pr\{\varepsilon_i = -1\} = \frac{1}{2}$. Take this sequence independent of $\{X_i\}$. Then, by Lemma 1.2.6 and (1.4.21)

$$\mathbb{E}|S_n| \approx \mathbb{E}\Big|\sum_{i=1}^{n} X_i \varepsilon_i\Big| \approx K_{X\varepsilon}(n) = K_X,$$

where the equality above follows from the definition of the K-function in (1.4.13). Indeed, for i.i.d. mean zero random variables one gets

$$\frac{1}{8e} K_X(n) \leq \mathbb{E}\Big|\sum_{i=1}^{n} X_i\Big| \leq 4K_X(n). \tag{2.1.14}$$

(An alternative approach would have been to use the L-function from Section 1.5.)

It is important to observe that (2.1.13) has a natural interpretation, where the sum is decoupled (independentized) from the stopping time. Let $(\{\tilde{X}_i\}, \tilde{T})$ be an independent copy of $(\{X_i\}, T)$. Then, $\tilde{T}$ is independent of $\{X_i\}$ and

$$\begin{aligned} \mathbb{E}a_T &= \sum_{n=1}^{\infty} \mathbb{E}|S_n| \Pr\{T = n\} = \sum_{n=1}^{\infty} \mathbb{E}|S_n| \Pr\{\tilde{T} = n\} \\ &= \sum_{n=1}^{\infty} \mathbb{E}|S_n| I_{\tilde{T}=n} = \sum_{n=1}^{\infty} \mathbb{E}|S_{\tilde{T}}| I_{\tilde{T}=n} = \mathbb{E}|S_{\tilde{T}}|. \end{aligned} \tag{2.1.15}$$

Therefore, condition (2.1.13) [and hence (2.1.12)] of the theorem can be replaced by $\mathbb{E}|S_{\tilde{T}}| < \infty$. This reinterpretation is important as a converse (in a broad sense) to Theorem 2.1.4 is obtained by observing that formally,

$$\mathbb{E}S_{\tilde{T}} = 0 \quad \text{implies} \quad \mathbb{E}|S_{\tilde{T}}| < \infty,$$

since a random variable has mean zero if its positive and negative parts have finite expectations and are equal. This implies that the absolute value of the random variable has finite expectation.

PROOF OF THEOREM 2.1.4. The proof of (2.1.12) is based on the bound

$$\mathbb{E}\max_{n \leq T} |S_n| \leq c\mathbb{E}K_X(T), \tag{2.1.16}$$

where $c < \infty$ is a universal constant, along with the dominated convergence theorem. In order to obtain this bound, we will use Davis' inequality for martingales [(1.4.29) with $p = 1$] to transform the problem into one involving only positive random variables. After this adjustment, we use the method of subsequences to trap the stopping time and decouple it from the Xs. Finally, we use the monotonicity properties of the K-function to recover the distributional properties of the random variables. We remark that in place of the K-function we could also use the L-function (Theorem 1.5.6) since it also has the desired monotonicity properties (Lemma 1.5.7).

Let $\kappa_T = \min\{k : 2^k > T\}$, $T^* = 2^{\kappa_T} - 1$. Then the following inequalities hold:

$$2^{\kappa_T - 1} \le T, \tag{2.1.17}$$

$$T^* \ge T, \tag{2.1.18}$$

and for all $k \ge 1$

$$I_{T^* \ge 2^k} = I_{T^* \ge 2^{k+1}-1}, \tag{2.1.19}$$

$$I_{\kappa_T - 1 \ge k} = I_{T^* \ge 2^k}. \tag{2.1.20}$$

To obtain (2.1.20) note that,

$$I_{\kappa_T - 1 \ge k} = I_{2^{\kappa_T} \ge 2^{k+1}} = I_{2^{\kappa_T} - 1 \ge 2^{k+1}-1} = I_{T^* \ge 2^{k+1}-1} = I_{T^* \ge 2^k},$$

where the last identity follows from (2.1.19).

We are now ready to proceed with the formal part of the proof. Throughout we will let the constant c denote a universal constant $0 < c < \infty$ that might change from line to line. Since $S_{T \wedge n} = \sum_{i=1}^{T \wedge n} X_i = \sum_{i=1}^{n} X_i I_{T \ge i}$, $(S_{T \wedge n}, \mathcal{F}_n)$ is a martingale, Davis' inequality (1.4.29) gives

$$\mathbb{E} \max_{n \le T} |S_n| = \mathbb{E} \sup_{n \le \infty} |S_{T \wedge n}| = \mathbb{E} \sup_{n \le \infty} \Big| \sum_{i=1}^{n} X_i I_{T \ge n} \Big|$$
$$\le c\mathbb{E}\Big(\sum_{j=1}^{\infty} X_i^2 I_{T \ge i} \Big)^{\frac{1}{2}} = c\mathbb{E}\Big(\sum_{j=1}^{T} X_i^2 \Big)^{\frac{1}{2}}.$$

The justification for the steps taken to bound this term are given at the end of the following string of inequalities:

$$c\mathbb{E}\Big(\sum_{j=1}^{T} X_i^2 \Big)^{\frac{1}{2}} = c\mathbb{E}\Big(\sum_{k=0}^{\infty} \sum_{2^k \le j < 2^{k+1}} X_i^2 I_{T \ge j} \Big)^{\frac{1}{2}}$$
$$\le c \sum_{k=0}^{\infty} \mathbb{E}\Big(\sum_{2^k \le j < 2^{k+1}} X_i^2 I_{T \ge j} \Big)^{\frac{1}{2}}$$
$$\le c \sum_{k=0}^{\infty} \mathbb{E}\Big(\sum_{2^k \le j < 2^{k+1}} X_i^2 \Big)^{\frac{1}{2}} I_{T \ge 2^k}$$
$$= c \sum_{k=0}^{\infty} \mathbb{E}\Big(\sum_{2^k \le j < 2^{k+1}} X_i^2 \Big)^{\frac{1}{2}} \Pr\{T \ge 2^k\} \qquad \text{(by A)}$$

$$\le c \sum_{k=0}^{\infty} \Pr\{T \ge 2^k\} \mathbb{E}|S_{2^k}| \qquad \text{(by B)}$$
$$\le c \sum_{k=0}^{\infty} \Pr\{T \ge 2^k\} K_X(2^k) \qquad \text{(by C)}$$
$$\le c\mathbb{E} \sum_{k=0}^{\infty} I_{T \ge 2^k} K_X(2^k) \qquad \text{(by D)}$$
$$\le c\mathbb{E} \sum_{k=0}^{\infty} I_{T^* \ge 2^k} K_X(2^k) \qquad \text{(by E)}$$
$$\le c\mathbb{E} \sum_{k=0}^{\infty} I_{\kappa_T - 1 \ge k} K_X(2^k) \qquad \text{(by F)}$$
$$\le c\mathbb{E} \sum_{k=0}^{\kappa_T - 1} \big(2^{-\frac{k}{2}} K_X(2^k)\big) 2^{\frac{k}{2}}$$
$$\le c\mathbb{E}\big(2^{-\frac{(\kappa_T - 1)}{2}} K_X(2^{\kappa_T - 1})\big) \sum_{k=0}^{\kappa_T - 1} 2^{\frac{k}{2}} \qquad \text{(by G)}$$
$$\le c\mathbb{E}K(2^{\kappa_T - 1}) 2^{-\frac{(\kappa_T - 1)}{2}} \frac{2^{\frac{\kappa_T - 1}{2}} - 1}{\sqrt{2} - 1}$$
$$= c\mathbb{E}K(2^{\kappa_T - 1})$$
$$\le c\mathbb{E}K(T), \qquad \text{(by H)}$$

where A through G can be justified as follows:

A) Use that $T \ge 2^k \in \mathcal{F}_{2^k - 1}$ and hence $\sum_{2^k \le j < 2^{k+1}} X_i^2$ is independent of $T \ge 2^k$.

B) Since the variables are i.i.d., using (1.4.30).

C) By (1.4.21) and (2.1.14).

D) By monotone convergence theorem.

E) By using (2.1.18).

F) By using (2.1.20).

G) by (1.4.23).

H) By using (2.1.17).

To obtain Theorem 2.1.4 from here, we observe that the condition

$$\mathbb{E} \sup_{n < \infty} |S_{T \wedge n}| \le C\mathbb{E}K_X(T) < \infty$$

along with the dominated convergence theorem gives

$$\mathbb{E}S_T = \mathbb{E} \lim_{n \to \infty} S_{T \wedge n} = \lim_{n \to \infty} \mathbb{E}S_{T \wedge n} = 0. \qquad (2.1.21) \qquad \square$$

2.2 Good-Lambda Inequalities

In this section we provide an introduction to the theory of good-lambda inequalities introduced by Burkholder and Gundy. This type of inequality is the theoretical foundation for the remaining sections of this chapter, and is typically used to derive inequalities for comparing the expectations of two variables that are highly correlated. The results of this section are also used in Chapters 6 and 7 to develop general decoupling and martingale inequalities.

We begin by introducing the class of functions that will be used throughout the chapter. For fixed $\alpha > 0$,

$$\mathcal{A}_\alpha = \big\{\Phi : R_+ \to R_+, \quad \Phi(0) = 0,\ \Phi \text{ nondecreasing, continuous, and } \Phi(cx) \le c^\alpha \Phi(x) \text{ for all } c \ge 2,\ x \ge 0\big\}. \tag{2.2.1}$$

Roughly speaking, a good-lambda inequality exploits the principle that whenever two variables U and V interact in such a way that if U is "big," it cannot be too big whenever V is "small," then, the expected value of U cannot exceed some multiple of the expected value of V.

LEMMA 2.2.1. *Let U and V be non-negative random variables, and let ε be a positive number. Suppose there exist positive reals β, γ, δ such that*

$$\beta^{-1} > \varepsilon$$

and

$$\Pr\{U \ge \beta y, \quad V < \delta y\} \le \varepsilon \Pr\{U \ge y\} \qquad \text{for all } y > 0. \tag{2.2.2}$$

Then,

$$\mathbb{E}U \le (\beta^{-1} - \varepsilon)^{-1}\delta^{-1}\mathbb{E}V. \tag{2.2.3}$$

The use of conditional probability helps in clarifying the meaning of the condition. For any β as above,

$$\Pr\{U \ge \beta y, V < \delta y | U \ge y\} \le \varepsilon < \frac{1}{\beta}.$$

Therefore, if U is "big," it cannot be "too big" whenever V is "small."

PROOF.

$$\begin{aligned}\Pr\{U \ge \beta y\} &= \Pr\{U \ge \beta y, V < \delta y\} + \Pr\{U \ge \beta y, V \ge \delta y\} \\ &\le \varepsilon \Pr\{U \ge y\} + \Pr\{V \ge \delta y\}.\end{aligned}$$

Integrating with respect to y:

$$\frac{1}{\beta}\mathbb{E}U \le \varepsilon\mathbb{E}U + \frac{1}{\delta}\mathbb{E}V.$$

Therefore,

$$\mathbb{E}U \le \frac{1}{\beta^{-1} - \varepsilon} \frac{1}{\delta} \mathbb{E}V. \qquad \square$$

We will use a slight generalization of this result stated in the following lemma.

LEMMA 2.2.2. *Fix $\alpha > 0$. Let U and V be non-negative random variables. Suppose there exist positive reals β, ε, δ for which (2.2.2) holds. Then, for all $\Phi \in \mathcal{A}_\alpha$,*

$$\mathbb{E}\Phi(U) \le (\gamma^{-1} - \varepsilon)^{-1} \mu \mathbb{E}\Phi(V), \tag{2.2.4}$$

whenever $\gamma^{-1} > \varepsilon$, where $\mu = (2 \vee \frac{1}{\delta})^\alpha$ and $\gamma = (\beta \vee 2)^\alpha$.

PROOF. Observe that under the conditions of the lemma,

$$\begin{aligned}\Pr\Big\{\Phi(U) > \gamma\Phi(y), \Phi(V) \le \frac{1}{\mu}\Phi(y)\Big\} &\le \Pr\big\{U > \beta y, V \le \delta y\big\} \\ &\le \varepsilon \Pr\big\{\Phi(U) > \Phi(y)\big\},\end{aligned}$$

and proceed as before taking $y' = \Phi(y)$. $\square$

LEMMA 2.2.3. *Let U and V be non-negative random variables. Suppose there exist positive reals β, γ, δ such that $\beta^{-1} > \varepsilon$ and (2.2.2) holds. Then,*

$$\sup_{y>0} y \Pr\{U \ge y\} \le \frac{\beta}{\delta(1 - \beta\varepsilon)} \sup_{y>0} y \Pr\{V \ge y\},$$

and

$$\limsup_{y>0} y \Pr\{U \ge y\} \le \frac{\beta}{\delta(1 - \beta\varepsilon)} \limsup_{y>0} y \Pr\{V \ge y\}. \tag{2.2.5}$$

PROOF. It follows from (2.2.2) that

$$\Pr\{U \ge \beta y\} \le \Pr\{U \ge \beta y, V < \delta y\} + \Pr\{V \ge \delta y\} \le \varepsilon \Pr\{U \ge y\} + \Pr\{V \ge \delta y\}.$$

Setting $U(y) = y \Pr\{U > y\}$, $V(y) = y \Pr\{V \ge y\}$, this inequality becomes

$$U(\beta y) \le \beta\varepsilon U(y) + \frac{\beta}{\delta} \sup_{y>0} V(y).$$

By a change of variables this can be modified to give

$$U(y) \le \beta\varepsilon U(y/\beta) + \frac{\beta}{\delta} \sup_{y>0} V(y).$$

Iterating the above one gets that

$$\begin{aligned}U(y) &\le (\beta\varepsilon)^n U(y/\beta^n) + (1 + \beta\varepsilon + \cdots + (\beta\varepsilon)^n) \frac{\beta \sup_{y>0} V(y)}{\delta} \\ &\le U(y/\beta^n) + \frac{1}{1 - \beta\varepsilon} \frac{\beta \sup_{y>0} V(y)}{\delta}.\end{aligned}$$

Letting $n \to \infty$ one gets the first inequality. To prove the second inequality, assuming $\limsup_{y>0} V(y) < \infty$ gives $\sup_{y>0} V(y) < \infty$ and therefore, using the first result one gets $\limsup_{y>0} U(y) \le \sup_{y>0} U(y) < \infty$. Finally, using the first steps of the proof, one easily gets that $\limsup U(y) \le \beta\varepsilon \limsup U(y) + \frac{\beta}{\delta} \limsup V(y)$, completing the proof. □

Having finished our introduction to good-lambda inequalities, we proceed to obtain the main decoupling inequality of this section.

2.3 Randomly Stopped Sums of Independent Banach-Valued Variables

The key result in the proof of Theorem 2.1.4 was given in (2.1.16). Letting $a_n = \mathbb{E}|S_n|$, a use of the properties of the K-function [Proposition 1.4.10 and (2.1.14)] shows that (2.1.16) is equivalent to the inequality $\mathbb{E}\max_{n\le T}|S_n| \le c'\mathbb{E}a_T < \infty$, for some universal constant $c' < \infty$. In this chapter we extend this to the case of arbitrary functions in $\mathcal{A}_\alpha$ [see (2.2.1)].

THEOREM 2.3.1. *Fix any $\alpha > 0$. Let $\Phi \in \mathcal{A}_\alpha$. Let $X_1, X_2, \ldots$ be independent random variables with values in a separable Banach space $(B, \|\cdot\|)$. Set $S_n = X_1 + \cdots + X_n$. Let T be a stopping time adapted to $\sigma(X_1, X_2, \ldots, X_n, \ldots)$. Then*

$$c_\alpha \mathbb{E}a^*_{T,\Phi} \le \mathbb{E}\max_{1\le n\le T}\Phi(\|S_n\|) \le C_\alpha \mathbb{E}a^*_{T,\Phi}, \tag{2.3.1}$$

where $a^*_{n,\Phi} = \mathbb{E}\max_{1\le k\le n}\Phi(\|S_k\|)$, and c_α, C_α are constants depending on α only [indeed, $C_\alpha \le 20(18)^\alpha$].

Equivalently if $\{\tilde{X}_i\}$ is an independent copy of $\{X_i\}$, and $\tilde{S}_n = \sum_{i=1}^n \tilde{X}_i$, then,

$$c_\alpha \mathbb{E}\max_{1\le n\le T}\Phi(\|\tilde{S}_n\|) \le \mathbb{E}\max_{1\le n\le T}\Phi(\|S_n\|) \le C_\alpha \mathbb{E}\max_{1\le n\le T}\Phi(\|\tilde{S}_n\|). \tag{2.3.2}$$

The reason why (2.3.2) is valid is that since $(\{X_i\}, T)$ and $(\{\tilde{X}_i\}, \tilde{T})$ have the same distribution, then, as in (2.1.15),

$$\begin{aligned}\mathbb{E}a^*_{T,\Phi} = \mathbb{E}a^*_{\tilde{T},\Phi} &= \mathbb{E}\Big(\mathbb{E}\big(\max_{1\le k\le \tilde{T}}\Phi(\|S_n\|)\big|\tilde{T}\big)\Big)\\ &= \mathbb{E}\max_{1\le k\le \tilde{T}}\Phi(\|S_n\|) = \mathbb{E}\max_{1\le k\le T}\Phi(\|\tilde{S}_n\|),\end{aligned}$$

making more obvious the decoupling nature of Theorem 2.3.1.

In what follows we introduce notation that is important in the development of a proof of the right-hand side of Theorem 2.3.1. Let

$$\begin{aligned} S^*_{(m,n]} &= \max_{m\le j\le n}\|S_j - S_m\| = \max_{m\le j\le n}\|X_{m+1} + \cdots + X_j\|;\\ S^*_n &= S^*_{(0,n]} = \max_{j\le n}\|S_j\|; \\ X^*_n &= \max_{1\le j\le n}\|X_j\|; \quad n'_y = n' = \sup\{n : a^*_n \le \delta y\}; \quad a^*_{n,\Phi} = \mathbb{E}\max_{1\le j\le n}\Phi(\|S_j\|).\end{aligned} \tag{2.3.3}$$

Fix $y > 0$. Let

$$T_y = \begin{cases} \min\{1 \le m \le T : \Phi(S_m^*) \ge y\} & \text{if such } m \text{ exists} \\ \infty & \text{otherwise.} \end{cases} \tag{2.3.4}$$

We will first obtain the following lemma.

LEMMA 2.3.2. *Fix $\alpha > 0$. Under the assumptions of Theorem 2.3.1 for all $\Phi \in \mathcal{A}_\alpha$ and all $y > 0$,*

$$\begin{aligned}\Pr\big\{\Phi(S_T^*) \ge \beta y, \Phi(X_T^*) \vee a_{T,\Phi}^* < \delta y\big\} &\le \frac{3^\alpha \cdot 2^\alpha \mathbb{E}\Phi(S_{n^*}^*)}{y(\beta - 3^\alpha(1+\delta))} \Pr\{T_y < T\} \\ &\le \frac{2^\alpha \delta . 3^\alpha}{\beta - 3^\alpha(1+\delta)} \Pr\{T_y < T\} \\ &\le \frac{6^\alpha \delta}{\beta - 3^\alpha(1+\delta)} \Pr\{\Phi(S_T^*) \ge y\},\end{aligned} \tag{2.3.5}$$

whenever $\delta > 0$ and $\beta > 3^\alpha(1+\delta)$. This implies

$$\mathbb{E} \max_{n \le T} \Phi(\|S_n\|) = \mathbb{E}\Phi(S_T^*) \le c\mathbb{E}(\Phi(X_T^*) \vee a_{T,\Phi}^*)$$

for some $c < \infty$.

PROOF. We begin by stating some useful facts. Note that for any $a, b, c \ge 0$,

$$\begin{aligned}\Phi(a+b+c) \le \Phi(3\max\{a,b,c\}) &\le \Phi(3a) + \Phi(3b) + \Phi(3c) \\ &\le 3^\alpha\{\Phi(a) + \Phi(b) + \Phi(c)\}.\end{aligned}$$

Observe that $\beta > 1$ implies that $T_y \le T$ on the set where $\Phi(S_T^*) \ge \beta y$. Furthermore, on the set where $T_y \le T$,

$$\begin{aligned}S_T^* &= \max_{j \le T_y} \|S_j\| \vee \max_{T_y < j \le T} \|S_j\| \\ &\le \|S_{T_y}\| \vee \Big\{\max_{T_y < j \le T} \|S_j - S_{T_y}\| + \|S_{T_y}\|\Big\} \\ &\le \max_{T_y < j \le T} \|S_j - S_{T_y}\| + \|S_{T_y - 1}\| + \|X_{T_y}\|.\end{aligned} \tag{2.3.6}$$

Putting these observations together one gets that, whenever $\Phi(S_T^*) \ge \beta y$,

$$\Phi(S_T^*) \le 3^\alpha\big[\Phi(S_{T_y-1}^*) + \Phi(\|X_{T_y}\|) + \Phi(S_{(T_y,T]}^*)\big].$$

Using the definitions of T_y and n' one can see that $\Phi(S_{T_y-1}^*) \le y$ and that $a_{T,\Phi}^* < \delta y$ implies that $T \le n'$. An application of all of the above facts gives that

$$\begin{aligned}&\Pr\big\{\Phi(S_T^*) \ge \beta y, \Phi(X_T^*) \vee a_{T,\Phi}^* < \delta y\big\} \\ &\quad \le \Pr\big\{3^\alpha\big(\Phi(S_{T_y-1}^*) + \Phi(\|X_{T_y}\|) + \Phi(S_{(T_y,T]}^*)\big) \ge \beta y, \Phi(X_T^*) < \delta y, T \le n'\big\} \\ &\quad \le \Pr\big\{3^\alpha\big(y + \delta y + \Phi(S_{(T_y,T]}^*)\big) \ge \beta y, \Phi(X_T^*) < \delta y, T \le n'\big\} \\ &\quad \le \Pr\big\{3^\alpha \Phi(S_{(T_y,T\wedge n']}^*) \ge y(\beta - 3^\alpha(1+\delta))\big\}\end{aligned}$$

This and Markov's inequality give

$$\Pr\big\{\Phi(S_T^*) \geq \beta y,\, \Phi(X_T^*) \vee a_{T,\Phi}^* < \delta y\big\} \\ \leq 3^\alpha y^{-1}(\beta - 3^\alpha(1+\delta))^{-1}\mathbb{E}\Phi(S^*_{(T_y,T\wedge n']}).$$

In what follows we bound the expectation:

$$\begin{aligned}
&\mathbb{E}\Phi(S^*_{(T_y,T\wedge n']}) \\
&= \sum_{j=1}^{n'-1} \mathbb{E}\Phi(S^*_{(j,T\wedge n']})I(T_y = j, T > j) \qquad \text{(since } (n', n'] \text{ is empty)} \\
&\leq \sum_{j=1}^{n'-1} \mathbb{E}\Phi(S^*_{(j,n']})I(T_y = j, T > j) \\
&= \sum_{j=1}^{n'-1} \mathbb{E}\Phi(S^*_{(j,n']}) \Pr\{T_y = j, T > j\} \qquad \text{(by independence, where decoupling occurs)} \\
&\leq \sum_{j=1}^{n'-1} \mathbb{E}\Phi(2S^*_{n'}) \Pr\{T_y = j, T > j\} \\
&\leq 2^\alpha \mathbb{E}\Phi(S^*_{n'}) \Pr\{T_y < T\} \\
&\leq 2^\alpha \delta y \Pr\{\Phi(S_T^*) \geq y\},
\end{aligned} \tag{2.3.7}$$

where the last line follows from the definition of T_y. Putting the two bounds together gives the proof of the lemma. □

PROOF OF UPPER-BOUND OF THEOREM 2.3.1. Letting $\varepsilon = 6^\alpha\delta(\beta - 3^\alpha(1+\delta))^{-1}$ with $\beta\varepsilon < 1$ in Lemma 2.2.1 gives

$$\mathbb{E}\Phi(S_T^*) \leq \delta^{-1}(\beta^{-1} - \gamma)^{-1}\mathbb{E}\big(\Phi(X_T^*) \vee a_{T,\Phi}^*\big). \tag{2.3.8}$$

Since $\mathbb{E}(\Phi(X_T^*) \vee a_{T,\Phi}^*) \leq \mathbb{E}\Phi(X_T^*) + \mathbb{E}a_{T,\Phi}^*$, to finish the proof of Theorem 2.3.1, we must bound $\mathbb{E}\Phi(X_T^*)$. We will do this by means of a lemma of independent interest.

LEMMA 2.3.3. *Let $\{d_i\}$, $\{e_i\}$ be two sequences of real random variables adapted to an increasing sequence of σ-fields $\{\mathcal{F}_i\}$. Assume that for all i, $\mathcal{L}(d_i|\mathcal{F}_{i-1}) = \mathcal{L}(e_i|\mathcal{F}_{i-1})$ (the sequences are said to be tangent). Then, for all $t > 0$,*

$$\Pr\Big\{\sup_{j\leq\infty} d_j > t\Big\} \leq 2\Pr\Big\{\sup_{j\leq\infty} e_j > t\Big\}. \tag{2.3.9}$$

PROOF. Let $\sigma = \inf\{n : e_n > t\}$. Then

$$\begin{aligned}
\Pr\Big\{\sup_{j\leq\infty} d_j > t\Big\} &= \Pr\Big\{\sup_{j\leq\infty} d_j > t, \sigma = \infty\Big\} + \Pr\Big\{\sup_{j\leq\infty} d_j > t, \sigma < \infty\Big\} \\
&= \mathbb{E}1_{\sup_{j\leq\infty} d_j > t} I_{\sigma=\infty} + \Pr\Big\{\sup_{j\leq\infty} d_j > t, \sigma < \infty\Big\}
\end{aligned}$$

$$
\begin{aligned}
&\le \mathbb{E}\sum_{j=1}^{\infty} I_{d_j>t} I_{\sigma=\infty} + \Pr\{\sigma < \infty\}\\
&\le \mathbb{E}\sum_{j=1}^{\infty} I_{d_j>t} I_{\sigma\ge j} + \Pr\{\sigma < \infty\}\\
&= \sum_{j=1}^{\infty} \mathbb{E}\Big(\mathbb{E}\big(I_{d_j>t} 1_{\sigma\ge j} \big| \mathcal{F}_{j-1}\big)\Big) + \Pr\{\sigma < \infty\}\\
&= \sum_{j=1}^{\infty} \mathbb{E} I_{(\sigma\ge j)} \mathbb{E}\big(I_{d_j>t} \big| \mathcal{F}_{j-1}\big) + \Pr\{\sigma < \infty\}\\
&= \sum_{j=1}^{\infty} \mathbb{E} I_{\sigma\ge j} \mathbb{E}\big(I_{e_j>t} \big| \mathcal{F}_{j-1}\big) + \Pr\{\sigma < \infty\} \qquad (\{d_i\}, \{e_i\} \text{ tangent})\\
&= \mathbb{E}\sum_{j=1}^{\sigma} I_{e_j>t} + \Pr\{\sigma < \infty\}\\
&= \mathbb{E} I_{e_\sigma>t} + \Pr\{\sigma < \infty\}\\
&\le \mathbb{E} I_{\sup_{j\le\infty} e_j>t} + \Pr\Big\{\sup_{j<\infty} e_j > t\Big\} \qquad (\text{def. of } \sigma)\\
&\le 2\Pr\Big\{\sup_{j\le\infty} e_j > t\Big\}.
\end{aligned}
$$

□

Letting $d_i = \Phi(\|X_i\|) I_{T\ge i}$, $e_i = \Phi(\|\tilde{X}_i\|) I_{T\ge i}$, and

$$\mathcal{F}_i = \sigma(X_1, \ldots, X_i; \tilde{X}_1, \ldots, \tilde{X}_i),$$

then d_i and e_i are tangent since, given $\mathcal{F}_{i-1}$, $I_{T\ge i}$ is a constant and the only additional randomness into d_i and e_i comes through X_i and $\tilde{X}_i$, which are i.i.d. Furthermore, observe that $\Phi(X_T^*) = \sup_i \Phi(\|X_i\|)1(T \ge i)$ and similarly for $\Phi(\tilde{X}_T^*)$. Therefore, applying Lemma 2.3.3 one gets

$$\Pr\big\{\Phi(X_T^*) > t\big\} \le 2\Pr\big\{\Phi(\tilde{X}_T^*) > t\big\}, \tag{2.3.10}$$

for all $t \ge 0$. Integrating over t we have,

$$
\begin{aligned}
\mathbb{E}\Phi(X_T^*) &= \mathbb{E}\Phi\big(\max_{n\le T} \|X_n\|\big) \le 2\mathbb{E}\Phi\big(\max_{n\le T} \|\tilde{X}_n\|\big) \le 2\mathbb{E}\Phi\big(\max_{n\le T} \|\tilde{S}_n - \tilde{S}_{n-1}\|\big)\\
&\le 2\mathbb{E}\Phi\big(\max_{n\le T} \|\tilde{S}_n\| + \max_{n\le T} \|\tilde{S}_{n-1}\|\big) \le 2\mathbb{E}\Phi\big(2\max_{n\le T} \|\tilde{S}_n\|\big)\\
&\le 2^{\alpha+1}\mathbb{E}\Phi\Big(\max_{n\le T} \Big\|\sum_{i=1}^{n} \tilde{X}_i\Big\|\Big) = \mathbb{E}a_{T,\Phi}^* = \mathbb{E}a_{\tilde{T},\Phi}^*.
\end{aligned} \tag{2.3.11}
$$

To complete the proof of the right-hand side of Theorem 2.3.1 observe that since $\mathbb{E}a_{T,\Phi} = \mathbb{E}\Phi\big(\max_{n\le T} \|\tilde{S}_n\|\big)$, (2.3.8) plus (2.3.11) give

$$\mathbb{E}\Phi(S_T^*) \le A_\alpha\Big[\mathbb{E}\Phi(X_T^*) + \mathbb{E}\Phi\big(\max_{n\le T} \|\tilde{S}_n\|\big)\Big] \le A_\alpha(2^{\alpha+1} + 1)\mathbb{E}\Phi(\max_{n\le T} \|\tilde{S}_n\|).$$

The proof of the left-hand side is very long and complicated. We devote the following section to it.

2.4 Proof of the Lower Bound of Theorem 2.3.1

To set the stage, recall that $S^*_{(m,n]} = \max_{m<j\le n} \|X_{m+1} + \cdots + X_j\|$. Let $\gamma_\alpha = 1 + (\frac{1}{16})^{\alpha+1}$, $m_0 = 0$, $m_1 = \inf\{m > m_0 : \Pr\{S^*_m > 0\} > 0\}$. Assuming that there is an m_1 with the above property, we define all other m_j by the iterative relation:

$$m_{j+1} = \inf\{n > m_j : \mathbb{E}\Phi(S^*_{(m_j,n]}) \ge \gamma_\alpha \mathbb{E}\Phi(S^*_{(m_{j-1},m_j]})\} \tag{2.4.1}$$

with $m_{j+1} = \infty$ if no such n exists. The expectation of the maximal increments of the process will be denoted by

$$b_j = \mathbb{E}\Phi\big(S^*_{(m_{j-1},m_j]}\big). \tag{2.4.2}$$

The proof of the lower bound is immediate once the following two auxiliary results are established.

PROPOSITION 2.4.1. *There exists a universal constant $0 < c_\alpha < \infty$ such that,*

$$\mathbb{E}a^*_{T,\Phi} \le c_\alpha \sum_{j=1}^{\infty} b_j \Pr\{T \ge m_j\}. \tag{2.4.3}$$

And

PROPOSITION 2.4.2. *There exists a universal constant $0 < c_\alpha < \infty$ such that,*

$$\sum_{j=1}^{\infty} b_j \Pr\{T \ge m_j\} \le c_\alpha \mathbb{E}\Phi(S^*_T). \tag{2.4.4}$$

The proof of Proposition 2.4.1 is obtained by using the fact (derived from the construction of the m_js) that for all m with $m_j \le m < m_{j+1}$, $a^*_{m,\Phi} \approx b_j$. That is, the sequence $\{m_j\}$ ties the growth of $a^*_{n,\Phi}$ to that of its increments. The proof of this is easy once Lemmas 2.4.3 through 2.4.5 below are established.

The proof of Proposition 2.4.2, is more complicated. One of the difficulties is that the growth of $U = \sum_{j=1}^{\infty} b_j I_{T\ge m_j}$ is not as directly linked to that of $V = \Phi(S^*_T)$ as to imply a good-lambda inequality. A second problem is that $S_{(m_{j-1},m_j]}$ and $I_{T\ge m_j}$ are not independent and hence we cannot use Wald's equation as was used in our first application to derive the good-lambda inequality [fourth line of (2.3.7)]. In order to solve these problems, several truncation arguments and rearrangement of summations are applied. In particular the surprisingly simple fact that, for a random variable $M > 0$,

$$\mathbb{E}M I_{M>\frac{1}{2}\mathbb{E}M} \ge \frac{1}{2}\mathbb{E}M \tag{2.4.5}$$

is used to create a situation where the growth rate of V does constrain that of a variant of U. The rearranging of summations is used to help in solving the second problem. A series of lemmas (Lemma 2.4.6 through Lemma 2.4.11) followed in order gives the proof of Proposition 2.4.2. We advise the reader to first review the statements of Lemmas 2.4.6 through 2.4.11 and check how they imply the main result. This will be helpful in providing the "big picture" before dealing with the intricate details.

Without further delay, let us continue with the proof. Since it is more convenient to work with a different class of functions than $\mathcal{A}_\alpha$, the following result allows the transfer needed. Fix $\alpha > 0$ and let $\mathcal{D}_\alpha$ be the set of all nondecreasing continuous functions Φ on $[0, \infty)$ with $\Phi(0) = 0$ such that

$$\Phi(cx) \le c^\alpha \Phi(x) \quad \text{for all } x \ge 0,\ c \ge 1. \tag{2.4.6}$$

LEMMA 2.4.3. *Let Φ be in $\mathcal{A}_\alpha$ and set*

$$\Phi_a(x) = x^{-1} \int_0^x \Phi(y)dy, \quad \text{for } x > 0 \text{ and } \Phi_a(0) = 0. \tag{2.4.7}$$

Taking $\alpha_a = 2^{2\alpha+1} - 1$ gives that $\Phi_a(\,\cdot\,)$ is a nondecreasing function, which is continuous and satisfies the condition $\Phi_a(0) = 0$. Moreover,

$$\Phi_a(x) \le \Phi(x) \le 2^{\alpha+1}\Phi_a(x) \quad \text{for all } x \ge 0 \tag{2.4.8}$$

and

$$\Phi_a(cx) \le c^{\alpha_a}\Phi_a(x) \quad \text{for all } c \ge 1 \text{ and } x \ge 0. \tag{2.4.9}$$

PROOF. If $x = 0$, the result is trivially valid. For $x > 0$, $\Phi_a(x) = x^{-1}\int_0^x \Phi(y)dy \le \int_0^x \Phi(x)dy = \Phi(x)$. Moreover,

$$\Phi_a(x) \ge x^{-1} \int_{\frac{x}{2}}^x \Phi(\frac{x}{2})dy = \frac{\Phi(\frac{x}{2})}{2} \ge \frac{\Phi(x)}{2^{\alpha+1}}.$$

It remains to establish (2.4.9). This new type of function is needed only in the case $1 \le c \le 2$, since for the case $c > 2$ the functions Φ in $\mathcal{A}_\alpha$ suffice, assuming that $1 \le c \le 2$ and $x > 0$,

$$\begin{aligned}
\Phi_a(xc) &= c^{-1}x^{-1}\int_0^{cx} \Phi(y)dy \\
&= c^{-1}\Phi_a(x) + c^{-1}x^{-1}\int_x^{cx} \Phi(y)dy \\
&\le c^{-1}\Phi_a(x) + c^{-1}(c-1)\Phi(2x) \qquad (\text{since } 1 \le c \le 2) \\
&\le c^{-1}\Phi_a(x) + c^{-1}(c-1)2^\alpha\Phi(x) \\
&\le c^{-1}(1 + (c-1)2^{2\alpha+1})\Phi_a(x) \qquad [\text{by } (2.4.8)].
\end{aligned}$$

The final result in (2.4.9) can be obtained using the fact that $(1+(c-1)y) \le c^y$ for $y \ge 1$. This follows easily by noting that the function $f_y(c) = c^y - (1+(c-1)y)$ is positive as it satisfies $f_y(1) = 0$ and $\frac{d}{dc}f_y(c) \ge 0$ for $1 \le c \le 2$ and $y \ge 1$. □

Let $\Phi \in \mathcal{F}_\alpha$ and assume that Theorem 2.3.1 is valid for all functions in $\mathcal{D}_\alpha$. Then using (2.4.8), (2.4.9), and (2.3.1) gives

$$\mathbb{E}a_{T,\Phi} \leq 2^{\alpha+1}\mathbb{E}a_{T,\Phi_a} \leq \frac{2^{\alpha+1}}{c_{\alpha_a}}\mathbb{E}\Phi_a(S_T^*) \leq \frac{2^{\alpha+1}}{c_{\alpha_a}}\mathbb{E}\Phi(S_T^*). \tag{2.4.10}$$

Hence, it suffices to prove Theorem 2.3.1 for Φ in $\mathcal{D}_\alpha$ to have it for all Φ in $\mathcal{F}_\alpha$.

From now until the end of this section we assume that Φ is in $\mathcal{D}_\alpha$ unless otherwise stated.

We will need the following quantities and random variables in addition to the ones introduced in (2.3.3). Let

$$S^*_{(m,n)} = S^*_{(m,n-1]} = \max_{m<j<n} \|S_j - S_m\| \tag{2.4.11}$$

and

$$\gamma_\alpha = 1 + (1/16)^{\alpha+1}. \tag{2.4.12}$$

Let

$$L = \sup\{j : m_{j-1} < \infty\} \qquad \text{[with } m_j \text{ from (2.4.1)]}, \tag{2.4.13}$$

$$\Phi_j = \Phi(S^*_{(m_{j-1},m_j)}), \tag{2.4.14}$$

$$\Phi'_j = \Phi_j I_{\Phi_j \geq 2^{-\alpha-2}b_j}, \tag{2.4.15}$$

$$X'_{m_j} = X_{m_j} I_{\Phi(\|X_{m_j}\|) \geq 2^{-\alpha-2}\varepsilon_\alpha b_j}, \tag{2.4.16}$$

for some $0 < \varepsilon_\alpha < 1$.

LEMMA 2.4.4. *Let U and V be two random variables with values in a separable Banach space $(B, \|\cdot\|)$. Let $\Phi : [0,\infty) \to [0,\infty)$ be a nondecreasing continuous function satisfying the growth property*

$$\Phi(cx) \leq c^\alpha \Phi(x) \quad \textit{for all } x \geq 0,\ c \geq 1. \tag{2.4.17}$$

Assume that for some $\lambda > 0$ and some $y \geq 0$,

$$\mathbb{E}\Phi(\|U\|) \leq y \quad \textit{and} \quad \mathbb{E}\Phi(\|V\|) \leq \lambda y. \tag{2.4.18}$$

Then

$$\mathbb{E}\max\{\Phi(\|U\|), \Phi(\|U+V\|)\} \leq h_\alpha(\lambda)y, \tag{2.4.19}$$

with

$$h_\alpha(x) = (1 + x^{1/2})(1 + x^{\frac{1}{2\alpha}})^\alpha. \tag{2.4.20}$$

PROOF. Taking $c = \lambda^{\frac{1}{2\alpha}}$ we have

$$\begin{aligned}
&\mathbb{E}\max\{\Phi(\|U\|), \Phi(\|U+V\|)\} \\
&\quad \leq \mathbb{E}\Phi(\|(1+c)U\|)I_{\|V\|\leq c\|U\|} + \mathbb{E}\Phi(\|(1+1/c)V\|)I_{\|V\|>c\|U\|} \\
&\quad \leq (1+c)^\alpha\mathbb{E}\Phi(\|U\|) + (1+1/c)^\alpha\mathbb{E}\Phi(\|V\|) \\
&\quad \leq (1+c)^\alpha(1+c^\alpha)y \\
&\quad = h_\alpha(\lambda)y.
\end{aligned}$$

□

LEMMA 2.4.5. *Take m_j, m_{j+1}, m so that $m_j \le m < m_{j+1}$. Then, there exists a positive finite constant c_α depending on α only such that*

$$2^{-\alpha} b_j \le \mathbb{E}\Phi(S_m^*) \le c_\alpha b_j. \tag{2.4.21}$$

PROOF. The idea of the proof is to use the structure of m_j to control the growth rate of $\mathbb{E}\Phi(S_m^*)$. Since $m \ge m_j$, $S^*_{(m_{j-1},m_j]} \le 2S^*_{m_j} \le 2S^*_m$. From this it follows that $b_j \le \mathbb{E}\Phi(2S_m^*) \le 2^\alpha \mathbb{E}\Phi(S_m^*)$. The other side is proved by dividing it into two cases. Let j^* be the first i such that $\mathbb{E}\Phi(S_{m_i}) \ge 2^{-1}\mathbb{E}\Phi(S^*_{m_j})$. Suppose first that

$$A = \left\{\mathbb{E}\Phi(S^*_{m_{j^*}-1}) > 2^{-\alpha-1}\mathbb{E}\Phi(S^*_{m_{j^*}})\right\}. \tag{2.4.22}$$

Since $h_\alpha(0) = 1$ for h_α as in the previous lemma, there exists a positive ε_α^* for which

$$\Pi_{i=0}^{\infty} h_\alpha(\varepsilon_\alpha^*(\gamma_\alpha)^{-i}) = 2. \tag{2.4.23}$$

The proof of this side is obtained by contradiction. Let

$$B = \left\{b_j \le 2^{-\alpha-2}\varepsilon_\alpha^*\mathbb{E}\Phi(S^*_{m_j})\right\}, \tag{2.4.24}$$

and assume that $A \cap B$ holds. If $j^* \le k \le j$ by using the definition of m_j, we have

$$\begin{aligned}
b_k &\le (\gamma_\alpha)^{k-j} b_j \\
&\le (\gamma_\alpha)^{k-j} 2^{-\alpha-2}\varepsilon_\alpha^*\mathbb{E}\Phi(S^*_{m_j}) \qquad \text{[by assumption in eq. (2.2.24)]} \\
&\le (\gamma_\alpha)^{k-j} 2^{-\alpha-1}\varepsilon_\alpha^*\mathbb{E}\Phi(S^*_{m_{j^*}}) \qquad \text{(by definition of } j^*) \\
&\le (\gamma_\alpha)^{k-j} \varepsilon_\alpha^*\mathbb{E}\Phi(S^*_{m_{j^*-1}}) \qquad \text{[by assumption in eq. (2.4.22)]} \\
&\le (\gamma_\alpha)^{k-j} \varepsilon_\alpha^*\mathbb{E}\Phi(S^*_{m_{k-1}}).
\end{aligned}$$

We use this observation in combination with Lemma 2.4.4 and the inequality

$$S^*_{m_k} \le S^*_{m_{k-1}} \vee \max_{m_{k-1}<j\le m_k} \|S_{m_{k-1}} + X_{m_{k-1}+1} + \cdots + X_j\| \le S^*_{m_{k-1}} + S^*_{(m_{k-1},m_k]} \tag{2.4.25}$$

to get

$$\mathbb{E}\Phi(S^*_{m_k}) \le \mathbb{E}\Phi(S^*_{m_{k-1}} + S^*_{(m_{k-1},m_k]}) \le h_\alpha(\varepsilon_\alpha^*(\gamma_\alpha)^{k-j})\mathbb{E}\Phi(S^*_{m_{k-1}}). \tag{2.4.26}$$

Iterating this inequality gives,

$$\begin{aligned}
\mathbb{E}\Phi(S^*_{m_j}) &\le \mathbb{E}\Phi(S^*_{m_{j^*}-1})\Pi_{k=j^*}^{j} h_\alpha(\varepsilon_\alpha^*(\gamma_\alpha)^{k-j}) \\
&< 2\mathbb{E}\Phi(S^*_{m_{j^*}-1}) \qquad \text{[by (2.4.23)]} \\
&< \mathbb{E}\Phi(S^*_{m_j}),
\end{aligned}$$

by the definition of j^*, contradicting $A \cap B$. Therefore, $A^c \cup B^c$ must hold. On B^c (2.4.21) holds since $\mathbb{E}\Phi(S_m^*) \le 2^\alpha\left[\mathbb{E}\Phi(S^*_{m_j}) + \mathbb{E}\Phi(S^*_{(m_j,m]})\right] \le (\frac{2^{2\alpha+2}}{\varepsilon_\alpha^*} + 2^\alpha\gamma_\alpha)b_j$. If

$$A^c = \left\{\mathbb{E}\Phi(S^*_{m_{j^*}-1}) \le 2^{-\alpha-1}\mathbb{E}\Phi(S^*_{m_{j^*}})\right\} \tag{2.4.27}$$

holds, then b_{j^*} should be comparable to $\mathbb{E}\Phi(S^*_{m_{j^*}})$ and hence to $\mathbb{E}\Phi(S^*_{m_j})$. This heuristic is formalized as follows. On A^c,

$$\begin{aligned}\mathbb{E}\Phi(S^*_{m_{j^*}}) &\le \mathbb{E}\Phi(2S^*_{m_{j^*-1}}) + \mathbb{E}\Phi(2S^*_{(m_{j^*-1},m_{j^*}]})\\ &\le 2^\alpha\big[\mathbb{E}\Phi(S_{m_{j^*-1}}) + b_{j^*}\big]\\ &\le 2^{-1}\mathbb{E}\Phi(S_{m_{j^*}}) + 2^\alpha b_{j^*} \qquad \text{[by (2.4.27)]}.\end{aligned}$$

Therefore, $\mathbb{E}\Phi(S^*_{m_{j^*}}) \le 2^{\alpha+1}b_{j^*}$. Using this we get,

$$\begin{aligned}\mathbb{E}\Phi(S^*_m) &\le \mathbb{E}\Phi(2S^*_{m_j}) + \mathbb{E}\Phi(2S^*_{(m_j,m_{j+1})})\\ &\le 2^\alpha\big[\mathbb{E}\Phi(S^*_{m_j}) + \mathbb{E}\phi_{j+1}\big]\\ &\le 2^{\alpha+1}\Phi(S^*_{m_{j^*}}) + 2^\alpha\gamma_\alpha b_j \qquad \text{(by def. of } j^* \text{ and } b_j)\\ &\le 2^{2\alpha+2}b_{j^*} + 2^\alpha\gamma_\alpha b_j\\ &\le (2^{2\alpha+2} + 2^\alpha\gamma_\alpha)b_j \qquad (\text{since } j \ge j^*),\end{aligned}$$

completing the proof. □

PROOF OF PROPOSITION 2.4.1. Using Lemma 2.4.5, it is easy to see that,

$$a_{T,\Phi}I_{m_j\le T<m_{j+1}} \le c_\alpha b_j I_{m_j\le T<m_{j+1}} \le c_\alpha b_j I_{T\ge m_j}. \tag{2.4.28}$$

Taking expectations and adding over j gives the result. □

The following lemmas will be used to prove Proposition 2.4.2.

LEMMA 2.4.6.

$$\sum_{j=1}^{\infty} b_j \Pr\{T \ge m_j\} \le 2^{\alpha+1}\Big[\mathbb{E}\sum_{j=2}^{\infty}\Phi'_j I_{T>m_{j-1}} + \mathbb{E}\sum_{j=1}^{\infty}\Phi(\|X'_{m_j}\|)I_{T\ge m_j}\Big]. \tag{2.4.29}$$

PROOF. Using (2.4.13) it follows that for $j \ge L$, $\Pr\{T \ge m_j\} = \Pr\{T \ge \infty\} = 0$. Therefore, the upper limits of summation in (2.4.29) can be replaced by $L-1$.

We will begin by showing that,

$$b_j \le 2^{\alpha+1}\big[\mathbb{E}\Phi'_j + \mathbb{E}\Phi(\|X'_{m_j}\|)\big]. \tag{2.4.30}$$

It is easy to see that,

$$\begin{aligned}b_j &\le \mathbb{E}\Phi(2S^*_{(m_{j-1},m_j)}) + \mathbb{E}\Phi(2\|X_{m_j}\|)\\ &\le 2^\alpha\big[\mathbb{E}\Phi_j + \mathbb{E}\Phi(\|X_{m_j}\|)\big]\\ &\le 2^\alpha\big[\mathbb{E}\Phi_j I_{\Phi_j<2^{-\alpha-2}b_j} + \mathbb{E}\Phi'_j + \mathbb{E}\Phi(\|X_{m_j}\|)I_{\Phi(\|X_{m_j}\|)<2^{-\alpha-2}\varepsilon_\alpha b_j}\\ &\qquad + \mathbb{E}\Phi(\|X'_{m_j}\|)\big] \qquad \text{[by (2.4.15), (2.4.16)]}\\ &\le 2^{-2}b_j + 2^\alpha\mathbb{E}\Phi'_j + 2^{-2}b_j + 2^\alpha\mathbb{E}\Phi(\|X'_{m_j}\|),\end{aligned}$$

which gives (2.4.30) by solving the above for b_j. Using this we have,

$$\begin{aligned}
\sum_{j=1}^{\infty} b_j \Pr\{T \geq m_j\} &= \mathbb{E}\sum_{j=1}^{L-1} b_j I_{T\geq m_j} \\
&\leq 2^{\alpha+1}\mathbb{E}\Big[\sum_{j=1}^{L-1}\big(\mathbb{E}\Phi_j' + \mathbb{E}\Phi(\|X'_{m_j}\|)\big)I_{T\geq m_j}\Big] \\
&= 2^{\alpha+1}\Big[\sum_{j=1}^{L-1}\big(\mathbb{E}\Phi_j' + \mathbb{E}\Phi(\|X'_{m_j}\|)\big)\Pr\{T \geq m_j\}\Big] \\
&\leq 2^{\alpha+1}\Big[\sum_{j=1}^{L-1}\mathbb{E}\Phi_j' \Pr\{T > m_{j-1}\} + \sum_{j=1}^{L-1}\mathbb{E}\Phi(\|X'_{m_j}\|)\Pr\{T \geq m_j\}\Big] \\
&= 2^{\alpha+1}\Big[\sum_{j=1}^{L-1}\mathbb{E}\Phi_j' I_{T>m_{j-1}} + \sum_{j=1}^{L-1}\mathbb{E}\Phi(\|X'_{m_j}\|)I_{T\geq m_j}\Big] \qquad \text{(independence)} \\
&\leq 2^{\alpha+1}\Big[\mathbb{E}\sum_{j=2}^{L-1}\Phi_j' I_{T>m_{j-1}} + \mathbb{E}\sum_{j=1}^{L-1}\Phi(\|X'_{m_j}\|)I_{T\geq m_j}\Big],
\end{aligned}$$

since L is fixed, $\Phi_1' = 0$ and the last line follows by using Wald's identity. □

LEMMA 2.4.7. *There exists a positive finite constant $c'_{\alpha,1}$ depending on α only so that the following inequality holds:*

$$\mathbb{E}\sum_{j=1}^{\infty}\Phi(\|X'_{m_j}\|)I_{T\geq m_j} \leq c'_{\alpha,1}\mathbb{E}\Phi(S_T^*). \tag{2.4.31}$$

PROOF. As in Lemma 2.4.6, the summation above extends only up to $L-1$. The proof of this result is done by obtaining a good-lambda inequality relating $U = \sum_{j=1}^{\infty}\Phi(\|X'_{m_j}\|)I_{T\geq m_j}$ to $V = \Phi(S_T^*)$. Interestingly, in this case the growth of V constrains the growth of U. Broadly speaking, this is because the point of truncation of $X'_{m_j} = X_{m_j} I_{\Phi(\|X_{m_j}\|)\geq 2^{-\alpha-2}\varepsilon_\alpha b_j}$ forces some portions of U to be small when V is small. More precisely, let $k_y = \max\{k < L : 2^{-\alpha-2}\varepsilon_\alpha b_k \leq 2^\alpha \delta y\}$. Observe that whenever $V = \Phi(S_T^*) \leq \delta y$, then $\Phi(\|X_{m_j}\|) \leq 2^\alpha\delta y$ whenever $m_j \leq T$. Therefore, $\Phi(\|X'_{m_j}\|) = 0$ when $m_{k_y} < m_j \leq T$. Therefore,

$$\begin{aligned}
\mathbb{E}\sum_{j=1}^{k_y}\Phi(\|X'_{m_j}\|) &\leq \mathbb{E}\sum_{j=1}^{k_y}\Phi(2S^*_{(m_{j-1},m_j]}) \\
&\leq 2^\alpha\sum_{j=1}^{k_y} b_j \leq 2^\alpha b_{k_y}\sum_{j=1}^{k_y}(\gamma_\alpha)^{j-k_y} \\
&\leq 2^\alpha(1-(\gamma_\alpha)^{-1})^{-1}b_{k_y} \leq \frac{2^{3\alpha+2}\delta y}{\varepsilon_\alpha(1-(\gamma_\alpha)^{-1})}.
\end{aligned}$$

Using this bound we will derive a good-lambda inequality.

Let $\tau_y = \min\{k : \sum_{j=1}^k \Phi(\|X'_{m_j}\|) \geq y\}$ if such $m_k \leq T$ exists, with $\tau_y = \infty$ otherwise. Using this we obtain for $\beta > 1 + \delta 2^\alpha$ and β, δ so that

$$\eta = \frac{2^{3\alpha+2}\delta}{\varepsilon_\alpha(\beta - 1 - \delta 2^\alpha)(1 - (\gamma_\alpha)^{-1})} < 1,$$

$$\Pr\Big\{\sum_{j=1}^{\infty} \Phi(\|X'_{m_j}\|) I_{T \geq m_j} \geq \beta y, \Phi(S_T^*) \leq \delta y\Big\}$$

$$= \Pr\Big\{\sum_{j=1}^{\tau_y - 1} \Phi(\|X'_{m_{\tau_y}}\|) + \Phi(\|X'_{m_{\tau_y}}\|)$$

$$+ \sum_{j:\tau_y < j \leq k_y, m_j \leq T} \Phi(\|X'_{m_j}\|) \geq \beta y, \tau_y < \infty, \Phi(S_T^*) \leq \delta y\Big\}$$

$$\leq \Pr\Big\{\sum_{j:\tau_y < j \leq k_y, m_j \leq m_{k_y} \wedge T} \Phi(\|X'_{m_j}\|) \geq (\beta - 1 - \delta 2^\alpha) y\Big\}$$

$$\leq \frac{1}{(\beta - 1 - \delta 2^\alpha) y} \mathbb{E} \sum_{j=2}^{k_y} \Phi(\|X'_{m_j}\|) I_{j > \tau_y, T \geq m_j}$$

$$= \frac{1}{(\beta - 1 - \delta 2^\alpha) y} \sum_{j=2}^{k_y} \mathbb{E}\Phi(\|X'_{m_j}\|) \Pr\{j > \tau_y, T \geq m_j\}$$

$$\leq \frac{1}{(\beta - 1 - \delta 2^\alpha) y} \Big[\sum_{j=2}^{k_y} \mathbb{E}\Phi(\|X'_{m_j}\|)\Big] \Pr\{m_{\tau_y} < T\}$$

$$\leq \frac{2^{3\alpha+2}\delta}{\varepsilon_\alpha(\beta - 1 - \delta 2^\alpha)(1 - (\gamma_\alpha)^{-1})} \times \Pr\Big\{\sum_{j=1}^{\infty} \Phi(\|X'_{m_j}\|) I_{T \geq m_j} \geq y\Big\}$$

$$= \eta \Pr\Big\{\sum_{j=1}^{\infty} \Phi(\|X'_{m_j}\|) I_{T \geq m_j} \geq y\Big\},$$

by the definitions of η, τ_y. To complete the proof pick β so large and δ so small so that $\frac{1}{\beta} > \eta$ and use Lemma 2.2.1. □

LEMMA 2.4.8. *There exists a positive finite constant $c'_{\alpha,2}$ depending on α only, so that the following inequality holds:*

$$\mathbb{E} \sum_{j=2}^{\infty} \Phi'_j I_{T \geq m_j} \leq c'_{\alpha,2} \mathbb{E}\Phi(S_T^*). \tag{2.4.32}$$

PROOF. As in Lemma 2.4.6, the summation above extends only up to $L - 1$. The proof is similar to the proof of Lemma 2.4.7, again using good-lambda inequalities. Let

$$k_y = \max\{k < L : 2^{-\alpha-2} b_k \leq 2^\alpha \delta y\}. \tag{2.4.33}$$

By construction $b_j \ge \gamma b_{j-1}$, hence $b_{k_y} \ge \gamma b_{k_y-1} \ge \cdots \ge \gamma^{k_y-j} b_j$. Therefore, for $j \le k_y$, $\mathbb{E}\Phi'_j \le b_j \le b_{k_y}(\gamma_\alpha)^{j-k_y}$. As a result,

$$\begin{aligned}\sum_{j\le k_y} \mathbb{E}\Phi'_j &\le \sum_{j\le k_y} b_{k_y}(\gamma_\alpha)^{j-k_y} \\ &\le 2^{2\alpha+2}\delta y \sum_{j=2}^{k_y}\Big(\frac{1}{\gamma_\alpha}\Big)^{k_y-j} \\ &\le 2^{2\alpha+2}\delta y\Big(\frac{1}{1-(1/\gamma_\alpha)}\Big).\end{aligned}$$

Let $\tau_y = \min\{k : \sum_{j=2}^k \Phi'_j \ge y\}$ if such $m_k \le T$ exists, with $\tau_y = \infty$ otherwise. In what follows we show that on the set

$$A = \big\{\tau_y < \infty, \quad \Phi(S_T^*) \le \delta y\big\}, \tag{2.4.34}$$

$$\Phi'_j I_{T\ge m_j} = 0, \quad \text{when} \quad k_y < j, \tag{2.4.35}$$

and

$$\Phi'_{\tau_y} \le 2^\alpha \delta y. \tag{2.4.36}$$

Observe that, on the intersection of A and $\{T \ge m_j,\ j > k_y\}$,

$$\begin{aligned}\Phi_j &= \Phi(S^*_{(m_{j-1},m_j)}) \\ &\le \Phi(2S_T^*) \qquad (\text{since } T \ge m_j) \\ &\le 2^\alpha \Phi(S_T^*) \\ &\le 2^\alpha \delta y \qquad [\text{by (2.4.34)}] \\ &< 2^{-\alpha-2} b_j \qquad (\text{since } j > k_y).\end{aligned}$$

Therefore,

$$\Phi'_j I_A I_{T\ge m_j} I_{j>k_y} \le \Phi_j I_{\Phi_j \ge 2^{-\alpha-2}b_j} I_{\Phi_j < 2^{-\alpha-2}b_j} = 0,$$

giving (2.4.35). Concerning (2.4.36), observing that on A, $\tau_y < \infty$ and $m_{\tau_y} \le T$ we have that

$$\Phi'_{\tau_y} \le \Phi(S^*_{(m_{\tau_y-1},m_{\tau_y})}) \le \Phi(2S_T^*) \le 2^\alpha \Phi(S_T^*) \le 2^\alpha \delta y. \tag{2.4.37}$$

Using these tools and assuming that $\{\tau_y < \infty,\ \Phi(S_T^*) \le \delta y\}$ holds we obtain,

$$\begin{aligned}U &= \sum_{j=1}^\infty \Phi'_j I_{T\ge m_j} \\ &= \sum_{j=2}^\infty \Phi'_j I_{T\ge m_j} \quad (\Phi'_j = 0)\end{aligned}$$

$$\le \sum_{j=2}^{\tau_y-1} \Phi'_j + \Phi'_{\tau_y} + \sum_{j:\tau_y<j\le k_y} \Phi'_j I_{T>m_{\tau_y}} \qquad \text{[by (2.4.35)]}$$

$$\le y + 2^\alpha \delta y + \sum_{j:\tau_y<j\le k_y} \Phi'_j I_{T>m_{\tau_y}} \qquad \text{[by definition of } \tau_y \text{ and (2.4.36)].}$$

Armed with this we are ready to obtain a proof of the good-lambda inequality:

$$\Pr\Big\{\sum_{j=1}^{\infty} \Phi'_j I_{T\ge m_j} \ge \beta y, \Phi(S_T^*) \le \delta y\Big\}$$

$$\le \frac{1}{(\beta - 1 - \delta 2^\alpha) y} \mathbb{E} \sum_{\tau_y<j\le k_y} \Phi'_j I_{T>m_{\tau_y}}$$

$$\le \frac{1}{(\beta - 1 - \delta 2^\alpha) y} \sum_{n=1}^{k_y-1} \Pr\big\{\tau_y = n, T > m_n\big\} \sum_{n<j\le k_y} \mathbb{E}\Phi'_j$$

$$\le \frac{\delta 4^{\alpha+1}}{(\beta - 1 - \delta 2^\alpha)(1-(1/\gamma_\alpha))} \sum_{n=1}^{k_y-1} \Pr\big\{\tau_y = n, T > m_n\big\}$$

[by approx. after (2.4.33)]

$$\le \frac{\delta 4^{\alpha+1}}{(\beta - 1 - \delta 2^\alpha)(1-(1/\gamma_\alpha))} \Pr\big\{\tau_y < \infty\big\}$$

$$= \frac{\delta 4^{\alpha+1}}{(\beta - 1 - \delta 2^\alpha)(1-(1/\gamma_\alpha))} \Pr\Big\{\sum_{j=1}^{\infty} \Phi'_j I_{T\ge m_j} \ge y\Big\}.$$

All these calculations make sense whenever $\beta > 1 + 2^\alpha \delta$ and $\delta > 0$ are such that $\varepsilon = \frac{\delta 4^{\alpha+1}}{(\beta-1-\delta 2^\alpha)(1-(1/\gamma_\alpha))} < 1$. Taking β, δ so that in addition $\frac{1}{\beta} > \varepsilon$, in combination with Lemma 2.2.1, gives the final result. □

LEMMA 2.4.9.

$$\mathbb{E}\sum_{j=2}^{\infty} \Phi'_j I_{m_{j-1}<T<m_j} \le 4^\alpha \Big[\mathbb{E}\Phi(S_T^*) + \sum_{j=2}^{\infty} \mathbb{E}\Phi_j \Pr\{m_{j-1} < T < m_j\}\Big] \tag{2.4.38}$$

PROOF. On the set $\{m_{j-1} < T < m_j\}$,

$$S^*_{(m_{j-1},m_j)} = \max_{m_{j-1}<r\le T} \|S_r - S_{m_{j-1}}\| \vee \max_{T<r<m_j} \|S_r - S_{m_{j-1}}\|$$

$$\le \Big\{\max_{m_{j-1}<r\le T} \|S_r\| + \max_{r\le T} \|S_r\|\Big\} \vee \max_{T<r<m_j} \|S_r - S_T - S_{m_{j-1}} + S_T\|$$

$$\le 2S_T^* \vee \{S^*_{(T,m_j)} + 2S_T^*\}$$

$$= S^*_{(T,m_j)} + 2S_T^*$$

$$\le \max\{4S_T^*, 2S^*_{(T,m_j)}\}.$$

Therefore,

$$\mathbb{E}\Phi'_j I_{m_{j-1}<T<m_j} \le 4^\alpha \mathbb{E}\Phi(S_T^*) I_{m_{j-1}<T<m_j} + 2^\alpha \mathbb{E}\Phi(S^*_{(T,m_j)}) I_{m_{j-1}<T<m_j}.$$

The second term above can be bounded by

$$\begin{aligned}\mathbb{E}\Phi(S^*_{(T,m_j)})I_{m_{j-1}<T<m_j} &= \sum_{m_{j-1}<n<m_j} \mathbb{E}\Phi(S^*_{(n,m_j)})\Pr\{T=n\}\\ &\le \sum_{m_{j-1}<n<m_j} \mathbb{E}\Phi(2S^*_{(m_{j-1},m_j)})\Pr\{T=n\}\\ &\le \sum_{m_{j-1}<n<m_j} 2^\alpha \mathbb{E}\Phi_j \Pr\{T=n\}\\ &= 2^\alpha \mathbb{E}\Phi_j \Pr\{m_{j-1}<T<m_j\}.\end{aligned}$$

Putting together the above observations and adding over j gives the result. □

LEMMA 2.4.10. *There exists a constant $0 < c'_{\alpha,3} < \infty$ depending on α only for which*

$$\mathbb{E}\Phi_{j+1} \le 2^{-3\alpha-2}\Big[(b_1+\cdots+b_j) + c'_{\alpha,3}\sum_{i=1}^{j}\mathbb{E}\Phi(\|X'_{m_i}\|)\Big]. \tag{2.4.39}$$

PROOF. We begin by showing that for any $1 \le k \le j$,

$$b_k \le \frac{(\gamma_\alpha-1)(b_1+\cdots+b_j)}{\gamma_\alpha^{j-k+1}-1}. \tag{2.4.40}$$

Using the fact that $b_{n+i} \ge (\gamma_\alpha)^i b_n$ for all $i \ge 0$, $n \ge 1$,

$$b_1+\cdots+b_j \ge b_k+\cdots+b_j \ge b_k\sum_{i=0}^{j-k}(\gamma_\alpha)^i = \frac{b_k(\gamma_\alpha^{j-k+1}-1)}{(\gamma_\alpha-1)}.$$

We will also need the following result from calculus. Let

$$d_\alpha = 1+[16^{\alpha+1}]. \tag{2.4.41}$$

Noting that $h_\alpha(0)=1$ and $h_\alpha(x)$ [from (2.4.20)] is strictly increasing to infinity, there is a unique constant $0<\varepsilon_\alpha<\infty$ for which

$$(h_\alpha(\varepsilon_\alpha/\gamma_\alpha))^{d_\alpha}(\gamma_\alpha)^{d_\alpha+1}(\gamma_\alpha-1)((\gamma_\alpha)^{d_\alpha+1}-1)^{-1} = 2^{-3\alpha-2}, \tag{2.4.42}$$

with γ_α as in (2.4.12). This can be verified by using the following bound:

$$\begin{aligned}(\gamma_\alpha)^{d_\alpha+1}(\gamma_\alpha-1)((\gamma_\alpha)^{d_\alpha+1}-1)^{-1} &= \frac{1}{(1-\gamma_\alpha^{-1-d_\alpha})16^{\alpha+1}}\\ &< \frac{1}{(1-\frac{1}{2})16^{\alpha+1}} < \frac{1}{2^{3(\alpha+1)}}.\end{aligned}$$

An important property of ε_α is that

$$\varepsilon_\alpha < 2-\gamma_\alpha. \tag{2.4.43}$$

Proving (2.4.43) is equivalent to showing

$$h_\alpha\Big(\frac{2-\gamma_\alpha}{\gamma_\alpha}\Big)^{d_\alpha} > (1-\gamma_\alpha^{-1-d_\alpha})2^{\alpha+2}. \tag{2.4.44}$$

To see this, the definition of h gives

$$\begin{aligned}
h_\alpha\Big(\frac{2-\gamma_\alpha}{\gamma_\alpha}\Big)^{d_\alpha} &= \frac{1}{\gamma_\alpha^{d_\alpha}}\Big[\Big(\Big(1+\frac{1}{16^{\alpha+1}}\Big)^{\frac{1}{2\alpha}}+\Big(1-\frac{1}{16^{\alpha+1}}\Big)^{\frac{1}{2\alpha}}\Big)^{\alpha d_\alpha} \\
&\qquad \times \Big(\Big(1+\frac{1}{16^{\alpha+1}}\Big)^{\frac{1}{2}}+\Big(1-\frac{1}{16^{\alpha+1}}\Big)^{\frac{1}{2}}\Big)^{d_\alpha}\Big] \\
&\geq \frac{1}{\gamma_\alpha^{d_\alpha}}2^{d_\alpha}\Big(1-\frac{1}{16^{\alpha+1}}\Big)^{d_\alpha} \\
&= 2^{d_\alpha}\Big(\frac{16^{\alpha+1}-1}{16^{\alpha+1}+1}\Big)^{d_\alpha}.
\end{aligned}$$

Noting that $\log 2 \geq \frac{1}{2}$ and

$$\log(1-x) = -x-\frac{x^2}{2}-\frac{x^3}{3}-\cdots \geq -x-x^2-x^3-\cdots = -\frac{x}{1-x},$$

we have

$$\begin{aligned}
\log\Big(2^{d_\alpha}\Big(\frac{16^{\alpha+1}-1}{16^{\alpha+1}+1}\Big)^{d_\alpha}\Big) &\geq d_\alpha\Big(\log 2-\frac{2}{16^{\alpha+1}-1}\Big) \\
&\geq (1+[16^{\alpha+1}])\Big(\frac{11}{30}\Big) \\
&> (\alpha+2)\log 2,
\end{aligned}$$

verifying (2.4.44) and hence (2.4.43).

If $\Phi \equiv 0$, Theorem 2.3.1 becomes trivial; hence, we may assume that Φ is not identically zero. But then, using the continuity and growth properties of $\Phi(x)$, it follows that $\Phi(x) > 0$ for all $x > 0$. To see this let $z = \inf\{x > 0 : \Phi(x) > 0\}$. If $z > 0$, the continuity of Φ implies that $\Phi(z) = 0$. Therefore, if $z > 0$, for any $x \geq 2z$ we have,

$$0 < \Phi(x) = \Phi\Big(\frac{x}{z}z\Big) \leq \Big(\frac{x}{z}\Big)^\alpha \Phi(z) = 0,$$

contradiction. In other words, we can assume that $\Phi(x) > 0$ for all $x > 0$. Therefore, $\mathbb{E}\Phi(\|X_{m_1}\|) > 0$. For $b_0 = 0$, define

$$k_j = \max\big\{k \leq j : \mathbb{E}\Phi(\|X_{m_k}\|) > \varepsilon_\alpha b_{k-1}\big\}. \tag{2.4.45}$$

The existence of such a k_j is guaranteed since the set of integers defining k_j includes $k = 1$. Our ultimate goal is to bound $\mathbb{E}\Phi_{j+1}$. This will be done in two parts, when $k_j \leq j - d_\alpha$ in which case enough of the b_is are of comparable size so that sum $\frac{1}{2^{3\alpha+2}}\sum_{i=1}^j b_i > \gamma_\alpha b_j \geq \mathbb{E}\Phi_{j+1}$. In the case $k_j > j - d_\alpha$, $\mathbb{E}\Phi(\|X_{m_{k_j}}\|)$ will be comparable to b_{k_j}, which will itself be comparable to $\mathbb{E}\Phi_{j+1}$. In order to substantiate these heuristics, we need some preliminary facts.

The following inequality holds:

$$\mathbb{E}\Phi(\|X_{m_{k_j}}\|) \leq 2\mathbb{E}\Phi(\|X'_{m_{k_j}}\|). \tag{2.4.46}$$

The proof of this is divided into two parts. If $b_{k_j} \le 2^{\alpha+1} b_{k_j-1}$, the definition of k_j gives

$$\mathbb{E}\Phi(\|X_{m_{k_j}}\|) > 2^{-\alpha-1}\varepsilon_\alpha b_{k_j}. \tag{2.4.47}$$

Using (2.4.5) and (2.4.47) we have,

$$\begin{aligned}\mathbb{E}\Phi(\|X_{m_{k_j}}\|)I_{\Phi(\|X_{m_{k_j}}\|)\ge 2^{-\alpha-2}\varepsilon_\alpha b_{k_j}} &\ge \mathbb{E}\Phi(\|X_{m_{k_j}}\|)I_{\Phi(\|X_{m_{k_j}}\|)\ge 2^{-1}\mathbb{E}\Phi(\|X_{m_{k_j}}\|)}\\ &\ge \frac{1}{2}\mathbb{E}\Phi(\|X_{m_{k_j}}\|).\end{aligned}$$

In the case $b_{k_j} > 2^{\alpha+1} b_{k_j-1}$,

$$\begin{aligned}b_{k_j} &\le 2^\alpha\big[\mathbb{E}\Phi_{k_j} + \mathbb{E}\Phi(\|X_{m_{k_j}}\|)\big]\\ &< 2^\alpha\big[\gamma_\alpha b_{k_j-1} + \mathbb{E}\Phi(\|X_{m_{k_j}}\|)\big]\\ &< 2^{-1}\gamma_\alpha b_{k_j} + 2^\alpha\mathbb{E}\Phi(\|X_{m_{k_j}}\|).\end{aligned}$$

Therefore, solving and using (2.4.43) we have,

$$\begin{aligned}\mathbb{E}\Phi(\|X_{m_{k_j}}\|) &> 2^{-\alpha}(1-2^{-1}\gamma_\alpha)b_{k_j}\\ &> 2^{-\alpha-1}\varepsilon_\alpha b_{k_j},\end{aligned}$$

which gives (2.4.47) and hence (2.4.46).

We continue by observing that whenever $k_j < k \le j$, $\mathbb{E}\Phi(\|X_{m_k}\|) \le \varepsilon_\alpha b_{k-1}$ and $\mathbb{E}\Phi_k \le \gamma_\alpha b_{k-1}$. A use of Lemma 2.4.4 gives

$$b_k \le h_\alpha\Big(\frac{\varepsilon_\alpha}{\gamma_\alpha}\Big)\gamma_\alpha b_{k-1}. \tag{2.4.48}$$

Iterating on (2.4.48) gives,

$$b_j \le \Big(h_\alpha\Big(\frac{\varepsilon_\alpha}{\gamma_\alpha}\Big)\gamma_\alpha\Big)^i b_{j-i} \qquad \text{for all} \quad k_j \le j-i \le j. \tag{2.4.49}$$

We are now ready to provide the announced bounds on $\mathbb{E}\Phi_{j+1}$. In the first case, when $k_j \le j - d_\alpha$,

$$\begin{aligned}\mathbb{E}\Phi_{j+1} &< \gamma_\alpha b_j \qquad \text{[by (2.4.1), (2.4.2) and (2.4.14)]}\\ &\le \gamma_\alpha\Big(h_\alpha\Big(\frac{\varepsilon_\alpha}{\gamma_\alpha}\Big)\gamma_\alpha\Big)^{d_\alpha} b_{j-d_\alpha} \qquad \text{[using (2.4.49)]}\\ &\le \gamma_\alpha\Big(h_\alpha\Big(\frac{\varepsilon_\alpha}{\gamma_\alpha}\Big)\gamma_\alpha\Big)^{d_\alpha}(\gamma_\alpha-1)(\gamma_\alpha^{d_\alpha+1}-1)^{-1}(b_1+\cdots+b_j)\\ &\qquad\qquad \text{[using (2.4.40)]}\\ &= \frac{b_1+\cdots+b_j}{2^{3\alpha+2}} \qquad \text{[by (2.4.42)]}.\end{aligned}$$

In the second case, when $k_j > j - d_\alpha$,

$$\begin{aligned}\mathbb{E}\Phi_{j+1} &< \gamma_\alpha b_j \\ &\le \gamma_\alpha \Big(h_\alpha\Big(\frac{\varepsilon_\alpha}{\gamma_\alpha}\Big)\gamma_\alpha\Big)^{j-k_j} b_{k_j} \qquad \text{[by (2.4.49) with } i = j - k_j] \\ &< \gamma_\alpha \Big(h_\alpha\Big(\frac{\varepsilon_\alpha}{\gamma_\alpha}\Big)\gamma_\alpha\Big)^{d_\alpha} b_{k_j}.\end{aligned}$$

It remains to bound b_{k_j}:

$$\begin{aligned}b_{k_j} &\le 2^\alpha\big[\mathbb{E}\Phi_{k_j} + \mathbb{E}\Phi(\|X_{m_{k_j}}\|)\big] \\ &\le 2^\alpha\big[\gamma_\alpha b_{k_j-1} + \mathbb{E}\Phi(\|X_{m_{k_j}}\|)\big] \\ &< 2^\alpha\Big(\frac{\gamma_\alpha}{\varepsilon_\alpha} + 1\Big)\mathbb{E}\Phi(\|X_{m_{k_j}}\|) \qquad \text{[by (2.4.45)]} \\ &< 2^{\alpha+1}\Big(\frac{\gamma_\alpha}{\varepsilon_\alpha} + 1\Big)\mathbb{E}\Phi(\|X'_{m_{k_j}}\|) \qquad \text{[by (2.4.46)]} \\ &\le 2^{\alpha+1}\Big(\frac{\gamma_\alpha}{\varepsilon_\alpha} + 1\Big)\sum_{i=1}^{j}\mathbb{E}\Phi(\|X'_{m_i}\|).\end{aligned}$$

Putting terms together,

$$\mathbb{E}\Phi_{j+1} \le 2^{-3\alpha-2}(b_1+\cdots+b_j)+\gamma_\alpha\Big(h_\alpha\Big(\frac{\varepsilon_\alpha}{\gamma_\alpha}\Big)\gamma_\alpha\Big)^{d_\alpha}2^{\alpha+1}\Big(1+\frac{\gamma_\alpha}{\varepsilon_\alpha}\Big)\sum_{i=1}^{j}\mathbb{E}\Phi(\|X'_{m_i}\|),$$

which is (2.4.39) with

$$c'_{\alpha,3} = 2^{4\alpha+3}\gamma_\alpha\Big(1+\frac{\gamma_\alpha}{\varepsilon_\alpha}\Big)\Big(\gamma_\alpha h_\alpha\Big(\frac{\varepsilon_\alpha}{\gamma_\alpha}\Big)\Big)^{d_\alpha}. \qquad \square$$

LEMMA 2.4.11.

$$\sum_{j=2}^{\infty}\mathbb{E}\Phi_j \Pr\{m_{j-1} < T < m_j\} \le 2^{-(3\alpha+2)}\Big[\sum_{j=1}^{\infty} b_j \Pr\{T \ge m_j\} + c'_{\alpha,1}c'_{\alpha,3}\mathbb{E}\Phi(S_T^*)\Big]. \tag{2.4.50}$$

PROOF. Using the previous lemma,

$$\begin{aligned}&\sum_{j=2}^{\infty}\mathbb{E}\Phi_j \Pr\{m_{j-1} < T < m_j\} \\ &\quad\le \frac{1}{2^{3\alpha+2}}\Big[\sum_{j=1}^{\infty}(b_1+\cdots+b_j)\Pr\{m_j \le T < m_{j+1}\} \\ &\qquad\qquad + c'_{\alpha,3}\sum_{j=1}^{\infty}\sum_{i=1}^{j}\mathbb{E}\Phi(\|X'_{m_i}\|)\Pr\{m_j \le T < m_{j+1}\}\Big]\end{aligned}$$

$$= \frac{1}{2^{3\alpha+2}}\Big[\sum_{j=1}^{\infty} b_j \Pr\{T \ge m_j\} + c'_{\alpha,3} \sum_{j=1}^{\infty} \mathbb{E}\Phi(\|X'_{m_j}\|) \Pr\{T \ge m_j\}\Big]$$

(summation by parts)

$$= \frac{1}{2^{3\alpha+2}}\Big[\sum_{j=1}^{\infty} b_j \Pr\{T \ge m_j\} + c'_{\alpha,3} \mathbb{E} \sum_{j=1}^{\infty} \Phi(\|X'_{m_j}\|) I_{T \ge m_j}\Big]$$

(by Wald's identity)

$$\le \frac{1}{2^{3\alpha+2}}\Big[\sum_{j=1}^{\infty} b_j \Pr\{T \ge m_j\} + c'_{\alpha,1} c'_{\alpha,3} \mathbb{E}\Phi(S^*_T)\Big],$$

where the last line follows from Lemma 2.4.7. □

PROOF OF PROPOSITION 2.4.2. Gathering Lemmas 2.4.6 through 2.4.11 in order one arrives at

$$\sum_{j=1}^{n} b_j \Pr\{T \ge m_j\} \le 2^{\alpha+1}(c'_{\alpha,2} + 4^{\alpha} + c'_{\alpha,1}) \mathbb{E}\Phi(S^*_T)$$
$$+ \frac{2^{3\alpha+1}}{2^{3\alpha+2}}\Big[\sum_{j=1}^{\infty} b_j \Pr\{T \ge m_j\} + c'_{\alpha,1} c'_{\alpha,3} \mathbb{E}\Phi(S^*_T)\Big].$$

Solving for $\sum_{j=1}^{n} b_j \Pr\{T \ge m_j\}$ gives the result. □

2.5 Continuous Time Processes

In this section we extend the results presented earlier to processes in continuous time. This extension is important for various reasons, one of which is that it provides an extension of the Burkholder–Gundy inequality to the case of processes with values in Banach spaces. (See Theorem 2.6.2 and Corollary 2.6.3 below.)

THEOREM 2.5.1. *Fix any $\alpha > 0$. Let $\Phi \in \mathcal{A}_\alpha$ as in (2.2.1). Let N_t, $t \ge 0$, be a B-valued process continuous on the right with limits from the left with independent increments, where $(B, \|\cdot\|)$ is a separable Banach space. Let T be a stopping time adapted to $\sigma(N_s, s \le t)$. Then,*

$$c_\alpha \mathbb{E} a^*_{T,\Phi} \le \mathbb{E} \sup_{0\le s\le T} \Phi(\|N_s\|) \le C_\alpha \mathbb{E} a^*_{T,\Phi}, \tag{2.5.1}$$

where $a^*_{t,\Phi} = \mathbb{E}\sup_{0\le s\le t} \Phi(\|N_s\|)$, and c_α, C_α are constants depending on α only [indeed, $C_\alpha \le 20(18)^\alpha$].

Equivalently if $\{\tilde{N}_t, t \ge 0\}$ is an independent copy of $\{N_t, t \ge 0\}$ and independent of T, then

$$c_\alpha \mathbb{E} \sup_{0\le s\le T} \Phi(\|\tilde{N}_s\|) \le \mathbb{E} \sup_{0\le s\le T} \Phi(\|N_s\|) \le C_\alpha \mathbb{E} \sup_{0\le s\le T} \Phi(\|\tilde{N}_s\|). \tag{2.5.2}$$

The proof of Theorem 2.5.1 is based on a limiting argument.

PROOF. For a fixed t, we take a subdivision of $[0, t]$, $\tau_n = \{s_i^n : 0 \le i \le n\}$, such that $0 = s_0^n < s_1^n < \cdots < s_{n-1}^n < s_n^n = t$, and $|\tau_n| \to 0$ as $n \to \infty$. For $n > 0$ we take

$$N_{s_j^n} = \sum_{i=1}^{j} (N_{s_i^n} - N_{s_{i-1}^n}) + N_{s_0^n}. \tag{2.5.3}$$

Observe that the restriction of N to τ_n is a discrete time process with independent increments. Therefore, one can choose $\{\tau_n\}$ such that τ_{n+1} contains τ_n. Setting $T_n(\omega) = s_{i+1}^n$ on $(s_i^n < T \wedge t \le s_{i+1}^n)$ makes T_n a stopping time with respect to $\{\mathcal{F}_{s_i^n}\}_{i\ge 0}$ where $\mathcal{F}_s = \sigma(X_u; u \le s)$. Observe also that $\{T_n\}_{n\ge 0}$ is a decreasing sequence converging to $T \wedge t$. Using Theorem 2.3.1 we get

$$c_\alpha \mathbb{E} \sup_{s\varepsilon\tau_n, s\le T_n} \Phi(\|\tilde{N}_s\|) \le \mathbb{E} \sup_{s\varepsilon\tau_n, s\le T_n} \Phi(\|N_s\|) \le C_\alpha \mathbb{E} \sup_{s\varepsilon\tau_n, s\le T_n} \Phi(\|\tilde{N}_s\|). \tag{2.5.4}$$

Using the right continuity assumption on $\{N_t, t \ge 0\}$ and the Lebesgue dominated convergence theorem we obtain

$$c_\alpha \mathbb{E} \sup_{s\le T\wedge t} \Phi(\|\tilde{N}_s\|) \le \mathbb{E} \sup_{s\le T\wedge t} \Phi(\|N_s\|) \le C_\alpha \mathbb{E} \sup_{s\le T\wedge t} \Phi(\|\tilde{N}_s\|). \tag{2.5.5}$$

The proof is completed by using Lebesgue's monotone convergence theorem. □

COROLLARY 2.5.2. *Under the assumptions of Theorem 2.5.1, and setting* $b_t = \mathbb{E}N_t$, $d_t = \mathbb{E}\|N_t - b_t\|$, *we have*

$$\mathbb{E}N_T = \mathbb{E}b_T = \mathbb{E}\tilde{N}_T \tag{2.5.6}$$

whenever $\mathbb{E}d_T < \infty$.

PROOF. Since $M_t = N_t - b_t$ is a mean-zero martingale with independent increments, the optional sampling theorem gives that for all fixed $t < \infty$, $\mathbb{E}M_{T\wedge t} = 0$. Let $\tilde{\tilde{N}}_t$ be a second independent copy of N_t. Then, using Theorem 2.5.1 and conditional Jensen's inequality, we have

$$\begin{aligned}
\mathbb{E} \sup_{s\le T} \|M_s\| &\le C\mathbb{E} \sup_{s\le T} \|\tilde{N}_s - b_t\| \\
&= C\mathbb{E} \sup_{s\le T} \|\tilde{N}_s - \mathbb{E}\tilde{\tilde{N}}_t\| \\
&\le C\mathbb{E} \sup_{s\le T} \|\tilde{N}_s - \tilde{\tilde{N}}_t\|.
\end{aligned}$$

Conditioning on T and using the independence of the stopping time from both $\tilde{N}_t$ and $\tilde{\tilde{N}}_t$ as well as an easy extension of Levy's inequality for sums of independent symmetric random variables to the case of continuous time processes with independent increments (a result similar to Proposition 2.6.1 below), one gets the bound

$$\begin{aligned} 2C\mathbb{E}\|\tilde{N}_T - \tilde{\tilde{N}}_T\| &= 2C\mathbb{E}\|(\tilde{N}_T - b_T) - (\tilde{\tilde{N}}_T - b_T)\| \\ &\leq 4C\mathbb{E}\|\tilde{N}_T - b_T\|. \end{aligned} \tag{2.5.7}$$

Therefore $\mathbb{E}\sup_t \|M_{T\wedge t}\| = \mathbb{E}\sup_{s\leq T}\|M_s\| \leq 4C\mathbb{E}d_T < \infty$. Now, the dominated convergence theorem gives the result. □

COROLLARY 2.5.3. *Under the assumptions of Theorem 2.5.1,*

$$\mathbb{E}\exp\big\{\sup_{t\leq T}\Phi(\|N_t\|)\big\} \leq 20\mathbb{E}\exp\big\{18^{\alpha}\Phi(\sup_{t\leq T}\|\tilde{N}_t\|)\big\}. \tag{2.5.8}$$

PROOF. This result follows easily from Theorem 2.5.1 by applying the right-hand side of this result to $\Phi_m(x) = \Phi(x)^m$, $m \geq 1$ and adding over m. □

2.6 Burkholder–Gundy Type Inequalities in Banach Spaces

As a preliminary result, in what follows we extend Theorem 1.1.5 to the case of continuous time processes with independent increments.

PROPOSITION 2.6.1. *Let $\{N_t, t \geq 0\}$, $N_0 = 0$, be a B-valued process, continuous on the right with limits from the left with increments that are stationary and independent, where $(B, \|\cdot\|)$ is a separable Banach space. Then, for all $y \geq 0$,*

$$\Pr\big\{\sup_{s\leq t}\|N_s\| > y\big\} \leq 9\Pr\big\{30\|N_t\| > y\big\}. \tag{2.6.1}$$

PROOF. For a fixed t, let $s_i^n = \frac{i}{2^n}t$ with $\tau_n = \{s_i^n : 0 \leq i \leq n\}$. Set $N_{s_j^n} = \sum_{i=1}^{j}(N_{s_i^n} - N_{s_{i-1}^n}) + N_{s_0^n}$. Observe that the restriction of $\{N_t\}$ to τ_n is a discrete time process with i.i.d. increments. Then a use of Theorem 1.1.5 gives that, for fixed $n \geq 1$, and all $y > 0$,

$$\Pr\big\{\sup_{s\in\tau_n, s\leq t}\|N_s\| > y\big\} \leq 9\Pr\big\{30\|N_t\| > y\big\}. \tag{2.6.2}$$

The result follows using the dominated convergence theorem, letting n tend to infinity. □

The following result contains as a special case the Burkholder–Gundy inequalities for randomly stopped Brownian motion and more generally a result for Bessel processes.

THEOREM 2.6.2. *Let $\{N_t, t \geq 0\}$, $N_0 = 0$, be a B-valued process continuous on the right with limits from the left with stationary and independent increments (in particular, $N_t - N_s$ has the same distribution as N_{t-s}, for all $t > s$), where $(B, \|\cdot\|)$*

is a separable Banach space. Let T be a stopping time adapted to $\sigma(N_s, s \le t)$. Let $\tilde{N}_t$ be an independent copy of N_t (hence, also independent of T). Fix $\alpha > 0$. Then, for any $\Phi \in \mathcal{A}_\alpha$ as in (2.2.1), there exist constants $0 < c_\alpha, C_\alpha < \infty$ depending on α only such that

$$c_\alpha \mathbb{E}\Phi(\|\tilde{N}_T\|) \le \mathbb{E}\sup_{t\le T}\Phi(\|N_t\|) \le C_\alpha \mathbb{E}\Phi(\|\tilde{N}_T\|). \tag{2.6.3}$$

PROOF. The left-hand side follows easily from the left-hand side of Theorem 2.5.1 since

$$\mathbb{E}\sup_{t\le T}\Phi(\|\tilde{N}_t\|) \ge \mathbb{E}\Phi(\|\tilde{N}_T\|). \tag{2.6.4}$$

The other bound follows by using the right-hand side of Theorem 2.5.1 and Proposition 2.6.1 (conditioning on T). One also needs to remember that for a non-negative variable X, $\mathbb{E}X = \int_0^\infty \Pr\{X > t\}dt$. Therefore, using the properties of Φ from (2.2.1), we have

$$\mathbb{E}\sup_{t\le T}\Phi(\|N_t\|) \le C_\alpha \mathbb{E}\sup_{t\le T}\Phi(\|\tilde{N}_t\|) \le 9C_\alpha \mathbb{E}\Phi(30\|\tilde{N}_T\|) \le 9C_\alpha 30^\alpha \mathbb{E}\Phi(\|\tilde{N}_T\|),$$

concluding the proof. □

Using the assumed independence between $\tilde{N}_t$ and N_t, if we take $N_t = B_t$, a standard Brownian motion, and apply the scaling properties of this process we get the Burkholder–Gundy inequality.

COROLLARY 2.6.3. *Let B_t, $t \ge 0$, be a standard Brownian motion and T a stopping time adapted to $\sigma(B_s, s > 0)$. Then, for all $p > 0$,*

$$c_p \mathbb{E}|B_1|^p \mathbb{E}T^{p/2} \le \mathbb{E}\sup_{t\le T}|B_t|^p \le C_1 C_2^p \mathbb{E}|B_1|^p \; ET^{p/2}, \tag{2.6.5}$$

where c_p is a constant depending on p only and C_1, C_2 are independent of p, the norm and the processes involved.

PROOF. Observe that

$$\mathbb{E}|\tilde{B}_T|^p = \mathbb{E}|T^{1/2}\tilde{B}_1|^p = \mathbb{E}T^{p/2}\mathbb{E}|\tilde{B}_1|^p = \mathbb{E}T^{p/2}\mathbb{E}|B_1|^p, \tag{2.6.6}$$

and apply Theorem 2.6.2. □

A more general result follows. Let $\{N_t = (B_t^{(1)}, \ldots, B_t^{(d)}), t \ge 0\}$, $1 \le d \le \infty$, be a vector of possibly dependent standard Brownian motions, with N_t taking values in a separable Banach space $(B, \|\cdot\|)$. Let T be a stopping time adapted to $\sigma(\{N_t\})$. Let $\{\tilde{N}_t\}$ be an independent copy of $\{N_t\}$ with $\{\tilde{N}_t\}$ independent of T as well. Letting $\|N_t\|_1 = \sup_{i\le d}|B_t^{(i)}|$ in Theorem 2.6.2 with $\Phi(x) = |x|^p$ we get that for each $p > 0$

$$c_p \mathbb{E}\sup_{i\le d}|\tilde{B}_T^{(i)}|^p \le \mathbb{E}\sup_{t\le T}\sup_{i\le d}|B_t^{(i)}|^p \le C_p \mathbb{E}\sup_{i\le d}|\tilde{B}_T^{(i)}|^p. \tag{2.6.7}$$

The independence between T and $\tilde{N}_t = (\tilde{B}_t^{(1)}, \ldots, \tilde{B}_t^{(d)})$ gives

$$\mathbb{E}\sup_{i\le d}|\tilde{B}_T^{(i)}|^p = \mathbb{E}\sup_{i\le d}|\sqrt{T}\tilde{B}_1^{(i)}|^p = \mathbb{E}T^{p/2}\mathbb{E}\sup_{i\le d}|\tilde{B}_1^{(i)}|^p = \mathbb{E}T^{p/2}\mathbb{E}\|N_1\|_1^p. \tag{2.6.8}$$

Likewise, letting $\|N_t\|_2 = \sqrt{(B_t^{(1)})^2 + \cdots + (B_t^{(d)})^2}$ in Theorem 2.6.2, we get for all $p > 0$

$$c_p\mathbb{E}\Big(\sum_{i=1}^d(\tilde{B}_T^{(i)})^2\Big)^{p/2} \le \mathbb{E}\sup_{t\le T}\Big(\sum_{i=1}^d(B_t^{(i)})^2\Big)^{p/2} \le C_p\mathbb{E}\Big(\sum_{i=1}^d(\tilde{B}_T^{(i)})^2\Big)^{p/2}. \tag{2.6.9}$$

After conditioning one obtains that

$$\mathbb{E}\Big(\sum_{i=1}^d(\tilde{B}_T^{(i)})^2\Big)^{p/2} = \mathbb{E}T^{p/2}\mathbb{E}\Big(\sum_{i=1}^d(\tilde{B}_1^{(i)})^2\Big)^{p/2} = \mathbb{E}T^{p/2}\mathbb{E}\|N_1\|_2^p, \tag{2.6.10}$$

which in the case the processes $\tilde{B}_t^{(i)}, i = 1, \ldots, d$ are independent Brownian motions gives $\mathbb{E}(\sum_{i=1}^d(\tilde{B}_T^{(i)})^2)^{p/2} = \mathbb{E}T^{p/2}\mathbb{E}(\mathcal{X}_d^2)^{p/2}$, where $\mathcal{X}_d^2$ is a chi-square random variable with d degrees of freedom. In this case $\|N_t\|_2$ is a Bessel process.

We remark that the constants c_p, C_p in (2.6.7) and (2.6.9) do not depend on d. The dependence on d is contained in $\mathbb{E}\|N_1\|^p$. It is easy to see that when $d = 1$, both (2.6.7) and (2.6.10) reduce to the Burkholder–Gundy inequalities of Corollary 2.6.3.

In order to streamline the presentation we will use the following notational convention: $A \approx_p B$ will mean that there exist constants $0 < c_p, C_p < \infty$ depending on p only such that

$$c_pA \le B \le C_pA. \tag{2.6.11}$$

Next, we will show how to use the above bounds to derive useful information related to boundary crossing times. Let

$$T_{1,b,d} = \inf\{t > 0 : \sup_{i\le d}|B_t^{(i)}| \ge b\} = \inf\{t > 0 : \|N_t\|_1 \ge b\}. \tag{2.6.12}$$

Using (2.6.7) and (2.6.8) we derive the following bounds on the average values of T_2^p. For all $p > 0$,

$$\mathbb{E}T_{1,b,d}^p \approx \frac{b^{2p}}{\mathbb{E}\sup_{i\le d}|B_1^{(i)}|^{2p}}. \tag{2.6.13}$$

Likewise, by setting

$$T_{2,b,d} = \inf\Big\{t > 0 : \sqrt{\sum_i^d(B_t^{(i)})^2} \ge b\Big\} = \inf\{t > 0 : \|N_t\|_2 \ge b\}, \tag{2.6.14}$$

a use of (2.6.9) and (2.6.10) gives

$$\mathbb{E}T_{2,b,d}^p \approx_p \frac{b^{2p}}{\mathbb{E}(\mathcal{X}_d^2)^p}, \tag{2.6.15}$$

for the first passage time through b of a Bessel process.

Since the constants involved are independent of the size of d, the above inequalities can be used to derive quantitative comparisons on the expected first passage times for processes with different values of d. That is, if we include the dependence on d and use (2.6.7) for the values $d = d_1$ and $d = d_2$ and take ratios, we obtain,

$$\mathbb{E}T^p_{1,b,d_2} \approx_p \frac{\mathbb{E}\sup_{i\le d_1}|B_1^{(i)}|^p}{\mathbb{E}\sup_{i\le d_2}|B_1^{(i)}|^p}\mathbb{E}T^p_{1,b,d_1}. \tag{2.6.16}$$

EXAMPLE 2.6.4. If we interpret the vector $N_t = (B_t^{(1)}, \ldots, B_t^{(d)})$ as the position (at time t) of a group of d agents in a market, the ratio

$$\frac{\mathbb{E}\sup_{i\le d_1}|B_1^{(i)}|^p}{\mathbb{E}\sup_{i\le d_2}|B_1^{(i)}|^p} \tag{2.6.17}$$

gives us information on how much longer (or less) we will need to wait before the boundary is hit by one of the agents if we use d_2 agents instead of d_1. A similar comparison can be made related to Bessel processes of different dimensions.

Concerning exponential powers we get the following corollary, which can be viewed as an exponential version of the Burkholder–Gundy inequality for randomly stopped Brownian motion.

COROLLARY 2.6.5. *Let $B_t, t \ge 0$ be a standard Brownian motion and T a stopping time adapted to $\sigma(B_s, s > 0)$. Let $(\tilde{B}_t, t \ge 0, \tilde{T})$ be an independent copy of the above process and stopping time. Then, for all λ, $p > 0$,*

$$\mathbb{E}\exp\{\lambda \sup_{t\le T}|B_t|^p\} \le C_1 \mathbb{E}\exp\{\lambda C_2^p |\tilde{B}_1|^p T^{p/2}\}, \tag{2.6.18}$$

where C_1, C_2 are universal constants.

PROOF. We note that $\sup_{s\le t}|\tilde{B}_s|$ is equal in distribution to $\sqrt{t}\sup_{s\le 1}|\tilde{B}_s|$. Therefore,

$$\mathbb{E}\exp\{\lambda \sup_{s\le t}|\tilde{B}_s|^p\} = \mathbb{E}\exp\{\lambda\sqrt{t}\sup_{s\le 1}|\tilde{B}_s|^p\} \le 2\mathbb{E}\exp\{\lambda\sqrt{t}|\tilde{B}_1|^p\}$$

by Levy's inequality (in continuous time). Using this and Corollary 2.5.3 (with $\Phi(x) = \lambda|x|^p$ and $N_t = B_t$) gives the result. More precisely, one uses a conditioning argument along with the fact that T and $\tilde{N}_t$ are independent. □

2.7 From Boundary Crossing of Nonrandom Functions to First Passage Times of Processes with Independent Increments

This section contains applications and refinements of the results from the previous section. The main application consists of a natural extension of the concept of boundary crossing by nonrandom functions to the case of random processes with independent increments. To motivate the development, suppose that a problem of interest can be studied by using a continuous increasing function $\eta(t)$ with the time of interest t_b equal to the first time $t \geq 0$ such that $\eta(t)$ reaches a fixed level $b > 0$. Then, it is easy to see that

$$\eta(t_b) = b \quad \text{and} \quad t_b = \eta^{-1}(b). \tag{2.7.1}$$

This model could be too simplistic for the problem at hand. A more realistic model might require that we introduce some randomness. Proceeding in this direction, let N_t be a process with continuous paths and independent increments. To make the framework more flexible we assume that N_t is B-valued where $(B, \|\cdot\|)$ is a separable Banach space. Taking $B = R^1$ and $\|x\| = |x|$ one retrieves the real valued case. We set

$$\eta(t) = \mathbb{E} \sup_{s \leq t} \|N_s\| \tag{2.7.2}$$

and

$$T_b = \inf\{s > 0 : \|N_s\| \geq b\}. \tag{2.7.3}$$

Then T_b is a random function. In this section we will show how to relate t_b to T_b. More precisely, if we assume that $\eta(t)$ is a strictly increasing function of t, then there exist universal constants $0 < c, C < \infty$ such that

$$cb \leq \mathbb{E}\eta(T_b) \leq Cb. \tag{2.7.4}$$

Moreover, if $\eta^{-1}(\cdot)$ is a convex function in $\mathcal{A}_\alpha$ for some $\alpha > 0$ [see (2.2.1)], then

$$c_\alpha \eta^{-1}(b) \leq \mathbb{E}T_b \leq C_\alpha \eta^{-1}(b), \tag{2.7.5}$$

for some constants $0 < c_\alpha, C_\alpha < \infty$ depending on α only. Observe that when N_t is deterministic, then $\eta(t) = \sup_{s \leq t} \|N_s\|$ and $T_b = t_b$, and hence the inequalities in (2.7.4) and (2.7.5) reduce to the equalities in (2.7.1). Therefore, our method produces an optimal extension of (2.7.1) up to multiplicative constants.

As in (2.6.11) we will use the expression $h(t) \approx_\alpha g(t)$ to denote that there are constants $0 < c_\alpha, C_\alpha < \infty$ depending on α only and such that

$$c_\alpha \leq \frac{h(t)}{g(t)} \leq C_\alpha. \tag{2.7.6}$$

If the constants are universal (numerical values not depending on anything), then we simply use $h(t) \approx g(t)$.

As will become apparent, it is possible to weaken (2.7.2) to

$$\eta(t) \approx_\alpha \mathbb{E} \sup_{s \le t} \|N_s\|. \tag{2.7.7}$$

We use this fact in the following example.

EXAMPLE 2.7.1. Let $N_t = B_t$, a standard Brownian motion. Then $\eta(t) = \mathbb{E}\sup_{s\le t}|B_s| = \sqrt{t}\,\mathbb{E}\sup_{s\le 1}|B_s|$ by the scaling properties of Brownian motion. Therefore,

$$\eta^{-1}(t) = \frac{t^2}{(\mathbb{E}\sup_{s\le 1}|B_s|)^2}$$

and hence, using (2.7.5),

$$\mathbb{E}T_b \approx \frac{b^2}{(\mathbb{E}\sup_{s\le 1}|B_s|)^2}.$$

More generally, if we take any process with continuous paths and independent increments for which $\eta(t) \approx t^r$ for some $0 < r \le 1$, then (2.7.5) gives $\mathbb{E}T_b \approx b^{\frac{1}{r}}$.

The theory of first passage times for random processes has been extensively developed in recent times. In particular, the distribution of the first hitting times of real Brownian motion has been studied through the use of the inverse Gaussian distribution. A similar approach to first hitting times involving Lévy processes is also available. However, the typical methods assume full knowledge of the distribution. In contrast, our approach is based on the approximate knowledge of L_p moments of the maximal process. In particular our method is well suited for dealing with problems in which a general Banach space structure is needed. This is the case in the example that follows and which might find applications in the natural sciences. It involves the study of the first passage time when the sphere of biggest radius among a set of d spheres reaches a fixed volume. This sphere could (perhaps) represent tumors in the body of a subject or a group of stars in a galaxy.

EXAMPLE 2.7.2. Let $\mathbf{p}_i = (p_i^{(1)}, p_i^{(2)}, p_i^{(3)})$, $i = 1, \ldots, d$, be a set of d points in three-dimensional space. At time $t = 0$, we start a three-dimensional Brownian motion with coordinate processes centered at each one of these points. For each i a sphere of radius $r_{t,i}$ is given, where the radius equals the distance of the location of the three-dimensional Brownian motion at time t to its starting point $\mathbf{p}_i$. We are interested in getting qualitative information on how long it will take before the radius (volume) of at least one of the spheres exceeds a fixed level (say b, for $b > 0$) as the size of d varies.

Let $\mathcal{X} = (\mathbf{X}_{t,i} = (X_{t,i}^{(1)}, X_{t,i}^{(2)}, X_{t,i}^{(3)}), i = 1, \ldots, d)$ where, for each i, $\mathbf{X}_{t,i} = \mathbf{p}_i + \mathbf{B}_{t,i}$ corresponds to the location of the three-dimensional Brownian motion $\mathbf{B}_{t,i}$ started at point $\mathbf{p}_i$ and $\mathcal{B}_t = (\mathbf{B}_{t,i} = (B_{t,i}^{(1)}, B_{t,i}^{(2)}, B_{t,i}^{(3)}), i = 1, \ldots, d)$. This radius of the largest sphere is given by $\|\mathcal{B}_t\| = \sup_{j\le d}\sqrt{\sum_{i=1}^3 (B_{t,j}^{(i)})^2}$.

Let $T_{b,d} = \inf\{s > 0 : \|\mathcal{B}_s\| = b\}$ be the stopping times of interest. If the coordinate processes at each point are independent from one another, using the scaling properties of Brownian motion we have

$$\eta(t) = \mathbb{E} \sup_{0 \le s \le t} \|\mathcal{B}_s\| = \sqrt{t}\mathbb{E} \sup_{0 \le s \le 1} \|\mathcal{B}_s\|,$$

which gives $\eta^{-1}(t) = \frac{t^2}{(\mathbb{E}\sup_{0\le s\le 1}\|\mathcal{B}_s\|)^2}$. Hence, by (2.7.5),

$$\mathbb{E}T_{b,d} \approx \frac{b^2}{\left(\mathbb{E}\sup_{0\le s\le 1}\sup_{j\le d}\sqrt{\sum_{i=1}^{3}(B_{s,j}^{(i)})^2}\right)^2}. \tag{2.7.8}$$

This can be further simplified by the extension of Lévy's inequality given in Proposition 2.6.1, since $\mathcal{B}_t$ is a process with stationary and independent increments. We therefore get

$$\mathbb{E}\sup_{s\le 1}\sup_{j\le d}\sqrt{\sum_{i=1}^{3}(B_{s,j}^{(i)})^2} \approx \mathbb{E}\sup_{j\le d}\sqrt{\sum_{i=1}^{3}(B_{1,j}^{(i)})^2} = \mathbb{E}\sup_{j\le d}\sqrt{\mathcal{X}_{3,j}^2}, \tag{2.7.9}$$

with $\mathcal{X}_{3,j}^2$ independent chi-square random variables with three degrees of freedom. Putting (2.7.8) and (2.7.9) together gives,

$$\mathbb{E}T_{b,d} \approx \frac{b^2}{\left(\mathbb{E}\sup_{j\le d}\sqrt{\mathcal{X}_{3,j}^2}\right)^2}.$$

The following growth properties of the functions in $\mathcal{A}_\alpha$ from (2.2.1) will be used extensively. Let $\alpha > 0$ be fixed. Then for all $c > 0$ and all $\Phi \in \mathcal{A}_\alpha$ there exists a constant $\gamma_{c,\alpha}$ depending on c and α only such that for all $x > 0$,

$$\frac{1}{\min(1, \gamma_{c,\alpha})}\Phi(x) \le \Phi(cx) \le \gamma_{c,\alpha}\Phi(x). \tag{2.7.10}$$

A useful corollary to (2.5.2) of Theorem 2.5.1 follows.

COROLLARY 2.7.3. *Let $\{N_t, t \ge 0\}$ be a B-valued process, continuous on the right with limits from the left with independent increments, where $(B, \|\cdot\|)$ is a separable Banach space. Let T be a stopping time adapted to $\mathcal{G}_t = \sigma(N_s, s \le t)$. Let $\{\tilde{N}_t, t \ge 0\}$ an independent copy of $\{N_t, t \ge 0\}$ (which is therefore independent of T as well). If Φ is a convex function in $\mathcal{A}_\beta$ for some $\beta > 0$, then*

$$\mathbb{E}\Phi\Big(\sup_{0\le s\le T}\|N_s\|\Big) \ge c_\beta \mathbb{E}\Phi\Big(\mathbb{E}\big[\big(\sup_{0\le s\le T}\|\tilde{N}_s\|\big)\big|T\big]\Big), \tag{2.7.11}$$

for a constant c_β depending on β only.

PROOF. From (2.5.2) and conditional Jensen's inequality,

$$\begin{aligned}\mathbb{E}\Phi\Big(\sup_{0\le s\le T}\|N_s\|\Big) &\ge c_\beta \mathbb{E}\Phi\Big(\sup_{0\le s\le T}\|\tilde N_s\|\Big)\\ &= c_\beta \mathbb{E}\Big[\mathbb{E}\big(\Phi\big(\sup_{0\le s\le T}\|\tilde N_s\|\big)\big|T\big)\Big]\\ &\ge c_\beta \mathbb{E}\Phi\Big(\mathbb{E}\big[\big(\sup_{0\le s\le T}\|\tilde N_s\|\big)\big|T\big]\Big).\end{aligned}$$

□

From this corollary, we can derive a useful bound on the moments of T.

PROPOSITION 2.7.4. *Let $\{N_t, t\ge 0\}$ be B-valued process continuous on the right with limits from the left with independent increments, where $(B, \|\cdot\|)$ is a separable Banach space. Let T be a stopping time adapted to $\mathcal{G}_t = \sigma(N_s, s\le t)$. Assume that $\mathbb{E}\sup_{s\le t}\|N_s\| \ge \eta(t)$ is a strictly increasing function of t (and therefore has a well-defined inverse). Let Φ be a function such that $\Phi(\eta^{-1}(x))$, $x>0$, is a convex function from $\mathcal{A}_\beta$ for some $\beta>0$. Then*

$$\mathbb{E}\Phi(T) \le \frac{1}{c_\beta}\mathbb{E}\Phi\Big(\eta^{-1}\big(\sup_{0\le s\le T}\|N_s\|\big)\Big), \tag{2.7.12}$$

for a constant c_β depending on β only.

PROOF. Using Corollary 2.7.3, we have

$$\begin{aligned}\mathbb{E}\Phi\Big(\eta^{-1}\big(\sup_{0\le s\le T}\|N_s\|\big)\Big) &\ge c_\beta \mathbb{E}\Phi\Big(\eta^{-1}\Big(\mathbb{E}\big[\big(\sup_{0\le s\le T}\|\tilde N_s\|\big)\big|T\big]\Big)\Big)\\ &\ge c_\beta\mathbb{E}\Phi\big(\eta^{-1}(\eta(T))\big) \qquad (2.7.13)\\ &= c_\beta\mathbb{E}\Phi(T).\end{aligned}$$

□

Recall that $T_b = \inf\{t>0 : \|N_t\|\ge b\}$. The following result provides information on the moments of first exit times from a fixed symmetric boundary by continuous processes.

COROLLARY 2.7.5. *Let $\{N_t, t\ge 0\}$ be a B-valued continuous time process with independent increments, where $(B, \|\cdot\|)$ is a separable Banach space. Assume that $\|N_t\|$ is a continuous function of t. Let T be a stopping time adapted to $\mathcal{G}_t = \sigma(N_s, s\le t)$. Assume that $\mathbb{E}\sup_{s\le t}\|N_s\|\ge\eta(t)$ is a strictly increasing function of t (and therefore has a well-defined inverse). Let Φ be a function such that $\Phi\big(\eta^{-1}(x)\big)$, $x>0$ is convex and in $\mathcal{A}_\beta$ for some $\beta>0$. Then*

$$\mathbb{E}\Phi(T_b) \le \frac{1}{c_\beta}\Phi(\eta^{-1}(b)), \tag{2.7.14}$$

for a constant c_β depending on β only.

In the case that the process has jumps, a weaker statement holds. A key result in developing this extension is the following decoupling inequality.

LEMMA 2.7.6. *Under the assumptions of Theorem 2.5.1, let* $\Delta N_t = N_t - N_{t^-}$, $\Delta N_t^* = \sup_{s\le t} \|\Delta N_s\|$ *and similarly* $\Delta\tilde{N}_t = \tilde{N}_t - \tilde{N}_{t^-}$, $\Delta\tilde{N}_t^* = \sup_{s\le t} \|\Delta\tilde{N}_s\|$. *Then*

$$\mathbb{E}\Phi(\Delta N_T^*) \le 4\mathbb{E}\Phi(\Delta\tilde{N}_T^*). \tag{2.7.15}$$

PROOF. Let $t_0 = 0$, $t_1 = \inf\{s : \mathbb{E}\Phi(\Delta N_s^*) > 0\}$. Define the sequence $t_2, t_3, \ldots$ as follows: $t_{k+1} = \inf\{s \ge t_k : \mathbb{E}\Phi(\Delta N_s^*) \ge 2^k \mathbb{E}\Phi(\Delta N_{t_1}^*)\}$ and $t_{k+1} = \infty$ if no such s exists. Let k^* be the last k such that $t_k \le T$, and $k^{**} = \sup\{k : t_k < \infty\}$. It can be easily shown that

$$2^{k^*-1}\mathbb{E}\Phi(\Delta N_{t_1}^*) \le \mathbb{E}\big(\Phi(\Delta\tilde{N}_T^*)\big|T\big) \le 2^{k^*}\mathbb{E}\Phi(\Delta N_{t_1}^*).$$

From this it follows that

$$\begin{aligned}
\mathbb{E}\Phi(\Delta N_T^*) &\le \mathbb{E}\sum_{k=1}^{k^{**}} \Phi\big(\sup_{t_k\le s<t_{k+1}} \|\Delta N_s\|\big) I_{T\ge t_k}\\
&\le \sum_{k=1}^{k^{**}} \mathbb{E}\Phi\big(\sup_{t_k\le s<t_{k+1}} \|\Delta N_s\|\big)\Pr\{T \ge t_k\}\\
&\le \sum_{k=1}^{k^{**}} \mathbb{E}\Phi\big(\sup_{s<t_{k+1}} \|\Delta N_s\|\big)\Pr\{T \ge t_k\}\\
&\le \sum_{k=1}^{k^{**}} 2^k \mathbb{E}\Phi(\Delta N_{t_1}^*)\Pr\{T \ge t_k\}\\
&\le \mathbb{E}\Phi(\Delta N_{t_1}^*)\sum_{k=1}^{k^*} 2^k \le 4\mathbb{E}2^{k^*-1}\Phi(\Delta N_{t_1}^*)\\
&\le 4\mathbb{E}(\mathbb{E}(\Phi(\Delta\tilde{N}_T^*)|T)) = 4\mathbb{E}\Phi(\Delta\tilde{N}_T^*).
\end{aligned}$$

□

As a consequence of the above we get the following asymptotic result.

THEOREM 2.7.7. *Under the assumptions of Proposition 2.7.4, let* $\Delta N_t = N_t - N_{t-}$. *Assume* $\sup_t \mathbb{E}\Phi(\eta^{-1}(\Delta N_t^*)) < \infty$. *Then*

$$\limsup_{b\to\infty} \mathbb{E}\frac{\Phi(T_b)}{\Phi(\eta^{-1}(b))} \le \frac{2^{\beta+2}}{c_\beta}. \tag{2.7.16}$$

PROOF. From Proposition 2.7.4,

$$\begin{aligned}
\mathbb{E}\Phi(T_b) &\le \frac{1}{c_\beta}\mathbb{E}\Phi\Big(\eta^{-1}\big(\sup_{0\le s\le T_b} \|N_t\|\big)\Big)\\
&= \frac{1}{c_\beta}\mathbb{E}\Phi\big(\eta^{-1}(\|N_{T_b}\|)\big)\\
&\le \frac{1}{c_\beta}\mathbb{E}\Phi\big(\eta^{-1}(\|N_{T_b}^-\| + \|\Delta N_{T_b}\|)\big)
\end{aligned}$$

$$
\begin{aligned}
&\le \frac{1}{c_\beta}\mathbb{E}\Phi\big(\eta^{-1}(2\{\|N_{T_b}^-\| \vee \|\Delta N_{T_b}\|\})\big) \\
&\le \frac{2^\beta}{c_\beta}\Big[\mathbb{E}\Phi\big(\eta^{-1}(\|N_{T_b}^-\|)\big) + \Phi\big(\eta^{-1}(\|\Delta N_{T_b}\|)\big)\Big] \\
&\le \frac{2^\beta}{c_\beta}[\Phi(\eta^{-1}(b)) + \mathbb{E}\Phi(\eta^{-1}(\Delta N_{T_b}^*))] \\
&\le \frac{2^{\beta+2}}{c_\beta}[\Phi(\eta^{-1}(b)) + \mathbb{E}\Phi(\eta^{-1}(\Delta \tilde{N}_{T_b}^*))] \qquad \text{(by Lemma 2.7.6)} \\
&= \frac{2^{\beta+2}}{c_\beta}\big[\Phi(\eta^{-1}(b)) + \mathbb{E}\big(\mathbb{E}\big(\Phi\big(\eta^{-1}(\Delta \tilde{N}_{T_b}^*)\big)\big|T_b\big)\big)\big] \\
&\le \frac{2^{\beta+2}}{c_\beta}\big[\Phi(\eta^{-1}(b)) + \mathbb{E}\big(\sup_t \mathbb{E}\big(\Phi\big(\eta^{-1}(\Delta \tilde{N}_{T_b}^*)\big)\big|T_b = t\big)\big)\big] \\
&= \frac{2^{\beta+2}}{c_\beta}\big[\Phi(\eta^{-1}(b)) + \sup_t \mathbb{E}\Phi(\eta^{-1}(\Delta N_t^*))\big].
\end{aligned}
$$

The proof is completed by dividing over by $\Phi(\eta^{-1}(b))$ and taking limits. □

The following result provides an alternate to the right-hand side of Theorem 2.5.1 in the case of continuous processes $\{N_t\}$. We will use it to obtain the reverse bound to (2.7.14). More generally, the results we will present in the remainder of this section hold for processes for which $\|N_t\|$ is continuous in t.

THEOREM 2.7.8. *Fix any $\alpha, \beta > 0$. Let Φ be a function in $\mathcal{A}_\alpha$, and Ψ a function in $\mathcal{A}_\beta$. Let $\{N_t, t \ge 0\}$ be a B-valued continuous process with independent increments, where $(B, \|\cdot\|)$ is a separable Banach space. Let T be a stopping time adapted to $\mathcal{G}_t = \sigma(N_s, s \le t)$. Then if $\{\tilde{N}_t, t \ge 0\}$ is an independent copy of $\{N_t, t \ge 0\}$ (hence independent of T as well),*

$$
\mathbb{E}\Phi\Big(\Psi\big(\sup_{0\le s\le T} \|N_s\|\big)\Big) \le C_{\alpha,\beta}\mathbb{E}\Phi\Big(\mathbb{E}\big[\Psi\big(\sup_{0\le s\le T} \|\tilde{N}_s\|\big)\big|T\big]\Big). \tag{2.7.17}
$$

If in addition Φ is assumed to be convex, then the reverse bound holds (as in Corollary 2.7.3), namely,

$$
\mathbb{E}\sup_{0\le s\le T} \Phi\big(\Psi(\|N_s\|)\big) \ge c_{\alpha,\beta}\mathbb{E}\Phi\Big(\mathbb{E}\big[\Psi\big(\sup_{0\le s\le T} \|\tilde{N}_s\|\big)\big|T\big]\Big). \tag{2.7.18}
$$

with $c_{\alpha,\beta}$, $C_{\alpha,\beta}$ constants depending on α, β only.

The proof of this result is deferred to introduce several applications. Even in the case $T = t$ is a fixed time (2.7.17) seems surprising.

COROLLARY 2.7.9. *Fix any $\alpha > 0$. Let Φ be a function in $\mathcal{A}_\alpha$, and Ψ a function in $\mathcal{A}_\beta$. Let N_t, $t \geq 0$, be a B-valued continuous process with independent increments, where $(B, \|\cdot\|)$ is a separable Banach space. Then, for all $t \geq 0$,*

$$\mathbb{E} \sup_{0\leq s\leq t} \Phi(\Psi(\|N_s\|)) \leq c_{\alpha,\beta}\Phi\Big(\mathbb{E}\Psi\big(\sup_{0\leq s\leq t} \|N_s\|\big)\Big), \tag{2.7.19}$$

with $c_{\alpha,\beta}$, $C_{\alpha,\beta}$ constants depending on α, β only.

We remark that (2.7.19) is (up to constants) a reversal of the bound obtained by applying Jensen's inequality in the case Φ is convex, that is,

$$\mathbb{E}\Phi\Big(\Psi\big(\sup_{s\leq t}\|N_s\|\big)\Big) \geq \Phi\Big(\mathbb{E}\Psi\big(\sup_{s\leq t}\|N_s\|\big)\Big). \tag{2.7.20}$$

This type of bound is typically encountered within the theory of hypercontractivity.

Several applications of Theorem 2.7.8 to boundary crossing problems follow. The first application deals with the nonstandard problem of approximating the expectation of a composite function of T_b. The advantage in this case is that minimal assumptions are therefore imposed on the growth rate of our process as will be seen next.

COROLLARY 2.7.10. *Fix any $\alpha, \beta > 0$. Let Φ be a function in $\mathcal{A}_\alpha$ and Ψ a function in $\mathcal{A}_\beta$. Let $\{N_t, t \geq 0\}$ be a B-valued continuous process with independent increments, where $(B, \|\cdot\|)$ is a separable Banach space. Set $\psi(t) = \mathbb{E}\Psi\big(\sup_{s\leq t}\|N_s\|\big)$. Then, if $\psi(t)$ is an increasing function of t,*

$$\mathbb{E}\Phi\big(\psi(T_b)\big) \geq \frac{1}{C_{\alpha,\beta}}\Phi\big(\Psi(b)\big). \tag{2.7.21}$$

If in addition Φ is assumed to be convex, then

$$\mathbb{E}\Phi\big(\psi(T_b)\big) \leq \frac{1}{c_{\alpha,\beta}}\Phi\big(\Psi(b)\big), \tag{2.7.22}$$

for constants $0 < c_{\alpha,\beta}, C_{\alpha,\beta} < \infty$ depending on α, β only.

Observe that in the case N_t is nonrandom, the above inequalities reduce to equalities, since then $\psi(t) = \Psi(t)$. This observation establishes the optimality of the above bounds up to the constants $c_{\alpha,\beta}$, $C_{\alpha,\beta}$.

PROOF. Observe in previous theorem that from the definition of ψ, for any stopping time T, $\mathbb{E}[\Psi(\sup_{0\leq s\leq T}\|\tilde{N}_s\|)|T] = \psi(T)$. Moreover, since $\|N_t\|$ is continuous, $\mathbb{E}\Phi\big(\Psi\big(\sup_{0\leq s\leq T_b}\|N_s\|\big)\big) = \Phi\big(\Psi(b)\big)$. □

The results that follow are of interest if one needs to bound a particular moment of T_b.

PROPOSITION 2.7.11. *Let $\{N_t, t \geq 0\}$ be a B-valued continuous process with independent increments, where $(B, \|\cdot\|)$ is a separable Banach space. Let T be a stopping time adapted to $\mathcal{G}_t = \sigma(N_s, s \leq t)$. Suppose $\mathbb{E}\sup_{s\leq t}\|N_s\| \leq \eta(t)$.*

Assume that $\eta(t)$ is a strictly increasing function of t (and therefore has a well-defined inverse). Let Φ be a function such that $\Phi(\eta^{-1}(x)),\ x > 0$ is in $\mathcal{A}_\alpha$ for some $\alpha > 0$. Then

$$\mathbb{E}\Phi(T) \geq \frac{1}{C_\alpha}\mathbb{E}\Phi\Big(\eta^{-1}\big(\sup_{0\leq s\leq T}\|N_t\|\big)\Big), \tag{2.7.23}$$

for a constant C_α depending on α only.

PROOF. Using Theorem 2.7.8 we have

$$\begin{aligned}\mathbb{E}\Phi\Big(\eta^{-1}\big(\sup_{0\leq s\leq T}\|N_s\|\big)\Big) &\leq C_\alpha\mathbb{E}\Phi\Big(\eta^{-1}\Big(\mathbb{E}\big[\big(\sup_{0\leq s\leq T}\|\tilde{N}_s\|\big)\big|T\big]\Big)\Big)\\ &= C_\alpha\mathbb{E}\Phi\big(\eta^{-1}(\eta(T))\big) \qquad (2.7.24)\\ &\leq C_\alpha\mathbb{E}\Phi(T). \qquad \square\end{aligned}$$

COROLLARY 2.7.12. *Let $\{N_t, t\geq 0\}$ be a B-valued continuous process with independent increments, where $(B, \|\cdot\|)$ is a separable Banach space. Let $T_b = \inf\{s\geq 0 : \|N_s\| = b\}$ and $\eta(t) = \mathbb{E}\sup_{s\leq t}\|N_s\|$. Suppose that $\eta(t)$ is an increasing function of t and that $\Phi\big(\eta^{-1}(t)\big)$ is a function in $\mathcal{A}_\alpha$. Then,*

$$\mathbb{E}\Phi(T_b) \geq \frac{1}{C_\alpha}\Phi\big(\eta^{-1}(b)\big), \tag{2.7.25}$$

for a constant C_α depending on α only.

The following Corollary provides an extension of (2.7.5).

COROLLARY 2.7.13. *Let $\{N_t, t\geq 0\}$ be a B-valued continuous process with independent increments, where $(B, \|\cdot\|)$ is a separable Banach space. Let $T_b = \inf\{s\geq 0 : \|N_s\| = b\}$ and $\eta(t) = \mathbb{E}\sup_{s\leq t}\|N_s\|$. Suppose that $\eta(t)$ is an increasing function and that $\Phi\big(\eta^{-1}(t)\big)$ is an increasing convex function in $\mathcal{F}_\alpha$. Then*

$$\mathbb{E}\Phi(T_b) \approx_\alpha \Phi\big(\eta^{-1}(b)\big). \tag{2.7.26}$$

An application of Corollary 2.7.13 provides a powerful tool for comparing the waiting times for the first passage of two (possibly different) processes over a boundary.

COROLLARY 2.7.14. *Let $\{N_{1t}, t\geq 0\}$ be a B_1-valued continuous process with independent increments and $\{N_{2t}, t\geq 0\}$ be a B_2-valued continuous process with independent increments, where $(B_1, \|\cdot\|_1)$ and $(B_2, \|\cdot\|_2)$, are two separable Banach spaces. Let $T_{1,b} = \inf\{s\geq 0 : \|N_{1s}\|_1 = b\}$, $T_{2,b} = \inf\{s\geq 0 : \|N_{2s}\|_2 = b\}$. Let $\eta_1(t) = \mathbb{E}\sup_{s\leq t}\|N_{1s}\|_1$, $\eta_2(t) = \mathbb{E}\sup_{s\leq t}\|N_{2s}\|_2$. Then, if $\eta_1^{-1} \in \mathcal{A}_{\alpha_1}$ and $\eta_2^{-1} \in \mathcal{A}_{\alpha_2}$ with η_1, η_2 convex increasing functions,*

$$\mathbb{E}T_{2,b} \approx_{\alpha_1,\alpha_2} \frac{\eta_2^{-1}(b)}{\eta_1^{-1}(b)}\mathbb{E}T_{1,b}. \tag{2.7.27}$$

EXAMPLE 2.7.15. Fix $p \geq 1$. Assume that $\mathbb{E}\sup_{s\leq t}\|N_s\| \geq \eta(t) = e^t - 1$. Taking $\Phi(t) = (e^t - 1)^p$, $\Phi(\eta^{-1}(t)) \geq t^p$ is a convex function in $\mathcal{A}_p$. One gets from Corollary 2.7.5, $\mathbb{E}(e^{T_b} - 1)^p \leq C_p b^p$. The process $N_t = e^t - 1 + B_t$ satisfies the above conditions with B_t a standard Brownian motion. Assume now that N_t is a homogeneous Poisson process with rate λ. Then, $\eta(t) = \mathbb{E}\sup_{s\leq t} N_s = \mathbb{E}N_t = \lambda t$. Moreover, since the jumps of the process are bounded, the conditions of Proposition 2.7.4 and Theorem 2.7.7 are satisfied. Therefore, for all $\Phi \in \mathcal{A}_\beta$,

$$\limsup_{b\to\infty} \frac{\mathbb{E}\Phi(T_b)}{\Phi(\frac{b}{\lambda})} \leq \frac{2^{\beta+1}}{c_\beta}.$$

Assume that $\mathbb{E}\sup_{s\leq t}\|N_s\| \leq \eta(t) = e^{\sqrt{t}} - 1$. Using Proposition 2.7.11 and since $\eta^{-1}(t) = (\log(t+1))^2$ is in $\mathcal{A}_2$, we get $\mathbb{E}T_b \geq c(\log(b+1))^2$. When B_t is again standard Brownian motion, then the process

$$N_t = \frac{e^{\sqrt{t}} - 1 + (B_t/\mathbb{E}\sup_{s\leq 1}|B_s|)}{2}$$

satisfies the above conditions. In the context of Corollary 2.7.13 with $\eta(t) \approx t^r$ for some $0 < r \leq s < \infty$, and $\Phi(t) = t^s$, we get $\mathbb{E}T_b^s \approx_{\frac{s}{r}} b^{\frac{s}{r}}$. When $\eta(t) \approx \log(t+1)$, then taking $\Phi(t) = (\log(t+1))^p$ gives $\mathbb{E}(\log(T_b+1))^p \approx_p b^p$, since $\Phi(\eta^{-1}(t)) = t^p$ is a convex function in $\mathcal{A}_p$.

An application of Corollary 2.7.10 gives,

COROLLARY 2.7.16. *In the context of Corollary 2.7.10 let $\Phi(t)$ be a function in $\mathcal{A}_\beta$, and $\eta(t)$ an increasing function. Then*

$$\mathbb{E}\Phi\big(\eta(T_b)\big) \geq c_\beta \Phi(b). \tag{2.7.28}$$

If in addition Φ is assumed to be convex, then

$$\mathbb{E}\Phi\big(\eta(T_b)\big) \approx c_\beta \Phi(b). \tag{2.7.29}$$

Observe that in this context the only condition imposed on $\eta(t)$ is that it should be increasing (so that an inverse exists)!

The following good-lambda inequality will be used to obtain the proof of Theorem 2.7.8, which is given at the end of the section.

LEMMA 2.7.17. *Let $\{N_t, t \geq 0\}$ be a B-valued continuous process with independent increments, where $(B, \|\cdot\|)$ is a separable Banach space. Let T be a stopping time adapted to $\mathcal{G}_t = \sigma(N_s, s \leq t)$. Let $\Psi(x)$, $x \geq 0$, be a function in $\mathcal{A}_\beta$. Let $N_t^* = \sup_{s\leq t}\|N_s\|$ and $a_{t,\Psi}^* = \mathbb{E}\sup_{s\leq t}\Psi(\|N_s\|)$. Then, for all $\rho > 1, \delta > 0$,*

$$\Pr\{\Psi(N_T^*) > \rho y, a_{T,\Psi}^* \leq \delta y\} \leq \frac{4^\beta \delta}{\rho - 2^\beta}\Pr\{\Psi(N_T^*) > y\}. \tag{2.7.30}$$

PROOF. For later reference we let $\{\tilde{N}_t, t \geq 0\}$ be an independent copy of $\{N_t, t \geq 0\}$ with $\{\tilde{N}_t, t \geq 0\}$. Let

$$T_y = \inf\{t > 0 : \Psi(\|N_t\|) > y\}, \qquad r' = \sup\{s > 0 : a^*_{s,\Psi} \leq \delta y\}. \quad (2.7.31)$$

Let $\mathcal{G}_{T_y} = \sigma\{N_s, s \leq T_y\}$. The fact that $\{N_t, t \geq 0\}$ has independent increments implies that conditional on $\mathcal{G}_{T_y}$ the joint distribution of $\|N_s - N_{T_y}\|$ is equal to that of $\|\tilde{N}_s - \tilde{N}_{T_y}\|$ for all fixed $s > T_y$. Therefore, for any fixed $\delta > 0$, $\rho > 1$ and all $y > 0$,

$$\begin{aligned}
&\Pr\{\Psi(N^*_T) > \rho y, a^*_{T,\Psi} \leq \delta y\} \\
&\quad \leq \Pr\Big\{2^\beta\big[\Psi\big(\sup_{T_y < s \leq T} \|N_s - N_{T_y}\|\big) + \Psi\big(\|N_{T_y}\|\big)\big] > \rho y, a^*_{T,\Psi} \leq \delta y, T \geq T_y\Big\} \\
&\quad \leq \Pr\Big\{2^\beta \sup_{T_y < s \leq T} \Psi(\|N_s - N_{T_y}\|) > (\rho - 2^\beta)y, T \leq r', T \geq T_y\Big\}. \quad (2.7.32)
\end{aligned}$$

The first inequality above follows from the growth property of the function Ψ, which gives $\Psi(a+b) \leq \Psi(2(a \vee b)) \leq \Psi(2a) + \Psi(2b) \leq 2^\beta\{\Psi(a) + \Psi(b)\}$, together with the following chain of inequalities:

$$\begin{aligned}
\sup_{s \leq T} \|N_s\| &= \sup_{s \leq T_y} \|N_s\| \vee \sup_{T_y < s \leq T} \|N_s\| \\
&\leq \sup_{s \leq T_y} \|N_s\| \vee \big(\sup_{T_y < s \leq T} \|N_s - N_{T_y}\| + \|N_{T_y}\|\big) \\
&\leq \sup_{T_y < s \leq T} \|N_s - N_{T_y}\| + \|N_{T_y}\|.
\end{aligned}$$

The second inequality follows from the definition of r'. Using Markov's inequality we bound (2.7.32) by

$$\begin{aligned}
&\frac{1}{(\rho - 2^\beta)y} 2^\beta \mathbb{E}\Psi\big(\sup_{T_y < s \leq T} \|N_s - N_{T_y}\| \; I_{T \leq r', T \geq T_y}\big) \\
&\quad \leq \frac{2^\beta}{(\rho - 2^\beta)y} \mathbb{E}\Psi\big(\sup_{T_y < s \leq r'} \|N_s - N_{T_y}\|\big) I_{T \geq T_y} \\
&\quad = \frac{2^\beta}{(\rho - 2^\beta)y} \mathbb{E}\Big[\mathbb{E}\big(\Psi\big(\sup_{T_y < s \leq r'} \|N_s - N_{T_y}\|\big) | \mathcal{G}_{T_y}\big) I_{T \geq T_y}\Big] \\
&\quad = \frac{2^\beta}{(\rho - 2^\beta)y} \mathbb{E}\Big[\mathbb{E}\big(\Psi\big(\sup_{T_y < s \leq r'} \|\tilde{N}_s - \tilde{N}_{T_y}\|\big) | \mathcal{G}_{T_y}\big) I_{T \geq T_y}\Big]. \quad (2.7.33)
\end{aligned}$$

Equation (2.7.33) above is obtained by using the equality in distribution of the difference processes discussed in the beginning of the proof. The decoupling happens in that line and is used in the second line below, exploiting the fact that $\{\tilde{N}_t\}$ and $(\{N_t\}, T)$ are independent. Using the triangle inequality, (2.7.33) is bounded by

$$\frac{2^\beta}{(\rho - 2^\beta)y} \mathbb{E}\Psi\big(\sup_{T_y < s \leq r'} \|\tilde{N}_s\| + \|\tilde{N}_{T_y}\|\big) I_{T \geq T_y}$$

$$
\begin{aligned}
&\le \frac{4^\beta}{(\rho - 2^\beta)y} \mathbb{E}\Psi\big(\sup_{s\le r'} \|\tilde{N}_s\|\big) \Pr\{T \ge T_y\} \\
&\le \frac{4^\beta \delta y}{(\rho - 2^\beta)y} \Pr\{T \ge T_y\} \\
&= \frac{4^\beta \delta}{(\rho - 2^\beta)} \Pr\big\{\sup_{s\le T} \Psi(\|N_s\|) > y\big\}, \qquad (2.7.34)
\end{aligned}
$$

where the last two lines follow from the definition of r' and of T_y. □

PROOF OF THEOREM 2.7.8. To obtain (2.7.17) just use Lemma 2.2.2. The proof of (2.7.18) uses the convexity of Φ and is an easy extension of the proof of Corollary 2.7.3. □

2.8 References for Chapter 2

Theorem 2.1.1 comes from Wald (1945), Blackwell (1946) (Wald's first equation). We followed Chow and Teicher (1988) for Wald's second equation. Theorem 2.1.3 (due to Stein) can be found in Proposition 2.19 of Siegmund (1985). The inequalities in (2.1.8) and (2.1.9) are from de la Peña and Govindarajulu (1992). The one in (2.1.10) is from Hitczenko (1994b). The result in (2.1.11) is from Chow, Robbins, and Siegmund (1971) extending work of Burkholder and Gundy (1970), Gordon [unpublished, referred to in Chung (1974) page 343], Theorem 2.1.4 and equation (2.1.16) from Klass (1988). Section 2.2 gathers the work of Burkholder and Gundy (1970) (Lemma 2.2.1, Lemma 2.2.2), while Lemma 2.2.3 is due to Burkholder (1979) and Azéma, Gundy, and Yor (1979). Section 2.3 introduces Theorem 2.3.1 which is a two-sided inequality of Klass (1988) and Klass (1990). The proof of the right-hand side given in Section 2.3 initially follows the approach taken by Klass, however, one part of the proof is done using Lemma 2.3.3, which comes from Kwapień and Woyczynski (1991), which was available in 1988 in preprint form [see also Hitczenko (1988)]. The left-hand side of Theorem 2.3.1 is proved in Section 2.4. Theorem 2.5.1 comes from de la Peña and Eisenbaum (1997) [see also de la Peña (1998a)]. de la Peña and Eisenbaum (1994) contains extensions of this result, with constants added, for certain stochastic integrals, and in particular martingales with respect to the filtration of linear Brownian motion. Corollary 2.5.2 is from de la Peña (1998a), Section 2.6 is mainly based on de la Peña (1998a) except for Corollary 2.6.5, which comes form de la Peña and Eisenbaum (1997), which contains further exponential results. The results of Section 2.7 are taken from de la Peña (1997a).

3

Decoupling of U-Statistics and U-Processes

U-statistics, first considered by Halmos (1946) in connection with unbiased statistics, and formally introduced by Hoeffding (1948), are defined as follows: Given an i.i.d. sequence of random variables $\{X_i\}_{i=1}^{\infty}$ with values in a measurable space $(S, \mathcal{S})$, and a measurable function $h : S^m \to \mathbb{R}$, the U-statistics of order m and kernel h based on the sequence $\{X_i\}$ are

$$U_n(h) = \frac{(n-m)!}{n!} \sum_{(i_1,\dots,i_m)\in I_n^m} h(X_{i_1}, \dots, X_{i_m}), \quad n \geq m, \tag{3.0.1}$$

where

$$I_n^m = \left\{(i_1, \dots, i_m) : i_j \in \mathbb{N}, 1 \leq i_j \leq n, i_j \neq i_k \quad \text{if} \quad j \neq k\right\}.$$

These objects appear often in statistics either as unbiased estimators of parameters of interest or, perhaps more often, as components of higher-order terms in expansions of smooth statistics (von Mises expansion, delta-method). Particularly in connection with von Mises expansions, it is sometimes convenient to also consider U-processes indexed by families $\mathcal{H}$ of kernels, that is, collections of U-statistics $\{U_n(h) : h \in \mathcal{H}\}$. Decoupling inequalities have had a major role in recent advances on the asymptotic theory of U-statistics and U-processes. In this chapter we present the decoupling theory and Chapters 4 and 5 are devoted to its applications to limit theorems.

By a *decoupling* result for U-statistics we mean a (usually two-sided) inequality between the quantities

$$\mathbb{E}\Phi\Big(\Big|\sum_{I_n^m} h(X_{i_1}, \ldots, X_{i_m})\Big|\Big) \tag{3.0.2}$$

and

$$\mathbb{E}\Phi\Big(\Big|\sum_{I_n^m} h(X_{i_1}^1, \ldots, X_{i_m}^m)\Big|\Big), \tag{3.0.3}$$

possibly multiplied by constants that depend only on m, where the sequences $\{X_i^k\}$, $k = 1, \ldots, m$, are independent copies of the original sequence $\{X_i\}$, and Φ is a non-negative function. Two quite different types of functions Φ will be considered: Φ convex (thus including $\Phi(|x|) = |x|^p$, $p \geq 1$) and $\Phi(|x|) = I_{|x|\geq t}$.

In fact one can decouple much more complicated expressions than (3.0.2): the variables $\big|\sum_{I_n^m} h(X_{i_1}, \ldots, X_{i_m})\big|$ in (3.0.2) can be replaced by

$$\sup_{h_{i_1\ldots i_m} \in \mathcal{H}_{i_1\ldots i_m}} \Big\|\sum_{I_n^m} h_{i_1\ldots i_m}(X_{i_1}, \ldots, X_{i_m})\Big\|$$

where $\mathcal{H}_{i_1\ldots i_m}$ are collections of B-valued kernels and B is a Banach space; then, of course, the variables in (3.0.3) are replaced by the corresponding decoupled expressions (decoupling in one direction requires some symmetry of $\mathcal{H}$ and h with respect to the multi-indices). The inequalities hold even if B is not separable or if the classes of functions $\mathcal{H}$ are not countable, with very few measurability requirements. In particular, decoupling applies to U-processes.

Whereas decoupling of convex functions of U-processes can be achieved by means of fairly elementary techniques, decoupling of tail probabilities requires specific tools from several different areas. These are hypercontractivity of Rademacher chaos variables (a sort of "reverse Hölder inequality" for polynomials in Rademacher variables extending Khinchin's inequalities), Paley–Zygmund-type lower bounds for tail probabilities of hypercontractive variables, and the extension of Lévy's maximal inequality to sums of i.i.d. not necessarily symmetric random vectors (Section 1.1.1). These three topics are important per se and their usefulness is not limited to decoupling (in fact, some of them are also used elsewhere in this book). So, we have chosen to develop them in a slightly more complete form than needed for decoupling. The third is presented in Chapter 1, Section 1.1, and the other two are the object of Sections 3.2 and 3.3 below. Sections 3.1 and 3.4 are devoted to decoupling of U-processes.

Randomization is one of the (several) reasons decoupling is useful. A randomization inequality compares moments or tail probabilities of the original U-statistic or process, $\sum_{I_n^m} h(X_{i_1}, \ldots, X_{i_m})$, with the corresponding moments or tail probabilities of

$$\sum_{I_n^m} \varepsilon_{i_1} \cdots \varepsilon_{i_r} h(X_{i_1}, \ldots, X_{i_m})$$

for some $1 \le r \le m$ that depends on the degree of degeneracy (meaning centering) of the kernel h, where ε_i are independent Rademacher variables independent from the X variables. Whereas randomization inequalities for moments work in both directions, randomization inequalities for tail probabilities are only possible in one direction (the randomized statistics have basically smaller tail probabilities than the original). These inequalities are examined in Section 5.

3.1 Decoupling of U-Processes: Convex Functions

First, we state and prove the main result of this section for kernels taking values in a separable Banach space. The proof for real-valued kernels is not easier, and presenting it in B (as opposed to $\mathbb{R}$) only changes absolute values by norms. Then we indicate how this result also holds true for U-processes, that is, when the norm $\|\cdot\|$ is replaced by the supremum over an often uncountable family of kernels. It is interesting to point out that the measurability problems that arise in this case can be handled in a very straightforward way. As a first, illustrative example—and we note that the next two chapters can be considered, in large part, also as examples of the results in this chapter—we show how to reduce the first absolute moment of a quadratic form in i.i.d. centered random variables to the (square of the) corresponding moment of a sum of independent variables.

THEOREM 3.1.1. *For natural numbers $n \ge m$, let $\{X_i\}_{i=1}^n$ be n independent random variables with values in a measurable space $(S, \mathcal{S})$, and let $\{X_i^k\}_{i=1}^n$, $k = 1, \ldots, m$, be m independent copies of this sequence. Let B be a separable Banach space and, for each $(i_1, \ldots, i_m) \in I_n^m$, let $h_{i_1 \ldots i_m} : S^m \mapsto B$ be measurable functions such that $\mathbb{E}\big(\|h_{i_1\ldots i_m}(X_{i_1}, \ldots, X_{i_m})\|\big) < \infty$. Let $\Phi : [0, \infty) \to [0, \infty)$ be a convex nondecreasing function such that $\mathbb{E}\Phi\big(\|h_{i_1\ldots i_m}(X_{i_1}, \ldots, X_{i_m})\|\big) < \infty$ for all $(i_1, \ldots, i_m) \in I_n^m$. Then,*

$$\mathbb{E}\Phi\Big(\Big\|\sum_{I_n^m} h_{i_1\ldots i_m}(X_{i_1}, \ldots, X_{i_m})\Big\|\Big) \le \mathbb{E}\Phi\Big(C_m\Big\|\sum_{I_n^m} h_{i_1\ldots i_m}(X_{i_1}^1, \ldots, X_{i_m}^m)\Big\|\Big), \tag{3.1.1}$$

where $C_m = 2^m(m^m - 1)((m-1)^{(m-1)} - 1) \times \cdots \times 3$. If, moreover, the functions $h_{i_1\ldots i_m}$ are symmetric in the sense that, for all $x_1, \ldots, x_m \in S$ and all permutations s of $\{1, \ldots, m\}$,

$$h_{i_1\ldots i_m}(x_1, \ldots, x_m) = h_{i_{s_1}\ldots i_{s_m}}(x_{s_1}, \ldots, x_{s_m}), \tag{3.1.2}$$

then inequality (3.1.1) can be reversed, that is,

$$\mathbb{E}\Phi\Big(\frac{1}{2^{(2m-2)}(m-1)!}\Big\|\sum_{I_n^m} h_{i_1\ldots i_m}(X_{i_1}^1, \ldots, X_{i_m}^m)\Big\|\Big) \le \mathbb{E}\Phi\Big(\Big\|\sum_{I_n^m} h_{i_1\ldots i_m}(X_{i_1}, \ldots, X_{i_m})\Big\|\Big). \tag{3.1.3}$$

PROOF FOR $m = 2$. For clarity of exposition we present first the proof for $m = 2$: this case is notationally much less involved than the general case and already contains the main idea. Replacing $h_{i,j}$ by h amounts only to formal changes in the proof, so we assume, for extra notational simplicity, $h_{i,j} = h$. In this part of the proof, we replace X_i^1, X_i^2, respectively, by X_i, X_i'. Let $\{\varepsilon_i\}_{i=1}^n$ be independent Rademacher variables, independent of $\{X_i, X_i'\}_{i=1}^n$, and let Z_i and Z_i', $i = 1, \ldots n$, be defined as follows:

$$Z_i = \begin{cases} X_i & \text{if } \varepsilon_i = 1 \\ X_i' & \text{if } \varepsilon_i = -1 \end{cases}, \quad Z_i' = \begin{cases} X_i' & \text{if } \varepsilon_i = 1 \\ X_i & \text{if } \varepsilon_i = -1 \end{cases}. \tag{3.1.4}$$

[That is, conditionally on the X variables, the vector (Z_i, Z_i') takes the "values" (X_i, X_i') and (X_i', X_i), each with probability $1/2$.] If for each $1 \le i \le n$, P_i is the law of X_i, then the law of the vector $(Z_1, \ldots, Z_n, Z_1', \ldots, Z_n')$ is $(P_1 \times \cdots \times P_n)^2$ since for each fixed $\varepsilon_1, \ldots, \varepsilon_n$ the coordinates of this vector are just $2n$ independent variables such that P_i is the law of the ith and the $(n+i)$th, $i = 1, \ldots, n$. That is,

$$\mathcal{L}(Z_1, \ldots, Z_n, Z_1', \ldots, Z_n') = \mathcal{L}(X_1, \ldots, X_n, X_1', \ldots, X_n').$$

Likewise, $\mathcal{L}(Z_1, \ldots, Z_n) = \mathcal{L}(X_1, \ldots, X_n)$. Therefore, for any $(P_1 \times \cdots \times P_n)^2$-integrable functions f and $(P_1 \times \cdots \times P_n)$-integrable functions g, we have

$$\begin{aligned} \mathbb{E} f(Z_1, \ldots, Z_n, Z_1', \ldots, Z_n') &= \mathbb{E} f(X_1, \ldots, X_n, X_1', \ldots, X_n'), \\ \mathbb{E} g(Z_1, \ldots, Z_n) &= \mathbb{E} g(X_1, \ldots, X_n). \end{aligned} \tag{3.1.5}$$

Note also that, if $\mathcal{Z}$ is the σ-algebra generated by the X variables,

$$\mathcal{Z} = \sigma(X_i, X_i' : i = 1, \ldots, n),$$

then conditional integration with respect to $\mathcal{Z}$ of any function of the Z variables is simply integration with respect to the ε_i variables only. In particular, for all $i \neq j$,

$$\begin{aligned} \mathbb{E}(h(Z_i, Z_j)|\mathcal{Z}) &= \mathbb{E}(h(Z_i, Z_j')|\mathcal{Z}) = \mathbb{E}(h(Z_i', Z_j)|\mathcal{Z}) = \mathbb{E}(h(Z_i', Z_j')|\mathcal{Z}) \\ &= \frac{1}{4}\Big(h(X_i, X_j) + h(X_i, X_j') + h(X_i', X_j) + h(X_i', X_j')\Big). \end{aligned} \tag{3.1.6}$$

[To see this, e.g., for $h(Z_i, Z_j)$, just note that $h(Z_i, Z_j) = h(X_i, X_j) I_{\varepsilon_i = \varepsilon_j = 1} + \cdots + h(X_i', X_j') I_{\varepsilon_i = \varepsilon_j = -1}$ and integrate with respect to the ε_i variables.] These observations, i.e., equations (3.1.5) and (3.1.6), together with the convexity and monotonicity of Φ, the integrability of the functions involved, and Jensen's inequality, justify the following two strings of inequalities, which, together, prove the theorem. (We leave to the reader the easy task of figuring out which of these properties are used at each step.)

1) For $h(= h_{ij} = h_{ji})$ symmetric in its entries,

$$
\begin{aligned}
\mathbb{E}\Phi\Big(\Big\|\sum_{I_n^2} h(X_i, X_j')\Big\|\Big) &= \mathbb{E}\Phi\Big(\frac{1}{2}\Big\|\sum_{I_n^2} h(X_i, X_j') + \sum_{I_n^2} h(X_i', X_j)\Big\|\Big) \\
&\le \frac{1}{2}\mathbb{E}\Phi\Big(\Big\|\sum_{I_n^2}\big[h(X_i, X_j') + h(X_i', X_j) + h(X_i, X_j) + h(X_i', X_j')\big]\Big\|\Big) \\
&\qquad + \frac{1}{4}\mathbb{E}\Phi\Big(2\Big\|\sum_{I_n^2} h(X_i, X_j)\Big\|\Big) + \frac{1}{4}\mathbb{E}\Phi\Big(2\Big\|\sum_{I_n^2} h(X_i', X_j')\Big\|\Big) \\
&= \frac{1}{2}\mathbb{E}\Phi\Big(4\Big\|\sum_{I_n^2} \mathbb{E}(h(Z_i, Z_j)|\mathcal{Z})\Big\|\Big) + \frac{1}{2}\mathbb{E}\Phi\Big(2\Big\|\sum_{I_n^2} h(X_i, X_j)\Big\|\Big) \\
&\le \frac{1}{2}\mathbb{E}\Phi\Big(4\Big\|\sum_{I_n^2} h(Z_i, Z_j)\Big\|\Big) + \frac{1}{2}\mathbb{E}\Phi\Big(2\Big\|\sum_{I_n^2} h(X_i, X_j)\Big\|\Big) \\
&= \frac{1}{2}\mathbb{E}\Phi\Big(4\Big\|\sum_{I_n^2} h(X_i, X_j)\Big\|\Big) + \frac{1}{2}\mathbb{E}\Phi\Big(2\Big\|\sum_{I_n^2} h(X_i, X_j)\Big\|\Big) \\
&\le \mathbb{E}\Phi\Big(4\Big\|\sum_{I_n^2} h(X_i, X_j)\Big\|\Big), \qquad (3.1.7)
\end{aligned}
$$

proving (3.1.3). Note that symmetry is essential for the first identity.

2) For $h(= h_{ij})$ general, letting $\mathcal{X} = \sigma(X_i : i = 1, \ldots, n)$,

$$
\begin{aligned}
&\mathbb{E}\Phi\Big(\Big\|\sum_{I_n^2} h(X_i, X_j)\Big\|\Big) \\
&\quad \le \frac{1}{2}\mathbb{E}\Phi\Big(2\Big\|\sum_{I_n^2} \mathbb{E}\big[h(X_i, X_j) + h(X_i', X_j) + h(X_i, X_j') + h(X_i', X_j')|\mathcal{X}\big]\Big\|\Big) \\
&\quad\qquad + \frac{1}{2}\mathbb{E}\Phi\Big(2\Big\|\sum_{I_n^2} \mathbb{E}\big[h(X_i, X_j') + h(X_i', X_j) + h(X_i', X_j')|\mathcal{X}\big]\Big\|\Big) \\
&\quad \le \frac{1}{2}\mathbb{E}\Phi\Big(2\Big\|\sum_{I_n^2}\big[h(X_i, X_j) + h(X_i', X_j) + h(X_i, X_j') + h(X_i', X_j')\big]\Big\|\Big) \\
&\quad\qquad + \frac{1}{6}\mathbb{E}\Phi\Big(6\Big\|\sum_{I_n^2} \mathbb{E}(h(X_i', X_j)|\mathcal{X})\Big\|\Big) + \frac{1}{6}\mathbb{E}\Phi\Big(6\Big\|\sum_{I_n^2} \mathbb{E}(h(X_i, X_j')|\mathcal{X})\Big\|\Big) \\
&\quad\qquad + \frac{1}{6}\mathbb{E}\Phi\Big(6\Big\|\sum_{I_n^2} \mathbb{E}(h(X_i', X_j')|\mathcal{X})\Big\|\Big)
\end{aligned}
$$

$$
\begin{aligned}
&\le \frac{1}{2}\mathbb{E}\Phi\Big(8\Big\|\sum_{I_n^2}\mathbb{E}(h(Z_i, Z_j')|\mathcal{Z})\Big\|\Big) + \frac{1}{3}\mathbb{E}\Phi\Big(6\Big\|\sum_{I_n^2}h(X_i, X_j')\Big\|\Big)\\
&\quad + \frac{1}{6}\Phi\Big(6\Big\|\sum_{I_n^2}\mathbb{E}h(X_i', X_j')\Big\|\Big)\\
&\le \frac{1}{2}\mathbb{E}\Phi\Big(8\Big\|\sum_{I_n^2}h(Z_i, Z_j')\Big\|\Big) + \frac{1}{3}\mathbb{E}\Phi\Big(6\Big\|\sum_{I_n^2}h(X_i, X_j')\Big\|\Big)\\
&\quad + \frac{1}{6}\Phi\Big(6\Big\|\sum_{I_n^2}\mathbb{E}h(X_i', X_j')\Big\|\Big)\\
&= \frac{1}{2}\mathbb{E}\Phi\Big(8\Big\|\sum_{I_n^2}h(X_i, X_j')\Big\|\Big) + \frac{1}{3}\mathbb{E}\Phi\Big(6\Big\|\sum_{I_n^2}h(X_i, X_j')\Big\|\Big)\\
&\quad + \frac{1}{6}\Phi\Big(6\Big\|\sum_{I_n^2}\mathbb{E}h(X_i, X_j')\Big\|\Big)\\
&\le \mathbb{E}\Phi\Big(8\Big\|\sum_{I_n^2}h(X_i, X_j')\Big\|\Big),
\end{aligned}
\tag{3.1.8}
$$

proving (3.1.1), even with a better constant. □

COMMENTS ON THE PROOF FOR $m > 2$. The general case is proved by recurrence. We simply indicate the proof for $m = 3$. Let r_i, $i = 1, \dots, n$, be i.i.d. random variables uniformly distributed over a set with three elements, say $\{1, 2, 3\}$, independent of the variables X_i^j, $i = 1, \dots, n$, $j = 1, 2, 3$. The role of Z_i, Z_i' above is now played by the new variables

$$
Y_i^1 = \begin{cases} X_i^1 & \text{if } r_i = 1 \\ X_i^2 & \text{if } r_i = 2 \\ X_i^3 & \text{if } r_i = 3 \end{cases}, \quad
Y_i^2 = \begin{cases} X_i^2 & \text{if } r_i = 1 \\ X_i^3 & \text{if } r_i = 2 \\ X_i^1 & \text{if } r_i = 3 \end{cases}, \quad
Y_i^3 = \begin{cases} X_i^3 & \text{if } r_i = 1 \\ X_i^1 & \text{if } r_i = 2 \\ X_i^2 & \text{if } r_i = 3 \end{cases}
$$

(note the circular permutation of the superindices of the X variables in the definition of the Ys), and the place of the σ-algebra $\mathcal{Z}$ is taken by the σ-algebra

$$\mathcal{Y} = \sigma(X_i^1, X_i^2, X_i^3 : i = 1, \dots, n).$$

Identities analogous to (3.1.5) and (3.1.6) now hold for these variables, in particular (3.1.6) becomes, for all $(i_1, i_2, i_3) \in I_n^3$ and $\ell_1, \ell_2, \ell_3 = 1, 2, 3$,

$$\mathbb{E}\big(h(Y_{i_1}^{\ell_1}, Y_{i_2}^{\ell_2}, Y_{i_3}^{\ell_3})|\mathcal{Y}\big) = \frac{1}{3^3}\sum_{j_1, j_2, j_3 \le 3} h(X_{i_1}^{j_1}, X_{i_2}^{j_2}, X_{i_3}^{j_3}). \tag{3.1.9}$$

For h symmetric in its entries the first three steps in the development (3.1.7) now yield

$$\mathbb{E}\Phi\Big(\Big\|\sum_{I_n^3} h(X_{i_1}^1, X_{i_2}^2, X_{i_3}^3)\Big\|\Big) \le \frac{1}{2}\mathbb{E}\Phi\Big(\frac{2\cdot 3^3}{3!}\Big\|\sum_{I_n^3}\mathbb{E}(h(Y_{i_1}^1, Y_{i_2}^1, Y_{i_3}^1)|\mathcal{Y})\Big\|\Big)$$
$$+\frac{1}{4}\mathbb{E}\Phi\Big(\frac{4}{3!}\Big\|\sum_{\substack{j_1,j_2,j_3\le 3\\ \text{two } j\text{s different}}}\sum_{I_n^3} h(X_{i_1}^{j_1}, X_{i_2}^{j_2}, X_{i_3}^{j_3})\Big\|\Big)$$
$$+\frac{1}{4}\mathbb{E}\Phi\Big(\frac{4}{3!}\Big\|\sum_{j=1}^{3}\sum_{I_n^3} h(X_{i_1}^{j}, X_{i_2}^{j}, X_{i_3}^{j})\Big\|\Big). \tag{3.1.10}$$

As in the case $m = 2$, the sum of the first and third terms on the right is dominated by

$$\frac{3}{4}\mathbb{E}\Phi\Big(\frac{2\cdot 3^3}{3!}\Big\|\sum_{I_n^3} h(X_{i_1}^1, X_{i_2}^1, X_{i_3}^1)\Big\|\Big),$$

which is what we want. Now we must also show that the second term is dominated by a similar expression. To prove this we first observe that

$$\mathbb{E}\Phi\Big(\frac{4}{3!}\Big\|\sum_{\substack{j_1,j_2,j_3\le 3\\ \text{two } j\text{s different}}}\sum_{I_n^3} h(X_{i_1}^{j_1}, X_{i_2}^{j_2}, X_{i_3}^{j_3})\Big\|\Big)$$
$$\le \mathbb{E}\Phi\Big(\frac{12}{3!}\Big\|\sum_{\substack{j_1,j_2,j_3\le 2\\ \text{two } j\text{s different}}}\sum_{I_n^3} h(X_{i_1}^{j_1}, X_{i_2}^{j_2}, X_{i_3}^{j_3})\Big\|\Big).$$

This follows by decomposing the sum over $j_1, j_2, j_3 \le 3$ with two js different into three sums according as to whether the two js different are $\{1, 2\}$, $\{1, 3\}$, or $\{2, 3\}$, and then using the convexity of Φ. We are now closer to the situation in the proof for $m = 2$. Then, if we relabel Z_i^1, Z_i^2 the variables Z_i, Z_i' defined in (3.1.4) (with X_i^1, X_i^2 instead of X_i, X_i'), we have that, as in (3.1.6),

$$\mathbb{E}\big(h(Z_{i_1}^1, Z_{i_2}^1, Z_{i_3}^1)|\mathcal{Z}\big) = \frac{1}{2^3}\sum_{j_1,j_2,j_3\le 2} h(X_{i_1}^{j_1}, X_{i_2}^{j_2}, X_{i_3}^{j_3})$$
$$= \frac{1}{2^3}\sum_{\substack{j_1,j_2,j_3\le 2\\ \text{two } j\text{s different}}} h(X_{i_1}^{j_1}, X_{i_2}^{j_2}, X_{i_3}^{j_3}) + \frac{1}{2^3}\sum_{j=1}^{3} h(X_{i_1}^{j}, X_{i_2}^{j}, X_{i_3}^{j}).$$

Then, using (3.1.5) and convexity and monotonicity of Φ and conditional Jensen, we obtain that, for any $c > 0$,

$$\begin{aligned}
&\mathbb{E}\Phi\Big(c\Big\|\sum_{\substack{j_1,j_2,j_3\le 2\\ \text{two } j\text{s different}}}\sum_{I_n^3} h(X_{i_1}^{j_1}, X_{i_2}^{j_2}, X_{i_3}^{j_3})\Big\|\Big)\\
&= \mathbb{E}\Phi\Big(c\Big\|2^3\sum_{I_n^3}\mathbb{E}\big(h(Z_{i_1}^1, Z_{i_2}^1, Z_{i_3}^1)|\mathcal{Z}\big) - c\sum_{I_n^3}\sum_{j=1}^{3} h(X_{i_1}^{j}, X_{i_2}^{j}, X_{i_3}^{j})\Big\|\Big)\\
&\le \frac{1}{2}\mathbb{E}\Phi\Big(2^4 c\Big\|\sum_{I_n^3}\mathbb{E}\big(h(Z_{i_1}^1, Z_{i_2}^1, Z_{i_3}^1)|\mathcal{Z}\big)\Big\|\Big) + \frac{1}{2}\mathbb{E}\Phi\Big(6c\Big\|\sum_{I_n^3} h(X_{i_1}^1, X_{i_2}^1, X_{i_3}^1)\Big\|\Big)\\
&\le \frac{1}{2}\mathbb{E}\Phi\Big(2^4 c\Big\|\sum_{I_n^3} h(X_{i_1}^1, X_{i_2}^1, X_{i_3}^1)\Big\|\Big) + \frac{1}{2}\mathbb{E}\Phi\Big(6c\Big\|\sum_{I_n^3} h(X_{i_1}^1, X_{i_2}^1, X_{i_3}^1)\Big\|\Big)\\
&\le \mathbb{E}\Phi\Big(2^4 c\Big\|\sum_{I_n^3} h(X_{i_1}^1, X_{i_2}^1, X_{i_3}^1)\Big\|\Big).
\end{aligned}$$

The conclusion from the last three sets of inequalities is then

$$\mathbb{E}\Phi\Big(\frac{4}{3!}\Big\|\sum_{\substack{j_1,j_2,j_3\le 3\\ \text{two } j\text{s different}}}\sum_{I_n^3} h(X_{i_1}^{j_1}, X_{i_2}^{j_2}, X_{i_3}^{j_3})\Big\|\Big) \le \mathbb{E}\Phi\Big(\frac{12\cdot 2^4}{3!}\Big\|\sum_{I_n^3} h(X_{i_1}^1, X_{i_2}^1, X_{i_3}^1)\Big\|\Big).$$

Combining this bound with inequality (3.1.10) gives

$$\mathbb{E}\Phi\Big(\Big\|\sum_{I_n^3} h(X_{i_1}^1, X_{i_2}^2, X_{i_3}^3)\Big\|\Big) \le \mathbb{E}\Phi\Big(2^5\Big\|\sum_{I_n^3} h(X_{i_1}^1, X_{i_2}^1, X_{i_3}^1)\Big\|\Big),$$

which is (3.1.3) for $m = 3$.

To prove (3.1.1) we proceed in analogy with the first three steps of the chain of inequalities (3.1.8), to obtain

$$\begin{aligned}
\mathbb{E}\Phi\Big(\Big\|\sum_{I_n^3} h(X_{i_1}^1, X_{i_2}^1, X_{i_3}^1)\Big\|\Big) \le{}& \frac{1}{2}\mathbb{E}\Phi\Big(2\cdot 3^3\Big\|\sum_{I_n^3}\mathbb{E}\big(h(Y_{i_1}^1, Y_{i_2}^2, Y_{i_3}^3)|\mathcal{Y}\big)\Big\|\Big)\\
&+\frac{1}{2\cdot 26}\sum_{\substack{j_1,j_2,j_3\le 3\\ \text{not all } j\text{s equal}}}\mathbb{E}\Phi\Big(52\Big\|\sum_{I_n^3} h(X_{i_1}^{j_1}, X_{i_2}^{j_2}, X_{i_3}^{j_3})\Big\|\Big)\\
&+\frac{2}{2\cdot 26}\Phi\Big(52\Big\|\sum_{I_n^3}\mathbb{E}h(X_{i_1}, X_{i_2}, X_{i_3})\Big\|\Big).
\end{aligned}$$

By (3.1.9), Jensen, and convexity and monotonicity of Φ, the sum of the first term at the right, the third, and the summands in the second term with all the js different, is dominated by

$$\Big(\frac{1}{2} + \frac{4}{26}\Big)\mathbb{E}\Phi\Big(2\cdot 3^3\Big\|\sum_{I_n^3} h(X_{i_1}^1, X_{i_2}^2, X_{i_3}^3)\Big\|\Big),$$

which is what is needed. Only the quantity

$$\frac{1}{52}\sum_{\substack{j_1,j_2,j_3\le 3\\ \text{two } js \text{ different}}} \mathbb{E}\Phi\Big(52\Big\|\sum_{I_n^3} h(X_{i_1}^{j_1},X_{i_2}^{j_2},X_{i_3}^{j_3})\Big\|\Big)$$

requires additional consideration. A typical term in this sum is

$$\mathbb{E}\Phi\Big(52\Big\|\sum_{I_n^3} h(X_{i_1}^{1},X_{i_2}^{1},X_{i_3}^{3})\Big\|\Big).$$

Conditioning on the variables X_i^3, $i=1,\dots,n$, this term reduces to the case $m=2$ for the kernels

$$g_{i_1i_2}(X_{i_1}^1,X_{i_2}^1)=\sum_{\substack{i_3\le n\\ i_3\ne i_1,i_2}} h(X_{i_1}^{1},X_{i_2}^{1},X_{i_3}^{3}).$$

Applying inequality (3.1.1) for $m=2$ [better, (3.1.8)] yields

$$\begin{aligned}
&\mathbb{E}\Phi\Big(52\Big\|\sum_{I_n^3} h(X_{i_1}^{1},X_{i_2}^{1},X_{i_3}^{3})\Big\|\Big)\\
&\quad=\mathbb{E}\mathbb{E}\Big[\Phi\Big(52\Big\|\sum_{I_n^2} g_{i_1i_2}(X_{i_1}^{1},X_{i_2}^{1})\Big\|\Big)\Big|X_1^3,\dots,X_n^3\Big]\\
&\quad\le\mathbb{E}\mathbb{E}\Big[\Phi\Big(416\Big\|\sum_{I_n^2} g_{i_1i_2}(X_{i_1}^{1},X_{i_2}^{2})\Big\|\Big)\Big|X_1^3,\dots,X_n^3\Big]\\
&\quad=\mathbb{E}\Phi\Big(416\Big\|\sum_{I_n^3} h(X_{i_1}^{1},X_{i_2}^{2},X_{i_3}^{3})\Big\|\Big).
\end{aligned}$$

Collecting terms yields (3.1.1) for $m=3$ (with a slightly better constant, namely, $2C_m/3$). □

There are other (in appearance simpler) ways to perform the main conditioning in the proof of the previous theorem. For instance, instead of introducing the new variables Z_i, Z_i' (this discussion is done only for $m=2$ but is valid for any m), one may just take the σ-algebra $\tilde{\mathcal{Z}}$ generated by the random variables

$$h_{ij}(X_i,X_j)+h_{ij}(X_i,X_j')+h_{ij}(X_i',X_j)+h_{ij}(X_i',X_j'),\qquad i\ne j=1,\dots,n,$$

and observe that

$$\begin{aligned}
\mathbb{E}(h_{ij}(X_i,X_j)|\tilde{\mathcal{Z}})&=\mathbb{E}(h_{ij}(X_i,X_j')|\tilde{\mathcal{Z}})=\mathbb{E}(h_{ij}(X_i',X_j)|\tilde{\mathcal{Z}})=\mathbb{E}(h_{ij}(X_i',X_j')|\tilde{\mathcal{Z}})\\
&=\frac{1}{4}\Big[h_{ij}(X_i,X_j)+h_{ij}(X_i,X_j')+h_{ij}(X_i',X_j)+h_{ij}(X_i',X_j')\Big].
\end{aligned}$$

These last identities can be used in the previous proof in place of the the identities (3.1.6), with only formal changes. However the conditioning used in the proof of Theorem 3.1.1 has at least two advantages. One is that it will also be used for decoupling tail probabilities. The other advantage, more important, is that if we take the variables X_i^j, ε_i to be the coordinate variables in the product probability space

$$\Big(S^{2n} \times \{-1, 1\}^n, \mathcal{S}^{2n} \otimes 2^{\{-1,1\}^n}, (P_1 \times \cdots \times P_n)^2 \times (\mathcal{L}(\varepsilon_1))^n\Big),$$

then all the conditional expectations used in the above proof become iterated integrals with respect to coordinates in a product probability space, and these are simpler objects than general conditional expectations. Obviously, in the situation of Theorem 3.1.1, i.e., for the B-valued kernels with B separable (hence, with a measurable norm), there is no loss of generality in assuming that the variables X_i^j are coordinates on the product space just described, although this assumption is not needed in this case. What is interesting is that this extra assumption allows extension of Theorem 3.1.1 to Banach spaces B which are not necessarily separable (hence, with not necessarily measurable norm), a situation that arises with U-processes. We briefly summarize in the next paragraph the necessary background for the nonseparable case.

This requires introducing the concepts of *measurable envelope* and *outer integral* of a nonmeasurable function. Given a probability space (Ω, Σ, μ) and $g : \Omega \mapsto [-\infty, +\infty]$ not necessarily measurable, the measurable envelope g^* of g is defined as the a.s. unique function $g^* \geq g$ such that if $h \geq g$ a.s. is measurable, then $g^* \leq h$ a.s. We refer to Dudley and Philipp (1983) or to van der Vaart and Wellner (1996) for existence as well as properties of measurable envelopes. The outer integral of g, $\mathbb{E}^*g$, is defined as

$$\mathbb{E}^*g = \inf\{\mathbb{E}h : h \geq g, h \text{ integrable}\}$$

(as usual, we take the infimum of the null set to be infinite). If $-\infty < \mathbb{E}g^* < \infty$, it is easy to see that $\mathbb{E}^*g = \mathbb{E}g^*$. Regarding the calculus of nonmeasurable functions, we further require the following one-sided version of Fubini's theorem, which follows immediately from the definition: If $\mathbb{E}_1$, $\mathbb{E}_2$ denote partial integation in a product probability space, then

$$\mathbb{E}_1^*(\mathbb{E}_2^*g) \leq (\mathbb{E}_1 \times \mathbb{E}_2)^*g. \tag{3.1.11}$$

Our final requirement on this subject is Jensen's inequality in the nonmeasurable setting that we will encounter in the next theorem. If $\Phi : [0, \infty) \to [0, \infty)$ is convex and nondecreasing and if, for a possibly *uncountable* index set A, f_α, $\alpha \in A$, are measurable functions, then

$$\Phi\big(\sup_{\alpha \in A} |\mathbb{E}f_\alpha|\big) \leq \mathbb{E}^*\Phi\big(\sup_{\alpha \in A} |f_\alpha|\big). \tag{3.1.12}$$

The proof follows by expressing Φ as the supremum of the supporting lines of its graph, $\Phi(t) = \sup_{\lambda \geq 0}(a_\lambda + b_\lambda t)$, since then (note the slopes b_λ are non-negative)

$$\begin{aligned}
\Phi\big(\sup_{\alpha \in A} |\mathbb{E}f_\alpha|\big) &= \sup_\lambda \sup_\alpha (a_\lambda + b_\lambda |\mathbb{E}f_\alpha|) \\
&\leq \mathbb{E}^* \sup_\lambda \sup_\alpha (a_\lambda + b_\lambda |f_\alpha|) \\
&= \mathbb{E}^*\Phi\big(\sup_{\alpha \in A} |f_\alpha|\big).
\end{aligned}$$

In the theorem that follows, we consider collections $\mathcal{H}_{i_1\dots i_m}$ of measurable functions $h : S^m \mapsto \mathbb{R}$ (these functions can also be Banach space valued, but this would not actually change the level of generality of the result) and it is convenient to have the following definition: an envelope (or a measurable envelope) of a class of functions $\mathcal{H}_{i_1\dots i_m}$ is any measurable function $H_{i_1\dots i_m}$ such that $\sup_{h\in\mathcal{H}_{i_1\dots i_m}} |h(x_1,\dots,x_m)| \le H_{i_1\dots i_m}(x_1,\dots,x_m)$ for all $x_1,\dots,x_m \in S$. All the classes of functions considered here will have everywhere finite envelope. The version of Theorem 3.1.1 for U-processes is as follows:

THEOREM 3.1.2. *For natural numbers $n \ge m$, let X_i, X_i^k, $i = 1,\dots,n$, $k = 1,\dots,m$, be the coordinate functions of the product probability space $\big((S^n)^{m+1}, (\mathcal{S}^n)^{(m+1)}, (P_1\times\cdots\times P_n)^{m+1}\big)$, in particular the variables $\{X_i\}_{i=1}^n$ are independent S-valued random variables, X_i with probability law P_i, $i \le n$, and the sequences $\{X_i^k\}_{i=1}^n$, $k \le m$, are i.i.d. copies of the sequence $\{X_i\}_{i=1}^n$. For each $(i_1,\dots,i_m) \in I_n^m$, let $\mathcal{H}_{i_1\dots i_m}$ be a collection of measurable functions $h_{i_1\dots i_m} : S^m \mapsto \mathbb{R}$ admitting an everywhere finite measurable envelope $H_{i_1\dots i_m}$ such that $\mathbb{E}H_{i_1\dots i_m}(X_{i_1},\dots,X_{i_m}) < \infty$. Let $\Phi : [0,\infty) \to [0,\infty)$ be a convex nondecreasing function such that $\mathbb{E}\Phi\big(H_{i_1\dots i_m}(X_{i_1},\dots,X_{i_m})\big) < \infty$ for all $(i_1,\dots,i_m) \in I_n^m$. Then*

$$\mathbb{E}^*\Phi\Big(\sup_{h_{i_1\dots i_m}\in\mathcal{H}_{i_1\dots i_m}} \Big|\sum_{I_n^m} h_{i_1\dots i_m}(X_{i_1},\dots,X_{i_m})\Big|\Big)$$
$$\le \mathbb{E}^*\Phi\Big(C_m \sup_{h_{i_1\dots i_m}\in\mathcal{H}_{i_1\dots i_m}} \Big|\sum_{I_n^m} h_{i_1\dots i_m}(X_{i_1}^1,\dots,X_{i_m}^m)\Big|\Big), \qquad (3.1.13)$$

where $C_m = 2^m(m^m-1)((m-1)^{(m-1)}-1)\times\cdots\times 3$. If, moreover, the classes $\mathcal{H}_{i_1\dots i_m}$ satisfy that for all $h_{i_1\dots i_m} \in \mathcal{H}_{i_1\dots i_m}$, $x_1,\dots,x_m \in S$ and permutations s of $\{1,\dots,m\}$,

$$h_{i_1\dots i_m}(x_1,\dots,x_m) = h_{i_{s_1}\dots i_{s_m}}(x_{s_1},\dots,x_{s_m}), \qquad (3.1.14)$$

then

$$\mathbb{E}^*\Phi\Big(\frac{1}{2^{(2m-2)}(m-1)!} \sup_{h_{i_1\dots i_m}\in\mathcal{H}_{i_1\dots i_m}} \Big|\sum_{I_n^m} h_{i_1\dots i_m}(X_{i_1}^1,\dots,X_{i_m}^m)\Big|\Big)$$
$$\le \mathbb{E}^*\Phi\Big(\sup_{h_{i_1\dots i_m}\in\mathcal{H}_{i_1\dots i_m}} \Big|\sum_{I_n^m} h_{i_1\dots i_m}(X_{i_1},\dots,X_{i_m})\Big|\Big). \qquad (3.1.15)$$

The proof is omitted because it follows with only formal changes from the proof of Theorem 3.1.1: e.g., for $m = 2$, enlarge the probability space to

$$\Big(S^{2n}\times\{-1,1\}^n, \mathcal{S}^{2n}\otimes 2^{\{-1,1\}^n}, (P_1\times\cdots\times P_n)^2\times(\mathcal{L}(\varepsilon_1))^n\Big), \qquad (3.1.16)$$

redefine the X and X' variables in the natural way as the S-coordinates, and define the variables ε_i as the $\{-1,1\}$ coordinates; then, as mentioned above, all the conditional expectations in the previous proof become integrals with respect to a number of coordinates in this product probability space and, owing to equations (3.1.11) and (3.1.12), all the arguments in the proof are valid if $\mathbb{E}$ is replaced by $\mathbb{E}^*$ at the obvious places.

It is worth noting that the symmetry conditions (3.1.2) and (3.1.14) are not superfluous: for instance, in the extreme case $h(x, y) = x - y$ the right side of equation (3.1.3) [and (3.1.15)] is zero whereas the left side may not vanish.

When choosing the generality in which to state Theorems 3.1.1 and 3.1.2, we tried to reach an equilibrium between simplicity of statements and readiness for use. There is thus the need to describe some apparent extensions to situations that we will encounter in Chapters 4 and 5.

1) Since the constants in the inequalities of these theorems do not depend on n, it follows that these inequalities also hold for $n = \infty$, provided the quantities at their right sides converge as $n \to \infty$.

2) The functions $h_{i_1 \ldots i_m} \in \mathcal{H}_{i_1 \ldots i_m}$ in Theorem 3.1.2 can take values in a not necessarily separable Banach space, instead of being just real valued. In order for Theorem 3.1.2 to hold in this situation, the only extra assumption to be made is that the random variables $x^*\big(h_{i_1 \ldots i_m}(X_1, \ldots, X_m)\big)$ be measurable for all x^* in the dual space B^* of B and, of course, absolute values should be replaced by norms both in the definition of the envelopes $H_{i_1 \ldots i_m}$ and in inequalities (3.1.12) and (3.1.15). This seemingly more general result is just Theorem 3.1.2 applied to the classes of real functions $\tilde{\mathcal{H}}_{i_1 \ldots i_m} = \{x^* \circ h : x^* \in B^*, \|x^*\| \le 1, h \in \mathcal{H}_{i_1 \ldots i_m}\}$.

3) The classes $\mathcal{H}$ may also be made dependent on n and then $\sup_{h_{i_1 \ldots i_m} \in \mathcal{H}_{i_1 \ldots i_m}}$ be replaced by $\sup_{n \in A} \sup_{h^n_{i_1 \ldots i_m} \in \mathcal{H}^n_{i_1 \ldots i_m}}$ for any finite set A of natural numbers (or even for infinite sets A if the suprema at the right of the inequalities exist). This can be seen by either checking the steps in the proof or by directly applying the theorem to suitably redefined classes of functions $\mathcal{H}$. We illustrate the technique of redefining the classes $\mathcal{H}$ with a simple example that will be of use in the next chapter. We show how Theorem 3.1.1 implies that for $a_n \to \infty$ and $h(x, y)$ symmetric in its entries and such that $\mathbb{E}|h(X_1, X_2)|^p < \infty$ for some $p \ge 1$,

$$\mathbb{E} \sup_{n \in \mathbb{N}} \frac{1}{a_n} \Big|\sum_{I_n^2} h(X_i, X_j)\Big|^p \approx \mathbb{E} \sup_{n \in \mathbb{N}} \frac{1}{a_n} \Big|\sum_{I_n^2} h(X_i, X'_j)\Big|^p \tag{3.1.17}$$

(where we take the equivalence in the sense that there is inequality in both directions up to multiplicative constants). For $k \in \mathbb{N}$ define new functions $\tilde{h}_k$ taking values in c_0, the space of real sequences that converge to zero, equipped with the sup norm, as follows:

$$\tilde{h}_k = \Big(0, \overset{(k-1)}{\ldots}, 0, \frac{h}{a_k}, \frac{h}{a_{k+1}}, \ldots\Big), \qquad k \ge 1,$$

(i.e., the first $k - 1$ entries of the vector $\tilde{h}_k$ are zero and the ℓth entry is h/a_ℓ for all $\ell \ge k$). Then, it is obvious that

$$\sup_{n \in \mathbb{N}} \frac{1}{a_n} \Big|\sum_{I_n^2} h(X_i, X_j)\Big| = \Big\| \sum_{\substack{i, j < \infty \\ i \ne j}} \tilde{h}_{i \vee j}(X_i, X_j) \Big\|$$

and that the corresponding decoupled identity is also true. So, in view of the previous comments 1) and 2), Theorem 3.1.1 (and also Theorem 3.1.2) applies and gives the equivalence (3.1.17). Even more, suppose $a_n \to \infty$ as above, but that h, symmetric, is not real valued but instead now takes values in a not necessarily separable Banach space B. Suppose h satisfies that $x^* \circ h(X_1, X_2)$ is measurable for all $x^* \in B^*$ and that $\mathbb{E}^*\|h\|^p < \infty$. Then the functions $\tilde{h}_{i\vee j}$ can still be formally defined as above, but now they take values in the nonseparable Banach space $c_0(B)$ of sequences of elements in B converging to zero. We cannot apply Theorem 3.1.1 but Theorem 3.1.2 certainly applies [again, by comments 1) and 2)] and gives

$$\mathbb{E}^* \sup_{n\in\mathbb{N}} \frac{1}{a_n} \Big\|\sum_{I_n^2} h(X_i, X_j)\Big\|^p \approx \mathbb{E}^* \sup_{n\in\mathbb{N}} \frac{1}{a_n} \Big\|\sum_{I_n^2} h(X_i, X_j')\Big\|^p. \tag{3.1.17$'$}$$

These observations are useful in the proof of the law of large numbers, central limit theorem, and law of the iterated logarithm for U-processes, as we see in Chapter 5.

In order to illustrate the usefulness of Theorem 3.1.1 we conclude with the following example, which deals with the simplest of kernels, $h(x, y) = xy$, in the simplest Banach space, $\mathbb{R}$.

EXAMPLE 3.1.3. We will prove the following: *Let X_i be a finite collection of centered independent identically distributed random variables. Then, there are universal constants c_1, c_2 such that*

$$c_1\Big(\mathbb{E}\Big|\sum_i X_i\Big|\Big)^2 \le \mathbb{E}\Big|\sum_{i\neq j} X_iX_j\Big| \le c_2\Big(\mathbb{E}\Big|\sum_i X_i\Big|\Big)^2, \tag{3.1.18}$$

thus reducing the L_1 norm of a quadratic form in independent variables to the L_1 norm of a sum. For the right-hand side inequality we recall that, by the Marcinkiewicz inequalities (Lemma 1.4.13),

$$\mathbb{E}\Big(\sum X_i^2\Big)^{1/2} \le \sqrt{2}\mathbb{E}\Big|\sum \varepsilon_i X_i\Big| \le 2\sqrt{2}\mathbb{E}\Big|\sum X_i\Big|.$$

Then, introducing an independent copy $\{X_i'\}$ of $\{X_i\}$ and applying Theorem 3.1.1 [actually, (3.1.8)] at the first step and Cauchy–Schwarz at the third, we have

$$\begin{aligned}
\mathbb{E}\Big|\sum_{i\neq j} X_iX_j\Big| &\le 8\mathbb{E}\Big|\sum_{i\neq j} X_iX_j'\Big| = 8\mathbb{E}\Big|\sum_{i,j} X_iX_j' - \sum_i X_iX_i'\Big| \\
&\le 8\mathbb{E}\Big|\sum_{i,j} X_iX_j'\Big| + 8\mathbb{E}\Big|\sum_i X_iX_i'\Big| \\
&\le 8\mathbb{E}\Big[\Big|\sum_i X_i\Big|\Big|\sum_i X_i'\Big|\Big] + 8\mathbb{E}\Big[\big(\sum_i X_i^2\big)^{1/2}\big(\sum_i X_i'^2\big)^{1/2}\Big] \\
&= 8\big(\mathbb{E}\Big|\sum_i X_i\Big|\big)^2 + 8\big[\mathbb{E}\big(\sum_i X_i^2\big)^{1/2}\big]^2 \le 72\big(\mathbb{E}\Big|\sum_i X_i\Big|\big)^2.
\end{aligned}$$

This argument works even if the variables are not identically distributed. For the converse, letting $S_n' = \sum_{i=1}^n X_i'$, and observing that $\mathbb{E}(X_j'|S_n') = S_n'/n$ (note that this does not hold if the X_i' variables are not i.i.d.), and using conditional Jensen and again Theorem 3.1.1 [inequality (3.1.7)], we have

$$\begin{aligned}\mathbb{E}\Big|\sum_{i\neq j\leq n} X_iX_j\Big| &\geq \frac{1}{4}\mathbb{E}\Big|\sum_{i\neq j\leq n} X_iX_j'\Big| \geq \frac{1}{4}\mathbb{E}\Big|\sum_{i\neq j\leq n} X_i\mathbb{E}(X_j'|S_n')\Big| \\ &= \frac{1}{4}\mathbb{E}\Big|\sum_{i\neq j} X_iS_n'/n\Big| = \frac{1}{4}\frac{n-1}{n}\mathbb{E}\Big|\sum_{i=1}^n X_i\Big|\mathbb{E}|S_n'| \\ &= \frac{1}{4}\frac{n-1}{n}\Big(\mathbb{E}\Big|\sum_{i=1}^n X_i\Big|\Big)^2,\end{aligned}$$

concluding the proof of inequality (3.1.18).

3.2 Hypercontractivity of Rademacher Chaos Variables

Let F be a normed vector space and let $\{\varepsilon_i\}_{i=1}^\infty$ be a Rademacher sequence (that is, the variables ε_i are i.i.d. and $\Pr\{\varepsilon_i = 1\} = \Pr\{\varepsilon_i = -1\} = 1/2$). An *$F$-valued Rademacher chaos of order d, X*, is a polynomial of order d in the Rademacher variables ε_i and with coefficients in F, i.e.,

$$X = x + \sum_{i=1}^n x_i\varepsilon_i + \sum_{i_1<i_2\leq n} x_{i_1i_2}\varepsilon_{i_1}\varepsilon_{i_2} + \cdots + \sum_{i_1<\cdots<i_d\leq n} x_{i_1\ldots i_d}\varepsilon_{i_1}\cdots\varepsilon_{i_d}, \quad (3.2.1)$$

with $n < \infty$ and $x, x_i, \ldots, x_{i_1\ldots i_d} \in F$. X is a homogeneous Rademacher chaos of order d if only the monomials of degree d in the variables ε_i in (3.2.1) are not zero. The Khinchin–Kahane inequalities for Rademacher sums introduced in Section 1.3,

$$\Big(\mathbb{E}\Big\|\sum x_i\varepsilon_i\Big\|^q\Big)^{\frac{1}{q}} \leq C_{p,q}\Big(\mathbb{E}\Big\|\sum x_i\varepsilon_i\Big\|^p\Big)^{\frac{1}{p}}, \; 0 < p < q < \infty, \quad (3.2.2)$$

with $C_{p,q}$ depending only on p and q, extend to the polynomials (3.2.1), and this fact has consequences for integrability of chaoses (Gaussian as well as Rademacher), decoupling in general, asymptotics of U-statistics and processes, etc. We should mention that the proof of the general decoupling theorem of Section 3.4 uses only hypercontractivity of *real* Rademacher chaos variables, in fact, only equation (3.2.12) below, and this is easier to prove than the more refined and more general result given as Theorem 3.2.1. Here is the main result of this section.

THEOREM 3.2.1. *Let $1 < p \leq q < \infty$ and let*

$$\gamma = \Big(\frac{p-1}{q-1}\Big)^{\frac{1}{2}}.$$

Then, for all $d \geq 1$,

$$\Big[\mathbb{E}\Big\|x+\gamma\sum_{i=1}^{n}x_i\varepsilon_i+\gamma^2\sum_{i_1<i_2\leq n}x_{i_1i_2}\varepsilon_{i_1}\varepsilon_{i_2}+\cdots+\gamma^d\sum_{i_1<\cdots<i_d\leq n}x_{i_1\ldots i_d}\varepsilon_{i_1}\cdots\varepsilon_{i_d}\Big\|^q\Big]^{\frac{1}{q}}$$
$$\leq\Big[\mathbb{E}\Big\|x+\sum_{i=1}^{n}x_i\varepsilon_i+\sum_{i_1<i_2\leq n}x_{i_1i_2}\varepsilon_{i_1}\varepsilon_{i_2}+\cdots+\sum_{i_1<\cdots<i_d\leq n}x_{i_1\ldots i_d}\varepsilon_{i_1}\cdots\varepsilon_{i_d}\Big\|^p\Big]^{\frac{1}{p}}. \tag{3.2.3}$$

PROOF. The proof is by induction on the number n of different Rademacher variables present in (3.2.3), and the basic step is the case $n = 1$.

1) *Proof of*

$$\big(\mathbb{E}|x+\gamma\varepsilon y|^q\big)^{\frac{1}{q}}\leq\big(\mathbb{E}|x+\varepsilon y|^p\big)^{\frac{1}{p}},\qquad x,y\in\mathbb{R}. \tag{3.2.4}$$

(Here ε denotes a Rademacher variable.) There is nothing to prove if either $x = 0$ or $y = 0$, so we can assume that neither x nor y are zero. Letting $u = y/x$, inequality (3.2.4) becomes

$$\Big[\frac{|1+\gamma u|^q+|1-\gamma u|^q}{2}\Big]^{\frac{1}{q}}\leq\Big[\frac{|1+u|^p+|1-u|^p}{2}\Big]^{\frac{1}{p}}. \tag{3.2.4$'$}$$

Consider first the case $1 < p \leq q \leq 2$ and $|u| \leq 1$ (which reduces by symmetry to $0 < u \leq 1$). Then, using the facts that $\binom{q}{2k}\gamma^2 \leq \frac{q}{p}\binom{p}{2k}$, that $(1+a)^\alpha \leq 1+\alpha a$ for $a \geq 0$ and $0 \leq \alpha \leq 1$, and that $\gamma \leq 1$, we have

$$\begin{aligned}\Big[\frac{|1+\gamma u|^q+|1-\gamma u|^q}{2}\Big]^{\frac{p}{q}} &= \Big[1+\sum_{k=1}^{\infty}\binom{q}{2k}\gamma^{2k}u^{2k}\Big]^{\frac{p}{q}}\\ &\leq\Big[1+\sum_{k=1}^{\infty}\frac{q}{p}\binom{p}{2k}u^{2k}\Big]^{\frac{p}{q}}\\ &\leq 1+\sum_{k=1}^{\infty}\binom{p}{2k}u^{2k}\\ &=\frac{(1+u)^p+(1-u)^p}{2}.\end{aligned}$$

The case $u \geq 1$ follows from the case $u \leq 1$ applied to $1/u$, because, for all $u \geq 1$, $|1\pm\gamma u| \leq |u\pm\gamma| = u|1\pm\gamma/u|$. The proof for $2 \leq p \leq q < \infty$ will follow by duality: On a two-point set, say $A = \{-1, 1\}$, consider the uniform measure $\lambda\{-1\} = \lambda\{1\} = 1/2$, and let id_A be the identity function on A. Consider the map $T : L_p(A,\lambda) \mapsto L_q(A,\lambda)$, $1 < p \leq q \leq 2$ given by

$$Tf=\int f d\lambda+\gamma\Big(\int f\cdot id_A d\lambda\Big)id_A.$$

The inequality just proved is equivalent to the fact that T is a contraction, i.e., $\|T\| \leq 1$ [take $f(1) = 1+u$, $f(-1) = 1-u$]. Then, the adjoint operator is also a contraction and this implies inequality (3.2.4′) for $2 \le p \le q < \infty$. The case $1 < p < 2 < q < \infty$ follows from the previous two cases by applying the inequality successively for q and 2, and for 2 and p.

2) *Proof of*

$$\left(\mathbb{E}\|x + \gamma\varepsilon y\|^q\right)^{\frac{1}{q}} \le \left(\mathbb{E}\|x + \varepsilon y\|^p\right)^{\frac{1}{p}}, \qquad x, y \in F. \tag{3.2.5}$$

Taking

$$z_1 = x + y, \quad z_2 = x - y, \quad u = \frac{\|z_1\| + \|z_2\|}{2}, \quad v = \frac{\|z_1\| - \|z_2\|}{2},$$

we obtain that, by convexity and inequality (3.2.4) [or, what is the same, (3.2.4′)],

$$\begin{aligned}
&\left[\frac{\|x + \gamma y\|^q + \|x - \gamma y\|^q}{2}\right]^{\frac{1}{q}} \\
&\quad \le \left[\frac{\left(\frac{1+\gamma}{2}\|z_1\| + \frac{1-\gamma}{2}\|z_2\|\right)^q + \left(\frac{1-\gamma}{2}\|z_1\| + \frac{1+\gamma}{2}\|z_2\|\right)^q}{2}\right]^{\frac{1}{q}} \\
&\quad = \left[\frac{|u + \gamma v|^q + |u - \gamma v|^q}{2}\right]^{\frac{1}{q}} \\
&\quad \le \left[\frac{|u + v|^p + |u - v|^p}{2}\right]^{\frac{1}{p}} \\
&\quad = \left[\frac{\|x + y\|^p + \|x - y\|^p}{2}\right]^{\frac{1}{p}},
\end{aligned}$$

proving (3.2.5).

3) *The induction.* To avoid notational complexity, we prove only that the theorem for $n = 1$, that is, inequality (3.2.5), implies the theorem for $n = 2$, that is,

$$\left[\mathbb{E}\left\|x + \gamma\sum_{i=1}^{2} x_i\varepsilon_i + \gamma^2 x_{12}\varepsilon_1\varepsilon_2\right\|^q\right]^{\frac{1}{q}} \le \left[\mathbb{E}\left\|x + \sum_{i=1}^{2} x_i\varepsilon_i + x_{12}\varepsilon_1\varepsilon_2\right\|^p\right]^{\frac{1}{p}}. \tag{3.2.6}$$

We should recall the generalized Minkowski's inequality, which is the main new ingredient in this part of the proof. If X and Y are independent and $f(X, Y) \in F$ is such that $\|f(X, Y)\|$ is measurable, we set $\|f\|_{r,X} = \left(\int \|f(x, y)\|^r d(\mathcal{L}(X))(x)\right)^{1/r}$ for $r \ge 1$, where $\mathcal{L}(X)$ denotes the probability law of X. $\|f\|_{r,Y}$ is defined by analogy. Then, the generalized Minkowski's inequality [which is just Minkowski replacing sum by integral: see, e.g., Jones (1993, page 271)] asserts

$$\|\,\|f\|_{1,X}\,\|_{r,Y} \le \|\,\|f\|_{r,Y}\,\|_{1,X}$$

for all $r \geq 1$. Let $\mathbb{E}_1$ (resp. $\mathbb{E}_2$) denote integration with respect to the variable ε_1 only (resp. ε_2 only) that is, $\mathbb{E}_1(\cdot) = \mathbb{E}(\cdot|\varepsilon_2)$, $\mathbb{E}_2(\cdot) = \mathbb{E}(\cdot|\varepsilon_1)$. The proof of (3.2.6) follows at once by application of Fubini (twice), generalized Minkowski with $r = q/p$, and twice the induction hypothesis (in our case, (3.2.5)):

$$\begin{aligned}
\Big[\mathbb{E}\Big\|x + \gamma\sum_{i=1}^{2} x_i\varepsilon_i + \gamma^2 x_{12}\varepsilon_1\varepsilon_2\Big\|^q\Big]^{\frac{1}{q}}
&= \Big[\mathbb{E}_1\mathbb{E}_2\Big\|x + \gamma\sum_{i=1}^{2} x_i\varepsilon_i + \gamma^2 x_{12}\varepsilon_1\varepsilon_2\Big\|^q\Big]^{\frac{1}{q}} \\
&\leq \Big[\mathbb{E}_1\big(\mathbb{E}_2\big\|x + \gamma x_1\varepsilon_1 + x_2\varepsilon_2 + \gamma x_{12}\varepsilon_1\varepsilon_2\big\|^p\big)^{\frac{q}{p}}\Big]^{\frac{1}{q}} \\
&\leq \Big[\mathbb{E}_2\big(\mathbb{E}_1\big\|x + \gamma x_1\varepsilon_1 + x_2\varepsilon_2 + \gamma x_{12}\varepsilon_1\varepsilon_2\big\|^q\big)^{\frac{p}{q}}\Big]^{\frac{1}{p}} \\
&\leq \Big[\mathbb{E}_2\mathbb{E}_1\big\|x + x_1\varepsilon_1 + x_2\varepsilon_2 + x_{12}\varepsilon_1\varepsilon_2\big\|^p\Big]^{\frac{1}{p}} \\
&= \Big[\mathbb{E}\big\|x + x_1\varepsilon_1 + x_2\varepsilon_2 + x_{12}\varepsilon_1\varepsilon_2\big\|^p\Big]^{\frac{1}{p}}.
\end{aligned}$$

□

Inequality (3.2.5) is referred to as the hypercontractivity property of Rademacher variables (it indicates that certain operators between L_p and L_q spaces are contractions). Inequality (3.2.3) shows that Rademacher products and, in general, homogeneous Rademacher polynomials, are also hypercontractive (with γ replaced by γ^d).

Theorem 3.2.1 takes a particularly simple form for homogeneous chaos variables:

THEOREM 3.2.2. *Let X be an F-valued homogeneous chaos of order d, $X = \sum_{i_1<\cdots<i_d\leq n} x_{i_1\ldots i_d}\varepsilon_{i_1}\cdots\varepsilon_{i_d}$, $x_{i_1,\ldots,i_d} \in F$, and let $1 < p < q < \infty$. Then the following holds:*

$$\big(\mathbb{E}\|X\|^q\big)^{\frac{1}{q}} \leq \Big(\frac{q-1}{p-1}\Big)^{\frac{d}{2}}\big(\mathbb{E}\|X\|^p\big)^{\frac{1}{p}}. \tag{3.2.7}$$

Theorem 3.2.2 reduces to Theorem 1.3.1 for $d = 1$.

Inequality (3.2.7), with a different constant at the right-hand side, extends to $0 < p < q < \infty$ (even to $p = 0$ as we will see in the next section). For instance, here is how to obtain it for $p \leq 1$ and $q = 2$ using Hölder's inequality, as in Littlewood's proof of Khinchin's inequality, and Theorem 3.2.2: for $0 < \alpha < 1$,

$$\begin{aligned}
\mathbb{E}\|X\|^2 &= \mathbb{E}\big(\|X\|^{\alpha p}\|X\|^{2-\alpha p}\big) \\
&\leq \big(\mathbb{E}\|X\|^p\big)^{\alpha}\big(\mathbb{E}\|X\|^{\frac{2-\alpha p}{1-\alpha}}\big)^{1-\alpha} \\
&\leq \big(\mathbb{E}\|X\|^p\big)^{\alpha}\Big(\frac{1+\alpha(1-p)}{1-\alpha}\Big)^{\frac{d(2-\alpha p)}{2}}\big(\mathbb{E}\|X\|^2\big)^{\frac{2-\alpha p}{2}}.
\end{aligned}$$

This gives

$$\left(\mathbb{E}\|X\|^2\right)^{\frac{1}{2}} \le \left(\frac{1+\alpha(1-p)}{1-\alpha}\right)^{\frac{d(2-\alpha p)}{2\alpha p}} \left(\mathbb{E}\|X\|^p\right)^{\frac{1}{p}}$$

and, letting $\alpha \to 0$,

$$\left(E\|X\|^2\right)^{\frac{1}{2}} \le e^{\frac{(2-p)d}{p}} \left(E\|X\|^p\right)^{\frac{1}{p}}. \tag{3.2.8}$$

The constant is not the best possible; for instance, it is not for $d = 1$ and $p = 1$ (Theorem 1.3.2).

Extending inequalities (3.2.7) and (3.2.8) to not necessarily homogeneous Rademacher chaos variables of order d requires some interesting arguments, namely, a sort of *contraction principle* for chaoses and comparison of the moments of a general chaos with those of its homogeneous components. This is the content of the next two lemmas.

For the general Rademacher chaos X of order d, as in (3.2.1), and for $t \in \mathbb{R}$, define $X(t)$ to be the random variable obtained from X by replacing ε_i with $t\varepsilon_i$, $i \le n$, that is,

$$X(t) = x + t\sum_{i=1}^{n} x_i\varepsilon_i + t^2 \sum_{i_1<i_2\le n} x_{i_1i_2}\varepsilon_{i_1}\varepsilon_{i_2} + \cdots + t^d \sum_{i_1<\cdots<i_d\le n} x_{i_1\ldots i_d}\varepsilon_{i_1}\cdots\varepsilon_{i_d}. \tag{3.2.9}$$

LEMMA 3.2.3. *Let X be a F-valued Rademacher chaos of order d, as in (3.2.1), and let $X(t)$ be defined by (3.2.9). Let $\phi : F \mapsto \mathbb{R}$ be a convex function. Then, for $|t| \le 1$,*

$$\mathbb{E}\phi(X(t)) \le \mathbb{E}\phi(X). \tag{3.2.10}$$

PROOF. For $x, y \in F$, the function $h(t) = \mathbb{E}\phi(x + ty\varepsilon) = [\phi(x + ty) + \phi(x - ty)]/2$, is convex and symmetric, and therefore, nondecreasing on $[0, \infty)$. Thus,

$$\mathbb{E}\phi(x + ty\varepsilon) \le \mathbb{E}\phi(x + y\varepsilon), \ |t| \le 1. \tag{3.2.10'}$$

In the general case, conditionally on the set of variables $\{\varepsilon_i : i \ne n\}$, $X(t)$ has the form $X(t) = x' + ty'\varepsilon_n$ for some $x', y' \in F$. This observation and Fubini make it possible to obtain (3.2.10) from (3.2.10′) by induction. Here is how to proceed for $n = 2$:

$$\begin{aligned}
\mathbb{E}\phi(x + t\sum_{i=1}^{2} x_i\varepsilon_i + t^2x_{12}\varepsilon_1\varepsilon_2) &= \mathbb{E}_1\big[\mathbb{E}_2\phi(x + t\sum_{i=1}^{2} x_i\varepsilon_i + t^2x_{12}\varepsilon_1\varepsilon_2)\big]\\
&\le \mathbb{E}_1\big[\mathbb{E}_2\phi(x + tx_1\varepsilon_1 + x_2\varepsilon_2 + tx_{12}\varepsilon_2\varepsilon_1)\big]\\
&= \mathbb{E}_2\big[\mathbb{E}_1\phi(x + tx_1\varepsilon_1 + x_2\varepsilon_2 + tx_{12}\varepsilon_2\varepsilon_1)\big]\\
&\le \mathbb{E}_2\mathbb{E}_1\phi(x + x_1\varepsilon_1 + x_2\varepsilon_2 + x_{12}\varepsilon_1\varepsilon_2). \qquad \square
\end{aligned}$$

General Rademacher chaos variables X of order d uniquely decompose into the sum of $d + 1$ homogeneous Rademacher chaos variables respectively of orders

$0, 1, \ldots, d$. The following lemma shows that the moments of these components are dominated by (some contants times) the moments of X.

LEMMA 3.2.4. *Let X be an F-valued Rademacher chaos of order d, as in (3.2.1), and let X_k be its homogeneous component of degree k, $0 \le k \le d$. Let $\phi : F \mapsto \mathbb{R}$ be a non-negative symmetric convex function. Then, there exists a universal constant K_d such that*

$$\mathbb{E}\phi(X_k) \le \mathbb{E}\phi(K_d X), \ 0 \le k \le d. \tag{3.2.11}$$

PROOF. We have $X = \sum_{k=0}^{d} X_k$ and $X(t) = \sum_{k=0}^{d} t^k X_k$. The computation that will give this lemma uses the following obvious properties of the function ϕ, that follow by convexity and symmetry: $\phi(tx)$ is nondecreasing in $t \in [0, \infty)$ for all $x \in F$; for any finite set of points $x_i \in F$ and real numbers λ_i with $\Lambda := \sum |\lambda_i| > 0$, we have

$$\phi\left(\sum \lambda_i x_i\right) \le \Lambda^{-1} \sum |\lambda_i| \phi(\Lambda x_i);$$

and the restriction of ϕ to any finite-dimensional subspace of F, in particular to the subspace generated by the vectors $x_{i_1 \ldots i_k}$ defining X_k, $k = 0, \ldots, d$, is continuous.

Since the functions $1, t, \ldots, t^d$, are linearly independent, there exist polynomials g_k, $0 \le k \le d$, such that $\int_{-1}^{1} t^r g_k(t)dt = 0$ for $r \ne k$ and $\int_{-1}^{1} t^k g_k(t)dt = 1$. Let $C_k = \int_{-1}^{1} |g_k(t)|dt$ and let $K_d = \max_{k \le d} C_k$.

Then, Lemma 3.2.3 and the previous observations readily give

$$\begin{aligned}
\mathbb{E}\phi(X_k) &= \mathbb{E}\phi\left(\int_{-1}^{1} X_k t^k g_k(t)dt\right) = \mathbb{E}\phi\left(\int_{-1}^{1} \sum_{r=0}^{d} X_r t^r g_k(t)dt\right) \\
&= \mathbb{E}\phi\left(\int_{-1}^{1} X(t) g_k(t)dt\right) \le C_k^{-1} \int_{-1}^{1} \mathbb{E}\phi\big(C_k X(t)\big)|g_k(t)|dt \\
&\le \mathbb{E}\phi\big(C_k X\big) \le \mathbb{E}\phi(K_d X).
\end{aligned}$$ □

Upper bounds for K_d can be easily computed.

Here is the extension of the Khinchin–Kahane inequality to not necessarily homogeneous Rademacher chaos of order d:

THEOREM 3.2.5. *For all $d \in \mathbb{N}$ there exists a constant $C_d \in (0, \infty)$ such that, if X is a (not necessarily homogeneous) Rademacher chaos of order d with values in any normed vector space F and $1 < p < q < \infty$, then*

$$\left(\mathbb{E}\|X\|^q\right)^{\frac{1}{q}} \le C_d \left(\frac{q-1}{p-1}\right)^{d/2} \left(\mathbb{E}\|X\|^p\right)^{\frac{1}{p}}. \tag{3.2.12}$$

PROOF. Lemma 3.2.4 and Theorem 3.2.2 [i.e., inequalities (3.2.7) and (3.2.11)], give that, for $1 < p < q < \infty$,

$$\left(\mathbb{E}\|X\|^q\right)^{\frac{1}{q}} \le \sum_{k=0}^{d} \left(\mathbb{E}\|X_k\|^q\right)^{\frac{1}{q}}$$

$$\le \sum_{k=0}^{d}\Big(\frac{q-1}{p-1}\Big)^{\frac{k}{2}}\big(\mathbb{E}\|X_k\|^p\big)^{\frac{1}{p}}$$

$$\le \Big[\sum_{k=0}^{d} K_d\Big]\Big(\frac{q-1}{p-1}\Big)^{\frac{d}{2}}\big(\mathbb{E}\|X\|^p\big)^{\frac{1}{p}}. \qquad \square$$

Just like (3.2.7), these inequalities also extend to $0 < p < q < \infty$, with different constants (see (3.2.8)).

These inequalities imply exponential integrability properties for Rademacher chaos variables. Let X be a Rademacher chaos of order d. Then, by Hölder and Theorem 3.2.5 there exists a universal constant $L_d < \infty$ such that

$$\mathbb{E}\|X\|^{\frac{2k}{d}} \le (L_d k)^k\big(\mathbb{E}\|X\|^2\big)^{\frac{k}{d}}$$

for all $k \in \mathbb{N}$. Combining this estimate with the expansion of the exponential immediately gives:

COROLLARY 3.2.6. *For every $d \in \mathbb{N}$ and $c > 1$ there exists $\lambda = \lambda(c,d) \in (0,\infty)$ such that, if X is an F-valued (not necessarily homogeneous) Rademacher chaos of order d, then*

$$\mathbb{E}\exp\Big[\Big(\frac{\|X\|}{\lambda\|X\|_2}\Big)^{\frac{2}{d}}\Big] \le c, \tag{3.2.13}$$

where $\|X\|_2 = (\mathbb{E}\|X\|^2)^{1/2}$.

The following consequence of Corollary 3.2.6 is sometimes useful:

COROLLARY 3.2.7. *For every $d \in \mathbb{N}$ and $0 < \alpha < 2/d$, there exist finite positive constants c_1, c_2 such that, if X is as in Corollary 3.2.6, then*

$$\mathbb{E}e^{t\|X\|^\alpha} \le c_1 \exp\Big(c_2(\|X\|_2^\alpha t)^{1/(1-\alpha d/2)}\Big), \quad t > 0. \tag{3.2.14}$$

PROOF (SKETCH). The inequality $|ab| \le p^{-1}|a|^p + q^{-1}|b|^q$ for $p^{-1}+q^{-1} = 1$, gives that for any $c > 0$,

$$t\|X\|^\alpha \le \frac{\alpha d}{2}\Big(\frac{c\|X\|}{\|X\|_2}\Big)^{\frac{2}{d}} + \Big(1-\frac{\alpha d}{2}\Big)\Big(\frac{t\|X\|_2^\alpha}{c^\alpha}\Big)^{1/(1-\alpha d/2)},$$

and (3.2.14) follows from (3.2.13) by taking c sufficiently small. $\square$

REMARK 3.2.8. *Comments on the case $F = \mathbb{R}$.* 1) It should be noted that for $F = \mathbb{R}$, if X is the Rademacher chaos (3.2.1), then

$$\mathbb{E}X^2 = x^2 + \sum x_i^2 + \cdots + \sum x_{i_1\ldots i_d}^2 := s^2(X).$$

So, Theorem 3.2.5 shows that for all $p > 0$ and $d \in \mathbb{N}$ there exist finite positive constants $c_{d,p}, C_{d,p}$, such that, for any real Rademacher chaos X of order d,

$$c_{d,p}s^p(X) \le \mathbb{E}|X|^p \le C_{d,p}s^p(X). \tag{3.2.15}$$

It is worthwhile to observe that this statement has a much more elementary proof than that of Theorem 3.2.5 (but the constants obtained are much worse), as follows. Direct, easy (but cumbersome) computation gives $\mathbb{E}X^4 \le C_{4,d}s^4(X)$ for some $C_{4,d} < \infty$. Then one notes that X^2 is also a Rademacher chaos so that this inequality can be rewritten as $s^2(X^2) \le C_{4,d}s^4(X)$. Then, recurrence gives

$$\mathbb{E}X^{2^k} \le C_{4,d}^{2^{k-1}-1}s^{2^k}(X)$$

[e.g., $\mathbb{E}X^8 = s^2(X^4) \le c_{d,4}s^4(X^2) = C_{d,4}[s^2(X^2)]^2 \le C_{d,4}^3 s^8(X)$]. By Hölder, this implies the right side of (3.2.15). The left side follows from the right side and Hölder, as in the derivation of (3.2.8) above. 2) Khinchin's inequality (3.2.7) for $F = \mathbb{R}$ and $d = 1$, that is, for real linear combinations of Rademacher variables, with constants of the same order $q^{1/2}$ as $q \to \infty$, has also a simple proof, e.g., based on Hoeffding's (1963) subgaussian tail estimate for these variables. By expanding the exponential we have that for all $s > 0$ and $x_i \in \mathbb{R}$,

$$\mathbb{E}\exp\Big(s\sum a_i\varepsilon_i\Big) = \prod \mathbb{E}e^{sa_i\varepsilon_i} = \prod \frac{e^{sa_i} + e^{-sa_i}}{2} \le \exp\Big(s^2\sum a_i^2/2\Big),$$

which, applying Chebyshev's inequality to the random variable $\exp\big(t\sum a_i\varepsilon_i/\sum a_i^2\big)$, gives that, for all $t > 0$,

$$\Pr\Big\{\sum a_i\varepsilon_i > t\Big\} \le \exp\Big(-\frac{t^2}{2\sum a_i^2}\Big), \tag{3.2.16}$$

which is Hoeffding's inequality. Now, integration by parts gives that, for $q \ge 2$,

$$\begin{aligned}\mathbb{E}\Big|\sum a_i\varepsilon_i\Big|^q &= q\int_0^\infty t^{q-1}\Pr\Big\{\Big|\sum a_i\varepsilon_i\Big| > t\Big\}dt\\ &\le 2q\Big(\int_0^\infty u^{q-1}e^{-u^2/2}du\Big)\Big(\sum a_i^2\Big)^{q/2}.\end{aligned}$$

REMARK 3.2.9. *Gaussian chaos.* Let $\mathbf{g} = \{g_i : 1 \le i \le n\}$ be a sequence of independent $N(0, 1)$ random variables. An *F-valued Gaussian chaos variable of degree d* is a polynomial of degree d in the variables g_i, with coefficients in F. Let $Q(\mathbf{g}) := Q(g_1, \ldots, g_n)$ be such a polynomial. Let $\boldsymbol{\epsilon} = \{\varepsilon_{i,k} : i \in \mathbb{N}, k \le d\}$, be i.i.d. Rademacher variables, and let, for a given polynomial Q, $Q_m(\boldsymbol{\epsilon}) = Q\big(\sum_{i=1}^m \varepsilon_{i,1}/\sqrt{m}, \ldots, \sum_{i=1}^m \varepsilon_{i,d}/\sqrt{m}\big)$. Then, we have

i) $\mathcal{L}\big(Q_m(\boldsymbol{\epsilon})\big) \to \mathcal{L}\big(Q(\mathbf{g})\big)$ weakly, and

ii) $\sup_m E\|Q_m(\boldsymbol{\epsilon})\|^p < \infty$ for all $p > 0$.

(i) follows from the central limit theorem in $\mathbb{R}^n$ and (ii) is an obvious consequence of Theorem 3.2.2 (this theorem implies $\sup_m \mathbb{E}\big|\sum_{i=1}^m \varepsilon_i/\sqrt{m}\big|^p < \infty$ for all $p > 0$). Then, by uniform integrability, (i) and (ii) yield that

$$\lim_{m\to\infty}\mathbb{E}\|Q_m(\boldsymbol{\epsilon})\|^p = \mathbb{E}\|Q(\mathbf{g})\|^p \tag{3.2.17}$$

for all $p > 0$. Therefore, the inequalities of Theorem 3.2.5 extend to general Gaussian polynomials of any degree. Whereas all F-valued polynomials in Rademacher variables have the form (3.2.1), F-valued polynomials in Gaussian variables may contain monomials of the form $xg_1^{p_1}\cdots g_n^{p_n}$ with one or more of the exponents p_i larger than 1. In this connection it is convenient to make the following definition: a polynomial Q in the variables $z_1,\ldots,z_n$ is *tetrahedral* if 1 is the largest exponent of any variable z_i occurring in Q. The limit (3.2.17) makes it possible to transfer to Gaussian F-valued tetrahedral polynomials the inequalities in Theorem 3.2.1 and the Lemmas 3.2.3 and 3.2.4. If Q is tetrahedral and homogeneous, then $Q_m(\boldsymbol{\epsilon})$ is a homogeneous Rademacher chaos, and therefore we can extend Theorem 3.2.2 to $Q(\mathbf{g})$. Collecting these observations, we have

THEOREM 3.2.10. *(i) The conclusions of Theorem 3.2.5 and Corollaries 3.2.6 and 3.2.7 are valid for general F-valued Gaussian chaos variables $Q(\mathbf{g})$ of order d, with the same constants; (ii) Theorem 3.2.1 and Lemmas 3.2.3 and 3.2.4 are also satisfied by $Q(\mathbf{g})$ if Q is tetrahedral, and so is Theorem 3.2.2 if Q is tetrahedral and homogeneous.*

REMARK 3.2.11. *Better exponential inequalities.* The integrability property (3.2.13) of homogeneous Rademacher and Gaussian chaos does not produce the best exponential bounds for the tail probabilities of these variables. Recent more refined isoperimetric methods produce better results. Ledoux and Talagrand (1991) contains an excellent presentation of this topic although not with the last word on the subject. For comparison purposes, here is the best known inequality for Rademacher chaos variables of order 2 (Talagrand, 1996, Theorem 1.2). Let $X = \sum_{(i,j)\in I_n^2}\varepsilon_i\varepsilon_j x_{ij}$, where $x_{ij} = x_{ji} \in B$, a Banach space, and n is an arbitrary natural number (it can be infinite if there is convergence). Let D be the unit ball of the dual B^* of B. Let

$$\begin{aligned}
M &= \text{a median of } \|X\|,\\
m^2 &= \mathbb{E}\sup_{|h|\le 1}\sup_{f\in D}\Big|\sum_{(i,j)\in I_n^2}\varepsilon_i h_j f(x_{ij})\Big|^2,\\
\sigma &= \sup_{|h|\le 1,|h'|\le 1}\sup_{f\in D}\Big|\sum_{(i,j)\in I_n^2} h_i h_j' f(x_{ij})\Big|,
\end{aligned}$$

where $h = (h_1,\ldots,h_n)$ and $|h|^2 = \sum_{i=1}^n h_i^2$ and likewise for h'. Then there exists a universal constant K, such that

$$\Pr\big\{\big|\|X\| - M\big| \ge t\big\} \le 2\exp\Big(-\frac{1}{K}\min\Big(\frac{t^2}{m^2},\frac{t}{\sigma}\Big)\Big). \tag{3.2.18}$$

This inequality shows, in particular, that the tail probabilities of $\big|\|X\| - M\big|$ are Gaussian-like in the range $t \lesssim m^2/\sigma$, a fact that Theorem 3.2.2 cannot detect.

3.3 Minorization of Tail Probabilities: The Paley–Zygmund Argument and a Conditional Jensen's Inequality

The arguments in this section, which originate in Paley–Zygmund (1932), yield lower bounds for tail probabilities of hypercontractive variables, thus extending the equivalence of their L_p-norms, $p > 0$, to $p = 0$ (that is, to medians and other quantiles). These bounds play an important role in decoupling, in particular, in the form of inequalities that can be viewed as *conditional Khinchin and Jensen inequalities for increasing not necessarily convex functions, in particular, for tail probabilities* of hypercontractive variables.

PROPOSITION 3.3.1. *Let ξ be a non-negative random variable such that $0 < \mathbb{E}\xi^q < \infty$ for some $q > 0$. Then, the following inequality is satisfied for all $t \geq 0$ and $0 < p < q$:*

$$\Pr\{\xi > t\} \geq \left[\frac{(\mathbb{E}\xi^p - t^p) \vee 0}{(\mathbb{E}\xi^q)^{\frac{p}{q}}}\right]^{\frac{q}{q-p}}. \tag{3.3.1}$$

PROOF. By Hölder,

$$\begin{aligned}\mathbb{E}\xi^p &\leq t^p + \mathbb{E}\xi^p I_{\xi > t} \\ &\leq t^p + (\mathbb{E}\xi^q)^{p/q}\big(\Pr\{\xi > t\}\big)^{\frac{q-p}{q}},\end{aligned}$$

hence,

$$\big(\Pr\{\xi > t\}\big)^{\frac{q-p}{q}} \geq \frac{(\mathbb{E}\xi^p - t^p) \vee 0}{(\mathbb{E}\xi^q)^{\frac{p}{q}}},$$

proving (3.3.1). □

As in previous sections, let us set $\|\xi\|_p = \big(\mathbb{E}|\xi|^p\big)^{1/p}$ for all $p > 0$. For X F-valued, we also set $\|X\|_p = \big(\mathbb{E}\|X\|^p\big)^{1/p}$, $p > 0$; with some abuse of notation we call this the L_p-norm of X if $p \geq 1$, and we refer to $\|X - Y\|_p^{1\wedge p}$ as the L_p distance between the random vectors X and Y. Here is an immediate, interesting consequence of Proposition 3.3.1:

COROLLARY 3.3.2. (PALEY–ZYGMUND'S INEQUALITY) *Let ξ, p, q, be as in Proposition 3.3.1, and let us set $c_{p,q} := \|\xi\|_p / \|\xi\|_q$. Then, for all $0 \leq \lambda \leq 1$,*

$$\Pr\{\xi > \lambda\|\xi\|_p\} \geq \left[(1 - \lambda^p)c_{p,q}^p\right]^{\frac{q}{q-p}}. \tag{3.3.2}$$

PROOF. Set $t = \lambda\|\xi\|_p$ in (3.3.1). □

Let $\mathfrak{R}_{d,0}$ be the set of F-valued (not necessarily homogeneous) Rademacher chaos variables of order d associated to a given Rademacher sequence. By Theorem 3.2.5 and the comment following its proof,

$$\sup_{X \in \mathfrak{R}_{d,0} \setminus \{0\}} \frac{\|X\|_p}{\|X\|_q} > 0$$

for all $p, q > 0$. Combining this with (3.3.2) shows that if a sequence $\{X_n\} \subset \mathfrak{R}_{d,0}$ is Cauchy in probability, then it is also Cauchy in L_p for all p. If F is a Banach space, this implies that the sequence $\{X_n\}$ has a limit X in L_p for all p (and hence, also in probability), and obviously the moments of X inherit inequality (3.2.12) of Theorem 3.2.5, with the same constants. The same comments apply to Gaussian chaos variables. As a consequence, we have the following theorem:

THEOREM 3.3.3. *Let F be a Banach space, let $\{\varepsilon_i\}$ be a Rademacher sequence, and, for each $d \in \mathbb{N}$, let $\mathfrak{R}_{d,0}$ be the set of all F-valued Rademacher chaos variables of order d associated to $\{\varepsilon_i\}$. Let $\mathfrak{R}_d$ be the closure of $\mathfrak{R}_{d,0}$ for the topology of convergence in probability. Then*

a) *for every $p > 0$, the closure $\mathfrak{R}_{d,0}$ for the L_p distance coincides with $\mathfrak{R}_d$, and*

b) *in $\mathfrak{R}_d$, convergence in probability is equivalent to convergence in L_p, $p > 0$.*

2) For $d \in \mathbb{N}$ let $\mathfrak{G}_{d,0}$ be the set of all B-valued Gaussian chaos variables of order d associated to a given i.i.d. sequence of $N(0, 1)$ random variables, and let $\mathfrak{G}_d$ be its closure in probability. Then statements (a) and (b) also hold for $\mathfrak{G}_{d,0}$ and $\mathfrak{G}_d$.

Since random vectors in $\mathfrak{R}_{d,0}$ or in $\mathfrak{G}_{d,0}$ take values in finite-dimensional subspaces of F, their norms are measurable and therefore Theorem 3.3.3 does not require separability of F.

Here is another corollary of Proposition 3.3.1 that will extend Khinchin's inequality to tail probabilities.

COROLLARY 3.3.4. *Let ξ, p, q, and $c_{p,q}$ be as in Corollary 3.3.2. Then,*

$$I_{\|\xi\|_p \geq 2^{1/p} t} \leq \left(\frac{2}{c_{p,q}^p}\right)^{\frac{q}{q-p}} \Pr\{\xi > t\} \tag{3.3.3}$$

for all $t > 0$. As a consequence, if f is a nondecreasing non-negative real function on $[0, \infty)$ and if f_- is its left-continuous modification, then, for all $c > 2$,

$$f\left(\|\xi\|_p / c^{1/p}\right) \leq \left(\frac{2}{c_{p,q}^p}\right)^{\frac{q}{q-p}} \mathbb{E} f_-(\xi). \tag{3.3.4}$$

In particular, if, moreover, f is either right-continuous or left-continuous, then

$$f\left(\|\xi\|_p / 2^{1/p}\right) \leq \left(\frac{2}{c_{p,q}^p}\right)^{\frac{q}{q-p}} \mathbb{E} f(\xi). \tag{3.3.4'}$$

PROOF. Since, by inequality (3.3.1),

$$\Pr\left\{\xi > \frac{\|\xi\|_p}{2^{1/p}}\right\} \geq \left(\frac{1}{2}\frac{\|\xi\|_p^p}{\|\xi\|_q^p}\right)^{\frac{q}{q-p}} = \left(\frac{c_{p,q}^p}{2}\right)^{\frac{q}{q-p}},$$

inequality (3.3.3) follows upon observing that

$$\Pr\{\xi > t\} \geq \Pr\left\{\xi > \frac{\|\xi\|_p}{2^{1/p}}\right\} I_{\|\xi\|_p/2^{1/p} \geq t}.$$

By monotonicity, $f\left(\|\xi\|_p/c^{1/p}\right) \leq f_-\left(\|\xi\|_p/2^{1/p}\right)$. So, (3.3.4) is a consequence of (3.3.4′) for f left-continuous. Proving (3.3.4′) for f left-continuous consists of a standard approximation argument, as follows. In this case, if we define the inverse of f as $f^{-1}(y) := \inf\{x : f(x) > y\}$, we have that $f(x) > y$ if and only if $x > f^{-1}(y)$. Then, (3.3.3) together with this observation give

$$\begin{aligned} f\left(\|\xi\|_p^p/2^{1/p}\right) &\leq \frac{1}{n} + \frac{1}{n}\sum_{k=1}^{\infty} I_{f(\|\xi\|_p^p/2^{1/p})>k/n} = \frac{1}{n} + \frac{1}{n}\sum_{k=1}^{\infty} I_{\|\xi\|_p^p/2^{1/p} > f^{-1}(k/n)} \\ &\leq \frac{1}{n} + \frac{1}{n}\left(\frac{2}{c_{p,q}^p}\right)^{\frac{q}{q-p}} \sum_{k=1}^{\infty} \Pr\{\xi > f^{-1}(k/n)\} \\ &= \frac{1}{n} + \frac{1}{n}\left(\frac{2}{c_{p,q}^p}\right)^{\frac{q}{q-p}} \sum_{k=1}^{\infty} \Pr\{f(\xi) > k/n\} \\ &\leq \frac{1}{n} + \left(\frac{2}{c_{p,q}^p}\right)^{\frac{q}{q-p}} \mathbb{E} f(\xi), \end{aligned}$$

proving inequality (3.3.4′) for f left-continuous and hence, also (3.3.4). Now, (3.3.4′) for f right-continuous follows from (3.3.4) since, in this case, $\mathbb{E} f_-(c\xi/2) \to \mathbb{E} f(\xi)$ as $c \searrow 2$. □

EXAMPLE 3.3.5. (*Extensions of Khinchin's inequality*) If we take $\xi = \|\sum x_i\varepsilon_i\|$, $x_i \in F$, $p = 1$ and $q = 2$, then Khinchin's inequality (1.3.2) gives $c_{1,2} \geq 1/\sqrt{2}$ and inequalities (3.3.3) and (3.3.4) extend Khinchin's inequality to arbitrary nondecreasing non-negative functions and to tail probabilities, as

$$I_{\left\|\sum x_i\varepsilon_i\right\|_1 \geq 2t} \leq 8\Pr\left\{\left\|\sum x_i\varepsilon_i\right\| > t\right\} \tag{3.3.5}$$

for $t > 0$, and

$$f\left(\left\|\sum x_i\varepsilon_i\right\|_1/c\right) \leq 8\mathbb{E} f\left(\left\|\sum x_i\varepsilon_i\right\|\right) \tag{3.3.6}$$

for any $c > 2$ and f nondecreasing and non-negative. If $F = \mathbb{R}$, another application of Khinchin gives

$$f\left(\left(\sum x_i^2\right)^{1/2}/(c\sqrt{2})\right) \leq 8\mathbb{E} f\left(\left\|\sum x_i\varepsilon_i\right\|\right), \tag{3.3.7}$$

which is inequality (1.3.5) in Chapter 1. With different constants these inequalities extend to any hypercontractive set of random variables, in particular, to (not necessarily homogeneous) Rademacher chaos variables of bounded degree (Section 3.2) and also to Gaussian chaos if F is separable.

We can view equations (3.3.4) and (3.3.6) as *Khinchin inequalities for arbitrary nondecreasing non-negative functions and* in particular, *for tail probabilities* [(3.3.3) and (3.3.5)]. If, moreover, in (3.3.6) the constants x are random and independent of the Rademacher variables, another integration produces a conditional Khinchin (a Marcinkiewicz) inequality. If the variables $\{\varepsilon_i\}$ and $\{x_i\}$ are defined on different factors of a product probability space, then Fubini's inequality (3.1.11) for outer probabilities applies without any further measurability restrictions to give, e.g., from (3.3.5) and (3.3.6),

$$\Pr^*\Big\{\mathbb{E}\big(\big\|\sum x_i\varepsilon_i\big\| \,\big|\, \{x_i\}\big) \geq 2t\Big\} \leq 8\Pr^*\Big\{\big\|\sum x_i\varepsilon_i\big\| > t\Big\},$$

$$\mathbb{E}^* f\Big(\mathbb{E}\big(\big\|\sum x_i\varepsilon_i\big\| \,\big|\, \{x_i\}\big)/c\Big) \leq 8\mathbb{E}^* f\Big(\big\|\sum x_i\varepsilon_i\big\|\Big),$$

respectively, for all $t \geq 0$ and all nondecreasing, non-negative functions on $[0, \infty)$. Here $\Pr^*$ of course refers to outer probability (see the end of Section 3.1.1, where outer expectation is defined). In the next section as well as in the next chapter, we require such a result for not necessarily homogeneous Rademacher chaos variables. We state it for reference:

THEOREM 3.3.6. *Let F be a normed linear space and let $n \in \mathbb{N}$. Let $\boldsymbol{\xi} = \{\xi_1, \ldots, \xi_n\}$ be n independent Rademacher variables or n independent standard normal variables, and let $\mathbf{X} = \{X_0; X_i, \ldots, X_{i_1,\ldots,i_d} : 1 \leq i, i_k \leq n\}$ be F-valued random variables such that $\boldsymbol{\xi}$ and $\mathbf{X}$ are defined on different factors of a product probability space (in particular, they are independent). Then, for every $p \in (0, \infty)$ there is a constant $C_{d,p}$ depending only on d and p (but not on F, n or $\mathbf{X}$) such that the conditional Rademacher or Gaussian chaos*

$$X = X_0 + \sum_{i=1}^{n} X_i\xi_i + \cdots + \sum_{I_n^d} X_{i_1,\ldots,i_d}\xi_{i_1}\cdots\xi_{i_d} \tag{3.3.8}$$

satisfies

$$\mathbb{E}^* f\Big(\big[\mathbb{E}(\|X\|^p|\mathbf{X})/c\big]^{1/p}\Big) \leq C_{d,p}\mathbb{E}^* f_-(\|X\|) \tag{3.3.9}$$

for all f non-negative, nondecreasing on $\mathbb{R}_+$ and $c > 2$, with $c = 2$ if f is either right or left continuous, and in particular,

$$\Pr^*\big\{\mathbb{E}(\|X\|^p|\mathbf{X}) \geq 2t^p\big\} \leq C_{d,p}\Pr^*\big\{\|X\| > t\big\} \tag{3.3.10}$$

for all $t \geq 0$.

PROOF. If $p > 1$, Corollary 3.3.4 and Theorem 3.2.5 or Theorem 3.2.10 with $q = 2p$ give

$$I_{\mathbb{E}(\|X\|^p|\mathbf{X})\geq 2t^p} \leq 4C_d^{2p}\left(\frac{2p-1}{p-1}\right)^{dp}\Pr\Big\{\|X\| > t\big|\mathbf{X}\Big\}.$$

Then, integrating with respect to $\mathbf{X}$, using Fubini's inequality (3.1.11) for outer expectations, yields (3.3.10) for $p > 1$. For $p \le 1$ we use the extension to chaos variables of inequality (3.2.8) for, e.g., $p \le 1$ and $q = 2$ instead of Theorem 3.2.5 (see the comment immediately below the proof of that theorem). Inequality (3.3.9) follows from (3.3.10) by the approximation argument in the proof of Corollary 3.3.4. □

The inequalities in the previous theorem apply not only to Rademacher and Gaussian chaos, but to more general *conditionally hypercontractive* random variables, as follows: *Let $\xi \ge 0$ be a non-negative real random variable and let $\mathcal{G}$ be a σ-algebra of Borel subsets of $\mathbb{R}$. If for some $0 < p < q$ there is $c_{p,q} > 0$ such that*

$$\left[\mathbb{E}(\xi^p|\mathcal{G})\right]^{1/p} \Big/ \left[\mathbb{E}(\xi^q|\mathcal{G})\right]^{1/q} \ge c_{p,q} \quad \text{a.s.},$$

then

$$\Pr\left\{\mathbb{E}(\xi^p|\mathcal{G}) \ge 2t^p\right\} \le \left(\frac{2}{c_{p,q}^p}\right)^{\frac{q}{q-p}} \Pr\{\xi > t\}$$

for all $t > 0$. This is obtained by integrating the equation

$$I_{\mathbb{E}(\xi^p|\mathcal{G}) \ge 2t^p} \le \left(\frac{2}{c_{p,q}^p}\right)^{\frac{q}{q-p}} \Pr\{\xi > t|\mathcal{G}\} \quad \text{a.s.},$$

which follows from (3.3.3) and the existence of regular conditional distributions [e.g., Dudley (1989, pages 269–270)]. It is obvious that the analog of (3.3.9) in this context also holds.

Inequality (3.3.10) for $p = 1$ implies, by the usual Jensen's inequality applied to the norm, that

$$\Pr{}^*\left\{\|\mathbb{E}(X|\mathbf{X})\| \ge 2t\right\} \le C_{d,1} \Pr{}^*\left\{\|X\| > t\right\} \tag{3.3.10$'$}$$

for all $t \ge 0$, a *conditional Jensen's inequality for tail probabilities*, and the same comment applies to inequality (3.3.9), giving a conditional Jensen inequality for nondecreasing, non-negative, not necessarily convex functions,

$$\mathbb{E}^* f\left(\|\mathbb{E}(X|\mathbf{X})\|/c\right) \le C_{d,1} \mathbb{E}^* f_-(\|X\|). \tag{3.3.9$'$}$$

It turns out that the weaker inequalities (3.3.9′) and (3.3.10′) do not require hypercontractivity of the vector X itself, but only of $x^*(X)$, for all x^* in the dual of F, and this, which is an irrelevant simplification for Rademacher or Gaussian chaos, may be important in other instances, as we will see in the next section. Here is the variation on Proposition 3.3.1 that will make such a result possible.

PROPOSITION 3.3.7. *Let ξ be a real mean zero random variable such that $\|\xi\|_p < \infty$ for some $p > 1$. Let $c_{1,p} := \|\xi\|_1/\|\xi\|_p$, with the convention $0/0 = 1$. Then*

$$\Pr\{\xi \ge 0\} > \left(\frac{c_{1,p}}{2}\right)^{\frac{p}{p-1}}. \tag{3.3.11}$$

PROOF. (3.3.11) is obvious if $\xi = 0$ a.s. If ξ is not a.s. zero, then, since $\mathbb{E}\xi I_{\xi>0} = -\mathbb{E}\xi I_{\xi<0}$, we have

$$\mathbb{E}|\xi| = \mathbb{E}|\xi| I_{|\xi|\neq 0} = 2\mathbb{E}\xi I_{\xi>0} < 2\|\xi\|_p \big[\Pr\{\xi > 0\}\big]^{\frac{p-1}{p}},$$

and therefore,

$$\Pr\{\xi > 0\} > \left(\frac{c_{1,p}}{2}\right)^{\frac{p}{p-1}}, \tag{3.3.11$'$}$$

an inequality slightly stronger than (3.3.11). □

COROLLARY 3.3.8. *Let $(\Omega, \Sigma, \Pr)$ be a probability space, let F be a normed linear space, and let $Y : \Omega \to F$ be an F-valued random element such that, for all $x^* \in F^*$, F^* the topological dual of F, the functions $x^*(Y)$ are measurable, centered, and square integrable. Then, for all $x \in F$,*

$$\Pr{}^*\big\{\|Y + x\| \geq \|x\|\big\} > \frac{1}{4} \inf_{x^* \in F^*} \left(\frac{\|x^*(Y)\|_1}{\|x^*(Y)\|_2}\right)^2. \tag{3.3.12}$$

PROOF. It is a consequence of the Hahn–Banach theorem that for every $x \in F$ there exists $x^* \in F^*$ such that $|x^*(y)| \leq \|y\|$ for all $y \in F$ and $x^*(x) = \|x\|$ [e.g., Rudin (1973, page 58)]. Then, Proposition 3.3.7 for $p = 2$ and $\xi = x^*(Y)$, together with this observation, implies

$$\begin{aligned}\Pr{}^*\big\{\|Y + x\| \geq \|x\|\big\} &\geq \Pr\big\{x^*(Y + x) \geq x^*(x)\big\} \\ &= \Pr\{x^*(Y) \geq 0\} > \frac{1}{4}\left(\frac{\|x^*(Y)\|_1}{\|x^*(Y)\|_2}\right)^2.\end{aligned}$$ □

The ratio between the L_1– and the L_2-norms of a random variable ξ, appearing in (3.3.12), is often more difficult to estimate than the ratio between its L_2– and L_4-norms (see, e.g., Remark 3.2.8 on real Rademacher chaos variables). But this second ratio can be used to estimate the first: if in the first part of the argument leading to (3.2.8) we take $p = 1$ and $\alpha = 2/3$ (so that $\frac{2-\alpha}{1-\alpha} = 4$), we obtain

$$\frac{\|\xi\|_1}{\|\xi\|_2} \geq \left(\frac{\|\xi\|_2}{\|\xi\|_4}\right)^2. \tag{3.3.13}$$

In particular, if Y satisfies the hypotheses of Corollary 3.3.8 and, moreover, the variables $x^*(Y)$ have finite fourth moment, then

$$\Pr{}^*\big\{\|Y + x\| \geq \|x\|\big\} > \frac{1}{4} \inf_{x^* \in F^*} \left(\frac{\|x^*(Y)\|_2}{\|x^*(Y)\|_4}\right)^4, \tag{3.3.14}$$

an inequality weaker, but handier, than (3.3.12). This observation applies also to the next result.

THEOREM 3.3.9. *Let F be a separable Banach space, let Y be an F-valued random variable, and let $\mathcal{G}$ be a σ-algebra such that*

$$c_{2,1,\mathcal{G}}^2 := \text{ess sup}\left[\sup_{x^*\in F^*} \frac{\mathbb{E}\left(\left[x^*(Y) - \mathbb{E}(x^*(Y)|\mathcal{G})\right]^2 \middle| \mathcal{G}\right)}{\left[\mathbb{E}\left(|x^*(Y) - \mathbb{E}(x^*(Y)|\mathcal{G})| \middle| \mathcal{G}\right)\right]^2}\right] < \infty. \tag{3.3.15}$$

Then,

$$\Pr\left\{\|\mathbb{E}(Y|\mathcal{G})\| > t\right\} \le 4c_{2,1,\mathcal{G}}^2 \Pr\{\|Y\| > t\} \tag{3.3.16}$$

for all $t > 0$.

PROOF. Since F is Polish, Y admits a regular conditional distribution $P_{Y|\mathcal{G}}(\cdot, \omega)$ with respect to $\mathcal{G}$ [e.g., Dudley (1989, pages 269–270)]. Hence we can apply inequality (3.3.12) for almost every ω as follows:

$$\begin{aligned}
&\Pr\left\{\|Y\| \ge \|\mathbb{E}(Y|\mathcal{G})\| \middle| \mathcal{G}\right\}(\omega)\\
&= P_{Y|\mathcal{G}}\left\{\left[\left\|y - \int y P_{Y|\mathcal{G}}(dy, \omega) + \int y P_{Y|\mathcal{G}}(dy, \omega)\right\| \ge \left\|\int y P_{Y|\mathcal{G}}(dy, \omega)\right\|\right], \omega\right\}\\
&> \frac{1}{4} \inf_{x^*\in F^*} \left(\frac{\left[\int \left|x^*(y) - \mathbb{E}(x^*(Y)|\mathcal{G})\right| P_{Y|\mathcal{G}}(dy, \omega)\right]^2}{\int \left(x^*(y) - \mathbb{E}(x^*(Y)|\mathcal{G})\right)^2 P_{Y|\mathcal{G}}(dy, \omega)}\right) \ge \frac{1}{4c_{2,1,\mathcal{G}}^2} \quad \text{a.s.}
\end{aligned}$$

Therefore,

$$\begin{aligned}
\Pr\left\{\|\mathbb{E}(Y|\mathcal{G})\| > t\right\} &\le \mathbb{E}\left[4c_{2,1,\mathcal{G}}^2 \Pr\left\{\|Y\| \ge \|\mathbb{E}(Y|\mathcal{G})\| \middle| \mathcal{G}\right\} I_{\|\mathbb{E}(Y|\mathcal{G})\| > t}\right]\\
&\le 4c_{2,1,\mathcal{G}}^2 \Pr\{\|Y\| > t\}.
\end{aligned}$$

□

Inequality (3.3.16), which is another *conditional Jensen inequality for tail probabilities* extends, just as (3.3.10), to arbitrary nondecreasing non-negative functions.

Separability or completeness of F are not required in Theorem 3.3.9 as long as the regular conditional distribution $P_{Y|\mathcal{G}}(\cdot, \omega)$ exists. In particular, it applies to the conditional chaos X defined in (3.3.8), which has coefficients in a general normed space (in that case, the conditional distributions are provided by Fubini's theorem). Thus, at least up to constants, inequality (3.3.10′) is contained in (3.3.16).

3.4 Decoupling of U-processes: Tail Probabilities

This section contains the main decoupling result in this chapter, (a version of) which is the following theorem.

THEOREM 3.4.1. *For natural numbers $n \ge m$, let $\{X_i\}_{i=1}^n$ be n independent random variables with values in a measurable space $(S, \mathcal{S})$, and let $\{X_i^k\}_{i=1}^n$, $k = 1, \dots, m$, be m independent copies of this sequence. Let B be a separable Banach space and, for each $(i_1, \dots, i_m) \in I_n^m$, let $h_{i_1 \dots i_m} : S^m \mapsto B$ be measurable* *functions. Then, there are constants $C_m \in (0, \infty)$, depending only on m, such that, for all $t > 0$,*

$$\Pr\Big\{\Big\|\sum_{I_n^m} h_{i_1\ldots i_m}(X_{i_1}, \ldots, X_{i_m})\Big\| > t\Big\}$$
$$\le C_m \Pr\Big\{C_m \Big\|\sum_{I_n^m} h_{i_1\ldots i_m}(X_{i_1}^1, \ldots, X_{i_m}^m)\Big\| > t\Big\}. \tag{3.4.1}$$

If, moreover, the functions $h_{i_1\ldots i_m}$ are symmetric in the sense that, for all $x_1, \ldots, x_m \in S$ and all permutations s of $\{1, \ldots, m\}$,

$$h_{i_1\ldots i_m}(x_1, \ldots, x_m) = h_{i_{s_1}\ldots i_{s_m}}(x_{s_1}, \ldots, x_{s_m}), \tag{3.4.2}$$

then the reverse inequality holds; that is, there are constants $D_m \in (0, \infty)$ depending only on m, such that, for all $t > 0$,

$$\Pr\Big\{\Big\|\sum_{I_n^m} h_{i_1\ldots i_m}(X_{i_1}^1, \ldots, X_{i_m}^m)\Big\| > t\Big\}$$
$$\le D_m \Pr\Big\{D_m \Big\|\sum_{I_n^m} h_{i_1\ldots i_m}(X_{i_1}, \ldots, X_{i_m})\Big\| > t\Big\}. \tag{3.4.3}$$

PROOF FOR $m = 2$. As in the proof of Theorem 3.1.1 for $m = 2$ we let $\{X_i'\}$ be an independent copy of $\{X_i\}$. As in that proof, we replace h_{ij} by h. Using Theorem 1.1.3, actually, the simple inequality (1.1.4′), we obtain

$$\Pr\Big\{\Big\|\sum_{I_n^2} h(X_i, X_j)\Big\| > t\Big\} \le 3\Pr\Big\{\Big\|\sum_{I_n^2}[h(X_i, X_j) + h(X_i', X_j')]\Big\| > \frac{2t}{3}\Big\}$$
$$\le 3\Pr\Big\{\Big\|\sum_{I_n^2}[h(X_i, X_j') + h(X_i', X_j) + h(X_i, X_j) + h(X_i', X_j')]\Big\| > \frac{t}{3}\Big\}$$
$$+ 6\Pr\Big\{\Big\|\sum_{I_n^2} h(X_i, X_j')\Big\| > \frac{t}{6}\Big\}. \tag{3.4.4}$$

As for the reverse inequality, if h is symmetric $[h_{ij}(x, y) = h_{ji}(y, x)]$, we obviously have

$$\Pr\Big\{\Big\|\sum_{I_n^2} h(X_i, X_j')\Big\| > t\Big\} = \Pr\Big\{\Big\|\sum_{I_n^2}[h(X_i, X_j') + h(X_i', X_j)]\Big\| > 2t\Big\}$$
$$\le \Pr\Big\{\Big\|\sum_{I_n^2}[h(X_i, X_j') + h(X_i', X_j) + h(X_i, X_j) + h(X_i', X_j')]\Big\| > t\Big\}$$
$$+ 2\Pr\Big\{\Big\|\sum_{I_n^2} h(X_i, X_j)\Big\| > \frac{t}{4}\Big\}. \tag{3.4.5}$$

Thus, the problem is reduced to estimating the tail probabilities of the polarized U-statistic

$$T_{n,2} = \sum_{I_n^2} \big[h(X_i, X_j) + h(X_i', X_j) + h(X_i, X_j') + h(X_i', X_j')\big], \tag{3.4.6}$$

in one case in terms of the tail probabilities of the decoupled U-statistic, and in terms of the original U-statistic in the other.

Let Z_i and Z_i', $i \le n$, be as defined in the proof of Theorem 3.1.1. To complete the proof of the first part of the theorem, we begin by observing that

$$\begin{aligned} 4h(Z_i, Z_j') =&(1+\varepsilon_i)(1-\varepsilon_j)h(X_i, X_j) \\ &+ (1+\varepsilon_i)(1+\varepsilon_j)h(X_i, X_j') \\ &+ (1-\varepsilon_i)(1-\varepsilon_j)h(X_i', X_j) \\ &+ (1-\varepsilon_i)(1+\varepsilon_j)h(X_i', X_j'), \end{aligned} \tag{3.4.7}$$

which is obvious from the definition of the variables Z_i, Z_j'. Therefore, setting

$$\begin{aligned} &X := 4\sum_{I_n^2} h(Z_i, Z_j'), \\ &\mathbf{X} := \{h(X_i, X_j), h(X_i', X_j), h(X_i, X_j'), h(X_i', X_j') : 1 \le i, j \le n\}, \end{aligned} \tag{3.4.8}$$

it follows that X, conditioned on $\mathbf{X}$, is a Rademacher chaos of order 2 and the conditional Jensen inequalities from Section 3.3 apply. Since

$$\mathbb{E}(X|\mathbf{X}) = T_{n,2} \tag{3.4.9}$$

inequality (3.3.10′) [or inequality (3.3.16)] gives

$$\begin{aligned} \Pr\big\{\|T_{n,2}\| > t\big\} &= \Pr\big\{\|\mathbb{E}(X|\mathbf{X})\| > t\big\} \\ &\le C_2 \Pr\big\{2\|X\| > t\big\} \\ &= C_2 \Pr\Big\{8\Big\|\sum_{I_n^2} h(X_i, X_j')\Big\| > t\Big\} \end{aligned} \tag{3.4.10}$$

for all $t > 0$. [For the last identity, recall form the proof of Theorem 3.1.1 that the random vectors $(Z_1, \dots, Z_n, Z_1', \dots, Z_n')$ and $(X_1, \dots, X_n, X_1', \dots, X_n')$ have the same probability law.] Combining (3.4.4) and (3.4.10) yields (3.4.1) for $m = 2$.

The second part of the theorem follows from (3.4.5) together with a similar estimation of the tail probabilities of $\|T_{n,2}\|$, now in terms of the decoupled U-statistic, based on the following analog of (3.4.7):

$$\begin{aligned} 4h(Z_i, Z_j) =&(1+\varepsilon_i)(1+\varepsilon_j)h(X_i, X_j) \\ &+ (1+\varepsilon_i)(1-\varepsilon_j)h(X_i, X_j') \\ &+ (1-\varepsilon_i)(1+\varepsilon_j)h(X_i', X_j) \\ &+ (1-\varepsilon_i)(1-\varepsilon_j)h(X_i', X_j'). \end{aligned} \tag{3.4.7′}$$

□

The previous proof uses Rademacher chaos hypercontractivity and elementary probability. To make the proof even easier, it should be noted that it does not require the most general hypercontractivity results from Section 3.2, but only the estimates in Remark 3.2.8, in fact, just the right-hand side of inequality (3.2.15) for $p = 4$; this inequality in combination with (3.3.13), is sufficient to give inequality (3.3.10), hence (3.3.10′), which is all that is used from hypercontractivity in the above proof.

PROOF OF THEOREM 3.4.1 FOR GENERAL m. PART 1: INEQUALITY (3.4.1). The general case is done by recurrence. For clarity of exposition we consider only the case $m = 3$, which is not different from the general case. Let us relabel the variables Z_i and Z_i' from the previous proof (and from the proof of Theorem 3.2.1), respectively, as Z_i^1 and Z_i^2 and, in analogy with (3.4.6), let us denote $T_{n,3}$ the polarized U-statistic, that is,

$$T_{n,3} = \sum_{(i_1,i_2,i_3)\in I_n^3} \sum_{1\le j_1,j_2,j_3\le 2} h(X_{i_1}^{j_1}, X_{i_2}^{j_2}, X_{i_3}^{j_3}).$$

Using inequality (1.1.4′), as in (3.4.4) above, we have

$$\begin{aligned}
&\Pr\Big\{\Big\|\sum_{I_n^3} h(X_{i_1}^1, X_{i_2}^1, X_{i_3}^1)\Big\| > t\Big\}\\
&\le 3\Pr\Big\{\Big\|\sum_{I_n^3}[h(X_{i_1}^1, X_{i_2}^1, X_{i_3}^1) + h(X_{i_1}^2, X_{i_2}^2, X_{i_3}^2)]\Big\| > \frac{2t}{3}\Big\}\\
&\le 3\Pr\Big\{\|T_{n,3}\| > \frac{t}{3}\Big\}\\
&\quad + 3\Pr\Big\{\Big\|\sum_{I_n^3}[h(X_{i_1}^1, X_{i_2}^1, X_{i_3}^1) + h(X_{i_1}^2, X_{i_2}^2, X_{i_3}^2) - T_{n,3}]\Big\| > \frac{t}{3}\Big\}\\
&= 3\Pr\Big\{\|T_{n,3}\| > \frac{t}{3}\Big\} + 3\Pr\Big\{\Big\|\sum_{\mathbf{i}\in I_n^3} \sum_{\substack{1\le j_1,j_2,j_3\le 2\\ \text{not all } j\text{s equal}}} h(X_{i_1}^{j_1}, X_{i_2}^{j_2}, X_{i_3}^{j_3})\Big\| > \frac{t}{3}\Big\}\\
&\le 3\Pr\Big\{\|T_{n,3}\| > \frac{t}{3}\Big\} + 3\sum_{\substack{1\le j_1,j_2,j_3\le 2\\ \text{not all } j\text{s equal}}} \Pr\Big\{\Big\|\sum_{\mathbf{i}\in I_n^3} h(X_{i_1}^{j_1}, X_{i_2}^{j_2}, X_{i_3}^{j_3})\Big\| > \frac{t}{18}\Big\}\\
&\le 3\Pr\Big\{\|T_{n,3}\| > \frac{t}{3}\Big\} + 18C_2\Pr\Big\{C_2\Big\|\sum_{\mathbf{i}\in I_n^3} h(X_{i_1}^1, X_{i_2}^2, X_{i_3}^3)\Big\| > \frac{t}{18}\Big\},
\end{aligned} \tag{3.4.11}$$

where the last inequality results from applying induction—that is, decoupling for $m = 2$—to each of the second summands, conditionally on the upper index j that occurs only once: this results in decoupling of the upper index j that repeats.

So, in analogy with the case $m = 2$, the problem reduces to estimating the tail probabilities of $\|T_{n,3}\|$. For this, we begin by noting that, as in (3.4.7), for any choice of $\ell_i = 1$ or $\ell_i = 2$, $1 \le i \le 3$, there is a choice of $\pm$ signs for which the identity

$$2^3 h(Z_{i_1}^{\ell_1}, Z_{i_2}^{\ell_2}, Z_{i_3}^{\ell_3}) = \sum_{1 \le j_1, j_2, j_3 \le 2} (1 \pm \varepsilon_{i_1})(1 \pm \varepsilon_{i_2})(1 \pm \varepsilon_{i_3}) h(X_{i_1}^{j_1}, X_{i_2}^{j_2}, X_{i_3}^{j_3})$$

holds. Let us take, for instance, $\ell_1 = 1$ and $\ell_2 = \ell_3 = 2$. Then, as in the case $m = 2$, the variable

$$X := 2^3 \sum_{I_n^3} h(Z_{i_1}^{\ell_1}, Z_{i_2}^{\ell_2}, Z_{i_3}^{\ell_3}) - T_{n,3},$$

conditioned on the sequences $\{X_i^j\}$, is a centered Rademacher chaos of order 3. Hence, applying Theorem 3.3.6 and Theorem 3.2.5 (or their weaker versions) as in the previous proof, we obtain [see (3.4.8)–(3.4.10)]

$$\Pr\Big\{\|T_{n,3}\| > t\Big\} \le C \Pr\Big\{2^4 \Big\|\sum_{\mathbf{i} \in I_n^3} h(X_{i_1}^1, X_{i_2}^2, X_{i_3}^2)\Big\| > t\Big\}$$

for some universal constant $C < \infty$. Now, again applying the result for $m = 2$, we can decouple with respect to the variables X_i^2 conditionally on the variables X_i^1 and obtain

$$\Pr\Big\{\|T_{n,3}\| > t\Big\} \le A \Pr\Big\{A \Big\|\sum_{\mathbf{i} \in I_n^3} h(X_{i_1}^1, X_{i_2}^2, X_{i_3}^3)\Big\| > t\Big\},$$

for some constant A that depends only on the fact that $m = 3$. Combined with (3.4.11), this inequality gives (3.4.1) thus concluding the proof of the first part of the theorem. □

The proof of the second part of the theorem requires some preparation. As we will see below, very much as with Theorem 3.1.1, the proof of (3.4.3) can be reduced to the following lemma:

LEMMA 3.4.2. *Under the hypotheses for the second part of Theorem 3.4.1, for all $1 \le \ell \le m \le n$, and for all $t > 0$, there is a constant $C_m \in (0, \infty)$, depending only on m, such that*

$$\Pr\Big\{\Big\|\sum_{\mathbf{i} \in I_n^m} \sum_{1 \le j_1, \dots, j_m \le \ell} h_{i_1 \dots i_m}(X_{i_1}^{j_1}, \dots, X_{i_m}^{j_m})\Big\| > t\Big\}$$
$$\le C_m \Pr\Big\{C_m \Big\|\sum_{I_n^m} h_{i_1 \dots i_m}(X_{i_1}, \dots, X_{i_m})\Big\| > t\Big\}. \tag{3.4.12}$$

PROOF. As usual, we replace $h_{i_1 \dots i_m}$ by h symmetric in its entries. Let

$$M_i = (\delta_{i1}, \dots, \delta_{i\ell}),\ i \le n,$$

be n independent Multinomial$(1; 1/\ell, \dots, 1/\ell)$ random vectors, that is, $\Pr\{\delta_{ij} = 1\} = \frac{1}{\ell} = 1 - \Pr\{\delta_{ij} = 0\}$, $\sum_{j=1}^{\ell} \delta_{ij} = 1$ for all i, and δs with different first subindices are independent. We further assume that the set of variables $\{M_i\}$ is independent from the set of variables $\{X_i^j\}$. Now we define new variables Y_i as

$$Y_i = \sum_{j=1}^{\ell} \delta_{ij} X_i^j, \quad i = 1, \ldots, n,$$

(that is, Y_i coincides with X_i^j with probability $1/\ell$). Obviously, the variables Y_i are independent and $\mathcal{L}(Y_i) = \mathcal{L}(X_i)$, $1 \le i \le n$. Note that, if $i_1, \ldots, i_m$ are all different,

$$h(Y_{i_1}, \ldots, Y_{i_m}) = \sum_{1 \le j_1, \ldots, j_m \le \ell} \delta_{i_1 j_1} \cdots \delta_{i_m j_m} h(X_{i_1}^{j_1}, \ldots, X_{i_m}^{j_m})$$

and that, if we let

$$\mathcal{X}_m = \sigma[X_i^j : 1 \le i \le n, 1 \le j \le m],$$

then we also have

$$\mathbb{E}\big[h(Y_{i_1}, \ldots, Y_{i_m}) | \mathcal{X}_m\big] = \Big(\frac{1}{\ell}\Big)^m \sum_{1 \le j_1, \ldots, j_m \le \ell} h(X_{i_1}^{j_1}, \ldots, X_{i_m}^{j_m}).$$

Let us set

$$\begin{aligned} U_n &= \sum_{\mathbf{i} \in I_n^m} h(Y_{i_1}, \ldots, Y_{i_m}) \\ &= \sum_{\mathbf{i} \in I_n^m} \sum_{1 \le j_1, \ldots, j_m \le \ell} \delta_{i_1 j_1} \cdots \delta_{i_m j_m} h(X_{i_1}^{j_1}, \ldots, X_{i_m}^{j_m}) \end{aligned} \tag{3.4.13}$$

and

$$T_n = \mathbb{E}(U_n | \mathcal{X}_m) = \Big(\frac{1}{\ell}\Big)^m \sum_{\mathbf{i} \in I_n^m} \sum_{1 \le j_1, \ldots, j_m \le \ell} h(X_{i_1}^{j_1}, \ldots, X_{i_m}^{j_m}). \tag{3.4.14}$$

Since the vectors $(Y_1, \ldots Y_n)$ and $(X_1, \ldots X_n)$ have the same probablity law, the lemma will result from comparison of the tail probabilities of the variables $\|U_n\|$ and $\|T_n\|$.

Suppose we proved that, for all $x^* \in B^*$,

$$\Big[\mathbb{E}\big(x^*(U_n - T_n) | \mathcal{X}_m\big)^4\Big]^{\frac{1}{4}} \le D_{m,\ell}\Big[\mathbb{E}\big(x^*(U_n - T_n) | \mathcal{X}_m\big)^2\Big]^{\frac{1}{2}} \tag{3.4.15}$$

for some finite constant $D_{m,\ell}$ depending only on m and ℓ. Then we could apply the conditional Jensen inequality (3.3.16) in conjunction with (3.3.14), conditionally on $\mathcal{X}_m$, and conclude

$$\Pr\big\{\|T_n\| > t\big\} \le 4D_{m,\ell}^4 \Pr\big\{\|U_n\| > t\big\},$$

which, by the equidistribution of the Ys and the Xs, is just inequality (3.4.12). Thus, *modulo inequality (3.4.15)*, Lemma 3.4.2 is proved. □

It is clear from the definitions of U_n and T_n, (3.4.13) and (3.4.14), respectively, that inequality (3.4.15) is a direct consequence of the following lemma.

LEMMA 3.4.3. *Let $1 \le \ell \le m \le n$. Let $M_i = (\delta_{i1}, \ldots, \delta_{i\ell})$, $i \le n$, be n independent Multinomial$(1; 1/\ell, \ldots, 1/\ell)$ random vectors and let $a_{i_1\ldots i_m}^{j_1\ldots j_m}$, $1 \le j_1, \ldots, j_m \le \ell$, $(i_1, \ldots, i_m) \in I_n^m$, be a collection of real numbers such that, for each $(j_1, \ldots, j_m)$ and $(i_1, \ldots, i_m)$, and for each permutation s of $\{1, \ldots, m\}$,*

$$a_{i_1\ldots i_m}^{j_1\ldots j_m} = a_{i_{s(1)}\ldots i_{s(m)}}^{j_{s(1)}\ldots j_{s(m)}}. \tag{3.4.16}$$

Let

$$\mathcal{U}_n = \sum_{\mathbf{i}\in I_n^m} \sum_{1\le j_1,\ldots,j_m\le\ell} \delta_{i_1 j_1}\cdots\delta_{i_m j_m} a_{i_1\ldots i_m}^{j_1\ldots j_m}$$

and

$$\mathcal{T}_n = \mathbb{E}\mathcal{U}_n = \Big(\frac{1}{\ell}\Big)^m \sum_{\mathbf{i}\in I_n^m} \sum_{1\le j_1,\ldots,j_m\le\ell} a_{i_1\ldots i_m}^{j_1\ldots j_m}.$$

Then, there exist constants $D_{m,\ell} \in (0, \infty)$, depending only on ℓ and m, such that

$$\Big[\mathbb{E}\big(\mathcal{U}_n - \mathcal{T}_n\big)^4\Big]^{\frac14} \le D_{m,\ell}\Big[\mathbb{E}\big(\mathcal{U}_n - \mathcal{T}_n\big)^2\Big]^{\frac12}. \tag{3.4.17}$$

PROOF. The proof combines techniques from Section 3.2 (hypercontractivity) with the main result in Section 3.1 (decoupling of moments of U-statistics). We begin with a simple inequality (as in the proof of hypercontractivity for Rademacher chaos variables).

CLAIM 1. *Let $a = (a_1, \ldots, a_\ell) \in \mathbb{R}^\ell$, let $M = (\delta_1, \ldots, \delta_\ell)$ be a Multinomial$(1; 1/\ell, \ldots, 1/\ell)$ random vector and let $V = a\cdot M = \sum_{j=1}^\ell a_j\delta_j$. Then, for all $y \in \mathbb{R}$,*

$$\|y + V\|_4 \le \|y + \mathbb{E}V + c_\ell(V - \mathbb{E}V)\|_2, \tag{3.4.18}$$

where $\|\cdot\|_p$ denotes L_p-norm, as usual, and where $c_\ell = (3 + 2\ell^{1/2})^{1/2}$.

To prove the claim, let us set $W = V - EV$ and $b_i = a_i - \frac{1}{\ell}\sum_{j=1}^\ell a_j$. Then

$$\mathbb{E}|W|^3 = \frac{1}{\ell}\sum_{j=1}^\ell |b_j|^3 \le \frac{1}{\ell}\Big(\sum_{j=1}^\ell b_j^2\Big)^{3/2} = \ell^{1/2}\big(\mathbb{E}W^2\big)^{3/2}.$$

Likewise, $\mathbb{E}W^4 \le \ell\big(\mathbb{E}W^2\big)^2$. Hence, setting $x = y + EV$,

$$\begin{aligned}
\mathbb{E}(y+V)^4 &= \mathbb{E}(x+W)^4 \le x^4 + 6x^2\mathbb{E}W^2 + 4|x|\ell^{1/2}(\mathbb{E}W^2)^{3/2} + \ell(\mathbb{E}W^2)^2 \\
&\le x^4 + (6 + 4\ell^{1/2})x^2\mathbb{E}W^2 + (\ell + 4\ell^{1/2})(\mathbb{E}W^2)^2 \\
&\le \big[x^2 + (3 + 2\ell^{1/2})\mathbb{E}W^2\big]^2 \le \Big[\mathbb{E}\big(x + (3 + 2\ell^{\frac12})^{\frac12}W\big)^2\Big]^2 \\
&= \Big[\mathbb{E}\big(y + \mathbb{E}V + c_\ell(V - \mathbb{E}V)\big)^2\Big]^2,
\end{aligned}$$

where we use $\mathbb{E}W = 0$ in the first and fourth steps, and, in the second, we consolidate (a bound for) the third term with the second or with the fourth, according as to whether $|x|$ exceeds or does not exceed $(\mathbb{E}W^2)^{1/2}$. Claim 1 is thus proved.

CLAIM 2. *Let $a_i = (a_{i1}, \ldots, a_{i\ell}) \in \mathbb{R}^\ell$ and let $V_i = a_i \cdot M_i = \sum_{j=1}^{\ell} a_{ij}\delta_{ij}$, $i = 1, \ldots, n$. Then, for all $y \in \mathbb{R}$,*

$$\Big\| y + \sum_{i=1}^{n} V_i \Big\|_4 \le c_\ell \Big\| y + \sum_{i=1}^{n} V_i \Big\|_2,$$

that is,

$$\Big\| y + \sum_{i=1}^{n} \sum_{j=1}^{\ell} a_{ij}\delta_{ij} \Big\|_4 \le c_\ell \Big\| y + \sum_{i=1}^{n} \sum_{j=1}^{\ell} a_{ij}\delta_{ij} \Big\|_2. \tag{3.4.19}$$

Inequality (3.4.18), with an induction as in the proof of Theorem 3.2.1, part 3, using the generalized Minkowski inequality, yields

$$\Big\| y + \sum_{i=1}^{n} V_i \Big\|_4 \le \Big\| y + \mathbb{E}\big(\sum_{i=1}^{n} V_i\big) + c_\ell \sum_{i=1}^{n} (V_i - \mathbb{E}V_i) \Big\|_2.$$

Then, on account of $c_\ell \ge 1$, developing the square gives

$$\begin{aligned} &\Big\| y + \mathbb{E}\big(\sum_{i=1}^{n} V_i\big) + c_\ell \sum_{i=1}^{n} (V_i - \mathbb{E}V_i) \Big\|_2 \\ &\qquad = \Big[c_\ell^2 \mathbb{E}\big(y + \sum_{i=1}^{n} V_i\big)^2 - (c_\ell^2 - 1)\Big(\mathbb{E}\big(y + \sum_{i=1}^{n} V_i\big)\Big)^2 \Big]^{1/2} \\ &\qquad \le c_\ell \Big\| y + \sum_{i=1}^{n} V_i \Big\|_2, \end{aligned}$$

proving (3.4.19).

CLAIM 3. *Let $\{M_i^j = (\delta_{i1}^j, \ldots, \delta_{i\ell}^j),\ i \le n\}$, $j = 1, \ldots, m$, be m independent copies of the sequence $\{M_i : i \le n\}$. Let*

$$\mathcal{U}_n^{\mathrm{dec}} = \sum_{\mathbf{i} \in I_n^m} \sum_{1 \le j_1, \ldots, j_m \le \ell} \delta_{i_1 j_1}^1 \cdots \delta_{i_m j_m}^m a_{i_1 \ldots i_m}^{j_1 \ldots j_m},$$

be the decoupled version of $\mathcal{U}_n$. Then, (3.4.17) holds for $\mathcal{U}_n^{\mathrm{dec}}$ with $D_{m,\ell} = c_\ell^m$, that is,

$$\Big[E\big(\mathcal{U}_n^{\mathrm{dec}} - \mathcal{T}_n\big)^4 \Big]^{\frac{1}{4}} \le c_\ell^m \Big[E\big(\mathcal{U}_n^{\mathrm{dec}} - \mathcal{T}_n\big)^2 \Big]^{\frac{1}{2}}. \tag{3.4.17$'$}$$

The proof consists in applying Claim 2 for one sequence $\{M_i^j : i = 1, \ldots, n\}$ at a time (that is, conditionally on the other sequences), with the help of the generalized Minkowski inequality, as in Part 3 of the proof of Theorem 3.2.1. Here is how it goes for $m = 2$:

$$\begin{aligned}
\|\mathcal{U}_n^{\mathrm{dec}} - \mathcal{T}_n\|_4 &= \Big(E_2 E_1\Big[\sum_{i_1=1}^{n}\sum_{j_1=1}^{\ell}\delta^1_{i_1 j_1}\Big(\sum_{\substack{i_2\le n\\ i_2\ne i_1}}\sum_{j_2=1}^{\ell}\delta^2_{i_2 j_2}a^{j_1 j_2}_{i_1 i_2}\Big) - \mathcal{T}_n\Big]^4\Big)^{\frac14}\\
&\le c_\ell\Big(E_2\Big\{E_1\Big[\sum_{i_1=1}^{n}\sum_{j_1=1}^{\ell}\delta^1_{i_1 j_1}\Big(\sum_{\substack{i_2\le n\\ i_2\ne i_1}}\sum_{j_2=1}^{\ell}\delta^2_{i_2 j_2}a^{j_1 j_2}_{i_1 i_2}\Big) - \mathcal{T}_n\Big]^2\Big\}^2\Big)^{\frac14}\\
&\le c_\ell\Big(E_1\Big\{E_2\Big[\sum_{i_1=1}^{n}\sum_{j_1=1}^{\ell}\delta^1_{i_1 j_1}\Big(\sum_{\substack{i_2\le n\\ i_2\ne i_1}}\sum_{j_2=1}^{\ell}\delta^2_{i_2 j_2}a^{j_1 j_2}_{i_1 i_2}\Big) - \mathcal{T}_n\Big]^4\Big\}^{\frac12}\Big)^{\frac12}\\
&\le c_\ell^2\Big[E_1 E_2\big(\mathcal{U}_n^{\mathrm{dec}} - \mathcal{T}_n\big)^2\Big]^{\frac12}.
\end{aligned}$$

Now the lemma follows from Claim 3 and Theorem 3.1.1. To see this, we let

$$g_{i_1\dots i_m}(M_{i_1},\dots,M_{i_m}) = \sum_{1\le j_1,\dots,j_m\le\ell}\delta_{i_1 j_1}\cdots\delta_{i_m j_m}a^{j_1\dots j_m}_{i_1\dots i_m} - \frac{1}{\binom{n}{m}}\mathcal{T}_n$$

with $a^{j_1\dots j_m}_{i_1\dots i_m} = h(X^{j_1}_{i_1},\dots,X^{j_m}_{i_m})$, which satisfy condition (3.4.16) by symmetry of h, and observe that

$$\mathcal{U}_n - \mathcal{T}_n = \sum_{\mathbf{i}\in I_n^m} g_{i_1\dots i_m}(M_{i_1},\dots,M_{i_m})$$

and

$$\mathcal{U}_n^{\mathrm{dec}} - \mathcal{T}_n = \sum_{\mathbf{i}\in I_n^m} g_{i_1\dots i_m}(M^1_{i_1},\dots,M^m_{i_m}).$$

Moreover, hypothesis (3.4.16) on the numbers $a^{j_1\dots j_m}_{i_1\dots i_m}$ implies that the functions g satisfy the symmetry hypothesis (3.1.2) in Theorem 3.1.1. Hence, Theorem 3.1.1 implies that the L_p-norms of these two statistics are comparable, that is, there exists a constant $C_m \in (0,\infty)$, depending only on m such that

$$\frac{1}{C_m^p}\mathbb{E}|\mathcal{U}_n^{\mathrm{dec}} - \mathcal{T}_n|^p \le \mathbb{E}|\mathcal{U}_n - \mathcal{T}_n|^p \le C_m^p\mathbb{E}|\mathcal{U}_n^{\mathrm{dec}} - \mathcal{T}_n|^p,$$

for all $p \ge 1$, in particular for $p = 2$ and $p = 4$. These inequalities and claim 3 prove inequality (3.4.17) with $D_{m,\ell} = c_\ell^m C_m^2$. □

Lemma 3.4.2 is thus completely proved. We can now proceed to complete the proof of Theorem 3.4.1.

PROOF OF THEOREM 3.4.1 FOR GENERAL m. PART 2: INEQUALITY (3.4.3). As with the first part, we only prove this inequality for $m = 3$ and we replace the functions $h_{i_1,\dots,i_m}$ by h symmetric. Let us consider the following string of inequalities:

$$\begin{aligned}
&\Pr\Big\{\Big\|\sum_{I_n^3} h(X_{i_1}^1, X_{i_2}^2, X_{i_3}^3)\Big\| > t\Big\} \\
&\quad = \Pr\Big\{\frac{1}{3!}\Big\|\sum_{\substack{j_1,j_2,j_3\le 3 \\ \text{all } j\text{s different}}}\sum_{I_n^3} h(X_{i_1}^{j_1}, X_{i_2}^{j_2}, X_{i_3}^{j_3})\Big\| > t\Big\} \qquad \text{(by symmetry of } h)\\
&\quad \le \Pr\Big\{\frac{1}{3!}\Big\|\sum_{j_1,j_2,j_3\le 3}\sum_{I_n^3} h(X_{i_1}^{j_1}, X_{i_2}^{j_2}, X_{i_3}^{j_3})\Big\| > \frac{t}{3}\Big\} \\
&\qquad + \Pr\Big\{\frac{1}{3!}\Big\|\sum_{\substack{j_1,j_2,j_3\le 3 \\ \text{two } j\text{s different}}}\sum_{I_n^3} h(X_{i_1}^{j_1}, X_{i_2}^{j_2}, X_{i_3}^{j_3})\Big\| > \frac{t}{3}\Big\} \\
&\qquad + \Pr\Big\{\frac{1}{3!}\Big\|\sum_{j=1}^{3}\sum_{I_n^3} h(X_{i_1}^{j}, X_{i_2}^{j}, X_{i_3}^{j})\Big\| > \frac{t}{3}\Big\} \\
&\quad \le \Pr\Big\{\frac{1}{3!}\Big\|\sum_{j_1,j_2,j_3\le 3}\sum_{I_n^3} h(X_{i_1}^{j_1}, X_{i_2}^{j_2}, X_{i_3}^{j_3})\Big\| > \frac{t}{3}\Big\} \\
&\qquad + 3\Pr\Big\{\Big\|\sum_{\substack{j_1,j_2,j_3\le 2 \\ \text{two } j\text{s different}}}\sum_{I_n^3} h(X_{i_1}^{j_1}, X_{i_2}^{j_2}, X_{i_3}^{j_3})\Big\| > \frac{2t}{3}\Big\} \\
&\qquad + 3\Pr\Big\{\frac{1}{3!}\Big\|\sum_{I_n^3} h(X_{i_1}, X_{i_2}, X_{i_3})\Big\| > \frac{2t}{3}\Big\}.
\end{aligned} \tag{3.4.20}$$

The first summand in the last term is dominated by a tail probability of the undecoupled statistic by Lemma 3.4.2 ($m = \ell = 3$), as wanted, and the third summand is already such a tail probability. The second summand undecouples by another application of Lemma 3.4.2 (for $m = 3,\ \ell = 2$) because, obviously,

$$\begin{aligned}
&\Pr\Big\{\Big\|\sum_{\substack{j_1,j_2,j_3\le 2 \\ \text{two } j\text{s different}}}\sum_{I_n^3} h(X_{i_1}^{j_1}, X_{i_2}^{j_2}, X_{i_3}^{j_3})\Big\| > t\Big\} \\
&\quad \le \Pr\Big\{\Big\|\sum_{j_1,j_2,j_3\le 2}\sum_{I_n^3} h(X_{i_1}^{j_1}, X_{i_2}^{j_2}, X_{i_3}^{j_3})\Big\| > \frac{t}{2}\Big\} \\
&\qquad + 2\Pr\Big\{\Big\|\sum_{I_n^3} h(X_{i_1}, X_{i_2}, X_{i_3})\Big\| > \frac{t}{4}\Big\}.
\end{aligned} \tag{3.4.21}$$

Inequality (3.4.3) is thus proved for $m = 3$. Inequalities (3.4.20) and (3.4.21) can easily be turned into a recursive argument that proves the theorem for any m. □

REMARK 3.4.4. *U-processes.* Just like Theorem 3.1.1, Theorem 3.4.1 has also a version for U-processes indexed by arbitrary families of kernels, analogous to Theorem 3.1.2. The details are left to the reader.

REMARK 3.4.5. *Tetrahedral polynomials in independent random variables.* One of the first results on decoupling (McConnell and Taqqu, 1986) consisted in comparison of moments of homogeneous multilinear forms

$$\sum_{\mathbf{i}\in I_n^m} a_{i_1\ldots i_m} X_{i_1}\cdots X_{i_m} \tag{3.4.22}$$

and of their decoupled expressions, where $a_{\mathbf{i}}$ is a symmetric array of B-valued coefficients and $X_1, \ldots, X_n$ are independent symmetric real random variables. We show next that the general result in this section on decoupling of U-statistics applies to *comparison of tail probabilities of not necessarily homogeneous tetrahedral B-valued polynomials in not necessarily symmetric independent random variables* X_i. Let $Q_{(m)}$ be a B-valued tetrahedral polynomial of degree m on $\mathbb{R}^n$, that is, if $\mathbf{x} = (x_1, \ldots, x_n) \in \mathbb{R}^n$ denotes a generic point in $\mathbb{R}^n$, then

$$Q_{(m)}(\mathbf{x}) = a_0 + \sum_{k=1}^{m}\sum_{\mathbf{i}\in I_n^k} a_{i_1,\ldots,i_k} x_{i_1}\cdots x_{i_k}, \tag{3.4.23}$$

where $a_{\mathbf{i}} \in B$ and B is a Banach space. Moreover, we assume that the coefficients $a_{\mathbf{i}}$ are *symmetric* in their subindices, that is, $a_{i_1,\ldots,i_k} = a_{i_{s(1)},\ldots,i_{s(k)}}$ for any permutation s of $1, \ldots, k$. Then, $Q_{(m)}(\mathbf{x})$ can be written in the form of a B-valued U-statistic of order m,

$$Q_{(m)}(\mathbf{x}) = \sum_{\mathbf{i}\in I_n^m} H^{(n)}_{i_1,\ldots,i_m}(x_{i_1}, \ldots, x_{i_m}) \tag{3.4.23$'$}$$

with

$$\begin{aligned} &H^{(n)}_{i_1,\ldots,i_m}(x_1, \ldots, x_m) \\ &\quad := \frac{(n-m)!}{n!}a_0 + \sum_{k=1}^{m}\frac{(n-m)!}{(n-k)!}\frac{(m-k)!}{m!}\sum_{\mathbf{r}\in I_m^k} a_{i_{r_1},\ldots,i_{r_k}} x_{r_1}\cdots x_{r_k}, \end{aligned} \tag{3.4.24}$$

as can be easily verified. Moreover, the symmetry of the coefficients $a_{\mathbf{i}}$ implies that

$$H^{(n)}_{i_1,\ldots,i_m}(x_1, \ldots, x_m) = H^{(n)}_{i_{s(1)},\ldots,i_{s(m)}}(x_{s(1)}, \ldots, x_{s(m)})$$

for all permutations s of $1, \ldots, m$, i.e., these kernels satisfy conditions (3.4.2). Hence, by Theorem 3.4.1, the tail probabilities of the random polynomial

$$Q_{(m)}(\mathbf{X}) = a_0 + \sum_{k=1}^{m}\sum_{\mathbf{i}\in I_n^k} a_{i_1,\ldots,i_k} X_{i_1}\cdots X_{i_k} = \sum_{\mathbf{i}\in I_n^m} H^{(n)}_{i_1,\ldots,i_m}(X_{i_1}, \ldots, X_{i_m}) \tag{3.4.25}$$

are comparable to the tail probabilities of its decoupled version

$$\tilde{Q}_{(m)}(\mathbf{X}^1, \ldots, \mathbf{X}^m) := \sum_{\mathbf{i}\in I_n^m} H^{(n)}_{i_1,\ldots,i_m}(X^1_{i_1}, \ldots, X^m_{i_m}), \tag{3.4.26}$$

which develops as

$$\tilde{Q}_{(m)}(\mathbf{X}^1, \ldots, \mathbf{X}^m) = a_0 + \sum_{k=1}^{m} \frac{(m-k)!}{m!} \frac{(n-m)!}{(n-k)!} \sum_{\mathbf{i} \in I_n^m} \sum_{\mathbf{r} \in I_m^k} a_{i_{r_1}, \ldots, i_{r_k}} X_{i_{r_1}}^{r_1} \cdots X_{i_{r_k}}^{r_k}$$

$$= a_0 + \sum_{k=1}^{m} \frac{(m-k)!}{m!} \sum_{\mathbf{i} \in I_n^k} \sum_{\mathbf{r} \in I_m^k} a_{i_1, \ldots, i_k} X_{i_1}^{r_1} \cdots X_{i_k}^{r_k}. \qquad (3.4.27)$$

In particular, if Q is homogeneous of degree m, then (3.4.27) becomes

$$\sum_{\mathbf{i} \in I_n^m} a_{i_1 \ldots i_m} X_{i_1}^1 \cdots X_{i_m}^m, \qquad (3.4.28)$$

which is the natural decoupling of the multilinear form (3.4.22). We then have

THEOREM 3.4.6. *For every $m \in \mathbb{N}$, there exists $C_m \in (0, \infty)$ such that if $\mathbf{X} = (X_1, \ldots X_n)$ is a vector of n independent random variables with $n > m$, $\mathbf{X}^j$, $j = 1, \ldots, m$, are m independent copies of $\mathbf{X}$, $Q_{(m)}(\mathbf{X})$, $N > m$, is a (not necessarily homogeneous) tetrahedral polynomial in $\mathbf{X}$ of degree m with coefficients in a Banach space B, as in (3.4.25), and if $\tilde{Q}_{(m)}(\mathbf{X}^1, \ldots, \mathbf{X}^m)$ is its decoupled version, as in (3.4.27), then, for all $t > 0$,*

$$\frac{1}{C_m} \Pr\Big\{ \|\tilde{Q}_{(m)}(\mathbf{X}^1, \ldots, \mathbf{X}^m)\| > C_m t \Big\}$$

$$\leq \Pr\Big\{ \|Q_{(m)}(\mathbf{X})\| > t \Big\} \qquad (3.4.29)$$

$$\leq C_m \Pr\Big\{ \|\tilde{Q}_{(m)}(\mathbf{X}^1, \ldots, \mathbf{X}^m)\| > \frac{t}{C_m} \Big\}.$$

3.5 Randomization

In this section we prove several randomization inequalities. We show that moments of U-statistics and U-processes are equivalent to the same moments of the randomized (or decoupled and randomized) U-statistics or U-processes. Regarding tail probabilities, we only show that the decoupled and randomized U-statistics are stochastically smaller than the original (the converse does not even hold for $m = 1$). This one-sided decoupling and randomization for tail probabilities is more elementary in nature, but still quite useful.

As usual, we let $(S, \mathcal{S}, P)$ be a probability space and let X_i, X_i^j ε_i, ε_i^j be the coordinate functions on the product of countably many copies of $(S, \mathcal{S}, P)$ and countably many copies of $(\{-1, 1\}, (\delta_1 + \delta_{-1})/2)$. In particular these variables are all independent, the Xs have law P, and the εs are Rademacher variables.

The amount of randomization one can perform on statistics of the form $\sum_{(i_1, \ldots, i_m) \in I_n^m} f(X_{i_1}, \ldots, X_{i_m})$ and still obtain a statistic of the same order of magnitude depends, as we shall see, on the *degree of P-degeneracy* of the kernel f, that we now define.

Definition 3.5.1. *A P^m-integrable symmetric function of m variables, $f : S^m \mapsto \mathbb{R}$, is P-degenerate of order $r-1$, $1 < r \le m$, if*

$$\int f(x_1, \dots, x_m) dP^{m-r+1}(x_r, \dots, x_m) = \int f dP^m \quad \text{for all } x_1, \dots, x_{r-1} \in S$$

whereas

$$\int f(x_1, \dots, x_m) dP^{m-r}(x_{r+1}, \dots, x_m)$$

is not a constant function. If f is P^m-centered and is P-degenerate of order $m-1$, that is, if

$$\int f(x_1, \dots, x_m) dP(x_1) = 0 \quad \text{for all } x_2, \dots, x_m \in S,$$

then f is said to be canonical or completely degenerate with respect to P. If f is not degenerate of any positive order, we say it is nondegenerate or degenerate of order zero.

In this definition the identities are usually taken in the almost everywhere sense. However, when dealing with uncountable families of functions (and only then), we need them to hold pointwise.

With the notation $P_1 \times \cdots \times P_m f = \int f d(P_1 \times \cdots \times P_m)$, the *Hoeffding projections* of $f : S^m \mapsto \mathbb{R}$ symmetric are defined as

$$\pi_k f(x_1, \dots, x_k) := \pi_{k,m}^P f(x_1, \dots, x_k) := (\delta_{x_1} - P) \times \cdots \times (\delta_{x_k} - P) \times P^{m-k} f$$

for $x_i \in S$ and $0 \le k \le m$. Note that $\pi_0 f = P^m f$ and that, for $k > 0$, $\pi_k f$ is a completely degenerate function of k variables. For f integrable these projections induce a decomposition of the U-statistic

$$U_n(f) := U_n^{(m)}(f) := U_n^{(m)}(f, P) := \frac{1}{\binom{n}{m}} \sum_{1 \le i_1 < \cdots < i_m \le n} f(X_{i_1}, \dots, X_{i_m})$$

into a sum of U-statistics of orders $k \le m$, which are orthogonal if $P^m f^2 < \infty$ and whose kernels are completely degenerate, namely, the *Hoeffding decomposition*:

$$U_n^{(m)}(f) = \sum_{k=0}^{m} \binom{m}{k} U_n^{(k)}(\pi_k f) \tag{3.5.1}$$

(here the superindex P and the subindex m of $\pi_{k,m}^P$ are not displayed; they will be dropped whenever no confusion is possible). This decomposition follows easily by expanding

$$f(x_1, \dots, x_m) = \delta_{x_1} \times \cdots \times \delta_{x_m} f = \big((\delta_{x_1} - P) + P\big) \times \cdots \times \big((\delta_{x_m} - P) + P\big) f$$

into terms of the form $(\delta_{x_{i_1}} - P) \times \cdots \times (\delta_{x_{i_k}} - P) \times P^{m-k} f$. It is very simple to check that f symmetric is P-degenerate of order $r-1$ iff $r = \min\{k > 0 : \pi_{k,m}^P f \not\equiv 0\}$.

Therefore, f is degenerate of order $r-1 \geq 0$ iff its Hoeffding expansion, except for the constant term, starts at term r, that is,

$$U_n(f) - P^m f = \sum_{k=r}^{m} \binom{m}{k} U_n^{(k)}(\pi_k f). \tag{3.5.2}$$

Hoeffding's decomposition is a basic tool in the analysis of U-statistics and in particular it will be put to use in the proof of the randomization theorem.

We are interested in the behavior of $\|U_n(f) - P^m f\|_{\mathcal{F}} := \sup_{f\in\mathcal{F}} |U_n(f) - P^m f|$ for possibly uncountable families $\mathcal{F}$ of symmetric functions $f : S^m \mapsto \mathbb{R}$. Whereas the *measurability* requirements for decoupling are minimal, randomization requires (or at least would not be useful without) the possibility of using Fubini's theorem on expressions of the form $\sup_{f\in\mathcal{F}} |\sum \varepsilon_{i_1}\cdots\varepsilon_{i_k} f(X_{i_1},\ldots,X_{i_m})|$, whose integrals one wants to compute by first integrating over the εs and then over the Xs or vice versa. In particular these expressions should be measurable. If the class $\mathcal{F}$ of measurable functions is countable, there are no measurability problems. A quite general situation for which one can work without measurability problems, as if the class were countable, is when the sigma algebra $\mathcal{S}$ is countably generated and contains the singletons [$(S, \mathcal{S})$ is a *separable measurable space*] and the class of functions $\mathcal{F}$ is *image-admissible Suslin*, that is, when there is a map from a Polish space Y onto $\mathcal{F}$, T, such that the composition of T and the evaluation map, $(y, x_1, \ldots, x_m) \mapsto T(y)(x_1, \ldots, x_m)$, is jointly measurable. Often the classes of functions of interest are parametrized by G_δ subsets $\Theta \subset \mathbb{R}^d$ and the evaluation map is jointly measurable in the arguments and the parameter, thus the usefulness of the image-admissible Suslin concept. The arguments in the proof of Theorem 10.3.2 in Dudley (1984) show that if $\mathcal{F}$ is image-admissible Suslin, then, for any constants as and b, the function

$$\|\sum a_{i_1,\ldots,i_m} f(x_{i_1},\ldots,x_{i_m}) + bP^m f\|_{\mathcal{F}}$$

is universally measurable (for any P). For $\mathcal{F}$ image-admissible Suslin the map $(y, x_1, \ldots, x_m) \mapsto \int T(y)(x_1, \ldots, x_m)dP^r(x_1, \ldots, x_r)$ is also jointly measurable for all $r \leq m$ by a monotone class argument and therefore the classes $\{\pi_k f : f \in \mathcal{F}\}$ are also image-admissible Suslin; in particular, we can replace f by $\pi_k f$ in the previous sup. We will also need to replace $\mathcal{F}$ in this sup by $\mathcal{F}'_\delta := \{f - g : f, g \in \mathcal{F}, d(f, g) \leq \delta\}$, where d is, e.g., the $L_2(P^m)$ distance. It follows from the definition that if $\mathcal{F}$ is image-admissible Suslin, so is $\mathcal{F}'_\delta$ and therefore the replacement is also possible. These observations are all we need on measurablity, and we refer for proofs to Dudley (1984) or to the soon to appear new book of Dudley (1998). For simplicity, image-admissible Suslin classes of functions over separable measurable spaces will simply be denoted as *measurable* classes.

Also, we will assume that all the classes of functions $\mathcal{F}$ considered here admit *everywhere finite envelopes*: $F(x_1, \ldots, x_m) := \sup_{f\in\mathcal{F}} |f(x_1, \ldots, x_m)| < \infty$, for all $x_i \in S$ and, likewise, $\sup_{f\in\mathcal{F}} |\pi_k f(x_1, \ldots, x_k)| < \infty$ for all $x_i \in S$ and $k \leq m$.

NOTATION. The symbol $\simeq$ between two expressions means two-sided inequality up to multiplicative constants that depend only on the order m of the U-processes and on the exponent p. Likewise, the symbols $\lesssim$ and $\gtrsim$ are used for one-sided inequalities up to multiplicative constants. We will also use, but only in proofs, the abbreviations $\mathbf{i}$ for the multiindex $(i_1, \ldots, i_m)$, $\mathbf{X_i}$ for the vector $(X_{i_1}, \ldots, X_{i_m})$, $\mathbf{X}_\mathbf{i}^{\text{dec}}$ for $(X^1_{i_1}, \ldots, X^m_{i_m})$, $\boldsymbol{\epsilon}_\mathbf{i}$ for the product $\varepsilon_{i_1} \cdots \varepsilon_{i_m}$, and $\boldsymbol{\epsilon}_\mathbf{i}^{\text{dec}}$ for the product $\varepsilon^1_{i_1} \cdots \varepsilon^m_{i_m}$. Here and in the rest of this section, $\{X^j_i\}$, $\{\varepsilon^j_i\}$ are independent copies of $\{X_i\}$, $\{\varepsilon_i\}$, defined as coordinate functions on the product proabability space (3.1.16).

3.5.1 *Moment Inequalities for Randomized U-Statistics*

We begin by recalling the simple randomization principle for sums of independent random variables from Lemma 1.2.6. In the present setting the proof requires a little extra care.

LEMMA 3.5.2. *Let $\mathcal{H}$ be a measurable class of P-centered functions $h : S \mapsto \mathbb{R}$ such that, for some $p \geq 1$, the envelope H of the class satisfies $PH^p < \infty$. Then, for all $n < \infty$,*

$$2^{-p}\mathbb{E}\Big\|\sum_{i=1}^n \varepsilon_i h(X_i)\Big\|^p_{\mathcal{H}} \leq \mathbb{E}\Big\|\sum_{i=1}^n h(X_i)\Big\|^p_{\mathcal{H}} \leq 2^p\mathbb{E}\Big\|\sum_{i=1}^n \varepsilon_i h(X_i)\Big\|^p_{\mathcal{H}}. \tag{3.5.3}$$

PROOF. Both inequalitites are easy and have similar proofs. We sketch the first and leave the second to the reader. Letting $Y_i = X_{n+i}$, owing to the measurability hypothesis on $\mathcal{H}$, the monotonicity of the integral, the invariance of $P^{\mathbb{N}}$ under permutations, and Jensen's inequality, we have

$$\begin{aligned}
\mathbb{E}\Big\|\sum \varepsilon_i h(X_i)\Big\|^p_{\mathcal{H}} &= \mathbb{E}_\varepsilon\Big(\mathbb{E}_X\Big\|\sum \varepsilon_i h(X_i) - \mathbb{E}_Y \sum \varepsilon_i h(Y_i)\Big\|^p_{\mathcal{H}}\Big) \\
&\qquad\qquad (\text{as } \mathbb{E}h(Y_i) = 0) \\
&\leq \mathbb{E}_\varepsilon\Big[\mathbb{E}_X\Big(\mathbb{E}_Y\Big\|\sum \varepsilon_i (h(X_i) - h(Y_i))\Big\|_{\mathcal{H}}\Big)^p\Big] \\
&\qquad\qquad (\text{by monotonicity of } \mathbb{E}_Y) \\
&\leq \mathbb{E}_\varepsilon\Big(\mathbb{E}_{X,Y}\Big\|\sum \varepsilon_i (h(X_i) - h(Y_i))\Big\|^p_{\mathcal{H}}\Big) \\
&\qquad\qquad (\text{by Jensen and Fubini}) \\
&= \mathbb{E}_{X,Y}\Big\|\sum (h(X_i) - h(Y_i))\Big\|^p_{\mathcal{H}} \\
&\qquad\qquad (\text{by permutation invariance of } P^{\mathbb{N}}) \\
&\leq 2^p\mathbb{E}\Big\|\sum_{i=1}^n h(X_i)\Big\|^p_{\mathcal{H}}.
\end{aligned}$$

□

In fact, exactly the same proof gives that if Ψ is a non-negative *convex* function on $[0, \infty)$ [such as x^p for $p \geq 1$, but also such as $\exp(x^p)$, $p \geq 1$], then

$$\mathbb{E}\Psi\Big(\frac{1}{2}\Big\|\sum_{i=1}^{n}\varepsilon_i h(X_i)\Big\|_{\mathcal{H}}\Big) \leq \mathbb{E}\Psi\Big(\Big\|\sum_{i=1}^{n}h(X_i)\Big\|_{\mathcal{H}}\Big) \leq \mathbb{E}\Psi\Big(2\Big\|\sum_{i=1}^{n}\varepsilon_i h(X_i)\Big\|_{\mathcal{H}}\Big). \tag{3.5.3$'$}$$

We can now prove the *randomization theorem*. This theorem is stated and proved in full generality although only the cases $r = m$ (the completely degenerate case), $r = 1$, and $r = 2$ are used in subsequent sections. The proof for $r = m$ is a straightforward consequence of the decoupling theorems of Section 3.1 and of Lemma 3.5.1 above. The proof for general r is equally easy but more complicated. We should remark that, although the theorem is stated only for $\Psi(|x|) = |x|^p$, it holds in fact for $\Psi(|x|)$ for all convex non-negative and nondecreasing Ψ.

THEOREM 3.5.3. *For $1 \leq r \leq m$ and $p \geq 1$, let $\mathcal{F}$ be a measurable class of real functions defined on S^m consisting of P-centered, P-degenerate functions of order at least $r-1$ such that $P^m F^p < \infty$. Then,*

$$\begin{aligned}
&\mathbb{E}\Big\|\sum_{(i_1,\ldots,i_m)\in I_n^m} f(X_{i_1},\ldots,X_{i_m})\Big\|_{\mathcal{F}}^p \\
&\qquad\simeq \mathbb{E}\Big\|\sum_{(i_1,\ldots,i_m)\in I_n^m} \varepsilon_{i_1}\cdots\varepsilon_{i_r} f(X_{i_1},\ldots,X_{i_m})\Big\|_{\mathcal{F}}^p \\
&\qquad\simeq \mathbb{E}\Big\|\sum_{(i_1,\ldots,i_m)\in I_n^m} \varepsilon_{i_1}^1\cdots\varepsilon_{i_r}^r f(X_{i_1},\ldots,X_{i_m})\Big\|_{\mathcal{F}}^p \\
&\qquad\simeq \mathbb{E}\Big\|\sum_{(i_1,\ldots,i_m)\in I_n^m} \varepsilon_{i_1}^1\cdots\varepsilon_{i_r}^r f(X_{i_1}^1,\ldots,X_{i_m}^m)\Big\|_{\mathcal{F}}^p.
\end{aligned} \tag{3.5.4}$$

PROOF. The case $r = m$ is immediate by decoupling and Lemma 3.5.2. Since in this case $\mathbb{E}f(X_1, x_2, \ldots, x_m) = 0$ for all $x_2, \ldots, x_m \in S$, letting $\mathbb{E}_r$ denote integration with respect to only the variables ε^r, X^r, Lemma 3.5.2 gives

$$\begin{aligned}
\mathbb{E}\Big\|\sum_{\mathbf{i}\in I_n^m} f(\mathbf{X}_{\mathbf{i}}^{\mathrm{dec}})\Big\|_{\mathcal{F}}^p &= \mathbb{E}\mathbb{E}_1\Big\|\sum_{i_1=1}^{n}\Big(\sum_{(i_2,\ldots,i_m):\mathbf{i}\in I_n^m} f(\mathbf{X}_{\mathbf{i}}^{\mathrm{dec}})\Big)\Big\|_{\mathcal{F}}^p \\
&\simeq \mathbb{E}\mathbb{E}_1\Big\|\sum_{\mathbf{i}\in I_n^m}\varepsilon_{i_1}^1 f(\mathbf{X}_{\mathbf{i}}^{\mathrm{dec}})\Big\|_{\mathcal{F}}^p \\
&= \mathbb{E}\mathbb{E}_2\Big\|\sum_{i_2=1}^{n}\Big(\sum_{(i_1,i_3,\ldots,i_m):\mathbf{i}\in I_n^m} \varepsilon_{i_1}^1 f(\mathbf{X}_{\mathbf{i}}^{\mathrm{dec}})\Big)\Big\|_{\mathcal{F}}^p \\
&\simeq \mathbb{E}\mathbb{E}_2\Big\|\sum_{\mathbf{i}\in I_n^m}\varepsilon_{i_1}^1\varepsilon_{i_2}^2 f(\mathbf{X}_{\mathbf{i}}^{\mathrm{dec}})\Big\|_{\mathcal{F}}^p \\
&\simeq \cdots \simeq \mathbb{E}\Big\|\sum_{\mathbf{i}\in I_n^m}\boldsymbol{\epsilon}_{\mathbf{i}}^{\mathrm{dec}} f(\mathbf{X}_{\mathbf{i}}^{\mathrm{dec}})\Big\|_{\mathcal{F}}^p.
\end{aligned}$$

Now the equivalences (3.5.4) for $r = m$ follow by several applications of the decoupling theorem in Section 3.1.

In what follows we drop the subindex $\mathcal{F}$ from the norm signs. For $r < m$ we use Hoeffding's decomposition to the effect that

$$\mathbb{E}\Big\|\sum_{I_n^m} f(\mathbf{X_i})\Big\|^p \le (m-r+1)^{p-1}\sum_{k=r}^{m}\Big(\frac{(n-k)!}{(n-m)!}\Big)^p \binom{m}{k}^p \mathbb{E}\Big\|\sum_{I_n^k}\pi_k f(\mathbf{X_i})\Big\|^p.$$

Since $\pi_k f$ is completely degenerate and $r \le k$, the previous argument shows

$$\mathbb{E}\Big\|\sum_{I_n^k}\pi_k f(\mathbf{X_i})\Big\|^p \simeq \mathbb{E}\Big\|\sum_{I_n^k}\varepsilon_{i_1}^1\cdots\varepsilon_{i_r}^r \pi_k f(\mathbf{X}_\mathbf{i}^{\mathrm{dec}})\Big\|^p.$$

Now, $\pi_k f(X_1^1, \dots, X_k^k)$ is a linear combination of 2^k terms of the form

$$P^{m-s} f(X_1^1, \dots, X_s^s, \cdot, \dots, \cdot) \quad \text{with} \quad s \le k,$$

and we have

$$\begin{aligned}
\mathbb{E}\Big\|\sum_{I_n^k}\varepsilon_{i_1}^1\cdots\varepsilon_{i_r}^r P^{m-s} f(X_{i_1}^1, \dots, X_{i_s}^s, \cdot, \dots, \cdot)\Big\|^p \\
&= \Big(\frac{(n-m)!}{(n-k)!}\Big)^p \mathbb{E}\Big\|\sum_{I_n^m}\varepsilon_{i_1}^1\cdots\varepsilon_{i_r}^r P^{m-s} f(X_{i_1}^1, \dots, X_{i_m}^m)\Big\|^p \\
&\le \Big(\frac{(n-m)!}{(n-k)!}\Big)^p \mathbb{E}\Big\|\sum_{I_n^m}\varepsilon_{i_1}^1\cdots\varepsilon_{i_r}^r f(X_{i_1}^1, \dots, X_{i_m}^m)\Big\|^p,
\end{aligned}$$

where the identity is obvious and the inequality follows from monotonicity of P^{m-s} and Jensen. We then conclude

$$\mathbb{E}\Big\|\sum_{I_n^m} f(\mathbf{X_i})\Big\|^p \lesssim \mathbb{E}\Big\|\sum_{I_n^m}\varepsilon_{i_1}^1\cdots\varepsilon_{i_r}^r f(\mathbf{X}_\mathbf{i}^{\mathrm{dec}})\Big\|^p. \tag{3.5.5}$$

We cannot undecouple in (ε, X) in the last expression because of lack of symmetry. In order to introduce symmetry we will use repeatedly and conditionally the following: if ξ is a symmetric random variable and a is a constant, then we have that for $p \ge 1$, $\mathbb{E}\|\xi\|^p \le \mathbb{E}\|\xi + a\|^p$, since

$$\mathbb{E}\|2\xi\|^p = \mathbb{E}\|(\xi+a)+(\xi-a)\|^p \le 2^{p-1}\big(\mathbb{E}\|a+\xi\|^p+\mathbb{E}\|a-\xi\|^p\big) = 2^p\mathbb{E}\|\xi+a\|^p.$$

Now, let

$$J_\ell = \big\{(j_1, \dots, j_r) : \\ 1 \le j_1 < \cdots < j_r \le m, \quad \ell = \max[i \le r : i \notin \{j_1, \dots, j_r\}]\big\},$$

$\ell = 1, \dots, r$. By applying the above inequality conditionally on all except the variables ε^1, we see that if we add to the last sum in (3.5.5), within the norm signs, the term $\sum_{\mathbf{i}\in I_n^m}\sum_{\mathbf{j}\in J_1}\varepsilon_{i(j_1)}^{j_1}\cdots\varepsilon_{i(j_r)}^{j_r} f(\mathbf{X_i}^{\mathrm{dec}})$, then the pth moment of the norm does not decrease; we can recursively add the analogous sums correponding to $J_2, \ldots, J_r$, and obtain

$$\mathbb{E}\Big\|\sum_{I_n^m} \varepsilon_{i_1}^1 \cdots \varepsilon_{i_r}^r f(X_{i_1}^1, \ldots, X_{i_m}^m)\Big\|^p$$
$$\le \mathbb{E}\Big\|\sum_{I_n^m}\Big(\sum_{1\le j_1<\cdots<j_r\le m} \varepsilon_{i(j_1)}^{j_1} \cdots \varepsilon_{i(j_r)}^{j_r} f(X_{i_1}^1, \ldots, X_{i_m}^m)\Big)\Big\|^p.$$

[For instance, if $r = 1$ and $m = 3$, then

$$\mathbb{E}\Big\|\sum_{I_n^3} \varepsilon_{i_1}^1 f(X_{i_1}^1, X_{i_2}^2, X_{i_3}^3)\Big\|^p \le \mathbb{E}\Big\|\sum_{I_n^3} (\varepsilon_{i_1}^1 + \varepsilon_{i_2}^2 + \varepsilon_{i_3}^3) f(X_{i_1}^1, X_{i_2}^2, X_{i_3}^3)\Big\|^p.]$$

The sum within the sup norm is the (unnormalized) decoupled U-statistic based on the symmetric kernel

$$h\big((\varepsilon_1, X_1), \ldots, (\varepsilon_m, X_m)\big) = \sum_{1<j_1<\cdots<j_r\le m} \varepsilon_{j_1} \cdots \varepsilon_{j_r} f(X_1, \ldots, X_m).$$

Therefore, we can now undecouple by Theorem 3.1.2 and obtain that the last expectation is dominated by a constant times

$$\mathbb{E}\Big\|\sum_{I_n^m} \varepsilon_{i_1} \cdots \varepsilon_{i_r} f(X_{i_1}, \ldots, X_{i_m})\Big\|^p.$$

Thus, we have proved

$$\mathbb{E}\Big\|\sum_{I_n^m} \varepsilon_{i_1}^1 \cdots \varepsilon_{i_r}^r f(\mathbf{X}_{\mathbf{i}}^{\text{dec}})\Big\|^p \lesssim \mathbb{E}\Big\|\sum_{I_n^m} \varepsilon_{i_1} \cdots \varepsilon_{i_r} f(\mathbf{X}_{\mathbf{i}})\Big\|^p. \tag{3.5.6}$$

By Theorem 2.1.2 the third term in (3.5.4) dominates a constant times the second and is dominated by a constant times the fourth and, moreover $\mathbb{E}\|\sum_{I_n^m} f(\mathbf{X}_{\mathbf{i}})\|^p \simeq \mathbb{E}\|\sum_{I_n^m} f(\mathbf{X}_{\mathbf{i}}^{\text{dec}})\|^p$. These considerations together with (3.5.5) and (3.5.6) reduce the proof of the theorem to showing

$$\mathbb{E}\Big\|\sum_{I_n^m} \varepsilon_{i_1}^1 \cdots \varepsilon_{i_r}^r f(\mathbf{X}_{\mathbf{i}}^{\text{dec}})\Big\|^p \lesssim \mathbb{E}\Big\|\sum_{I_n^m} f(\mathbf{X}_{\mathbf{i}}^{\text{dec}})\Big\|^p. \tag{3.5.7}$$

If f is centered and degenerate of order $r - 1$ then, as hinted immediately above equation (3.5.2),

$$f(x_1, \ldots, x_m) = \sum_{k=r}^{m} \sum_{1\le j_1<\cdots<j_k\le m} \pi_k f(x_{j_1}, \ldots, x_{j_k}),$$

and we have

$$\mathbb{E}\Big\|\sum_{I_n^m}\varepsilon_{i_1}^1\cdots\varepsilon_{i_r}^r f(\mathbf{X}_{\mathbf{i}}^{\mathrm{dec}})\Big\|^p$$

$$\lesssim \sum_{k=r}^{m}\mathbb{E}\Big\|\sum_{\mathbf{i}\in I_n^m}\varepsilon_{i_1}^1\cdots\varepsilon_{i_r}^r\sum_{1\le j_1<\cdots<j_k\le m}\pi_k f(X_{i(j_1)}^{j_1},\dots,X_{i(j_k)}^{j_k})\Big\|^p$$

$$\lesssim \sum_{k=r}^{m}\sum_{1\le j_1<\cdots<j_k\le m}\mathbb{E}\Big\|\sum_{\mathbf{i}\in I_n^m}\varepsilon_{i_1}^1\cdots\varepsilon_{i_r}^r\pi_k f(X_{i(j_1)}^{j_1},\dots,X_{i(j_k)}^{j_k})\Big\|^p.$$

If $\#\big[\{1,\dots,r\}\cap\{j_1,\dots,j_k\}\big]=\ell$, we have

$$\mathbb{E}\Big\|\sum_{I_n^m}\varepsilon_{i_1}^1\cdots\varepsilon_{i_r}^r\pi_k f(X_{i(j_1)}^{j_1},\dots,X_{i(j_k)}^{j_k})\Big\|^p$$

$$=\Big(\frac{(n-s)!}{(n-m)!}\Big)^p\mathbb{E}\Big\|\sum_{I_n^s}\varepsilon_{i_1}^1\cdots\varepsilon_{i_\ell}^\ell\varepsilon_{i_{k+1}}^{k+1}\cdots\varepsilon_{i_s}^s\pi_k f(X_{i_1}^1,\dots,X_{i_k}^k)\Big\|^p$$

$$:=V_N(k,r),$$

where $\ell+s-k=r$, $k\le s\le m$, and $N:=s-k=r-\ell$ is the number of Rademacher factors in the generic summand above, whose superindices are not shared by the Xs. So, (3.5.7) and hence the theorem will be proved if we show that

$$V_N(k,r)\lesssim\mathbb{E}\Big\|\sum_{I_n^m}f(\mathbf{X}_{\mathbf{i}}^{\mathrm{dec}})\Big\|^p \tag{3.5.8}$$

for all $r\le k\le m$ and $0\le N\le r\wedge(m-k)$. For each r and k such that $r\le k\le m$, we prove (3.5.8) by induction on N. For $N=0$, that is, for

$$V_0(k,r)=\mathbb{E}\Big\|\sum_{I_n^m}\varepsilon_{i_1}^1\cdots\varepsilon_{i_r}^r\pi_k f(X_{i_1}^1,\dots,X_{i_k}^k)\Big\|^p,$$

we can replace $\pi_k f(X_{i_1}^1,\dots,X_{i_k}^k)$ by its expansion as a linear combination of partial integrals of f, as in the proof of (3.5.5), then take the partial integrals out of the pth power of the sup norm (monotonicity of P^{m-s} and Jensen), and finally remove the εs using Lemma 3.5.2 conditionally several times, to obtain

$$V_0(k,r)\lesssim\mathbb{E}\|\sum_{I_n^m}f(\mathbf{X}_{\mathbf{i}}^{\mathrm{dec}})\|^p \tag*{$(3.5.8)_0$}$$

for all k and r with $r\le k\le m$. Suppose now (3.5.8) holds for $M<N$. If we try to take the Rademacher sums $\sum_{I_n^{s-k}}\varepsilon_{i_{k+1}}^{k+1}\cdots\varepsilon_{i_s}^s$ as common factors in $V_N(k,r)$, that is, if we try to compare $V_N(k,r)$ with

$$\bar V_N(k,r)$$

$$:=n^{(m-s)p}\mathbb{E}\Big\|\sum_{\mathbf{j}\in I_n^{s-k}}(\varepsilon_{j_{k+1}}^{k+1}\cdots\varepsilon_{j_s}^s)\Big(\sum_{\mathbf{i}\in I_n^k}\varepsilon_{i_1}^1\cdots\varepsilon_{i_\ell}^\ell\pi_k f(X_{i_1}^1,\dots,X_{i_k}^k)\Big)\Big\|^p,$$

we must introduce "diagonal terms" that is, terms for which some of the $\varepsilon^{(t)}$s, $k+1 \le t \le s$, say, α of them, have the same subindex as α of the ε^js, $1 \le j \le \ell$, and some *other* β $\varepsilon^{t'}$s, $k+1 \le t' \le s$, have the same subindices as β of the $X^{j'}$s, $\ell < j' \le k$. But the pth moment of the sup norm of such a term is dominated by a constant times V_M for $M = N - \alpha - \beta < N$ and therefore induction applies to it. [For instance, the diagonal term corresponding to $t = s$, $j = \ell$, $t' = s-1, s-2$, $j' = \ell+1, \ell+2$, is the sum of $(n-s+3)!/(n-s)!$ summands of the form

$$\sum_{I_n^{s-3}} \varepsilon_{i_1}^1 \cdots \varepsilon_{i_\ell}^\ell \varepsilon_{i_{k+1}}^{k+1} \cdots \varepsilon_{i_{s-3}}^{s-3} \varepsilon_{i_{\ell+2}}^{s-2} \varepsilon_{i_{\ell+1}}^{s-1} \varepsilon_{i_\ell}^{s} \pi_k f(X_{i_1}^1, \ldots, X_{i_k}^k).$$

Since $\{\varepsilon_i^\ell \varepsilon_i^s : i \le n\}$ is a Rademacher sequence independent from the others, and since we can relabel the superindices $(s-1)$, $(s-2)$, respectively, as $(\ell+1)$, $(\ell+2)$, the pth moment of the norm of these diagonal terms, multiplied by $n^{(m-s)p}$, is dominated by a constant times $V_{N-3}(k, r-1)$.] Thus, by the induction hypothesis, it suffices to prove

$$\bar{V}_N(k,r) \lesssim \mathbb{E}\Big\| \sum_{I_n^m} f(\mathbf{X}_{\mathbf{i}}^{\mathrm{dec}}) \Big\|^p. \tag{3.5.9}$$

Repeated application of Khinchin's inequality conditionally on all but one sequence of Rademacher variables shows

$$\mathbb{E}\Big| \sum_{(j_{k+1},\ldots,j_s)\in I_n^{s-k}} \varepsilon_{j_{k+1}}^{k+1} \cdots \varepsilon_{j_s}^{s} \Big|^p \simeq \big[\# I_n^{(s-k)}\big]^{\frac{p}{2}},$$

and therefore,

$$\begin{aligned}
\bar{V}_N(k,r) &= n^{(m-s)p} \mathbb{E}\Big\| \Big(\sum_{\mathbf{j}\in I_n^{s-k}} \varepsilon_{j_{k+1}}^{k+1} \cdots \varepsilon_{j_s}^{s}\Big)\Big(\sum_{\mathbf{i}\in I_n^k} \varepsilon_{i_1}^1 \cdots \varepsilon_{i_\ell}^\ell \pi_k f(X_{i_1}^1, \ldots, X_{i_k}^k)\Big)\Big\|^p \\
&\lesssim n^{(m-s)p} n^{\frac{(s-k)p}{2}} \mathbb{E}\Big\| \sum_{(i_1,\ldots,i_k)\in I_n^k} \varepsilon_{i_1}^1 \cdots \varepsilon_{i_\ell}^\ell \pi_k f(X_{i_1}^1, \ldots, X_{i_k}^k)\Big\|^p \\
&= n^{(m-s)p} n^{\frac{(s-k)p}{2}} n^{-(m-k)p} V_0(k,\ell) \\
&\lesssim \frac{1}{n^{\frac{(s-k)p}{2}}} \mathbb{E}\Big\| \sum_{I_n^m} f(\mathbf{X}_{\mathbf{i}}^{\mathrm{dec}}) \Big\|^p,
\end{aligned}$$

where in the last inequality we use $(3.5.8)_0$ with ℓ replacing r. So, (3.5.9) is proved, and therefore so are (3.5.8) and the theorem. □

REMARKS 3.5.4. 1) With only formal changes, the previous proof also shows that Theorem 3.5.3 holds as well for non-negative, nondecreasing, convex functions Ψ on $\mathbb{R}_+$, that is, under the assumptions of Theorem 3.5.3,

$$\begin{aligned}
&\mathbb{E}\Psi\Big(\Big\| \sum_{(i_1,\ldots,i_m)\in I_n^m} f(X_{i_1}, \ldots, X_{i_m})\Big\|_{\mathcal{F}}\Big) \\
&\quad \le \mathbb{E}\Psi\Big(C_1 \Big\| \sum_{(i_1,\ldots,i_m)\in I_n^m} \varepsilon_{i_1} \cdots \varepsilon_{i_r} f(X_{i_1}, \ldots, X_{i_m})\Big\|_{\mathcal{F}}\Big)
\end{aligned}$$

$$\leq \mathbb{E}\Psi\Big(C_2\Big\| \sum_{(i_1,\ldots,i_m)\in I_n^m} \varepsilon_{i_1}^1 \cdots \varepsilon_{i_r}^r f(X_{i_1},\ldots,X_{i_m})\Big\|_{\mathcal{F}}\Big)$$

$$\leq \mathbb{E}\Psi\Big(C_3\Big\| \sum_{(i_1,\ldots,i_m)\in I_n^m} \varepsilon_{i_1}^1 \cdots \varepsilon_{i_r}^r f(X_{i_1}^1,\ldots,X_{i_m}^m)\Big\|_{\mathcal{F}}\Big)$$

$$\leq \mathbb{E}\Psi\Big(C_4\Big\| \sum_{(i_1,\ldots,i_m)\in I_n^m} f(X_{i_1},\ldots,X_{i_m})\Big\|_{\mathcal{F}}\Big). \tag{3.5.4'}$$

2) The proof of Theorem 3.5.3, again with only formal changes, also shows that the class of functions $\mathcal{F}$ can be replaced by classes of functions $\mathcal{F}_{i_1,\ldots,i_m}$ with the same properties as $\mathcal{F}$ and with the additional permutation invariance property (3.1.14). Also, these classes of functions can be replaced by single functions $f_{i_1,\ldots,i_m}$ with values in a separable Banach space, degenerate of order $r-1$, with p-integrable norm, and satisfying (3.1.14); then, of course, the $\|\cdot\|_{\mathcal{F}}$ norm is replaced in the statement of the theorem by the Banach space norm (which can actually be thought of as the sup norm over the class $\mathcal{F}$ of functions in B', the dual of B, of norm one). This also applies to the above remark.

3.5.2 Randomization of Tail Probabilities for U-Statistics and Processes

Let F be a vector space, let the function $h : S^m \mapsto F$ be symmetric in its entries and let $I = \{(i_1,\ldots,i_m) \in \mathbb{N}^m : i_j \neq i_k \text{ for all } j \neq k\}$. For finite sets $A \subset \mathbb{N}$ we set

$$S_A = \sum_{\mathbf{i}\in I\cap A^m} h(\mathbf{x}_{\mathbf{i}}).$$

Now let $A = \cup_{i=1}^r A_i$ be a partition of a finite set $A \subset \mathbb{N}$ and let $m_1,\ldots m_r$ be non-negative integers whose sum is m; then we define $S_{A_1,\ldots,A_r}(m_1,\ldots,m_r)$ as $\sum h(x_{i_1},\ldots,x_{i_m})$ with the sum extended over all the multi-indices $(i_1,\ldots,i_m) \in I$ such that $\mathrm{Card}\big[\{i_1,\ldots,i_m\}\cap A_j\}\big] = m_j$ for $j = 1,\ldots,r$. Then, it is clear that

$$S_A = \sum_{(m_1,\ldots,m_r):\sum_{i=1}^r m_i = m} S_{A_1,\ldots,A_r}(m_1,\ldots,m_r). \tag{3.5.10}$$

Finally, on subsets C of $\{(m_1,\ldots,m_r) : \sum_{i=1}^r m_i = m\}$, we define the additive set function

$$S_{A_1,\ldots,A_r}(C) = \sum_{(m_1,\ldots,m_r)\in C} S_{A_1,\ldots,A_r}(m_1,\ldots,m_r).$$

The following elementary lemma will provide decoupling and randomization of tail probabilities.

LEMMA 3.5.5. *Let A_i, $i = 0, \ldots, m$, be $m+1$ finite disjoint sets of integers, $A_i \neq \emptyset$ if $i \neq 0$, and let $A = \cup_{i=0}^m A_i$. Then,*

$$m! \sum_{\mathbf{i} \in A_1 \times \cdots \times A_m} h(\mathbf{x_i}) = (-1)^m S_{A_0} + \sum_{k=1}^{m} (-1)^{m-k} \sum_{1 \leq i_1 < \cdots < i_k \leq m} S_{A_0 \cup A_{i_1} \cup \cdots \cup A_{i_k}}. \tag{3.5.11}$$

PROOF. Let $M = \{(m_0, m_1, \ldots, m_m) : m_i \in \mathbb{N} \cup \{0\}, \sum_{i=0}^m m_i = m\}$ and, for $i = 1, \ldots, m$, let $C_i = \{(m_0, \ldots, m_m) \in M : m_i = 0\}$. Then,

$$M = \{(0, 1, \ldots, 1)\} \cup C_1 \cup \cdots \cup C_m,$$

and $\{(0, 1, \ldots, 1)\}$ is disjoint from $C_1 \cup \cdots \cup C_m$. Therefore, equation (3.5.10), the additivity of the set function $S_{A_0, \ldots, A_m}(\cdot)$ and the inclusion-exclusion formula give

$$\begin{aligned} S_A &= S_{A_0, \ldots, A_m}(M) \\ &= S_{A_0, \ldots, A_m}(0, 1, \ldots, 1) \\ &\quad + \sum_{\ell=1}^{m} (-1)^{\ell+1} \sum_{1 \leq i_1 < \cdots < i_\ell \leq m} S_{A_0, \ldots, A_m}(C_{i_1} \cap \cdots \cap C_{i_\ell}). \end{aligned}$$

This is just equation (3.5.11) because

$$S_{A_0, \ldots, A_m}(C_{i_1} \cap \cdots \cap C_{i_\ell}) = S_{A \setminus (A_{i_1} \cup \cdots \cup A_{i_\ell})}$$

and, by symmetry,

$$S_{A_0, \ldots, A_m}(0, 1, \ldots, 1) = S_{A_1, \ldots, A_m}(1, \ldots, 1) = m! \sum_{\mathbf{i} \in A_1 \times \cdots \times A_m} h(\mathbf{x_i}). \qquad \square$$

THEOREM 3.5.6. *Let $\mathcal{F}$ be a measurable class of real functions on S^m, symmetric in their entries. Then,*

a) *for integers $0 \leq n_0 < n$, if D_j are subsets of $\{n_0 + 1, \ldots, n\}$, $j = 1, \ldots, m$, and $M = n_0 + \sum_{j=1}^m |D_j|$, then, for all $t > 0$,*

$$\begin{aligned} \Pr\Big\{\Big\| \sum_{\mathbf{i} \in D_1 \times \cdots \times D_m} f(X_{i_1}^1, \ldots, X_{i_m}^m) \Big\|_{\mathcal{F}} \geq \frac{2^m t}{m!}\Big\} \\ \leq 2^m \max_{n_0 < k \leq M} \Pr\Big\{\Big\| \sum_{\mathbf{i} \in I_k^m} f(X_{i_1}, \ldots, X_{i_m}) \Big\|_{\mathcal{F}} \geq t\Big\} \end{aligned} \tag{3.5.12}$$

(where $I_k^m = \emptyset$ if $k < m$);

b) *for integers $0 \leq n_0 < n$ and all $t > 0$,*

$$\begin{aligned} \Pr\Big\{\Big\| \sum_{n_0 < i_1, \ldots, i_m \leq n} \varepsilon_{i_1}^1 \cdots \varepsilon_{i_m}^m f(X_{i_1}^1, \ldots, X_{i_m}^m) \Big\|_{\mathcal{F}} \geq \frac{2^{2m} t}{m!}\Big\} \\ \leq 2^{2m} \max_{n_0 < k \leq nm} \Pr\Big\{\Big\| \sum_{\mathbf{i} \in I_k^m} f(X_{i_1}, \ldots, X_{i_m}) \Big\|_{\mathcal{F}} \geq t\Big\}. \end{aligned} \tag{3.5.13}$$

PROOF. Inequality (3.5.12) is trivially true if $D_j = \emptyset$ for some j. So, we can assume that D_j is not empty for any j. To prove (3.5.12) we now take $A_0 = \{1, \dots, n_0\}$, $A_j = \{1 + n_0 + \sum_{i=1}^{j-1} |D_i|, \dots, n_0 + \sum_{i=1}^{j} |D_i|\}$, $j = 1, \dots, m$, which are disjoint, and note that, by permutation of the factors in the infinite product of $(S, \mathcal{S}, P)$,

$$\Pr\Big\{\Big\|\sum_{\mathbf{i}\in D_1\times\cdots\times D_m} f(X_{i_1}^1, \dots, X_{i_m}^m)\Big\| \ge \frac{2^m t}{m!}\Big\}$$
$$= \Pr\Big\{\Big\|\sum_{\mathbf{i}\in A_1\times\cdots\times A_m} f(X_{i_1}, \dots, X_{i_m})\Big\| \ge \frac{2^m t}{m!}\Big\}.$$

Now, part (a) follows by direct application of Lemma 3.5.5.

Part (b) follows from part (a) and Fubini's theorem because

$$\sum_{n_0 \le i_1, \dots, i_m \le n} \varepsilon_{i_1}^1 \cdots \varepsilon_{i_m}^m f(X_{i_1}^1, \dots, X_{i_m}^m)$$

is a linear combination with coefficients ± 1 of 2^m terms of the form

$$\sum_{\mathbf{i}\in D_1\times\cdots\times D_m} f(X_{i_1}^1, \dots, X_{i_m}^m)$$

with $D_j = \{n_0 < i \le n : \varepsilon_i^j = 1\}$ or $D_j = \{n_0 < i \le n : \varepsilon_i^j = -1\}$. □

Of course, Theorem 3.5.6 can also be stated for Banach-space-valued kernels and Banach space norms (see Remark 3.5.4).

It should be noted that *there is no converse to inequality (3.5.13)*, in general, even for $m = 1$. For instance, if X is such that $\mathbb{E}X = 0$ and

$$\Pr\{X > t\} \simeq c_1 t^{-1}(\log t)^{-2}, \qquad \Pr\{X < -t\} \simeq c_2 t^{-1}(\log t)^{-2} \qquad (3.5.14)$$

as $t \to \infty$ with $0 \ne c_1 \ne c_2$ (it can be easily seen that such variables exist), then $\sum_{i=1}^n X_i = O_P\big(n(\log n)^{-1}\big)$, whereas $\sum_{i=1}^n \varepsilon_i X_i = O_P\big(n(\log n)^{-2}\big)$. To see this just note that X is in the domain of attraction of a 1-stable law with normings $a_n = n(\log n)^{-2}$ and centerings $b_n = -na_n^{-1}\mathbb{E}XI(|X| > a_n) \simeq \pm \log n$, that is, the variables $a_n^{-1}\sum_{i=1}^n X_i - b_n$ converge in distribution to a 1-stable variable and, on the other hand, the symmetric variables $a_n^{-1}\sum_{i=1}^n \varepsilon_i X_i$ also converge in law (to another 1-stable variable). [See, e.g., Feller (1966) or Araujo and Giné (1980) for definition and properties of domains of attraction.]

Part b) of Theorem 3.5.6 is all that is needed to show that stochastic boundedness of the sequence $\big\{\big\|\sum_{\mathbf{i}\in I_n^m} f(\mathbf{X_i})/b_n\big\|\big\}_{n=1}^\infty$ implies stochastic boundedness of the sequence of decoupled and randomized statistics, $\big\{\big\|\sum_{|\mathbf{i}|\le n} \boldsymbol{\epsilon}_{\mathbf{i}}^{\mathrm{dec}} f(\mathbf{X}_{\mathbf{i}}^{\mathrm{dec}})/b_n\big\|\big\}_{n=1}^\infty$, or to show that if the first converges to zero in probability, so does the second, at least if the normalizing constants b_n tend to infinity regularly enough. So, Theorem 3.5.6 should be useful for the central limit theorem and for the law of large numbers. However, more is needed for the law of the iterated logarithm, and we address this next.

The next theorem makes iterative use of Lévy's maximal inequality, Theorem 1.1.1 (this is one of the reasons decoupling and randomization is useful).

THEOREM 3.5.7. *Let $\mathcal{F}$ be as in Theorem 3.5.6. Let $\gamma(\cdot)$ be a positive nondecreasing function and $\{n_j\}$ be a sequence of positive integers satisfying $n_j \nearrow \infty$ and $\gamma\left(m\sum_{k\le j} n_k\right) \le c\gamma\left(n_j\right)$ for some $c < \infty$ and all $j \in \mathbb{N}$. Then,*

$$\limsup_{n\to\infty} \frac{1}{\gamma(n)} \Big\| \sum_{\mathbf{i}\in I_n^m} f(X_{i_1}, \dots, X_{i_m}) \Big\|_{\mathcal{F}} < c_0 < \infty \quad \text{a.s.} \tag{3.5.15}$$

implies

$$\sum_{j=1}^{\infty} \Pr\Big\{ \frac{1}{\gamma(n_j)} \max_{1\le n\le n_j} \Big\| \sum_{|\mathbf{i}|\le n} \varepsilon_{i_1}^1 \cdots \varepsilon_{i_m}^m f(X_{i_1}^1, \dots, X_{i_m}^m) \Big\|_{\mathcal{F}} > 4^m c c_0 \Big\} < \infty. \tag{3.5.16}$$

PROOF. Let $\tau_0 = 0$, $\tau_j = m\sum_{k=1}^{j} n_k$, $j \ge 1$. For each $j \in \mathbb{N}$ we decompose the interval $(\tau_{j-1}, \tau_j]$ into m equal subintervals $I_{j,1}, \dots, I_{j,m}$, and, for $1 \le r \le m$, relabel the sequence $\{X_i : i \in \cup_{j=1}^{\infty} I_{j,r}\}$ as $\{X_k^r : k = 1, 2, \dots\}$ [so that, if $i = m(n_1 + \cdots + n_s) + (r-1)n_{s+1} + t$, $0 < t \le n_{s+1}$, then $X_i = X^r_{n_1+\cdots+n_s+t}$]. The sequences $\{X_k^r : k \in \mathbb{N}\}$ are i.i.d. copies of the original sequence $\{X_i\}$. So, we can write $\mathbf{X}_{\mathbf{i}}^{\mathrm{dec}} = (X_{i_1}^1, \dots, X_{i_m}^m)$. Let $\{\varepsilon_i^r : i \in \mathbb{N}, 1 \le r \le m\}$ be a Rademacher array obtained from a Rademacher sequence $\{\varepsilon_i\}$ independent of $\{X_i\}$ in the same way. We can use these variables to construct a decoupled and randomized version of the original U-statistic. Obviously, for all $j \in \mathbb{N}$, the variable

$$\sum_{n_1+\cdots+n_{j-1} < i_1, \dots, i_m \le n_1+\cdots+n_j} \boldsymbol{\epsilon}_{\mathbf{i}}^{\mathrm{dec}} f(\mathbf{X}_{\mathbf{i}}^{\mathrm{dec}})$$

is the sum of 2^m terms of the form

$$\pm \sum_{\mathbf{i}\in A_{j,1}\times\cdots\times A_{j,m}} f(\mathbf{X}_{\mathbf{i}})$$

where, for each $1 \le r \le m$, $A_{j,r}$ is one of the two sets

$$A_{j,r} = \{i \in I_{j,r} : \varepsilon_i = 1\} \quad \text{or} \quad A_{j,r} = \{i \in I_{j,r} : \varepsilon_i = -1\}.$$

For $\{\varepsilon_i\}$ fixed, let us consider one (any) of the 2^m collections $\{A_{j,r} : j \in \mathbb{N}, 1 \le r \le m\}$ of such sets. Set $A_{j,0} = (0, \tau_{j-1}]$. By (3.5.11)

$$\begin{aligned} m! &\sum_{\mathbf{i}\in A_{j,1}\times\cdots\times A_{j,m}} f(\mathbf{X}_{\mathbf{i}}) \\ &= (-1)^m S_{A_{j,0}} + \sum_{r=1}^{m} (-1)^{m-r} \sum_{1\le s_1<\cdots<s_r\le m} S_{A_{j,0}\cup A_{j,s_1}\cup\cdots\cup A_{j,s_r}}. \end{aligned}$$

It follows from (3.5.15) that

$$\limsup_{j\to\infty} \frac{1}{\gamma(\tau_j)} \left| S_{B_j} \right| < c_0 \quad \text{a.s.}$$

for all sequences of sets of integers $\{B_j, j \geq 1\}$ such that $B_j \subseteq B_{j+1}$ and $B_j \subseteq (0, \tau_j]$, as $X, X_i, i \in \cup_j B_j$, are also i.i.d. random variables. By definition $A_{j,0} \cup A_{j,s_1} \cup \cdots \cup A_{j,s_r} \subseteq (0, \tau_j] = A_{j+1,0}$ for each subset $\{s_1 < s_2 < \cdots < s_r\}$ of $\{1, \ldots, m\}$, so that we can apply the previous limsup for $B_j := A_{j,0} \cup A_{j,s_1} \cup \cdots \cup A_{j,s_r}$, and obtain

$$\limsup_{j\to\infty} \frac{1}{\gamma(\tau_j)} \Big\| \sum_{\mathbf{i}\in A_{j,1}\times\cdots\times A_{j,m}} f(\mathbf{X}) \Big\|$$
$$\leq \sum_{r=0}^{m} \sum_{1\leq s_1<\cdots<s_r\leq m} \limsup_{j\to\infty} \frac{1}{\gamma(\tau_j)} \left\| S_{A_{j,0}\cup A_{j,s_1}\cup\cdots\cup A_{j,s_r}} \right\| < 2^m c_0 \quad \text{a.s.}$$

conditionally on the Rademacher array. Therefore, by Fubini's theorem,

$$\limsup_{j\to\infty} \frac{1}{\gamma(\tau_j)} \Big\| \sum_{n_1+\cdots+n_{j-1}<i_1,\ldots,i_m\leq n_1+\cdots+n_j} \epsilon_{\mathbf{i}}^{\text{dec}} f(\mathbf{X}_{\mathbf{i}}^{\text{dec}}) \Big\| < 2^m 2^m c_0 := M \quad \text{a.s.}$$

and then Borel–Cantelli gives

$$\sum_{j=1}^{\infty} \Pr\Big\{ \frac{1}{\gamma(\tau_j)} \Big\| \sum_{n_1+\cdots+n_{j-1}<i_1,\ldots,i_m\leq n_1+\cdots+n_j} \epsilon_{\mathbf{i}}^{\text{dec}} f(\mathbf{X}_{\mathbf{i}}^{\text{dec}}) \Big\| > M \Big\} < \infty.$$

In this last expression, the sum can be taken, instead, over the subindices $\mathbf{i}$ such that $|\mathbf{i}| \leq n_j$, and $\gamma(\tau_j)$ can be replaced by $c\gamma(n_j)$. Thus, the theorem will be proved as soon as we show that

$$\sum_{j=1}^{\infty} \Pr\Big\{ \frac{1}{\gamma(n_j)} \max_{n\leq n_j} \Big\| \sum_{|\mathbf{i}|\leq n} \epsilon_{\mathbf{i}}^{\text{dec}} f(\mathbf{X}_{\mathbf{i}}^{\text{dec}}) \Big\| > cM \Big\}$$
$$\leq \sum_{j=1}^{\infty} 2^m \Pr\Big\{ \frac{1}{\gamma(n_j)} \Big\| \sum_{|\mathbf{i}|\leq n_j} \epsilon_{\mathbf{i}}^{\text{dec}} f(\mathbf{X}_{\mathbf{i}}^{\text{dec}}) \Big\| > cM \Big\}. \tag{3.5.17}$$

This Lévy-type inequality can be obtained by conditionally applying Lévy's inequality (1.1.1) m times, as follows. For $1 \leq \ell \leq m \leq N$ and $1 \leq k_1, \ldots, k_\ell \leq N$, let $I_{N,m}(k_1, \ldots, k_\ell) = \{(i_1, \ldots, i_m) : i_j \leq k_j \text{ for } j \leq \ell, i_j \leq N \text{ for } j > \ell\}$, and let $\Pr_\ell$ denote conditional probability given $\{X_i^{(r)}, \varepsilon_i^{(r)} : r \neq \ell\}$. Also, for $1 \leq 1 \leq N$, let $\mathbf{v}_i$ be a vector in $\ell_{N^{\ell-1}}^{\infty}\big(\ell^\infty(\mathcal{F})\big)$ with coordinates all the possible sums

$$\Big\{ \sum_{\substack{i_j\leq k_j \text{ for } 1\leq j\leq \ell-1 \\ i_j\leq N \text{ for } \ell+1\leq j\leq N}} \varepsilon_{i_1}^1 \cdots \varepsilon_{i_{\ell-1}}^{\ell-1} \varepsilon_{i_{\ell+1}}^{\ell+1} \cdots \varepsilon_{i_m}^m f(X_{i_1}^1, \ldots, X_i^\ell, \ldots, X_{i_m}^m) : f \in \mathcal{F} \Big\},$$

with $1 \le k_j \le N$ and $1 \le j \le \ell - 1$. Then, Lévy's inequality, applied to the (conditionally) independent symmetric random vectors $\varepsilon_i^\ell \mathbf{v}_i \in \ell^\infty_{N^{\ell-1}}\big(\ell^\infty(\mathcal{F})\big)$, gives

$$\begin{aligned}
\Pr\Big\{ \max_{\substack{k_j \le N \\ \text{for } 1\le j\le \ell}} \Big\| \sum_{\mathbf{i}\in I_{N,m}(k_1,\dots,k_\ell)} \boldsymbol{\epsilon}_{\mathbf{i}}^{\text{dec}} f(\mathbf{X}_{\mathbf{i}}^{\text{dec}}) \Big\| > t \Big\}
&= \mathbb{E}\Pr_\ell \Big\{ \max_{k_\ell \le N} \Big\| \sum_{i=1}^{k_\ell} \varepsilon_i^\ell \mathbf{v}_i \Big\| > t \Big\} \\
&\le 2\mathbb{E}\Pr_\ell \Big\{ \Big\| \sum_{i=1}^{N} \varepsilon_i^\ell \mathbf{v}_i \Big\| > t \Big\} \\
&= 2\Pr\Big\{ \max_{\substack{k_j \le N \\ \text{for } 1\le j\le \ell-1}} \Big\| \sum_{\mathbf{i}\in I_{N,m}(k_1,\dots,k_{\ell-1})} \boldsymbol{\epsilon}_{\mathbf{i}}^{\text{dec}} f(\mathbf{X}_{\mathbf{i}}^{\text{dec}}) \Big\| > t \Big\},
\end{aligned}$$

and inequality (3.5.17) follows by iteration. □

By Borel–Cantelli, condition (3.5.15) is implied by convergence of the series

$$\sum_{j=1}^{\infty} \Pr\Big\{ \frac{1}{\gamma(n_j)} \max_{1\le n\le n_j} \Big\| \sum_{|\mathbf{i}|\le n} h(\mathbf{X}_{\mathbf{i}}) \Big\| > c_0 \Big\}.$$

Theorem 3.5.7 provides a sort of converse implication. For sums of independent random variables, similar statements about *blocking* follow directly from symmetrization, Lévy's inequality, and Borel–Cantelli, but for U-statistics, we achieve symmetrization (and decoupling) by means of the inclusion-exclusion formula (3.5.11).

It is natural to ask whether Corollary 1.1.5 in conjunction with decoupling of tail probabilities (Theorem 3.4.1) produces the same result as Lemma 3.5.5. However, the fact that the variables X_i in Corollary 1.1.5 must be identically distributed prevents its iterative application in the situation at hand.

3.6 References for Chapter 3

U-processes were introduced by Nolan and Pollard (1987) for $m = 2$ and by Arcones and Giné (1993) and Sherman (1994) in general. Nolan and Pollard's paper contains a randomization inequality whose proof uses methods close to decoupling. Decoupling of convex functions of tetrahedral quadratic forms $\sum_{I_n^2} X_i X_j$, and of tetrahedral multilinear forms in general, has a long history (Burkholder, 1983; McConnell and Taqqu, 1986, 1987; Zinn, 1985; Pisier (1985, unpublished) and de Acosta, 1987; Kwapień, 1987; Bourgain and Tzafriri, 1987; de la Peña and Klass, 1994, among others); the final result in the direction of U-statistics, Theorem 3.1.1 (and 3.1.2) above, was obtained by de la Peña (1992). The comments on measurability in Section 1 are based on remarks by D. Pollard sketched in de la Peña, loc. cit. The constants in Theorem 3.1.1 in the case of bivariate non-negative kernels can be drastically improved (de la Peña, 1992a). J. Cvitanic helped in obtaining the constants in the multivariate version of de la Peña (1992).

Nelson (1973) proved hypercontractivity for real Gaussian tetrahedral polynomials, Bonami (1970) and Gross (1975) proved it for real Rademacher polynomials and Borell (1979) extended it to Banach-valued Rademacher (and Gaussian) polynomials. Here we follow Beckner (1975) for the proof of the important inequality (3.2.4), and Kwapień and Szulga (1991) for the induction argument in the proof of Theorem 3.2.1. Lemmas 2.3 and 2.4 come from Kwapień (1987). We refer to Borell (1984), Krakowiak and Szulga (1988) and Kwapień and Szulga (1991) for extension of hypercontractivity to larger classes of variables. Lindenstrauss and Tzafriri (1979), pp. 75–77, use the same hypercontractivity methods to prove the Khinchin–Kahane inequalities.

Proposition 3.3.1 and Corollary 3.3.2 go back to Paley and Zygmund (1932). We learned these inequalities from Kahane (1968). In this generality, Theorem 3.3.3 is due to Borell (1984). Giné and Zinn (1994) contains inequalities (3.3.3) and (3.3.10), which are the basis for Corollary 3.3.4 and Theorem 3.3.6, "conditional Khinchin for tail probabilities"; they used these inequalities to prove the converse CLT for U-statistics. Proposition 3.3.7 and Corollary 3.3.8, "conditional Jensen for tail probailities" are due to Kwapień (1987), who used them for decoupling of tail probabilities of tetrahedral Gaussian multilinear forms. Theorem 3.3.9 is made explicit in this generality in Szulga (1993), building up on Kwapień's 3.3.8.

Kwapień (1987) was first in obtaining decoupling results for tail probabilities. He proved Theorem 3.4.6 in the particular case of tetrahedral multilinear forms in Gaussian variables, and he set up part of the present methodology by using results from Sections 2 and 3 above. A decoupling inequality for the tail probability of quadratic forms in independent symmetric random variables is found in Kwapień and Woyczynski (1992), Theorem 6.9.2. The main decoupling result of this chapter, Theorem 3.4.1, was obtained by de la Peña and Montgomery-Smith (1994, 1995). Theorem 3.4.1 contains, modulo constants, the decoupling result for convex functions in Section 1, which is however used in its proof. Theorem 3.4.1 can be seen as the culmination of a series of previous efforts on decoupling. For instance, modulo constants, it contains and improves upon previous decoupling inequalities for tetrahedral multilinear forms in symmetric variables due to to McConnell and Taqqu (1986), Pisier (1985, unpublished), de Acosta (1987), Kwapień (1987), Kwapień and Woyczynski (1992) and de la Peña, Montgomery-Smith, and Szulga (1994). Giné (1998) observed that Theorem 3.4.1 also applies to the decoupling of not necessarily homogeneous polynomials, as given in Theorem 3.4.6. For a variation including diagonals, see Klass and Nowicki (1996).

The randomization inequalities for moments of U-processes contained in Theorem 3.5.3 were first used in Arcones and Giné (1993) to prove the law of large numbers and the central limit theorem for U-processes. They are contained in de la Peña (1992) for the completely degenerate (canonical) case and in Arcones and Giné, loc. cit., for $r = 1, 2$. Nolan and Pollard (1987) for $m = 2$ and Sherman (1994) for general m, also have randomization inequalities for moments of suprema of canonical U-processes; they are quite different from those in Theorem 3.5.3, but they have similar uses. The randomization inequalities for tail probabilities (Theorem 3.5.6) were obtained for $n_0 = 1$ by Giné and Zinn (1994), and for general n_0 by Zhang (1996). Theorem 3.5.7, on decoupling in the LIL is in Giné and Zhang (1996).

4

Limit Theorems for U-Statistics

The decoupling results of Chapter 3 are very well suited for application in the asymptotic theory of U-statistics and U-processes, as we will see in this chapter and in the next. In this chapter we consider U-statistics. In general, we are interested in the "three gems" of classical probability theory, the law of large numbers, the central limit theorem, and the law of the iterated logarithm. Some inequalities, such as exponential and Hoffmann-Jørgensen-type inequalities, are also considered, mainly as first examples of application of decoupling and randomization, but also for their intrinsic interest.

The law of large numbers for U-statistics with integrable kernels is a simple consequence of the reverse martingale convergence theorem; however, integrability of the kernel h is not a necessary condition for the a.s. convergence of $U_n(h)$. There are other aspects of convergence a.s. and in probability of U-statistics, such as e.g., the analogs of the Marcinkiewicz–Zygmund law of large numbers, that we will not consider. We will only present an application of Hoffmann-Jørgensen's inequality to prove that some integrability of the kernel h is necessary for the law of large numbers (in fact, $\mathbb{E}h < \infty$ is necessary—as well as sufficient—if the kernel is non-negative). Section 4.1 contains the Hoffmann-Jørgensen and the exponential inequalities as well as comments on the law of large numbers.

Whereas the $\sqrt{n}$ central limit theorem for not necessarily degenerate U-statistics has a very simple proof and was obtained by Hoeffding in 1948, the CLT for degenerate U-statistics, more difficult, was obtained much later. The limit laws in the degenerate case are the laws of Gaussian chaos variables. Section 4.2 is devoted to the solution of the central limit theorem for completely degenerate U-statistics: the sequence $\{n^{k/2}U_n^{(k)}(h)\}$ converges in distribution if and only if

$\mathbb{E}h^2 < \infty$ and h is canonical. On the way to proving the direct part of this theorem we also construct the chaos decomposition of the L_2 space generated by a Gaussian process. This part of the theorem does not involve decoupling, but it applies to decoupling of Gaussian polynomials, a subject that is historically important for decoupling and that has recently had some applications in the analysis of Gaussian chaos (Heinrich, 1996; Marcus and Rosen, 1996; Marcus, 1997). The converse CLT, on the other hand, is strongly based on decoupling and randomization of tail probabilities and on the properties of Rademacher chaos.

The law of the iterated logarithm, presented in Section 4.3, is also proved for the components of the Hoeffding decomposition, that is, for U-statistics based on canonical kernels. The LIL for canonical U-statistics is not yet completely solved, and the situation is as follows: we know it holds if $\mathbb{E}h^2 < \infty$ and that it dos not hold if $\mathbb{E}h^2/(\log\log|h|)^{m-1+\delta} = \infty$ for some $\delta > 0$ [where $\log x$ is taken to mean $\log(x \vee 1)$], and we also know that $\mathbb{E}h^2 < \infty$ is not a necessary condition, but we know of no necessary and sufficient conditions for the validity of the LIL in general. The proofs of these results make extensive use of decoupling and randomization, as well as the properties of Rademacher chaos. The proof of the bounded LIL for canonical square integable kernels constitutes one of the neatest instances of application of decoupling and randomization of U-statistics and U-processes.

4.1 Some Inequalities; the Law of Large Numbers

As a first application of decoupling and randomization to U-statistics and processes, we derive in this section two types of inequalities. The first type constitute extensions of Hoffmann-Jørgensen's inequalities to canonical or non-negative U-processes, whereas the second consist of exponential inequalities for canonical U-statistics that extend some of the classical ones for sums of independent random variables. Hoffmann-Jørgensen's inequality is then put to work in the law of large numbers for U-statistics. It is also used in Section 3. The exponential inequalities, which will not be used in the sequel, are interesting in their own right as an important complement to the central limit theorem and as tools for other aspects of the asymptotics of U-statistics, such as moderate and large deviations, that we do not consider here.

4.1.1 Hoffmann-Jørgensen's Inequality for U-Processes

Decoupling will allow us to iterate Hoffmann-Jørgensen's inequality for sums of independent random vectors and obtain a similar inequality for U-statistics with positive or with canonical kernels. But the iteration must be applied to sums of independent random vectors with values in L_q, $0 < q < \infty$. Although the L_q spaces are not Banach for $0 < q < 1$, it turns out that the Hoffmann-Jørgensen inequalities are also true for these spaces, as follows. The L_q pseudodistance for

$q < 1, d(X, Y) = d(X - Y, 0) := \|X - Y\|_q^q$, which is translation invariant, fails to be homogeneous but satisfies instead $\|aX\|_q^q = |a|^q \|X\|_q^q$. This introduces a small change in Lévy's inequality (1.2.1), to wit: *if X_i are independent, symmetric random vectors in L_q of some probability (or measure) space, $0 < q < 1$, then*

$$\Pr\Big\{\max_{1\le k\le n} \|\sum_{i=1}^{k} X_i\|_q^q > t\Big\} \le 2\Pr\Big\{\|\sum_{i=1}^{n} X_i\|_q^q > 2^{q-1}t\Big\}. \tag{4.1.1}$$

The proof of Theorem 1.2.1, with $\|\cdot\|$ replaced by $\|\cdot\|_q^q$, works almost verbatim in this case, the only change being that now one can only prove

$$A_k = \big[A_k \cap \{\|S_n\|_q^q > 2^{q-1}t\}\big] \cup \big[A_k \cap \{\|S_n^{(k)}\|_q^q > 2^{q-1}t\}\big],$$

with factors of 2^{q-1} where before there were none. (This holds true because, otherwise,

$$2^q \|S_k(\omega)\|_q^q = \|S_n(\omega) + S_n^{(k)}(\omega)\|_q^q \le \|S_n(\omega)\|_q^q + \|S_n^{(k)}(\omega)\|_q^q \le 2 \cdot 2^{q-1}t$$

for some $\omega \in A$, contradiction.) And of course, if the random functions X_i are non-negative, then $\max_{1\le k\le n} \|\sum_{i=1}^{k} X_i\|_q^q = \|\sum_{i=1}^{n} X_i\|_q^q$. Inequality (4.1.1) has only the effect of changing some numerical constants in the proof of inequality (1.2.3) for $\|S_n\|_q^q$, as can be seen by going through the proof with some care. The conclusion is that *if Z_i are independent symmetric or independent non-negative random functions in L_q of a probability space* ($0 < q < 1$), *then there exist $K_{c,q}$ such that for all $c \le \alpha < \beta$,*

$$\mathbb{E}\|\sum Z_i\|_q^{q\beta} \le \Big(K_{c,q}\frac{\beta}{\alpha}\Big)^{\beta}\Big[\Big(\mathbb{E}\|\sum Z_i\|_q^{q\alpha}\Big)^{\beta/\alpha} + \mathbb{E}\max\|Z_i\|_q^{q\beta}\Big]. \tag{4.1.2}$$

We can now prove the following Hoffmann-Jørgensen-type inequality for U-statistics and processes. We state it first for non-negative kernels.

THEOREM 4.1.1. *Let h be a non-negative kernel symmetric in its entries, $h : S^m \mapsto \mathbb{R}$, and let X_i, $1 \le i \le n$, be independent S-valued random variables. Then, for each $0 < p < \infty$ there exist constants $C_{p,m,1}$, $C_{p,m,2}$, that depend only on p and m but not on h, S, X_i, or n, such that, if*

$$t_0 := \inf\Big[t > 0 : \Pr\Big\{\sum_{I_n^m} h(X_{i_1}, \dots, X_{i_m}) > t\Big\} \le C_{p,m,2}\Big],$$

then

$$\mathbb{E}\Big(\sum_{I_n^m} h(X_{i_1}, \dots, X_{i_m})\Big)^p \le C_{p,m,1}\Big[t_0^p + \mathbb{E}\max_{1\le i_1\le n}\Big(\sum_{I_n^m(i_1)} h(X_{i_1}, \dots, X_{i_m})\Big)^p\Big], \tag{4.1.3}$$

where $I_n^m(i_1) := \{(i_2, \dots, i_m) : (i_1, i_2, \dots, i_m) \in I_n^m\}$.

PROOF. First we will prove the decoupled version of (4.1.3) with t_0 replaced by the qth moment of the sum for any $q < p$, then we will use the decoupling inequalities from previous chapters to undecouple and, finally, Paley–Zygmund's argument (proof of Proposition 3.3.1) will take us from the qth moment of the sum to its quantile t_0.

Let $\{X_i^j : i \le n\}$, $j = 1, \dots, m$, be m independent copies of the sequence $\{X_i : i \le n\}$ and let $0 < q < p$. Let $\mathbb{E}_j$ denote expectation with respect to the X^j variables only and let $\mathbb{E}_{[r,s]}$ denote expectation with respect to the $X^r, \dots, X^s$ variables, for any $1 \le r < s \le m$. Then, Hoffmann-Jørgensen's inequality (1.2.7) for non-negative variables gives

$$\begin{aligned}\mathbb{E}\Big(\sum_{I_n^m} h(X_{i_1}^1, \dots, X_{i_m}^m)\Big)^p &= \mathbb{E}_{[2,m]}\mathbb{E}_1\Big[\sum_{i_1=1}^n\Big(\sum_{I_n^m(i_1)} h(X_{i_1}^1, \dots, X_{i_m}^m)\Big)\Big]^p \qquad (4.1.4)\\ &\le C\mathbb{E}_{[2,m]}\bigg\{\Big[\mathbb{E}_1\Big(\sum_{I_n^m} h(X_{i_1}^1, \dots, X_{i_m}^m)\Big)^q\Big]^{p/q}\\ &\qquad + \mathbb{E}_1 \max_{1\le i_1\le n}\Big(\sum_{I_n^m(i_1)} h(X_{i_1}^1, \dots, X_{i_m}^m)\Big)^p\bigg\},\end{aligned}$$

with a constant C depending only on p and q. The second summand is already the decoupled version of the second summand in (4.1.3) but the first must be further reduced. The reduction requires $m - 1$ analogous steps, and we just indicate the first. For fixed $X_{i_3}^3, \dots, X_{i_m}^m$, we consider the independent random vectors Z_{i_2} with values in $L_q(P^n)$ defined as

$$Z_{i_2}(x_1, \dots, x_n, X_{i_2}^2) = \sum_{I_n^m(i_2)} h(x_{i_1}, X_{i_2}^2, \dots, X_{i_m}^m), \qquad i_2 = 1, \dots, n,$$

with $I_n^m(i_2) := \{(i_1, i_3, \dots, i_m) : (i_1, i_2, i_3, \dots, i_m) \in I_n^m\}$. We note that the $L_q(P^n)$ distance to zero of the process $\sum_{i=2}^n Z_{i_2}$ is

$$\begin{aligned}&\int\Big(\sum_{i=2}^n Z_{i_2}(x_1, \dots, x_n, X_{i_2}^2)\Big)^q dP^n(x_1, \dots, x_n)\\ &\qquad = \mathbb{E}_1\Big(\sum_{I_n^m} h(X_{i_1}^1, \dots, X_{i_m}^m)\Big)^q = \mathbb{E}_1\Big(\sum_{i_2=2}^n Z_{i_2}(X_1^1, \dots, X_n^1, X_{i_2}^2)\Big)^q.\end{aligned}$$

Now, we apply Hoffmann-Jørgensen's inequality to these L_q-valued random vectors, to obtain

$$\mathbb{E}_2\Big[\mathbb{E}_1\Big(\sum_{i_2} Z_{i_2}\Big)^q\Big]^{p/q} \le C\Big[\mathbb{E}_{[1,2]}\Big(\sum_{i_2} Z_{i_2}\Big)^q\Big]^{p/q} + C\mathbb{E}_2 \max_{i_2}\big(\mathbb{E}_1 Z_{i_2}^q\big)^{p/q}. \quad (4.1.5)$$

For $q < 1$, (4.1.5) follows from (4.1.2) with $\alpha = 1$, and $\beta = p/q$, whereas for $q > 1$, it follows from inequality (1.2.7) for non-negative functions together with

Jensen's inequality. Rewriting inequality (4.1.5) in terms of h, applying Jensen to the second term and integrating the rest of the variables, gives

$$\mathbb{E}_{[2,m]}\Big[\mathbb{E}_1\Big(\sum_{I_n^m} h(X_{i_1}^1,\ldots,X_{i_m}^m)\Big)^q\Big]^{p/q}$$
$$\leq C\mathbb{E}_{[3,m]}\Big\{\Big[\mathbb{E}_{[1,2]}\Big(\sum_{I_n^m} h(X_{i_1}^1,\ldots,X_{i_m}^m)\Big)^q\Big]^{p/q} + \mathbb{E}_{[1,2]}\max_{1\leq i_2\leq n}\Big(\sum_{I_n^m(i_2)} h(X_{i_1}^1,\ldots,X_{i_m}^m)\Big)^p\Big\}.$$

This and (4.1.4) then give, by symmetry of h,

$$\mathbb{E}\Big(\sum_{I_n^m} h(X_{i_1}^1,\ldots,X_{i_m}^m)\Big)^p \leq C\mathbb{E}_{[3,m]}\Big\{\Big[\mathbb{E}_{[1,2]}\Big(\sum_{I_n^m} h(X_{i_1}^1,\ldots,X_{i_m}^m)\Big)^q\Big]^{p/q} + \mathbb{E}_{[1,2]}\max_{1\leq i_1\leq n}\Big(\sum_{I_n^m(i_1)} h(X_{i_1}^1,\ldots,X_{i_m}^m)\Big)^p\Big\},$$

Hence, it follows by iteration that

$$\mathbb{E}\Big(\sum_{I_n^m} h(X_{i_1}^1,\ldots,X_{i_m}^m)\Big)^p \leq C\Big\{\Big[\mathbb{E}\Big(\sum_{I_n^m} h(X_{i_1}^1,\ldots,X_{i_m}^m)\Big)^q\Big]^{p/q} + \mathbb{E}\max_{1\leq i_1\leq n}\Big(\sum_{I_n^m(i_1)} h(X_{i_1}^1,\ldots,X_{i_m}^m)\Big)^p\Big\}, \tag{4.1.6}$$

where the constant C depends only on p, q, and m.

Now we undecouple. By Theorem 3.4.1 the pth and qth moments of the sum $\sum_{I_n^m} h(X_{i_1}^1,\ldots,X_{i_m}^m)$ are equivalent to the corresponding moments of the original undecoupled variable $\sum_{I_n^m} h(X_{i_1},\ldots,X_{i_m})$, even if p or q are smaller than 1. Also, using the trick of incorporating the maximum into the norm, as in the proof of inequality (1.2.9) in Lemma 1.2.6, we also have, by the same theorem,

$$\mathbb{E}\max_{1\leq i_1\leq n}\Big(\sum_{I_n^m(i_1)} h(X_{i_1}^1,\ldots,X_{i_m}^m)\Big)^p \simeq \mathbb{E}\max_{1\leq i_1\leq n}\Big(\sum_{I_n^m(i_1)} h(X_{i_1}^1,X_{i_2},\ldots,X_{i_m})\Big)^p. \tag{4.1.7}$$

In order to complete the undecoupling of the maximum term we consider now the sigma algebras

$$\mathcal{F}_{i_1} := \sigma\{X_1^1,\ldots,X_{i_1}^1,X_1,\ldots,X_{i_1}\}, \qquad i_1 = 1,\ldots,n,$$

and note that

$$\mathbb{E}\Big[\sum_{I_n^m(i_1)} h(X_{i_1}^1,X_{i_2},\ldots,X_{i_m})\Big|\mathcal{F}_{i_1-1}\Big] = \mathbb{E}\Big[\sum_{I_n^m(i_1)} h(X_{i_1},X_{i_2},\ldots,X_{i_m})\Big|\mathcal{F}_{i_1-1}\Big],$$

that is, the two sequences $\{d_i\}$ and $\{e_i\}$ defined as

$$d_{i_1} = \sum_{I_n^m(i_1)} h(X_{i_1}^1, X_{i_2}, \ldots, X_{i_m}),\ e_{i_1} = \sum_{I_n^m(i_1)} h(X_{i_1}, X_{i_2}, \ldots, X_{i_m}),\ i_1 = 1, \ldots, n,$$

are tangent with respect to the σ-algebras $\mathcal{F}_{i_1}$, $i_1 = 1, \ldots, n$. Therefore the tail probabilities of their maxima over i_1 are comparable (both ways) by Lemma 2.3.3 on decoupling of maxima, thus obtaining,

$$\mathbb{E} \max_{1 \le i_1 \le n} \Big(\sum_{I_n^m(i_1)} h(X_{i_1}^1, X_{i_2}, \ldots, X_{i_m}) \Big)^p \simeq \mathbb{E} \max_{1 \le i_1 \le n} \Big(\sum_{I_n^m(i_1)} h(X_{i_1}, X_{i_2}, \ldots, X_{i_m}) \Big)^p. \tag{4.1.8}$$

Hence, by the previous observations and (4.1.7) and (4.1.8), we can completely undecouple inequality (4.1.6) to the effect that

$$\mathbb{E}\Big(\sum_{I_n^m} h(X_{i_1}, \ldots, X_{i_m})\Big)^p \le C\bigg\{\Big[\mathbb{E}\Big(\sum_{I_n^m} h(X_{i_1}, \ldots, X_{i_m})\Big)^q\Big]^{p/q} + \mathbb{E} \max_{1 \le i_1 \le n} \Big(\sum_{I_n^m(i_1)} h(X_{i_1}, \ldots, X_{i_m}) \Big)^p\bigg\}, \tag{4.1.9}$$

where C is a constant that depends only on p and q (and is possibly different from the constants in previous inequalities, also denoted as C).

Next, as a final step, we apply Paley–Zygmund's argument in conjunction with equation (4.1.9). For ease of notation we set

$$\xi = \sum_{I_n^m} h(X_{i_1}, \ldots, X_{i_m}).$$

As in the proof of Proposition 3.3.1, we have

$$\mathbb{E}\xi^{p/2} \le t_0^{p/2} + \big(\mathbb{E}\xi^p\big)^{1/2}\big(\Pr\{\xi > t_0\}\big)^{1/2}.$$

Taking t_0 such that

$$\Pr\{\xi > t_0\} \le \frac{1}{4C},$$

where C is the constant in inequality (4.1.9) corresponding to $q = p/2$, this gives

$$\big(\mathbb{E}\xi^{p/2}\big)^2 \le 2t_0^p + \frac{1}{2C}\mathbb{E}\xi^p,$$

and, combining with (4.1.9) for $q = p/2$,

$$\mathbb{E}\Big(\sum_{I_n^m} h(X_{i_1}, \ldots, X_{i_m})\Big)^p \le C\mathbb{E} \max_{1 \le i_1 \le n} \Big(\sum_{I_n^m(i_1)} h(X_{i_1}, \ldots, X_{i_m}) \Big)^p + 2Ct_0^p + \frac{1}{2}\mathbb{E}\Big(\sum_{I_n^m} h(X_{i_1}, \ldots, X_{i_m})\Big)^p.$$

Taking the last summand to the left side of the inequality proves (4.1.3). □

The same proof gives similar inequalities in the following situations: 1) if h non-negative takes values in a Banach space of functions for which $\|f\| \leq \|g\|$ whenever $0 \leq f \leq g$; 2) if $h(X_1, \ldots, X_n)$ is a symmetric random vector; and 3) if h is P-canonical B-valued and $p \geq 1$. For instance, in the last case, the fact that $\mathbb{E}h(X_1, x_2, \ldots, x_m) = 0$ allows us to use Theorem 1.2.5 for centered variables, and in the second case we must use equation (4.1.2) for $p \leq 1$. We only state the theorem for canonical kernels.

THEOREM 4.1.2. *Let h be a B-valued symmetric P-canonical kernel, $h : S^m \to B$, where B is a separable Banach space, and let X_i, $1 \leq i \leq n$, be independent S valued random variables with common probability law P. Then, for each $0 < p < \infty$ there exist universal constants $C_{p,m,1}, C_{p,m,2}$ (that is, constants that depend only on p and m but not on h, S, X_i, or n) such that, if*

$$t_0 := \inf\Big[t > 0 : \Pr\Big\{\Big\|\sum_{I_n^m} h(X_{i_1}, \ldots, X_{i_m})\Big\| > t\Big\} \leq C_{p,m,2}\Big],$$

then

$$\mathbb{E}\Big\|\sum_{I_n^m} h(X_{i_1}, \ldots, X_{i_m})\Big\|^p \leq C_{p,m,1}\Big[t_0^p + \mathbb{E}\max_{1\leq i_1\leq n}\Big\|\sum_{I_n^m(i_1)} h(X_{i_1}, \ldots, X_{i_m})\Big\|^p\Big]. \tag{4.1.10}$$

This theorem may be described in the following terms: when estimating moments of a multiple sum of U-statistic type, based on a canonical or a non-negative kernel, we may replace *one* of the sums by a max at the price of adding (a power of) a quantile. Then, a natural question is whether we are allowed to replace more than one sum by a max, and the answer is negative.

EXAMPLE 4.1.3. For each n, let X_i, $i \in \mathbb{N}$, be i.i.d. symmetric α-stable random variables for some $0 < \alpha < 2$, and let a_n satisfy $a_n^\alpha = n^{1+\delta}$ for some fixed $\delta \in (0, 1)$. Let ε_i, $i \in \mathbb{N}$, be independent Rademacher variables independent from the X variables. Then,

$$\mathbb{E}\Big(\sum_{I_n^m} \varepsilon_{i_1} \cdots \varepsilon_{i_m} I_{|X_{i_1}+\cdots+X_{i_m}|>a_n}\Big)^4 \gtrsim n^{2m-2-\delta},$$

$$\Big[\mathbb{E}\Big(\sum_{I_n^m} \varepsilon_{i_1} \cdots \varepsilon_{i_m} I_{|X_{i_1}+\cdots+X_{i_m}|>a_n}\Big)^2\Big]^2 \lesssim n^{2m-2-2\delta},$$

and

$$\mathbb{E}\max_{i_1\neq i_2\leq n}\Big(\sum_{I_n^m(i_1,i_2)} \varepsilon_{i_3} \cdots \varepsilon_{i_m} I_{|X_{i_1}+\cdots+X_{i_m}|>a_n}\Big)^4 \lesssim n^{2m-3-\delta},$$

so that the first quantity is of larger order than either of the other two. [Here and elsewhere, $A_n \lesssim B_n$ ($A_n \gtrsim B_n$) indicates that there exists a positive finite constant c independent of n such that $A_n \leq cB_n$ ($A_n \geq cB_n$), and $A_n \simeq B_n$ means that, both, $A_n \lesssim B_n$ and $A_n \gtrsim B_n$.] We prove these inequalities only for $m = 2$

since the proof for general m is not significantly different. We recall that if X is nondegenerate α-stable symmetric, then $t^\alpha \Pr\{X > t\} \to c$ as $t \to \infty$, where c is a positive constant. For the fourth moment, we have (Remark 3.2.8)

$$\begin{aligned}\mathbb{E}\Big(\sum_{I_n^2} \varepsilon_i \varepsilon_j I_{|X_i+X_j|>a_n}\Big)^4 &\simeq \mathbb{E}\Big(\sum_{I_n^2} I_{|X_i+X_j|>a_n}\Big)^2 \\ &\gtrsim n^3 \Pr\{|X_1 + X_2| > a_n, |X_1 + X_3| > a_n\} \\ &\gtrsim n^3 \Pr\{X_1 > a_n, X_2 > 0, X_3 > 0\} \gtrsim \frac{n^3}{a_n^\alpha} = n^{2-\delta};\end{aligned}$$

again by Remark 3.2.8, the square of the second moment is bounded by

$$\Big[\mathbb{E}\Big(\sum_{I_n^2} \varepsilon_i \varepsilon_j I_{|X_i+X_j|>a_n}\Big)^2\Big]^2 \simeq \Big[n^2 \Pr\{|X_1 + X_2| > a_n\}\Big]^2 \simeq \frac{n^4}{a_n^{2\alpha}} = n^{2-2\delta}$$

since $X_1 + X_2$ is also α-stable; finally, the maximum term obviously satisfies

$$\mathbb{E} \max_{1 \le i \ne j \le n} I_{|X_i+X_j|>a_n} \le n^2 \Pr\{|X_i + X_j| > a_n\} \simeq n^{1-\delta}.$$

4.1.2 *An Application to the Law of Large Numbers*

To prove the law of large numbers for U-statistics with an integrable kernel we can proceed by analogy with the martingale proof of the law of large numbers for sums of independent random variables, as follows. The random variables $U_n(h)$, for h integrable, form a reverse martingale with respect to the sigma fields $\mathcal{F}_n$ generated by $X_{(1,1)}, \ldots, X_{(1,n)}, X_{n+1}, \ldots$, where $X_{(1,1)}, \ldots, X_{(1,n)}$ are the order statistics corresponding to the i.i.d. sample $X_1, \ldots, X_n$. Hence, if $\mathbb{E}|h(X_1, \ldots, X_m)| < \infty$, the sequence $U_n(h)$ converges a.s. and in L_1. The limit being a constant by the Hewitt–Savage zero–one law, it can only be $\mathbb{E}h(X_1, \ldots, X_m)$. We have thus proved the following theorem.

THEOREM 4.1.4. *Let h be a real measurable function on S^m symmetric in its entries and let X_i, $i \in \mathbb{N}$, be i.i.d. S-valued random variables. If*

$$\mathbb{E}\big|h(X_1, \ldots, X_m)\big| < \infty, \tag{4.1.11}$$

then

$$\frac{(n-m)!}{n!} \sum_{I_n^m} h(X_{i_1}, \ldots, X_{i_m}) \to \mathbb{E}h(X_1, \ldots, X_m) \tag{4.1.12}$$

a.s. and in L_1.

It is natural to ask whether the integrability condition (4.1.11) on h can be relaxed in general. The answer is negative, as the following proposition shows.

PROPOSITION 4.1.5. *If $h \geq 0$ and if the sequence*

$$\left\{\frac{1}{n^m}\sum_{I_n^m} h(X_{i_1},\ldots,X_{i_m})\right\}_{n=1}^{\infty}$$

is stochastically bounded, then $\mathbb{E}h(X_1,\ldots,X_m) < \infty$.

PROOF. By positivity, the sequence of variables

$$\left\{\frac{1}{n^m}\sum_{I_n^m} h(X_{i_1},\ldots,X_{i_m})I_{h(X_{i_1},\ldots,X_{i_m})\leq n}\right\}_{n=1}^{\infty} \tag{4.1.13}$$

is also stochastically bounded. Hence, if $t_{0,n}$ is defined as

$$t_{0,n} = \inf\left[t > 0 : \Pr\left\{\frac{1}{n^m}\sum_{I_n^m} h(X_{i_1},\ldots,X_{i_m})I_{h(X_{i_1},\ldots,X_{i_m})\leq n} > t\right\} > C_{1,m,2}\right],$$

$C_{1,m,2}$ as in Hoffmann-Jørgensen's inequality above (Theorem 4.1.1), then

$$t_0 := \sup_n t_{0,n} < \infty.$$

This gives control of the first term at the right of inequality (4.1.3) uniformly over the sequence in (4.1.13). Moreover, the boundedness imposed on these variables implies

$$\frac{1}{n^m}\max_{1\leq i_1\leq n}\left[\sum_{I_n^m(i_1)} h(X_{i_1},\ldots,X_{i_m})I_{h(X_{i_1},\ldots,X_{i_m})\leq n}\right] \leq 1,$$

meaning that we also control the second term in Hoffmann-Jørgensen's inequality for the variables in (4.1.13) uniformly in n. So, if we apply Theorem 4.1.1 to the variables in the sequence (4.1.13) for $p = 1$, we obtain

$$\begin{aligned}
&\sup_n \mathbb{E}h(X_1,\ldots,X_1)I_{h(X_1,\ldots,X_m)\leq n}\\
&\qquad = \sup_n \mathbb{E}\left[\frac{1}{n^m}\sum_{I_n^m} h(X_{i_1},\ldots,X_{i_m})I_{h(X_{i_1},\ldots,X_{i_m})\leq n}\right]\\
&\qquad \leq C_{1,m,1}(t_0+1) < \infty.
\end{aligned}$$

Hence, $\mathbb{E}h < \infty$ by monotone convergence. □

Proposition 4.1.5 can be proved in a different way: instead of Hoffmann-Jørgensen's inequality, we could have applied the law of large numbers (4.1.12) to the truncated kernels $hI_{|h|\leq c}$ and reach the same conclusion. But the method of proof we have chosen gives other results as well, as we will see in Proposition 4.1.7 and in Theorem 4.3.10 below.

Since condition (4.1.11) cannot be relaxed in general, we may ask whether it is necessary for the LLN, and the answer is negative, by far:

EXAMPLE 4.1.6. For every $m \geq 2$ and $\varepsilon > 0$ there is a (symmetric) kernel h on $\mathbb{R}^m$ and a probability measure P on $\mathbb{R}$ such that, if X_i, $i \in \mathbb{N}$, are i.i.d. with law P and ε_i, $i \in \mathbb{N}$, are independent Rademacher variables independent of the X variables, then

$$\mathbb{E}|h(X_1, \ldots, X_m)|^{2/(m+1)+\varepsilon} = \infty, \tag{4.1.14}$$

whereas

$$\lim_{n\to\infty} \frac{(n-m)!}{n!} \sum_{I_n^m} \varepsilon_{i_1} \cdots \varepsilon_{i_m} h(X_{i_1}, \ldots, X_{i_m}) = 0 \quad \text{a.s.} \tag{4.1.15}$$

Note that, for each n, the random variable at the left of (4.1.15) is a U-statistic with kernel $f\big((x_1, y_1), \ldots, (x_m, y_m)\big) = y_1 \cdots y_m h(x_1, \ldots, x_m)$, canonical with respect to the variables $Z_i = (X_i, \varepsilon_i)$ and with the same absolute value as h. The example is as follows. Let $0 < \delta < \varepsilon$ be such that $p := \frac{2}{m+1} + \delta < 1$, and let X_i be i.i.d., non-negative and such that $\mathbb{E}X_i < \infty$ but $\mathbb{E}X_i^{1+\eta} = \infty$ for all $\eta > 0$ [e.g., $\Pr\{X > t\} = e^2 t^{-1} (\log t)^{-2}$ for $t \geq e$]. Then, the kernel

$$h(x_1, \ldots, x_m) = \sum_{I_m^m} x_{i_1}^{1/p} x_{i_2}^{(m-1/p)/(m-1)} \cdots x_{i_m}^{(m-1/p)/(m-1)}$$

satisfies the required properties. We prove this only for $m = 2$ because the proof for general m is similar but somewhat cumbersome. In this case, $p = \frac{2}{3} + \delta < 1$ and, obviously, $\mathbb{E}(X_1^{1/p} X_2^{2-1/p})^{\frac{2}{3}+\varepsilon} = \infty$. So, only the limit (4.1.15) must be verified. This limit will be a direct consequence of the Marcinkiewicz law of large numbers for sums of i.i.d. random variables ξ_i, so we recall it now (e.g., Stout, 1974): $n^{-1/r} \sum_{i=1}^n \xi_i \to 0$ a.s. if and only if $\mathbb{E}|\xi_1|^r < \infty$ whenever $0 < r < 1$, and if and only if both $\mathbb{E}|\xi_1|^r < \infty$ and $\mathbb{E}\xi_i = 0$ whenever $1 \leq r < 2$. Since $\mathbb{E}X < \infty$, $p < 2$ and $2 - 1/p > 1/2$, the Marcinkiewicz law of large numbers gives

$$\begin{aligned}
&\frac{1}{n^2} \sum_{1 \leq i \neq j \leq n} \varepsilon_i \varepsilon_j X_i^{1/p} X_j^{2-1/p} \\
&\quad = \Big(\frac{1}{n^{1/p}} \sum_{i=1}^n \varepsilon_i X_i^{1/p}\Big)\Big(\frac{1}{n^{2-1/p}} \sum_{i=1}^n \varepsilon_i X_i^{2-1/p}\Big) - \frac{1}{n^2} \sum_{i=1}^n X_i^2 \to 0 \quad \text{a.s.}
\end{aligned}$$

In light of this example, we may ask whether the law of large numbers for $U_n(h)$ does imply any integrability at all on the kernel h. Hoffmann-Jørgensen's inequality allows us to give an answer to this question that Example 4.1.6 shows to be essentially best possible (just "essentially": the exponent is only determined up to ε).

PROPOSITION 4.1.7. *If the sequence*

$$\Big\{\frac{1}{n^m} \sum_{I_n^m} h(X_{i_1}, \ldots, X_{i_m})\Big\}_{n=m}^{\infty}$$

is stochastically bounded, then

$$\mathbb{E}|h(X_1, \ldots, X_m)|^{2/(m+1)-\varepsilon} < \infty$$

for all $0 < \varepsilon \leq 2/(m+1)$.

PROOF. By Theorem 3.5.6 b), the decoupled and randomized sequence

$$\Big\{Z_n(\boldsymbol{\epsilon}, \mathbf{X}) := \frac{1}{n^m} \sum_{1 \leq i_1, \ldots, i_m \leq n} \varepsilon_{i_1}^1 \cdots \varepsilon_{i_m}^m h(X_{i_1}^1, \ldots, X_{i_m}^m)\Big\}_{n=m}^{\infty}$$

is also stochastically bounded. Conditionally on the sequence $\mathbf{X} = \{X_i\}$, the variable $Z_n(\boldsymbol{\epsilon}, \mathbf{X})$ is a nonhomogeneous Rademacher chaos of degree m. Then, we can apply Theorem 3.3.6 [concretely, inequality (3.3.10), conditional Khinchin for tail probabilities] for $p = 2$, and obtain

$$\Pr\{\mathbb{E}\big(Z_n^2(\boldsymbol{\epsilon}, \mathbf{X})|\mathbf{X}\big) > 2M^2\} \leq C_{m,2} \Pr\{|Z_n(\boldsymbol{\epsilon}, \mathbf{X})| > M\}$$

for all $M > 0$, where $C_{m,2}$ is a constant that depends only on m. Since, by stochastic boundedness,

$$\lim_{M \to \infty} \sup_n \Pr\{|Z_n(\boldsymbol{\epsilon}, \mathbf{X})| > M\} = 0,$$

and since

$$\mathbb{E}\big(Z_n^2(\boldsymbol{\epsilon}, \mathbf{X})|\mathbf{X}\big) = \frac{1}{n^{2m}} \sum_{1 \leq i_1, \ldots, i_m \leq n} h^2(X_{i_1}^1, \ldots, X_{i_m}^m), \qquad n \geq m,$$

it follows that this last sequence is also stochastically bounded. Then, by positivity, the collection of random variables

$$\Big\{\frac{1}{n^{2m}} \sum_{I_n^m} h^2(X_{i_1}^1, \ldots, X_{i_m}^m) I_{h^2(X_{i_1}^1, \ldots, X_{i_m}^m) \leq c} : n \geq m,\, c > 0\Big\}$$

is stochastically bounded as well. The decoupling inequality (3.4.1) then implies that the corresponding collection of "undecoupled" U-statistics is also stochastically bounded. So, there is a finite t_0 such that

$$\Pr\Big\{\frac{1}{n^{2m}} \sum_{I_n^m} h^2(X_{i_1}, \ldots, X_{i_m}) I_{h^2(X_{i_1}, \ldots, X_{i_m}) \leq c} \geq t_0\Big\} \leq C_{1,m,2}$$

for all n and c, where $C_{1,m,2}$ is an in Theorem 4.1.1 (Hoffmann-Jørgensen for non-negative kernels). Also, if for each n we truncate the kernel h^2 at the level $n^{2m-(m-1)} = n^{m+1}$, we have

$$\frac{1}{n^{2m}} \max_{1 \leq i_1 \leq n} \sum_{I_n^m(i_1)} h^2(X_{i_1}, \ldots, X_{i_m}) I_{h^2(X_{i_1}, \ldots, X_{i_m}) \leq n^{m+1}} \leq 1.$$

Hence, we can apply Hoffmann-Jørgensen's inequality (Theroem 4.1.1) to the sequence of random variables

$$\frac{1}{n^{2m}} \sum_{I_n^m} h^2(X_{i_1}, \ldots, X_{i_m}) I_{h^2(X_{i_1}, \ldots, X_{i_m}) \leq n^{m+1}},$$

just as in the proof of Proposition 4.1.5, and conclude that $\sup_n n^{-m}\mathbb{E}h^2 I_{h^2\le n^{m+1}} < \infty$, that is,

$$\mathbb{E}h^2 I_{h^2\le n^{m+1}} \le Cn^m, \qquad n \in \mathbb{N},$$

for a constant C independent of n. Then, taking $\beta = 2/(m+1)-\varepsilon$ in the following inequalities, these moment bounds give that, for some constant $C' < \infty$,

$$\begin{aligned}\mathbb{E}|h|^\beta &\le 1 + \sum_{k=0}^{\infty} \mathbb{E}|h|^\beta I_{2^k<h^2\le 2^{k+1}} \\ &\le 1 + \sum_{k=0}^{\infty} 2^{k(\beta-2)/2}\mathbb{E}h^2 I_{h^2\le 2^{k+1}} \simeq 1 + C'\sum 2^{-\varepsilon k/2} < \infty. \qquad \square\end{aligned}$$

There is no need to "undecouple" in the previous proof since Hoffman-Jørgensen's inequality also works for decoupled U-statistics (see the proof of Theorem 4.1.1).

It is clear from Proposition 4.1.5 and Example 4.1.6 that, contrary to the case of sums of independent random variables, *there cannot exist a characterization of the law of large numbers in terms of only moments of h*. The problem of finding necessary and sufficient condition on the random variable $h(X_1, \dots, X_m)$ equivalent to the law of large numbers for $\varepsilon_1 \cdots \varepsilon_m h(X_1, \dots, X_m)$ (and only depending on m integrations with respect to the law of X_1) has been very recently solved by Zhang (1997) and Latała and Zinn (1998).

4.1.3 Exponential Inequalities for Canonical U-Statistics

Hoeffding (1963) generalized to U-statistics with bounded nondegenerate kernels some of the classical exponential inequalities for sums of bounded independent random variables. The first term in the Hoeffding decomposition of $U_n(h) - P^m h$ for the nondegenerate case [see (3.5.1)] is just the sum of i.i.d. random variables expressed in linear functional notation as

$$\frac{m}{n}\sum_{i=1}^{n}\big(P^{m-1}h(X_i) - P^m h\big).$$

This term is $O_P(n^{-1/2})$ by the central limit theorem; in fact, it is asymptotically normal when multiplied by $\sqrt{n}$. As we will see in the next section, the remaining terms in Hoeffding's decomposition (the remainder after linearization of the U-statistic) are of smaller order, concretely, the kth term is $O_P(n^{-k/2})$, and its limit in distribution, when normalized by $n^{k/2}$, is a homogeneous Gaussian chaos variable of degree k. It is then natural to expect that exponential inequalities for nondegenerate U-statistics based on bounded kernels should be quite similar to the classical inequalities for sums of i.i.d. bounded random variables. And it is also natural that exponential inequalities for degenerate U-statistics should reflect their limiting behavior and thus be completely different from the nondegenerate case. It turns out that these inequalities for degenerate statistics were not obtained until very recently, when decoupling techniques were available, and their proofs use decoupling and randomization quite heavily (this is not to say that other proofs are not possible but, certainly, decoupling does simplify things). These inequalities are interesting in themselves and quite useful, e.g., in the study of large and moderate deviations. They may however be optimal only within a limited range of the argument t, and the problem of finding bounds that are optimal over the whole range remains open.

Hoeffding's inequalities are classical and there is no point in reproducing their proof here. However, we state them for the purpose of comparison.

THEOREM 4.1.8. (HOEFFDING (1963)) *Let h be a bounded measurable kernel on S^m and let X_i, $i \in \mathbb{N}$, be i.i.d. S-valued random variables such that $\mathbb{E}h(X_1, \dots, X_m) = 0$. Set $c := \|h\|_\infty$ and $\sigma^2 := \mathbb{E}h^2(X_1, \dots, X_m)$. Then, for all $t > 0$,*

a)

$$\Pr\{U_n(h) > t\} \le \exp\Big(-\frac{[n/m]t^2}{2\sigma^2 + (2/3)ct}\Big)$$

and

b)

$$\Pr\{U_n(h) > t\} \le \exp\Big(-\frac{[n/m]t^2}{2c^2}\Big).$$

Let now h be a P-canonical kernel symmetric in its entries (here symmetry is only a matter of convenience). The exponential inequalities for the variables $Y := \sum_{I_n^m} h(X_{i_1}, \dots, X_{i_m})/n^{m/2}$ that we will now obtain are based on estimates of $\mathbb{E}\exp(t|Y|^\alpha)$ for convenient values of $\alpha \le 1$, actually, for $\alpha = 2/(m+1)$. The function $\exp(x^\alpha)$, $x > 0$, is only convex in the range $x^\alpha \ge (1-\alpha)/\alpha$, but it can be made convex by modifying it in the range $0 \le x^\alpha \le (1-\alpha)/\alpha$: e.g., we can replace $y = \exp(x^\alpha)$ by $y = \exp\big((1-\alpha)/\alpha\big)$ for $0 \le x^\alpha \le (1-\alpha)/\alpha$. Then, by the randomization theorem for moments in Section 3.5 [see Remark 3.5.4, 1)],

$$\mathbb{E}\exp\big(t|Y|^\alpha\big) \simeq \mathbb{E}\exp\Big(ct\Big|\sum_{I_n^m} \epsilon_{\mathbf{i}} h(\mathbf{X}_{\mathbf{i}})/n^{m/2}\Big|^\alpha\Big),$$

and by the hypercontractivity properties of Rademacher chaos variables in Section 3.2,

$$\mathbb{E}\exp\Big(ct\Big|\sum_{I_n^m} \epsilon_{\mathbf{i}} h(\mathbf{X}_{\mathbf{i}})/n^{m/2}\Big|^\alpha\Big) \lesssim \mathbb{E}\exp\Big(c't^{2/\alpha}\sum_{I_n^m} h^2(\mathbf{X}_{\mathbf{i}})/n^m\Big).$$

The functions h^2 can be centered at the expense of multiplying by $\exp(-Kt^{2/\alpha}\mathbb{E}h^2)$, and then, an averaging procedure (that goes back to Hoeffding, 1963) together with convexity of the exponential reduce the problem to one on sums of independent centered random variables that can be handled by means of classical inequalities. Randomization and hypercontractivity of Rademacher chaos have the effect of doubling the power of n (from $n^{-m/2}$ to n^{-m}) and this allows for use of the averaging device. In order to carry out this argument, we begin by isolating the one-dimensional inequality, which is nothing but Bernstein's inequality in integral form.

LEMMA 4.1.9. (BERNSTEIN'S INEQUALITY) *Let ξ_i, $i \le n$, be independent random variables such that, for each i, $\mathbb{E}\xi_i = 0$, $\sigma_i^2 := \mathbb{E}\xi_i^2 < \infty$ and*

$$\mathbb{E}|\xi_i|^k \le \frac{k!}{2}\sigma_i^2 M^{k-2}, \qquad k \ge 2, \tag{4.1.16}$$

for some constant $M \in (0, \infty)$. Then, letting $\sigma^2 = \sum_{i=1}^n \sigma_i^2$, we have

$$\mathbb{E}\exp\Big(t\sum_{i=1}^n \xi_i\Big) \le \exp\Big(\frac{\sigma^2 t^2}{2 - 2M|t|}\Big), \qquad \text{for} \quad M|t| < 1, \tag{4.1.17}$$

and

$$\Pr\Big\{\sum_{i=1}^n \xi_i > u\Big\} \le \exp\Big(-\frac{u^2}{2\sigma^2 + 2Mu}\Big) \tag{4.1.18}$$

for all $u \ge 0$.

PROOF. Using hypothesis (4.1.16) in the development of the exponential, assuming $M|t| < 1$, we have

$$\begin{aligned}
\mathbb{E}e^{t\xi_i} &\le 1 + \frac{t^2\sigma_i^2}{2} + \frac{|t|^3\sigma_i^2 M}{2} + \frac{t^4\sigma_i^2 M^2}{2} + \cdots \\
&= 1 + \frac{t^2\sigma_i^2}{2}\Big[1 + |t|M + t^2M^2 + \cdots\Big] \\
&= 1 + \frac{t^2\sigma_i^2}{2}\frac{1}{1 - |t|M} \le \exp\Big(\frac{t^2\sigma_i^2}{2 - 2M|t|}\Big)
\end{aligned}$$

for all $1 \le i \le n$. Now (4.1.17) follows by independence. To obtain Bernstein's inequality (4.1.18) we first plug the bound (4.1.17) into Chebyshev's inequality for exponentials,

$$\begin{aligned}
\Pr\Big\{\sum_{i=1}^n \xi_i > u\Big\} &= \Pr\Big\{\exp\Big(t\sum_{i=1}^n \xi_i\Big) > \exp(tu)\Big\} \\
&\le \exp\Big(\frac{\sigma^2 t^2}{2 - 2Mt} - tu\Big)
\end{aligned}$$

for $0 < t < 1/M$, and then take $t = u/(\sigma^2 + Mu)$. □

REMARK 4.1.10. If the variables ξ_i are uniformly bounded by a constant c, then we can take $M = c/3$ in the previous lemma. In particular, *if ξ_i are i.i.d., centered, $\|\xi_i\|_\infty \le c \le \infty$ and $\sigma^2 = \mathbb{E}\xi_i^2$, then*

$$\mathbb{E}\exp\Big(t\frac{1}{\sqrt{n}}\sum_{i=1}^n \xi_i\Big) \le \exp\Big(\frac{\sigma^2 t^2}{2 - \frac{2}{3}c|t|/\sqrt{n}}\Big), \quad \text{for } |t| < 3\sqrt{n}/c, \tag{4.1.17$'$}$$

and

$$\Pr\Big\{\frac{1}{\sqrt{n}}\sum_{i=1}^{n}\xi_i > u\Big\} \le \exp\Big(-\frac{u^2}{2\sigma^2 + \frac{2}{3}cu/\sqrt{n}}\Big) \tag{4.1.18$'$}$$

for all $u \ge 0$.

REMARK 4.1.11. We can do a little better than inequality (4.1.18) if, instead of taking t such that $\sigma^2 t^2/(2-2Mt) = tu/2$, we minimize the exponent $\sigma^2 t^2/(2-2Mt) - tu$. The resulting inequality, one of *Bennett's inequalities*, which holds under the moment hypothesis (4.1.16), just as Bernstein's (4.1.18), is as follows:

$$\Pr\Big\{\sum_{i=1}^{n}\xi_i > u\Big\} \le \exp\Big(-\frac{u^2}{\sigma^2\big(1+\sqrt{1+2Mu/\sigma^2}\big)+Mu}\Big). \tag{4.1.19}$$

Back to U-statistics, it may be convenient to recall at this point that if Z is a homogeneous Rademacher chaos of degree m, that is, if

$$Z = \sum_{1\le i_1<\cdots<i_m\le n} \varepsilon_{i_1}\cdots\varepsilon_{i_m} a_{i_1,\dots,i_m}$$

for some n (finite or infinite) and real constants $a_{\mathbf{i}}$, then Z satisfies Corollaries 3.2.6 and 3.2.7. In fact, since in the homogeneous chaos case we can use Theorem 3.2.2 instead of Theorem 3.2.5, a simple computation along the lines of the proofs of these corollaries gives

$$\mathbb{E}\exp\Big(\big[|Z|/(4es)\big]^{2/m}\Big) \le 2,$$

where $s^2 := \mathbb{E}Z^2 = \sum_{1\le i_1<\cdots<i_m\le n} a^2_{i_1,\dots,i_m}$, and, from this,

$$\mathbb{E}\exp\Big(t|Z|^\alpha\Big) \le 2\exp\Big(K(s^\alpha t)^{2/(2-\alpha m)}\Big), \qquad 0\le\alpha<2/m, \tag{4.1.20}$$

where $K = (1-\alpha m/2)\big((4e)^\alpha(\alpha m/2)^{\alpha m/2}\big)^{2/(2-\alpha m)}$ is a constant that depends only on α and m.

We should also recall the part of the randomization inequalities (3.5.4$'$) that we will use below, namely, that if h is a kernel of m variables that is P-canonical and sufficiently integrable, X_i are i.i.d.(P), and Ψ is a nondecreasing convex function on $[0,\infty)$, then

$$\mathbb{E}\Psi\Big(\big|\sum_{I_n^m} h(X_{i_1},\dots,X_{i_m})\big|\Big) \le \mathbb{E}\Psi\Big(C_m\big|\sum_{I_n^m}\varepsilon_{i_1}\cdots\varepsilon_{i_k}h(X_{i_1},\dots,X_{i_m})\big|\Big) \tag{4.1.21}$$

for $1\le k\le m$ and constant C_m that only depends on m.

THEOREM 4.1.12. *Let $(S,\mathcal{S},P)$ be a probability space, let X_i be i.i.d.(P), and let h be a bounded P-canonical kernel of m variables that, for convenience, we assume symmetric in its entries. We set $c := \|h\|_\infty$ and $\sigma^2 = \mathbb{E}h^2(X_1,\dots,X_m)$. Then there exist finite positive constants c_i that depend only on m such that*

a)

$$\Pr\left\{\left|\frac{1}{n^{m/2}}\sum_{I_n^m} h(X_{i_1},\dots,X_{i_m})\right| \geq t\right\} \leq c_1 \exp\left(-\frac{c_2 t^{2/m}}{\sigma^{2/m} + \left(ct^{1/m}/n^{1/2}\right)^{2/(m+1)}}\right) \tag{4.1.22}$$

for all $t > 0$, and

b)

$$\Pr\left\{\left|\frac{1}{n^{m/2}}\sum_{I_n^m} h(X_{i_1},\dots,X_{i_m})\right| \geq t\right\} \leq c_3 \exp\{-c_4(t/c)^{2/m}\} \tag{4.1.23}$$

for all $t > 0$. These inequalities also hold for decoupled U-statistics.

PROOF. In this proof the constants c_i may vary from line to line and depend only on m. We take $\alpha = 2/(m+1)$ in (4.1.20) so that the exponent $2\alpha/(2-\alpha m)$ of s becomes 2. The convex modification Ψ of $\exp(x^\alpha)$ alluded to above satisfies $a_\alpha \Psi(|x|) \leq \exp(|x|^\alpha) \leq \Psi(|x|)$ with $a_\alpha = \exp((1-\alpha)/\alpha)$ (note that $a_1 = 1$). Hence, we can apply the randomization inequality (4.1.21) with $k = m$, to conclude

$$\begin{aligned}
&\mathbb{E}\exp\left(t\left|\frac{1}{n^{m/2}}\sum_{i_1<\cdots<i_m\leq n} h(X_{i_1},\dots,X_{i_m})\right|^{2/(m+1)}\right)\\
&\quad\leq c_1\mathbb{E}\exp\left(c_2 t\left|\frac{1}{n^{m/2}}\sum_{1\leq i_1<\cdots<i_m\leq n} \varepsilon_{i_1}\cdots\varepsilon_{i_m} h(X_{i_1},\dots,X_{i_m})\right|^{2/(m+1)}\right)
\end{aligned}$$

for all $t > 0$. We can now apply inequality (4.1.20) on Rademacher chaos integrability, conditionally on the X variables, to obtain

$$\begin{aligned}
&\mathbb{E}\exp\left(t\left|\frac{1}{n^{m/2}}\sum_{1\leq i_1<\cdots<i_m\leq n} h(X_{i_1},\dots,X_{i_m})\right|^{2/(m+1)}\right)\\
&\leq c_1\mathbb{E}\exp\left(c_2 t^{m+1}\frac{1}{n^m}\sum_{1\leq i_1<\cdots<i_m\leq n} h^2(X_{i_1},\dots,X_{i_m})\right)\\
&\leq c_1\exp\left(c_2\sigma^2 t^{m+1}\right) \qquad\qquad (4.1.24)\\
&\quad\times\mathbb{E}\exp\left(c_2 t^{m+1}\frac{1}{n^m}\sum_{1\leq i_1<\cdots<i_m\leq n}\left(h^2(X_{i_1},\dots,X_{i_m}) - \mathbb{E}h^2(X_1,\dots,X_m)\right)\right).
\end{aligned}$$

Next we describe and apply Hoeffding's averaging. If for a symmetric kernel $f(x_1,\dots,x_m)$, and for n points $x_1,\dots,x_n$, letting $k := [n/m]$, we define

$$W_f(x_1,\dots,x_n) := \frac{1}{k}\sum_{i=1}^{k} f(x_{(i-1)m+1},\dots,x_{im}),$$

the average of f over the k disjoint m-blocks of xs, $\{x_{(i-1)m+1},\dots,x_{im}\}$, $i = 1,\dots,k$, then it is clear that the U-statistic $U_n(f)$ is the average of the variables $W_f(X_{i_1},\dots,X_{i_n})$, taken over all the permutations $i_1,\dots,i_n$ of $1,\dots,n$. The exponent in the last term of (4.1.24) contains the U-statistic $U_n(h^2 - \mathbb{E}h^2)$; writing this U-statistic as the average of the $W_{h^2-\mathbb{E}h^2}$ variables and then using the convexity of the exponential function, gives

$$\begin{aligned}
&\mathbb{E}\exp\left(c_2 t^{m+1}\frac{1}{n^m}\sum_{1\le i_1<\cdots<i_m\le n}\left(h^2(X_{i_1},\ldots,X_{i_m}) - \mathbb{E}h^2(X_1,\ldots,X_m)\right)\right)\\
&\quad\le \mathbb{E}\exp\left(c_2 t^{m+1} W_{h^2-\mathbb{E}h^2}(X_1,\ldots,X_m)\right)\\
&\quad\le \mathbb{E}\exp\left(2mc_2 t^{m+1} n^{-1}\sum_{i=1}^{[n/m]}(h^2(X_{1+(i-1)m},\ldots,X_{im}) - \mathbb{E}h^2(X_1,\ldots,X_m))\right),
\end{aligned}$$

or, by (4.1.24),

$$\begin{aligned}
&\mathbb{E}\exp\left(t\left|\frac{1}{n^{m/2}}\sum_{i_1<\cdots<i_m\le n} h(X_{i_1},\ldots,X_{i_m})\right|^{2/(m+1)}\right)\\
&\quad\le c_1\exp\left(c_2\sigma^2 t^{m+1}\right) \qquad (4.1.25)\\
&\qquad\times\mathbb{E}\exp\left(2mc_2 t^{m+1} n^{-1}\sum_{i=1}^{[n/m]}\left(h^2(X_{1+(i-1)m},\ldots,X_{im}) - \mathbb{E}h^2(X_1,\ldots,X_m)\right)\right).
\end{aligned}$$

Now, we apply Lemma 4.1.9 [actually, inequality (4.1.17′)] to the random variables

$$\xi_i = h^2(X_{1+(i-1)m},\ldots,X_{im}) - \mathbb{E}h^2(X_1,\ldots,X_m), \qquad i = 1,\ldots,n,$$

which are independent, centered, bounded by c^2, and with variance dominated by $\sigma^2 c^2$, and we obtain

$$\begin{aligned}
&\mathbb{E}\exp\left(t\left|\frac{1}{n^{m/2}}\sum_{i_1<\cdots<i_m\le n} h(X_{i_1},\ldots,X_{i_m})\right|^{2/(m+1)}\right)\\
&\quad\le c_1\exp\left(c_2 t^{m+1}\sigma^2 + \frac{4m^2c_2^2\sigma^2c^2t^{2(m+1)}}{2n-(8/3)mc_2c^2t^{m+1}}\right) \qquad (4.1.26)
\end{aligned}$$

for $n > (2/3)c_2c^2t^{m+1}$.

The rest of the proof is standard: we apply Chebyshev's inequality with exponential moments using the bound (4.1.26) and then choose t so as to minimize the exponent. So, for the permissible values of t we have

$$\begin{aligned}
&\Pr\left\{\frac{1}{n^{m/2}}\left|\sum_{i_1<\cdots<i_m\le n} h(X_{i_1},\ldots,X_{i_m})\right| \ge u\right\}\\
&\quad\le c_1\exp\left(-tu^{2/(m+1)} + c_2t^{m+1}\sigma^2 + \frac{4m^2c_2^2\sigma^2c^2t^{2(m+1)}}{2n-(8/3)mc_2c^2t^{m+1}}\right). \qquad (4.1.27)
\end{aligned}$$

If $t = \left(\frac{u^{2/(m+1)}}{c_2(m+1)\sigma^2}\right)^{1/m}$, which is the value obtained by minimizing $-tu^{2/(m+1)} + c_2t^{m+1}\sigma^2$, then the first two terms in the exponent are of the order $-c(u/\sigma)^{2/m}$. This will be the order of the whole exponent if two conditions are met, one, that this value of t be permissible and second, that the third term be smaller than a small constant times $(u/\sigma)^{2/m}$. It turns out that both requirements are met if

$$\frac{u^{2/m}c^2}{n\sigma^{2(m+1)/m}} \le K \tag{4.1.28}$$

for small enough K, as is easy to check. That is, there are K and c_2' such that

$$\Pr\left\{\frac{1}{n^{m/2}}\left|\sum_{i_1<\cdots<i_m\le n} h(X_{i_1},\ldots,X_{i_m})\right| \ge u\right\} \le c_1 \exp\left(-c_2'\left(\frac{u}{\sigma}\right)^{2/m}\right) \tag{4.1.29}$$

for all u satisfying inequality (4.1.28). Finally, it is easy to check that there is L positive small enough and $c_2'' < \infty$ such that, letting $t^{m+1} = Ln/c^2$ in (4.1.27) gives

$$\Pr\left\{\frac{1}{n^{m/2}}\left|\sum_{i_1<\cdots<i_m\le n} h(X_{i_1},\ldots,X_{i_m})\right| \ge u\right\} \le c_1 \exp\left(-c_2''\left(\frac{nu^2}{c^2}\right)^{1/(m+1)}\right)$$

for all $u > 0$ and *not* satisfying inequality (4.1.28). This and (4.1.29) prove part (a) after some elementary estimates.

To prove part (b) of the theorem, we note that, as in (4.1.24),

$$\begin{aligned} &\mathbb{E}\exp\left(t\left|\frac{1}{n^{m/2}}\sum_{i_1<\cdots<i_m\le n} h(X_{i_1},\ldots,X_{i_m})\right|^{2/(m+1)}\right) \\ &\le c_1 \mathbb{E}\exp\left(c_2 t^{m+1}\frac{1}{n^m}\sum_{i_1<\cdots<i_m\le n} h^2(X_{i_1},\ldots,X_{i_m})\right) \\ &\le c_1\exp\left(c_2 t^{m+1}c^2\right). \end{aligned} \tag{4.1.30}$$

Then,

$$\Pr\left\{\frac{1}{n^{m/2}}\left|\sum_{i_1<\cdots<i_m\le n} h(X_{i_1},\ldots,X_{i_m})\right| \ge u\right\} \le c_1 \exp\left(-tu^{2/(m+1)} + c_2 t^{m+1}c^2\right),$$

and the inequality in part (b) is obtained by minimizing the exponent with respect to t, which gives $t = \left(u^{2m/(m+1)}/(c_2(m+1)c^2)\right)^{1/m}$.

The proof for decoupled U-statistics is very similar and is omitted. □

We see from this theorem that the tail probabilities of $n^{m/2}U_n(h)$ for h canonical and bounded, are of the order $\exp\left(-c_2(t/\sigma)^{2/m}\right)$ at least on the range $0 \le t \lesssim \sigma^{m+1}n^{m/2}/c^m$. As we will see in the next section the sequence $\{n^{m/2}U_n(h)\}$ converges in distribution and its limit law, a Gaussian chaos variable with variance σ^2, has the same tail probability behavior, possibly for a different constant c_2 and at least from some t on (it is better behaved for smaller values of t, but we will not consider this). Hence, the inequalities in part (a) are best possible on some range. In fact, there are examples of canonical bounded kernels h for which the inequality with the right side replaced by $c_1\exp\left(-c_2(t/\sigma)^{2/m}\right)$ cannot happen for $t \gtrsim \sigma^{m+1}n^{m/2}/c^m$ (Arcones and Giné, 1993). So, this "breakpoint" is optimal. The inequality may be suboptimal both for small and for large values of t.

Clearly, inequalities (4.1.22) and (4.1.23) are less precise than the inequalities of Hoeffding in Theorem 4.1.8 for not necessarily degenerate U-statistics.

The first term of the Hoeffding decomposition of $\sqrt{n}\big(U_n(h) - P^m(h)\big)$, namely $m\sum_{i=1}^n \pi_1 h(X_i)/\sqrt{n}$, where $\pi_1(h)(X_1) = (P^{m-1}h)(X_1) - P^m h$, has a normal limit with variance that of $\pi_1(h)$, which is smaller, in general, than that of h. Thus, since the remainder is of smaller order, we should be able to replace σ^2 by $m^2\mathrm{Var}(P^{m-1}h)$ in the Hoeffding–Bernstein inequality on part (a) of Theorem 4.1.8. This is what we do next.

THEOREM 4.1.13. *Let $(S, \mathcal{S}, P)$ be a probability space, let X_i be i.i.d.(P) and let h be a bounded P-centered kernel of m variables from S, that we assume symmetric in its entries. Let $c := \|h\|_\infty$, and let*

$$s^2 := P(P^{m-1}h)^2 - (P^m h)^2$$

be the variance of $P^{m-1}(h)$. Then, there exists a constant c_1 depending only on m such that

$$\Pr\Big\{n^{1/2}|U_n(h)| > t\Big\} \le 4\exp\Big(-\frac{t^2}{2m^2s^2 + c_1 ct/\sqrt{n}}\Big). \tag{4.1.31}$$

PROOF. By Hoeffding's decomposition (3.5.1), since h is centered, we have

$$n^{1/2}U_n(h) = \frac{m}{\sqrt{n}}\sum_{i=1}^n \pi_1(h)(X_i) + n^{1/2}\sum_{k=2}^m \binom{m}{k} U_n^{(k)}(\pi_k(h)), \tag{4.1.32}$$

that is, $n^{1/2}U_n(h)$ is the sum of a linear term and a "remainder." The size of the linear term is estimated by Bernstein's inequality: since $\pi_1 h = P^{m-1}h - P^m h$ is bounded by $2c$, inequality (4.1.18′) gives

$$\Pr\Big\{\frac{m}{\sqrt{n}}\sum_{i=1}^n \pi_1(h)(X_i) > t\Big\} \le \exp\Big(-\frac{t^2}{2m^2s^2 - \frac{4}{3}mct/\sqrt{n}}\Big). \tag{4.1.33}$$

Next we estimate the size of the remainder using decoupling, randomization, and hypercontractivity of Rademacher chaos variables, with several differences with respect to the previous proof. Convexity of $\exp|x|$ and decoupling and randomization [inequality (4.1.21)], give

$$\begin{aligned}
&\mathbb{E}\exp\Big(tn^{1/2}\Big|\sum_{k=2}^m \binom{m}{k} U_n^{(k)}(\pi_k(h))\Big|\Big)\\
&\le \frac{1}{m-1}\sum_{k=2}^m \mathbb{E}\exp\Big(tn^{1/2}(m-1)\binom{m}{k}\big|U_n^{(k)}(\pi_k(h))\big|\Big)\\
&\le \frac{1}{m-1}\sum_{k=2}^m \mathbb{E}\exp\Big(C_k tn^{1/2}(m-1)\binom{m}{k}\frac{(n-k)!}{n!}\\
&\qquad\qquad \times\Big|\sum_{I_n^k} \varepsilon_{i_1}\varepsilon_{i_2}\pi_k(h)(X_{i_1},\dots,X_{i_k})\Big|\Big)
\end{aligned}$$

$$\leq \frac{1}{m-1}\sum_{k=2}^{m}\mathbb{E}\exp\left(c_m\frac{t}{n^{1/2}}\left|\frac{1}{n^{k-1}}\sum_{I_n^k}\varepsilon_{i_1}\varepsilon_{i_2}\pi_k(h)(X_{i_1},\ldots,X_{i_k})\right|\right), \tag{4.1.34}$$

where the last inequality only consolidates constants, and C_k (c_m) are constants that depend only on k (m). The conditional variance of the conditional Rademacher chaos of order two in the exponent can be estimated as follows:

$$\begin{aligned}
&\frac{1}{n^{k-1}}\left[\mathbb{E}_\varepsilon\left(2\sum_{1\leq i_1\leq i_2\leq n}\varepsilon_{i_1}\varepsilon_{i_2}\sum_{(i_3,\ldots,i_k):(i_1,\ldots,i_k)\in I_n^k}\pi_k(h)(X_{i_1},\ldots,X_{i_k})\right)^2\right]^{1/2}\\
&\quad=\frac{1}{n^{k-1}}\left[\sum_{1\leq i_1\leq i_2\leq n}\left(2\sum_{(i_3,\ldots,i_k):(i_1,\ldots,i_k)\in I_n^k}\pi_k(h)(X_{i_1},\ldots,X_{i_k})\right)^2\right]^{1/2}\\
&\quad\leq n^{1-k}\left(n^2(4n^{k-2}c)^2\right)^{1/2}=4c.
\end{aligned}$$

Now, we use hypercontractivity of Rademacher chaos in the form of Corollary 3.2.6, which asserts in particular that if Y is a Rademacher chaos of degree 2, then there is a universal constant λ positive and finite such that $\mathbb{E}\exp\bigl(\lambda|Y|/\|Y\|_2\bigr)\leq 2$. (Exercise: show that $\lambda = 1/8$ works.) Hence, it follows that if $c_m t/n^{1/2}\leq\lambda/4c$, then

$$\mathbb{E}\exp\left(c_m\frac{t}{n^{1/2}}\left|\frac{1}{n^{k-1}}\sum_{I_n^k}\varepsilon_{i_1}\varepsilon_{i_2}\pi_k(h)(X_{i_1},\ldots,X_{i_k})\right|\right)\leq 2,\qquad 2\leq k\leq m,$$

and, by (4.1.34), that

$$\mathbb{E}\exp\left(tn^{1/2}\left|\sum_{k=2}^{m}\binom{m}{k}U_n^{(k)}(\pi_k(h))\right|\right)\leq 2.$$

Then, letting $t_0=\lambda n^{1/2}/(4c_m c)$, we have

$$\begin{aligned}
&\Pr\left\{n^{1/2}\left|\sum_{k=2}^{m}\binom{m}{k}U_n^{(k)}(\pi_k(h))\right|>u\right\}\\
&\quad=\Pr\left\{\exp\left(t_0n^{1/2}\left|\sum_{k=2}^{m}\binom{m}{k}U_n^{(k)}(\pi_k(h))\right|\right)>\exp(t_0u)\right\}\\
&\quad\leq 2\exp(-t_0u)=2\exp\left(-\frac{C_mn^{1/2}u}{c}\right)
\end{aligned}\tag{4.1.35}$$

for a constant C_m that depends only on m.

Now, we combine the estimates (4.1.33) and (4.1.35) with Hoeffding's decomposition (4.1.32) to obtain the desired result: for $0 < \tau < 1/m$,

$$\begin{aligned}\Pr\Big\{n^{1/2}|U_n(h)| > t\Big\} &\leq \Pr\Big\{\frac{m}{\sqrt{n}}\sum_{i=1}^{n}\pi_1(h)(X_i) > \big(1-m\tau\big)t\Big\}\\ &\quad + \Pr\Big\{n^{1/2}\Big|\sum_{k=2}^{m}\binom{m}{k}U_n^{(k)}(\pi_k(h))\Big| > m\tau t\Big\}\\ &\leq 2\exp\Big(-\frac{(1-2m\tau)t^2}{2m^2s^2+\frac{4}{3}mct/\sqrt{n}}\Big) + 2\exp\Big(\frac{C_m n^{1/2}m\tau t}{c}\Big);\end{aligned}$$

if we take τ so that these two exponents coincide, that is,

$$\tau = \frac{ct}{C_m m n^{1/2}\big(2m^2s^2+\frac{4}{3}mct/\sqrt{n}\big)+2mct}$$

we obtain the bound (4.1.31) with $c_1 = 4m/3 + 1/C_m$. □

We have not computed the constants c_i in the inequalities from this subsection because we do not know how sharp they would be, but it is not difficult to obtain reasonable constants from the above proofs.

4.2 Gaussian Chaos and the Central Limit Theorem for Canonical U-Statistics

The central limit theorem for sums of i.i.d. random variables can be described in the following terms. Let $L_2^c(S, \mathcal{S}, P)$ be the space of real-valued P-centered, P-square integrable functions on S. Let G_P be an isonormal Gaussian process on $L_2^c(S, \mathcal{S}, P)$, that is, a centered Gaussian process with parameter space $L_2^c(S, \mathcal{S}, P)$, such that $EG_P(f)G_P(g) = \int fg\,dP$. Isonormal processes exist by Kolmogorov's consistency theorem; alternatively, if $\{\phi_i\}_{i\in I}$ is an orthonormal basis of $L_2^c(S, \mathcal{S}, P)$ and if $\{g_i\}_{i\in I}$ is a family of independent $N(0,1)$ random variables, then the equation

$$G_P\Big(\sum_{i\in I} a_i\phi_i\Big) = \sum_{i\in I} a_i g_i$$

whenever $\sum_{i\in I} a_i^2 < \infty$, produces such a process. By identifying random variables that are a.s. equal, G_P becomes an isometry from $L_2^c(S, \mathcal{S}, P)$ onto the Hilbert space of jointly normal random variables generated by $\{G_P(f)\}$ (or, in the example, by the g_is). Then, the finite-dimensional central limit theorem simply asserts that the finite-dimensional distributions of the processes $\{\frac{1}{n^{1/2}}\sum_{i=1}^{n} f(X_i) : f \in L_2^c(S, \mathcal{S}, P)\}$ converge in law to the finite-dimensional distributions of $\{G_P(f) : f \in L_2^c(S, \mathcal{S}, P)\}$, that is, for every finite set of functions $f_1, \ldots, f_k$ in $L_2^c(S, \mathcal{S}, P)$,

$$\left(\frac{1}{n^{1/2}}\sum_{i=1}^{n} f_1(X_i), \ldots, \frac{1}{n^{1/2}}\sum_{i=1}^{n} f_k(X_i)\right) \to_{\mathcal{L}} \left(G_P(f_1), \ldots, G_P(f_k)\right), \quad (4.2.1)$$

with convergence of up to second moments (of any norm) as well. The central limit theorem for canonical U-statistics that we describe in this section may be viewed as the extension of the isometry G_P to an isometry K_P from the Hilbert space of all P-canonical square integrable kernels onto a Gaussian chaos Hilbert space (precisely, the Gaussian chaos corresponding to G_P) in such a way that the finite-dimensional distributions of properly normalized U-statistics converge to the corresponding finite-dimensional distributions of the process K_P. We present a proof of this result as well as of its converse. The chaos decomposition of the L_2 space generated by the process G_P is obtained as a byproduct of the proof of the direct part of the CLT. At the end of the section the CLT for canonical U-statistics is applied to obtain results on decoupling and hypercontractivity for Gaussian chaos that complement Theorem 3.4.6 and Theorem 3.2.10 for these processes.

Let $L_2^{c,k}(S, \mathcal{S}, P)$ [$L_2^{c,k}(P)$ for short] denote the Hilbert space of P-canonical functions of k variables. Let $\{\phi_i\}_{i\in J}$ be an orthonormal basis for $L_2^c(S, \mathcal{S}, P)$, where J is an index set. It then follows easily from basic Hilbert space and measure theory that the following set of functions is an orthonormal basis for $L_2^{c,k}(S, \mathcal{S}, P)$:

$$\left\{\frac{1}{\binom{k}{r_j : j\in J}^{\frac{1}{2}}} \sum_{\mathbf{i}\in J^k : j(\mathbf{i})=r_j} \phi_{i_1}(x_1)\cdots\phi_{i_k}(x_k) : \sum_{j\in J} r_j = k, r_j \in \mathbb{N}\right\}, \quad (4.2.2)$$

where, for any $j \in J$ and $\mathbf{i} \in J^k$, $j(\mathbf{i}) = \sum_{\ell=1}^k I_{i_\ell = j}$ is the number of occurrences of j among the coordinates of the multi-index $\mathbf{i}$, and where $\binom{k}{r_j : j\in J}$ denotes the combinatorial number $k!/\prod_{j\in J} r_j!$. The general term in (4.2.2) can be written as a sum of at most 2^k summands of the form

$$h_k^{\psi}(x_1, \ldots, x_k) = \psi(x_1)\cdots\psi(x_k), \quad (4.2.3)$$

$\psi \in L_2^c(P)$. This is a consequence of the following obvious *polarization identitity*: if $\varepsilon_1, \ldots, \varepsilon_k$ are i.i.d. Rademacher variables, then, for any k functions $f_1, \ldots, f_k$ of one variable, not necessarily different, and for any $x_1, \ldots, x_k$ in S, we have

$$\sum_{\sigma} f_{\sigma(1)}(x_1)\cdots f_{\sigma(k)}(x_k) = \mathbb{E}\left[\varepsilon_1\cdots\varepsilon_k\left(\sum_{i=1}^k \varepsilon_i f_i(x_1)\right)\cdots\left(\sum_{i=1}^k \varepsilon_i f_i(x_k)\right)\right], \quad (4.2.4)$$

where σ runs over all the permutations of $\{1, \ldots, k\}$. [So, in our case, the functions $\psi(x)$ are $\pm\phi_{i_1}(x) \pm \cdots \pm \phi_{i_k}(x)$.] Hence, if $h \in L_2^{c,k}(S, \mathcal{S}, P)$, then there exists a sequence $\{\psi_r\}$ of centered square integrable functions (for P) and a sequence of constants c_r such that

$$h = \sum_{r=1}^{\infty} c_r h_k^{\psi_r}, \tag{4.2.5}$$

with convergence in the sense of $L_2(P^k)$. These considerations will allow us to reduce the problem to weak convergence of U-statistics with kernel h_k^{ψ}. In order to handle this case, we recall *Newton's identities*: given $t_1, \ldots, t_n$ in $\mathbb{R}$, if, for $1 \le r \le n$, we let $p_r = \sum_{1 \le i_1 < \cdots < i_r \le n} t_{i_1} \cdots t_{i_r}$ and $s_r = \sum_{i=1}^n t_i^r$, then the identity

$$s_k - p_1 s_{k-1} + p_2 s_{k-2} - \cdots + (-1)^{k-1} s_1 p_{k-1} + (-1)^k k p_k = 0$$

holds for all $k < n$. These identities, which can be checked by induction, give, also by induction, that for every $k \in \mathbb{N}$ there is a real polynomial R_k of degreee k, in k variables, such that for all $n > k$,

$$\sum_{1 \le i_1 < \cdots < i_k \le n} t_{i_1} \cdots t_{i_k} = R_k\Big(\sum_{i=1}^n t_i, \sum_{i=1}^n t_i^2, \ldots, \sum_{i=1}^n t_i^k\Big).$$

Moreover, the polynomial $R_k(u_1, \ldots, u_k)$ is a sum of monomials of the form $c \prod_{i=1}^k u_i^{k_i}$ with $\sum_{i=1}^k i k_i = k$ and the coefficient of u_1^k is $1/k!$. (See, e.g., Jacobson, 1951.) Therefore,

$$\frac{1}{n^{k/2}} \sum_{1 \le i_1 < \cdots i_k \le n} h_k^{\psi}(X_{i_1}, \ldots, X_{i_k}) = \sum_{1 \le i_1 < \cdots < i_k \le n} \frac{\psi(X_{i_1})}{n^{\frac{1}{2}}} \cdots \frac{\psi(X_{i_k})}{n^{\frac{1}{2}}}$$

$$= R_k\Big(\frac{1}{n^{\frac{1}{2}}} \sum_{i=1}^n \psi(X_i), \frac{1}{n} \sum_{i=1}^n \psi^2(X_i), \ldots, \frac{1}{n^{\frac{k}{2}}} \sum_{i=1}^n \psi^k(X_i)\Big). \tag{4.2.6}$$

Then, the central limit theorem for i.i.d. random vectors [(4.2.1)], and the law of large numbers (both, Kolmogorov and Marzinkiewicz) give, by the continuous mapping theorem, that for any $r < \infty$, $\psi_1, \ldots, \psi_r$ in $L_2^c(P)$ and $k_1, \ldots, k_r$ in $\mathbb{N}$,

$$\bigg(\binom{n}{k_1}^{\frac{1}{2}} U_n(h_{k_1}^{\psi_1}), \ldots, \binom{n}{k_r}^{\frac{1}{2}} U_n(h_{k_r}^{\psi_r})\bigg)$$

$$\to_{\mathcal{L}} \Big((k_1!)^{\frac{1}{2}} R_{k_1}(G_P(\psi_1), \mathbb{E}\psi_1^2, 0, \ldots, 0),$$

$$\ldots, (k_r!)^{\frac{1}{2}} R_{k_r}(G_P(\psi_r), \mathbb{E}\psi_r^2, 0, \ldots, 0)\Big). \tag{4.2.7}$$

[Here and elsewhere, the symbol $\to_{\mathcal{L}}$ denotes weak convergence of the probability laws of the variables involved.] Moreover, there is convergence of up to the second moment of the norm (e.g., the Euclidean norm) in the limit (4.2.7), as we now prove. For this, by uniform integrability, it suffices to show that, for any ψ and $r \le n$,

$$\binom{n}{r} \mathbb{E}\big[U_n(h_r^{\psi})\big]^2 = r! \mathbb{E}\big[R_r(G_P(\psi), \mathbb{E}\psi^2, 0, \ldots, 0)\big]^2.$$

The left-hand side is obviously equal to $\mathbb{E}\big(h_r^\psi\big)^2 = \big(\mathbb{E}\psi^2\big)^r$. For the right-hand side, we observe that, by (4.2.7),

$$\frac{(\mathbb{E}\psi^2)^{r/2}}{\binom{n}{r}^{\frac{1}{2}}} \sum_{1 \le i_1 < \cdots < i_r \le n} \varepsilon_{i_1} \cdots \varepsilon_{i_r} \to_{\mathcal{L}} (r!)^{\frac{1}{2}} R_r(G_P(\psi), \mathbb{E}\psi^2, 0, \ldots, 0),$$

where $\{\varepsilon_i\}$ is a Rademacher sequence, and that there is convergence of all the moments in this limit because hypercontractivity (e.g. (3.2.15)) provides enough uniform integrability. Hence, we also have $r!\mathbb{E}\big[R_r(G_P(\psi), \mathbb{E}\psi^2, 0, \ldots, 0)\big]^2 = \big(\mathbb{E}\psi^2\big)^r$.

Convergence of up to the second moments implies that the left and right sides of the limit (4.2.7) both have the same covariance structure. The covariance structure of the vector at the left side is just that of the vector $(h_{k_1}^{\psi_1}, \ldots, h_{k_r}^{\psi_r})$. So, if we set

$$K_P(h_k^\psi) = (k!)^{\frac{1}{2}} R_k(G_P(\psi), \mathbb{E}\psi^2, 0, \ldots, 0) \tag{4.2.8}$$

and extend this map linearly and continuously (for the L_2-norm), then K_P defines an *isometry* between the Hilbert subspace of $L_2(S^{\mathbb{N}}, \mathcal{S}^{\mathbb{N}}, P^{\mathbb{N}})$ generated by the set of functions $\{h_k^\psi : k \in \mathbb{N}, \psi \in L_2^c(P)\}$, which, by (4.2.5), includes the set of all P-canonical kernels, and its image, which is the closure in L_2 of a set of polynomials in the random variables $G_P(\psi)$. In order to better describe K_P and extend the limit theorem (4.2.7) to arbitrary degenerate kernels (and more), it is now convenient to introduce *Hermite polynomials* and the *Gaussian chaos* decomposition of the Hilbert space spanned by the process G_P.

The system of Hermite polynomials, $\{\frac{1}{\sqrt{k!}} H_k(x) : k = 0, 1, \ldots\}$, is defined as the only orthonormal basis of $L_2(\mathbb{R}, \mathcal{B}, \frac{1}{\sqrt{2\pi}} e^{-x^2/2} dx)$ such that $H_k(x)$ is, for each k, a polynomial of degree k in x. Equivalently, Hermite polynomials are also defined by the relationship

$$e^{ux - \frac{u^2}{2}} = \sum_{k=0}^{\infty} \frac{u^k}{k!} H_k(x). \tag{4.2.9}$$

Obviously, the function H_k in (4.2.9) is a polynomial of degree k for each k. To see that both definitions of H_k coincide, we show that if H_k is defined by equation (4.2.9), and if g_1 and g_2 are jointly normal random variables with mean zero and variance 1, then

$$\mathbb{E}\Big(H_k(g_1) H_\ell(g_2)\Big) = \begin{cases} k!(\mathbb{E}g_1 g_2)^k & \text{if } k = \ell \\ 0 & \text{otherwise.} \end{cases} \tag{4.2.10}$$

For $u, v \in \mathbb{R}$ we have

$$\begin{aligned} \mathbb{E}\Big(\exp\Big(ug_1 - \frac{u^2}{2}\Big) \exp\Big(vg_2 - \frac{v^2}{2}\Big)\Big) &= \exp\Big(-\frac{u^2 + v^2}{2}\Big) \exp\Big(\frac{1}{2}\mathbb{E}(ug_1 + vg_2)^2\Big) \\ &= \exp\big(uv(\mathbb{E}g_1 g_2)\big), \end{aligned} \tag{4.2.11}$$

and (4.2.10) follows from (4.2.11) by just differentiating k times with respect to u and ℓ times with respect to v at $(u, v) = (0, 0)$.

If we compare the covariance structure of the Gaussian polynomials at the right side of equation (4.2.8) with that of Hermite polynomials, we conclude that, for all $k \geq 1$ and $\psi \in L_2^c(P)$,

$$K_P(h_k^\psi) = \frac{\sigma_\psi^k}{\sqrt{k!}} H_k\big(G_P(\psi)/\sigma_\psi\big), \tag{4.2.8'}$$

where $\sigma_\psi^2 = P(\psi^2)$.

Let $(\Omega, \Sigma, \Pr)$ be the probability space where the isonormal process G_P is defined, and let $\sigma(G_P)$ be the sub-σ-algebra of Σ generated by the random variables $\{G_P(\psi) : \psi \in L_2^c(P)\}$. We will show that the range of K_P, extended as the identity on constants, is precisely the set of all G_P-measurable square integrable random variables, $L_2(G_P) := L_2(\Omega, \sigma(G_P), \Pr)$. Since by (4.2.5) the kernels of the form h_k^ψ generate the set of all canonical kernels, we already know that the domain of K_P, extended as the identity on constants, is the Hilbert space generated by the P-canonical kernels of all orders together with the constants, that is, the space

$$\mathbb{R} \oplus L_2^{c,\mathbb{N}}(P) := \mathbb{R} \oplus \left(\oplus_{k=1}^\infty L_2^{c,k}(P)\right),$$

where $\oplus$ denotes orthogonal sum in $L_2(S^{\mathbb{N}}, \mathcal{S}^{\mathbb{N}}, P^{\mathbb{N}})$. We will also see that K_P translates this orthogonal decomposition into the Gaussian chaos decomposition of $L_2(G_P)$.

PROPOSITION 4.2.1. a) *The set*

$$\left\{\exp\big(G_P(\psi)\big) : \psi \in L_2^c(P)\right\}$$

generates $L_2(G_P)$.

b) *The set*

$$\left\{\exp^\sim(\psi) : \psi \in L_2^c(P)\right\},$$

where

$$\exp^\sim(\psi) := \sum_{k=0}^\infty \frac{h_k^\psi}{\sqrt{k!}}, \tag{4.2.12}$$

generates $\mathbb{R} \oplus L_2^{c,\mathbb{N}}(P)$.

c) *The isometry* K_P *determined on all of* $\mathbb{R} \oplus L_2^{c,\mathbb{N}}(P)$ *by equation (4.2.8′) [or (4.2.8)] and by* $K_P(1) = 1$, *is onto* $L_2(G_P)$, *and is also uniquely determined by the equation*

$$K_P\big(\exp^\sim(\psi)\big) = \exp\left(G_P(\psi) - \frac{\sigma_\psi^2}{2}\right). \tag{4.2.13}$$

PROOF. a) Suppose $Z \in L_2(G_P)$ is orthogonal to $\exp\big(G_P(\psi)\big)$ for all $\psi \in L_2^c(P)$ and consider the (finite total variation) measure $Q(A) = \int_A Z\, d\Pr$, $A \in$

Σ. Then, for any $n \in \mathbb{N}$ and $\psi_1, \ldots, \psi_n$ in $L_2^c(P)$, the measure $\mu = Q \circ (G_P(\psi_1), \ldots, G_P(\psi_n))^{-1}$, defined on $\mathbb{R}^n$, has moment generating function

$$\hat{\mu}(t_1, \ldots, t_n) = E\Big[Z \cdot \exp\big(G_P(\textstyle\sum t_i\psi_i)\big)\Big] = 0.$$

Hence $\mu = 0$. So, Q is zero on $\sigma\big(G_P(\psi_1), \ldots, G_P(\psi_n)\big)$, hence, also on $\sigma(G_P)$. Since $Z = dQ/d\Pr$ and Z is $\sigma(G_P)$-measurable, we have $Z = 0$ a.s., proving part a).

b) $\exp^{\sim}\psi \in \mathbb{R} \oplus L_2^{c,\mathbb{N}}(P)$ because $\mathbb{E}(\exp^{\sim}\psi)^2 = \sum_{k=0}^{\infty}(\mathbb{E}\psi)^k/k! < \infty$ (note that h_k^ψ and h_ℓ^ψ are orthogonal if $k \neq \ell$). Now let $h \in \mathbb{R} \oplus L_2^{c,\mathbb{N}}(P)$ be orthogonal to $\exp^{\sim}\psi$ for all $\psi \in L_2^c(P)$. Then, $h = c + \sum_{k=1}^{\infty} h_k$ with $h_k \in L_2^{c,k}(P)$, and, for all $s \in \mathbb{R}$,

$$0 = \mathbb{E}\big[h \cdot \exp^{\sim}(s\psi)\big] = c + \sum_{k=1}^{\infty} \frac{s^k}{\sqrt{k!}} \mathbb{E}(h_k h_k^\psi).$$

Taking $s = 0$, this identity gives $c = 0$, and differentiating k times at $s = 0$, $k = 1, 2, \ldots$, it gives $\mathbb{E}(h_k h_k^\psi) = 0$ for all k. Then, equation (4.2.5) implies $\mathbb{E}h_k^2 = 0$, hence $h = 0$ a.s.

c) By orthogonality of canonical kernels of different orders and by (4.2.11), for all $\psi, \phi \in L_2^c(P)$ we have

$$\begin{aligned}
\mathbb{E}\Big(\big(\exp^{\sim}\psi\big)\big(\exp^{\sim}\phi\big)\Big) &= 1 + \sum_{k=1}^{\infty} \frac{1}{k!}(\mathbb{E}\psi\phi)^k \\
&= \exp(\mathbb{E}\psi\phi) \\
&= \mathbb{E}\bigg(\exp\Big(G_P(\psi) - \frac{\sigma_\psi^2}{2}\Big)\exp\Big(G_P(\phi) - \frac{\sigma_\phi^2}{2}\Big)\bigg).
\end{aligned}$$

Therefore, by a) and b), equation (4.2.13) defines an isometry of $\mathbb{R} \oplus L_2^{c,\mathbb{N}}(P)$ *onto* $L_2(G_P)$, that we provisorily denote by $\tilde{K}_P$. To see that $\tilde{K}_P$ coincides with the isometry K_P determined by equation (4.2.8′), it is sufficient to show

$$\tilde{K}_P(h_k^\psi) = \frac{\sigma_\psi^k}{\sqrt{k!}} H_k\big(G_P(\psi)/\sigma_\psi\big)$$

for all ψ and k. But this is obvious from the following two expressions for the variable $\tilde{K}_P(\exp^{\sim} t\psi)$, which are valid for all $t \in \mathbb{R}$, and which follow from (4.2.9) and (4.2.12):

$$\tilde{K}_P(\exp^{\sim} t\psi) = 1 + \sum_{k=1}^{\infty} \frac{t^k}{k!}\sqrt{k!}\tilde{K}_P(h_k^\psi),$$

$$\tilde{K}_P(\exp^{\sim} t\psi) = \exp\Big(G_P(t\psi) - \frac{t^2\sigma_\psi^2}{2}\Big)$$

$$= \exp\Big(t\sigma_\psi \frac{G_P(\psi)}{\sigma_\psi} - \frac{(t\sigma_\psi)^2}{2}\Big)$$

$$= \sum_{k=0}^{\infty} \frac{t^k}{k!}\sigma_\psi^k H_k\Big(\frac{G_P(\psi)}{\sigma_\psi}\Big). \qquad \square$$

REMARK 4.2.2. *The process K_P in terms of the basis (4.2.2).* Computing $K_P\big(\exp^{\sim}(\sum_{\ell=1}^k t_\ell \phi_{i_\ell})\big)$ in two ways, just as done immediately above, one obtains

$$K_P\Bigg(\frac{1}{\binom{k}{r_j : j\in J}^{\frac{1}{2}}} \sum_{\mathbf{i}\in J^k : j(\mathbf{i})=r_j} \phi_{i_1}(x_1)\cdots\phi_{i_k}(x_k)\Bigg) = \prod_{j\in J} \frac{1}{\sqrt{r_j!}} H_{r_j}\big(G_P(\phi_j)\big), \quad (4.2.14)$$

where, we recall, $j(\mathbf{i})$ is the number of occurrences of j in the multi-index $\mathbf{i}$. Let now $h \in L_2^{c,k}(P)$. Then,

$$h(x_1, \dots, x_k) = \sum_{\mathbf{i}\in J^k} a_{\mathbf{i}} \phi_{i_1}(x_1)\cdots\phi_{i_k}(x_k), \qquad (4.2.15)$$

in the L_2 sense, with coefficients $a_{\mathbf{i}} = a_{i_1,\dots,i_k} = E\big[h(x_1,\dots,x_k)\prod_{r=1}^k \phi_{i_r}(x_r)\big]$, which are symmetric in their indices, and (4.2.14) gives

$$K_P(h) = \frac{1}{\sqrt{k!}} \sum_{\mathbf{i}\in J^k} a_{\mathbf{i}} \prod_{j\in J} H_{j(\mathbf{i})}\big(G_P(\phi_j)\big). \qquad (4.2.16)$$

Equation (4.2.16) constitutes the most explicit description of the map K_P.

Given an isonormal process G_P on $L_2^c(P)$, we let $\mathcal{P}_k(G_P)$ ($\mathcal{P}_k$ for short) be the Hilbert subspace of $L_2(G_P)$ generated by the polynomials of degree at most k in the variables $G_P(\psi)$, $\psi \in L_2^c(P)$, and let $\mathcal{H}_k(G_P)$ ($\mathcal{H}_k$ for short) be the orthogonal complement of $\mathcal{P}_{k-1}$ in $\mathcal{P}_k$, that is,

$$\mathcal{H}_k = \mathcal{P}_k \ominus \mathcal{P}_{k-1}.$$

THEOREM 4.2.3. *Let G_P be an isonormal process on $L_2^c(P)$ and let K_P be its extension to $\mathbb{R} \oplus L_2^{c,\mathbb{N}}(P)$ determined by equation (4.2.8′) [or alternatively, by (4.2.13) or (4.2.16)] and by $K_P(1) = 1$. Then,*

$$K_P(L_2^{c,k}(P)) = \mathcal{H}_k(P), \qquad k \in \mathbb{N}, \qquad (4.2.17)$$

and, in particular,

$$L_2(G_P) = \oplus_{k=0}^{\infty}\mathcal{H}_k(P) \qquad (4.2.18)$$

as an infinite orthogonal sum of Hilbert spaces.

PROOF. Since K_P is an isometry of $\mathbb{R} \oplus L_2^{c,\mathbb{N}}(P) = \mathbb{R} \oplus \big(\oplus_{k=1}^\infty L_2^{c,k}(P)\big)$ onto $L_2(G_P)$ (Proposition 4.2.1), only equation (4.2.17) requires a proof. But (4.2.17) follows immediately from the facts that (i) the set $\{1, h_\ell^\psi : \psi \in L_2^c(P), 1 \le \ell \le k\}$ generates the space $\mathbb{R} \oplus \big(\oplus_{\ell=1}^k L_2^{c,\ell}(P)\big)$ by (4.2.5), (ii) the set of images

$$\{1, K_P(h_\ell^\psi) = \sigma_\psi^\ell H_\ell\big(G_P(\psi)/\sigma_\psi\big)\sqrt{\ell!} : \psi \in L_2^c(P), 1 \le \ell \le k\}$$

generates $\mathcal{P}_k$ (as, by construction, the monomials $1, x, \ldots, x^k$ are linear combinations of $H_0(x), H_1(x), \ldots, H_k(x)$), and (iii) h_k^ψ is orthogonal to h_ℓ^ϕ for all $\ell < k$ and all $\psi, \phi \in L_2^c(P)$. □

By Remark 4.2.2, if $\{\phi_i : i \in J\}$ is an orthonormal basis for $L_2^c(P)$, then the set of Gaussian polynomials at the right of (4.2.14), for all possible maps $r : J \mapsto \{0\} \cup \mathbb{N}$ with $\sum_{j \in J} r_j = k$, form an orthonormal basis of $K_P\big(L_2^{c,k}(P)\big) = \mathcal{H}_k(G_P)$, $k \in \mathbb{N}$.

Identity (4.2.18) is known as the *Wiener chaos decomposition* of the Hilbert space $L_2(G_P)$ of the Gaussian process G_P. It holds for any Gaussian processes. (In fact the above theorem is general since all Gaussian processes are isonormal.) We call the process K_P determined by (4.2.8′) (or (4.2.13), or (4.2.16)), the *isonormal Gaussian chaos process* associated to G_P.

We can now state the *central limit theorem for canonical U-statistics*. However, in order to streamline the proof, it is convenient first to recall that the dual bounded Lipschitz distance between Borel probability measures μ, ν on $\mathbb{R}^d$ (and in general separable metric spaces: e.g. Dudley, 1989, Theorem 11.3.3),

$$d_{BL}(\mu, \nu) := \sup\left\{\left|\int f d(\mu - \nu)\right| : \|f\|_{\text{Lip}} \le 1, \|f\|_\infty \le 1\right\},$$

where $\|f\|_{\text{Lip}} = \sup_{x \ne y} |f(x) - f(y)|/|x - y|$, metrizes weak convergence and that, therefore, the pseudodistance

$$d(\mathbf{Y}, \mathbf{Z}) := d_{BL}\big(\mathcal{L}(\mathbf{Y}), \mathcal{L}(\mathbf{Z})\big) + \big|(\mathbb{E}\|\mathbf{Y}\|^2)^{1/2} - (\mathbb{E}\|\mathbf{Z}\|^2)^{1/2}\big|$$

metrizes convergence in law together with convergence of the second moments of the norm. Here is the CLT:

THEOREM 4.2.4. *For arbritrary natural numbers r_ℓ, $1 \le \ell \le k < \infty$, let h_ℓ be P-square integrable, P-canonical kernels in r_ℓ variables (that is, $h_\ell \in L_2^{c,r_\ell}(P)$), and let K_P be an isonormal Gaussian chaos process on $\oplus_{r=1}^\infty L_2^{c,r}(P)$. Then,*

$$\left(\binom{n}{r_1}^{\frac{1}{2}} U_n(h_1), \ldots, \binom{n}{r_k}^{\frac{1}{2}} U_n(h_k)\right) \to_{\mathcal{L}} \big(K_P(h_1), \ldots, K_P(h_k)\big) \qquad (4.2.19)$$

as $n \to \infty$, with convergence of up to second moments of the norm.

PROOF. Set $\mathbf{h} = (h_1, \ldots, h_k)$ and $S_n(\mathbf{h}) = \left(\binom{n}{r_1}^{\frac{1}{2}} U_n(h_1), \ldots, \binom{n}{r_k}^{\frac{1}{2}} U_n(h_k)\right)$. The vector $(h_1, \ldots, h_k)$ can be approximated in $L_2(P)$ by finite linear combinations of vectors of the form $(h_{r_1}^{\psi_1}, \ldots, h_{r_k}^{\psi_k})$, say $\tilde{\mathbf{h}}_m = (\tilde{h}_{m,1}, \ldots, \tilde{h}_{m,k})$, by (4.2.5). As an immediate consequence of the limit (4.2.7) [see also (4.2.8′)], we have

$$S_n(\tilde{\mathbf{h}}_m) \to_{\mathcal{L}} \big(K_P(\tilde{h}_{m,1}), \ldots, K_P(\tilde{h}_{m,k})\big) := \mathbf{K}_P(\tilde{\mathbf{h}}_m) \qquad (4.2.20)$$

as $n \to \infty$, with convergence of the second moments of the norm as well. Also, if $\|\cdot\|$ denotes the Euclidean norm of $\mathbb{R}^k$, we have

$$\begin{aligned} \mathbb{E}\|\mathbf{K}_P(\tilde{\mathbf{h}}_m) - \mathbf{K}_P(\mathbf{h})\|^2 &= \mathbb{E}\|\tilde{\mathbf{h}}_m(\mathbf{X}) - \mathbf{h}(\mathbf{X})\|^2 \\ &= \mathbb{E}\|S_n(\tilde{\mathbf{h}}_m) - S_n(\mathbf{h})\|^2. \end{aligned} \qquad (4.2.21)$$

Now, the theorem follows from (4.2.20) and (4.2.21) by a simple triangle inequality, e.g., using the metric d:

$$d\big(S_n(\mathbf{h}), K_P(\mathbf{h})\big) \\ \leq d\big(S_n(\mathbf{h}), S_n(\tilde{\mathbf{h}}_m)\big) + d\big(S_n(\tilde{\mathbf{h}}_m), K_P(\tilde{\mathbf{h}}_m)\big) + d\big(K_P(\tilde{\mathbf{h}}_m), K_P(\mathbf{h})\big);$$

by (4.2.21) the first and third terms tend to 0 as $m \to \infty$ uniformly in n (note that twice the L_2 distance dominates d), and the second tends to zero as $n \to \infty$ for all m by (4.2.20). □

REMARKS 4.2.5. 1. This limit theorem admits an extension to finite numbers of functions $h \in L_2^{c,\mathbb{N}}(P)$, with the same proof. For a single function $h \in L_2^{c,\mathbb{N}}(P)$ the result is that, if $h = \sum_{k=1}^{\infty} h_k$ with $h_k \in L_2^{c,k}(P)$, then

$$\sum_{k=1}^{\infty} \binom{n}{k}^{\frac{1}{2}} U_n(h_k) \to_{\mathcal{L}} K_P(h).$$

2. Combined with Hoeffding's expansion, Theorem 4.2.4 gives the central limit theorem for *any* U-statistics $U_n(h)$ whose kernel h (that can be assumed to be symmetric) satisfies $Eh^2 < \infty$: since there is joint convergence in law of the variables $\{n^{\ell/2}U_n(\pi_\ell^P h) : \ell = 0, \ldots, k\}$, only the first term of the Hoeffding expansion of $U_n(h) - P^k h$ contributes to the limit. For instance, if h is nondegenerate, we have

$$\mathcal{L}\Big(n^{\frac{1}{2}}\big(U_n(h) - P^k h\big)\Big) \to N\big(0, k^2 \operatorname{Var}_P P^{k-1}h\big).$$

It should be observed, however, that this limit holds under a less restrictive hypothesis, namely, that $P[P^{k-1}h]^2 < \infty$ and that, for $2 \leq \ell \leq k$, the kernels $\pi_\ell^P h$ satisfy $n^{1/2}U_n^{(\ell)}(\pi_\ell^P h) \to 0$ in probability. It can be proved (Giné and Zinn, 1992a, Theorem 4) that these limits hold if $t^{2\ell/(2\ell-1)} \Pr\{|\pi_\ell^P(h)| > t\} \to 0$ as $t \to \infty$ for $2 \leq \ell \leq k$. We do not pursue this here.

Next we show that Theorem 4.2.4 has a converse. This is done as an application of Theorem 3.3.6 (Khinchin's inequality for the tail probabilities of Rademacher chaos variables) together with decoupling and randomization for tail probabilities, Theorem 3.5.6, part b).

THEOREM 4.2.6. *Let $h : S^k \mapsto \mathbb{R}$ be a measurable symmetric function on $(S, \mathcal{S})$ and let X, X_i, $i \in \mathbb{N}$, be i.i.d. S-valued random variables with probability law P. If the sequence of random variables*

$$\Big\{\frac{1}{n^{k/2}} \sum_{I_n^k} h(X_{i_1}, \ldots, X_{i_k})\Big\}_{n=k}^{\infty} \tag{4.2.22}$$

is stochastically bounded, then $Eh^2(X_1, \ldots, X_k) < \infty$ and, moreover, h is P-canonical.

PROOF. Stochastic boundedness of the sequence (4.2.22) implies, by inequality (3.5.13) in Theorem 3.5.6, that the decoupled and randomized sequence

$$\left\{\frac{1}{n^{k/2}}\sum_{1\le i_1,\ldots,i_k\le n}\varepsilon^1_{i_1}\cdots\varepsilon^k_{i_k}h(X^1_{i_1},\ldots,X^k_{i_k})\right\}_{n=k}^{\infty} \tag{4.2.23}$$

is also stochastically bounded. To ease notation, let us denote by S_n and $\tilde{S}_n$, respectively, the general terms of the sequences (4.2.22) and (4.2.23). The random variables $\tilde{S}_n$ from (4.2.23) are conditionally Rademacher chaos variables of degree k and, therefore, Theorem 3.3.6 applies to them. If $\mathbb{E}_\varepsilon$ denotes conditional expectation given the variables X^j_i, or what is the same, integration with respect to the Rademacher variables only, then, by iteration of Khinchin's inequality [or by (3.2.8), with a worse constant],

$$\mathbb{E}_\varepsilon|\tilde{S}_n| \ge \frac{1}{2^{k/2}}\left[\mathbb{E}_\varepsilon|\tilde{S}_n|^2\right]^{1/2} = \frac{1}{2^{k/2}}\left[\frac{1}{n^k}\sum_{1\le i_1,\ldots,i_k\le n}h^2(X^1_{i_1},\ldots,X^k_{i_k})\right]^{1/2}$$

and therefore, setting

$$[S_n]^2 := \frac{1}{n^k}\sum_{1\le i_1,\ldots,i_k\le n}h^2(X^1_{i_1},\ldots,X^k_{i_k})$$

inequality (3.3.10) in Theorem 3.3.6 (conditional Khinchin for tail probabilities) gives

$$\Pr\left\{[S_n]^2 > 2t^2\right\} \le C_{k,2}\Pr\left\{|\tilde{S}_n| > t\right\} \tag{4.2.24}$$

for all $n \in \mathbb{N}$ and $t > 0$, where $C_{k,2}$ is a constant that only depends on k. Hence, since the sequence in (4.2.23), $\{\tilde{S}_n\}_{n=k}^\infty$, is stochastically bounded, (4.2.24) implies that so is the sequence $\{[S_n]\}_{n=k}^\infty$. By the law of large numbers for U-statistics (the decoupled version of Theorem 4.1.4, which has a similar proof) we have that, for any constant $C < \infty$,

$$\frac{1}{n^k}\sum_{1\le i_1,\ldots,i_k\le n}h^2I_{h^2\le C}(X^1_{i_1},\ldots,X^k_{i_k}) \to P^k\left(h^2I_{h^2\le C}\right) \quad \text{a.s.}$$

This limit and the facts that $[S_n]$ is non-negative and the sequence $\{[S_n]^2\}$ is stochastically bounded, yield that, for all $t > 0$,

$$\begin{aligned}
&\sup_C \Pr[P^k(h^2I_{h^2\le C}) > t] \\
&\qquad \le \sup_C\sup_n \Pr\left\{\frac{1}{n^k}\sum_{1\le i_1,\ldots,i_k\le n}(h^2I_{h^2\le C})(X^1_{i_1},\ldots,X^k_{i_k}) > t\right\} \\
&\qquad \le \sup_n \Pr\left\{[S_n]^2 > t\right\} \to 0 \quad \text{as} \quad t\to\infty.
\end{aligned}$$

Hence, there is $t_0 < \infty$ such that $\sup_C I[P^k(h^2I_{h^2\le C}) > t_0] = 0$ that is, $\sup_C P^k(h^2I_{h^2\le C}) \le t_0$, and therefore, $Eh^2(X_1,\ldots,X_k) \le t_0 < \infty$ by monotone convergence.

To show that h is P-canonical, we observe that Theorem 4.2.4, the central limit theorem, which applies to $\pi_\ell h$ because $P^k h^2 < \infty$, implies that the ℓth term in the Hoeffding expansion $U_n(h) = \sum_{\ell=0}^n \binom{k}{\ell} U_n(\pi_k^P h)$ is either exactly of the order $O_P(n^{-\ell/2})$ or $\pi_\ell^P h = 0$ a.s., $\ell = 0, 1, \ldots, k$. Since, by hypothesis, $U_n(h) = O_P(n^{-k/2})$, it follows that $\pi_\ell^P h = 0$ a.s. for $\ell = 0, 1, \ldots, k-1$, showing that h is P-canonical. □

Finally, we will use the central limit theorem (Theorem 4.2.4) together with Theorem 3.4.1 on decoupling of U-statistics to obtain a decoupling result for *tail probabilities of F-valued Gaussian chaos polynomials of degree $m < \infty$*, not necessarily tetrahedral, not necessarily homogeneous, where F is a normed vector space. Let $G := \{g_i : i \in \mathbb{N}\}$ be a sequence of i.i.d. $N(0, 1)$ random variables. As a consequence of Theorem 4.2.3, any polynomial $Q_{(m)}$ of degree m in the elements of G with coefficients in F admits the decomposition

$$Q_{(m)} = \sum_{k=0}^{m} Q_k \tag{4.2.25}$$

with $Q_0 = h_0 \in B$ and

$$Q_k = \frac{1}{(k!)^{1/2}} \sum_{1 \le i_1, \ldots, i_k \le N} x_{i_1, \ldots, i_k} \prod_{j \le N} H_{j(i_1, \ldots, i_k)}(g_j), \qquad k \ge 1, \tag{4.2.26}$$

for some $N < \infty$, where the coefficients $x_{\mathbf{i}}$ are *symmetric* in the multi-index and $j(\mathbf{i})$ is the number of occurrences of j among the coordinates of the multi-index $\mathbf{i}$.

Let ϕ_i, $i = 0, 1, \ldots$ be an orthonormal system of $L_2([0, 1], \mathcal{B}, \lambda)$ with $\phi_0 \equiv 1$ (e.g., the Haar system). Given $Q_{(m)}$ as in (4.2.25), (4.2.26), we consider the symmetric kernels

$$h_k(t_1, \ldots, t_k) = \sum_{1 \le i_1, \ldots, i_k \le N} x_{i_1, \ldots, i_k} \phi_{i_1}(t_1) \cdots \phi_{i_k}(t_k), \quad k \ge 1.$$

Note that, the functions ϕ_i being centered for $i > 1$, h_k is a canonical or completely degenerate kernel for $k \ge 1$. Here we need to keep track of the order of several U-statistics and therefore we will add a superindex to the usual notation, that is, given a symmetric kernel h of k variables and an i.i.d. sequence $\mathbf{X} = (X_i : i \in \mathbb{N})$, we denote

$$U_n^{(k)}(h) := U_n^{(k)}(h, \mathbf{X}) := \binom{n}{k}^{-1} \sum_{1 \le i_1 < \cdots < i_k \le n} h(X_{i_1}, \ldots, X_{i_k})$$

the U statistic of order k and kernel h based on $\mathbf{X}$, with the convention $U_n^{(0)}(c) = c$. In what follows $\mathbf{X}$ is a sequence of i.i.d. random variables uniform on $[0, 1]$. Then, the central limit theorem for degenerate U-statistics taking values in a finite-dimensional space (in our case, in the span of the vectors $x_{\mathbf{i}}$), that is, Theorem 4.2.4, gives

$$\sum_{k=0}^{m}\binom{n}{k}^{1/2}U_n^{(k)}(h_k)\to_d Q_{(m)}, \tag{4.2.27}$$

where, here and elsewhere, the symbol '$\to_d$' denotes convergence in distribution. The left side of (4.2.27) can be written as a U statistic of order m: clearly, if h_k is a symmetric function of k variables, $k\le m$, then $U_n^{(k)}(h)=U_n^{(m)}(H_{k,m})$ for the kernel

$$H_{k,m}(t_1,\ldots,t_m):=\frac{(m-k)!}{m!}\sum_{\mathbf{j}\in I_m^k}h_k(t_{j_1},\ldots,t_{j_k}),\qquad 1\le k\le m,$$

and, also letting $H_{0,m}:=h_0\in F$, (4.2.27) becomes

$$U_n^{(m)}\left(\sum_{k=0}^{m}\binom{n}{k}^{1/2}H_{k,m}\right)\to_d Q_{(m)}. \tag{4.2.27$'$}$$

The kernel in (4.2.27$'$),

$$H_{(n)}:=\sum_{k=0}^{m}\binom{n}{k}^{1/2}H_{k,m},$$

is symmetric so that it satisfies condition (3.4.2) of Theorem 3.4.1. Let now X_i^j, $i\in\mathbb{N}$, $1\le j\le m$, be i.i.d. uniform on [0, 1] random variables, and define

$$U_n^{(m),dec}(H_{(n)}):=\frac{(n-m)!}{n!}\sum_{\mathbf{i}\in I_n^m}H_{(n)}(X_{i_1}^1,\ldots,X_{i_m}^m). \tag{4.2.28}$$

Theorem 3.4.1 then asserts that *the tail probabilities of the variables* $\|U_n^{(m)}(H_{(n)})\|$ *and* $\|U_n^{(m),dec}(H_{(n)})\|$ *are comparable.* In order to translate this into a comparison result for Gaussian polynomials, it only remains to compute the limiting distribution of $U_n^{(m),dec}(H_{(n)})$. Setting $|\mathbf{i}|=\max_{\ell\le k}i_\ell$ and applying the previous definitions we obtain

$$\begin{aligned}
&U_n^{(m),dec}(H_{(n)})\\
&=h_0+\frac{(n-m)!}{n!}\sum_{k=1}^{m}\binom{n}{k}^{1/2}\frac{(m-k)!}{m!}\sum_{|\mathbf{i}|\le N}x_{\mathbf{i}}\sum_{\mathbf{j}\in I_m^k}\sum_{\mathbf{r}\in I_n^m}\phi_{i_1}(X_{r_{j_1}}^{j_1})\cdots\phi_{i_k}(X_{r_{j_k}}^{j_k})\\
&=h_0+\frac{(n-m)!}{n!}\sum_{k=1}^{m}\binom{n}{k}^{1/2}\frac{(m-k)!}{m!}\frac{(n-k)!}{(n-m)!}\\
&\qquad\times\sum_{|\mathbf{i}|\le N}x_{\mathbf{i}}\sum_{\mathbf{j}\in I_m^k}\sum_{\mathbf{r}\in I_n^k}\phi_{i_1}(X_{r_1}^{j_1})\cdots\phi_{i_k}(X_{r_k}^{j_k})\\
&\asymp h_0+\sum_{k=1}^{m}\frac{1}{n^{k/2}}\frac{(m-k)!}{(k!)^{1/2}m!}\sum_{|\mathbf{i}|\le N}x_{\mathbf{i}}\sum_{\mathbf{j}\in I_m^k}\sum_{\mathbf{r}\in I_n^k}\phi_{i_1}(X_{r_1}^{j_1})\cdots\phi_{i_k}(X_{r_k}^{j_k}),
\end{aligned} \tag{4.2.29}$$

where we use the symbol $\asymp$ to mean "asymptotically equivalent" as n tends to infinity. The first identity in (4.2.29) is obvious from the definitions; the second identity follows by consolidating repetitions in the sum over $\mathbf{r} \in I_n^m$, according to the equation

$$\sum_{\mathbf{r}\in I_n^m} g(X_{r_{j_1}}^{j_1}, \dots, X_{r_{j_k}}^{j_k}) = \frac{(n-k)!}{(n-m)!} \sum_{\mathbf{r}\in I_n^k} g(X_{r_1}^{j_1}, \dots, X_{r_k}^{j_k}),$$

which is easily verified; and in the last line of (4.2.29) we simply consolidate numerical coefficients. Now, since the functions ϕ_i are orthonormal for the uniform law and the variables X_r^j are i.i.d. uniform, it follows that if $\mathbf{r}, \mathbf{s}$ are two distinct points in $\{1, \dots, n\}^k$, then

$$\mathbb{E}\Big[\phi_{i_1}(X_{r_1}^1)\cdots\phi_{i_k}(X_{r_k}^k)\phi_{i_1}(X_{s_1}^1)\cdots\phi_{i_k}(X_{s_k}^k)\Big] = 0.$$

This observation together with the fact that $n^k - \#(I_n^k) = O\big(n^{k-1}\big)$ imply

$$\mathbb{E}\Big[\frac{1}{n^{k/2}}\Big(\sum_{|\mathbf{r}|\le n} - \sum_{\mathbf{r}\in I_n^k}\Big)\phi_{i_1}(X_{r_1}^{j_1})\cdots\phi_{i_k}(X_{r_k}^{j_k})\Big]^2 = \frac{n^k - \#(I_n^k)}{n^k} \to 0.$$

This and (4.2.29) then give

$$\begin{aligned} &U_n^{m,dec}(H_{(n)}) \\ &\asymp h_0 + \sum_{k=1}^m \frac{1}{n^{k/2}} \frac{(m-k)!}{(k!)^{1/2} m!} \sum_{|\mathbf{i}|\le N} x_{\mathbf{i}} \sum_{\mathbf{j}\in I_m^k} \sum_{|\mathbf{r}|\le n} \phi_{i_1}(X_{r_1}^{j_1})\cdots\phi_{i_k}(X_{r_k}^{j_k}) \\ &= h_0 + \sum_{k=1}^m \frac{(m-k)!}{(k!)^{1/2} m!} \sum_{|\mathbf{i}|\le N} x_{\mathbf{i}} \sum_{\mathbf{j}\in I_m^k} \Big(\frac{\sum_{r=1}^n \phi_{i_1}(X_r^{j_1})}{\sqrt{n}}\Big)\cdots\Big(\frac{\sum_{r=1}^n \phi_{i_k}(X_r^{j_k})}{\sqrt{n}}\Big). \end{aligned} \tag{4.2.30}$$

Let now g_i^j, $1 \le i \le N$, $1 \le j \le m$, denote i.i.d. $N(0,1)$ random variables. Since the variables X_r^j are independent and the functions ϕ_i are orthonormal for the law of the Xs, the vectors $(g_i^j : 1 \le i \le N, 1 \le j \le m)$ and $(\phi_i(X_1^j) : 1 \le i \le N, 1 \le j \le m)$ have the same covariance. Hence, the central limit theorem in $\mathbb{R}^{Nm}$ together with the continuous mapping theorem give, by (4.2.30), that

$$U_n^{(m),dec}(H_{(n)}) \to_d h_0 + \sum_{k=1}^m \frac{(m-k)!}{(k!)^{1/2} m!} \sum_{|\mathbf{i}|\le N} x_{\mathbf{i}} \sum_{\mathbf{j}\in I_m^k} g_{i_1}^{j_1}\cdots g_{i_k}^{j_k}. \tag{4.2.31}$$

The limit in (4.2.31) is a "decoupled" Gaussian chaos polynomial. If we define

$$Q_{(m)}^{dec} := h_0 + \sum_{k=1}^m \frac{(m-k)!}{(k!)^{1/2}} \sum_{|\mathbf{i}|\le N} x_{i_1,\dots,i_k} \sum_{\mathbf{j}\in I_m^k} g_{i_1}^{j_1}\cdots g_{i_k}^{j_k}, \tag{4.2.32}$$

then Theorem 3.4.1 and the limits (4.2.27′) and (4.2.31) readily give the following:

THEOREM 4.2.7. *For every $m \in \mathbb{N}$ there exists $C_m \in (0, \infty)$ such that if $Q_{(m)}$ is a (not necessarily homogeneous) Gaussian polynomial of degree m with coefficients* *in a normed space F, as in (4.2.25), (4.2.26), and if $Q^{dec}_{(m)}$ is its decoupled version, as in (4.2.32), then, for all $t > 0$,*

$$\frac{1}{C_m}\Pr\Big\{\|Q^{dec}_{(m)}\| > C_m t\Big\} \le \Pr\Big\{\|Q_{(m)}\| > t\Big\} \le C_m \Pr\Big\{\|Q^{dec}_{(m)}\| > \frac{t}{C_m}\Big\}. \tag{4.2.33}$$

The factors $1/(k!)^{1/2}$ in the defintions of Q and Q^{dec} are obviously superfluous.

The importance of this theorem resides in the fact that it allows one to treat any Gaussian polynomial of a finite degree as a conditionally Gaussian random variable.

REMARKS 4.2.8. 1) Theorem 4.2.7 extends to $N = \infty$, that is, to all of $\mathcal{P}_m(G; F)$, the closure in $L_2(\Omega, \Sigma, \Pr; F)$ of the set of F-valued polynomials of degree m in elements of G.

2) If $Q_{(m)}$ is homogeneous and tetrahedral, that is, if

$$Q_{(m)} = \sum_{\mathbf{i}\in I_N^m} x_{i_1,\dots,i_m} g_{i_1}\cdots g_{i_m},$$

then its decoupled version according to equation (4.2.32) is

$$Q^{dec}_{(m)} = \sum_{(i_1,\dots,i_m)\in I_N^m} x_{i_1,\dots,i_m} g^1_{i_1}\cdots g^m_{i_m}.$$

$Q^{dec}_{(m)}$ also takes a simple form in case $Q_{(m)}$ is in $\mathcal{H}_m(G; F)$, that is, when

$$Q_{(m)} = \sum_{1\le i_1,\dots,i_m\le N} x_{i_1,\dots,i_m} \prod_{j\le N} H_{j(i_1,\dots,i_m)}(g_j).$$

In this case,

$$Q^{dec}_{(m)} = \sum_{1\le i_1,\dots,i_m\le N} x_{i_1,\dots,i_m} g^1_{i_1}\cdots g^m_{i_m}.$$

3) The central limit theorem for U-statistics can also be used to show that the elements of the homogeneous Gaussian chaos of order m, $\mathcal{H}_m(G; F)$, as defined in the previous remark, satisfy the hypercontractivity inequality (3.2.7) in Theorem 3.2.2, as follows. Let $S = \{-1, 1\}^{\mathbb{N}}$, let $P = \left(\frac{1}{2}\delta_{-1} + \frac{1}{2}\delta_1\right)^{\mathbb{N}}$, let $\phi_i : S \mapsto \{-1, 1\}$ be the coordinate functions, let $X_i : S^{\mathbb{N}} \mapsto S$ be the coordinate functions, and let $\{X_i^j\}_{i=1}^\infty$, $j \in \mathbb{N}$, be independent copies of $\{X_i\}_{i=1}^\infty$. Then, the U-statistics $\binom{n}{m}^{1/2} U_n^m(h, \mathbf{X})$, with kernel

$$h(t_1, \dots, t_m) = \sum_{1\le i_1,\dots,i_m\le N} x_{i_1,\dots,i_m}\phi_{i_1}(t_1)\cdots\phi_{i_m}(t_m)$$

converge in law to

$$\sum_{1\le i_1,\dots,i_m\le N} x_{i_1,\dots,i_m} \prod_{j\le N} H_{j(i_1,\dots,i_m)}(g_j).$$

But these U-statistics are homogeneous Rademacher polynomials with uniformly bounded moments of any order due to hypercontractivity, and therefore inequality (3.2.7) in Theorem 3.2.2 for the moments of these U-statistics is inherited by the moments of the limiting Gaussian chaos.

4.3 The Law of the Iterated Logarithm for Canonical U-Statistics

As soon as the kernel h is integrable, as with the CLT, Hoeffding's decomposition reduces the law of the iterated logarithm (LIL) for U-statistics with kernel h to the LIL for canonical (completely degenerate) U-statistics. In order to guess the natural norming in the LIL for degenerate U-statistics, it is instructive to begin with the simplest example, namely, the kernel $h(x, y) = xy$ and random variables X_i i.i.d. with $EX_i = 0$ and $EX_i^2 = 1$. This example is appropriate because of, e.g., the representation (4.2.5). Then, the Hartman–Wintner LIL for sums of i.i.d. square integrable random variables and the law of large numbers readily show that

$$\begin{aligned}\limsup_n \frac{1}{2n\log\log n}\sum_{I_n^2} X_iX_j\\ = \limsup_n \left[\left(\frac{1}{(2n\log\log n)^{\frac{1}{2}}}\sum_{i=1}^n X_i\right)^2 - \frac{1}{2n\log\log n}\sum_{i=1}^n X_i^2\right] = 1 \quad \text{a.s.}\end{aligned}$$

In fact, by the Strassen LIL (see e.g. Ledoux and Talagrand, 1991, page 206) for almost every ω, the set of limit points of the sequence

$$\left\{\sum_{I_n^2} X_i(\omega)X_j(\omega)/2n\log\log n\right\}_{n=1}^{\infty}$$

is precisely the interval $[0, 1]$. This is just a particular case of a more general statement where the kernel xy is replaced by a general square integrable P-canonical kernel in m variables, the norming $2n\log\log n$ is replaced by $a_n = (2n\log\log n)^{m/2}$, and the limit set $[0, 1]$ becomes the set

$$\left\{\mathbb{E}\big[h(X_1, \dots, X_m)g(X_1)\cdots g(X_m)\big] : \mathbb{E}g^2(X_1) \le 1\right\}.$$

The main object of this section consists of proving such a law of the iterated logarithm. Decoupling and randomization, together with the hypercontractivity property of Rademacher chaos, will be seen to provide a relatively simple path towards this result.

Contrary to the case of sums of i.i.d. random variables, square integrability of the kernel is not a necessary condition for the LIL when $m \ge 2$. However, it will be seen to be necessary when h is restricted to be of a particular type. A necessary condition for the LIL in terms of integrability of h (different from square integrability only by a power of $\log\log|h|$) will also be derived. In fact we know, at present, of no sensible set of necessary and sufficient conditions (on the degenerate kernel h) for the LIL.

We begin with a sharp inequality for the estimation of the maximum of a sequence of random variables that enjoy good integrability properties. This inequality will be used in subsequent sections. It elaborates on the following simple argument. Let Φ be a positive, convex, strictly increasing function on $\mathbf{R}_+ \cup \{0\}$ and let $\{\xi_i : i = 1, \ldots, N\}$ be a finite collection of non-negative random variables satisfying

$$\mathbb{E}\Phi(\xi_i/c) \leq d$$

for some $0 < c, d < \infty$, and $E\xi_i < \infty$. Then, by convexity and monotonicity,

$$\begin{aligned}\Phi\big(\mathbb{E}\max_{i\leq N}\xi_i/c\big) &\leq \mathbb{E}\Phi\big(\max_{i\leq N}\xi_i/c\big)\\ &= \mathbb{E}\max_{i\leq N}\Phi\big(\xi_i/c\big)\\ &\leq \sum_{i=1}^{N}\mathbb{E}\Phi\big(\xi_i/c\big)\\ &\leq Nd.\end{aligned}$$

Inverting Φ gives, again by monotonicity,

$$\mathbb{E}\max_{i\leq N}\xi_i \leq c\Phi^{-1}(Nd). \tag{4.3.1}$$

This simple inequality can be quite sharp if the variables ξ_i are i.i.d. and if Φ is of exponential type [that is, if $\Phi(x) \simeq \exp(x^\alpha)$ for some $\alpha > 0$ as $x \to \infty$: this is explicitly defined below]. For example, if ξ_i are $N(0, 1)$, then $\mathbb{E}\exp(\xi_i^2/c^2) = 1/\sqrt{1 - 2/c^2} := d_c$ for $c > 1/2$, and inequality (4.3.1) gives $\mathbb{E}\max_{i\leq N}|\xi_i| \leq c\sqrt{\log(Nd_c)}$, an estimate of the right order of magnitude.

Inequality (4.3.1) can be sharpened, particularly for functions Φ of exponential type. A function ψ on $[0, \infty)$ is a *Young modulus* if $\psi(0) = 0$ and ψ is convex and strictly increasing to infinity. If ψ is a Young modulus, the space $L_\psi(\Omega, \Sigma, \Pr)$ of all the random variables ξ defined on Ω such that $\mathbb{E}\psi(|\xi|/c) < \infty$ for some $0 < c < \infty$, equipped with the norm

$$\|\xi\|_\psi = \inf\big\{c > 0 : \mathbb{E}\psi(|\xi|/c) \leq 1\big\},$$

is a Banach space (cf. Krasnoselsky and Rutitsky, 1961). For instance, if $\psi(x) = x^p$, $1 \leq p < \infty$, then $\|\xi\|_\psi = \|\xi\|_p$. We are more interested in the Young functions of exponential type, ψ_α, $0 < \alpha < \infty$, which are defined as follows:

$$\psi_\alpha(x) = \begin{cases}\exp(x^\alpha) - 1, & \text{if } \alpha \geq 1\\ \tau_\alpha(x) - \alpha\exp\big(\frac{1-\alpha}{\alpha}\big), & \text{if } 0 < \alpha < 1,\end{cases} \tag{4.3.2}$$

where $\tau_\alpha(x)$ is defined as $\exp(x^\alpha)$ if $x \geq \big(\frac{1-\alpha}{\alpha}\big)^{\frac{1}{\alpha}}$, and as the tangent line to the function $y = \exp(x^\alpha)$ at the point $\Big(\big(\frac{1-\alpha}{\alpha}\big)^{\frac{1}{\alpha}}, \exp\big(\frac{1-\alpha}{\alpha}\big)\Big)$ if $0 \leq x \leq \big(\frac{1-\alpha}{\alpha}\big)^{\frac{1}{\alpha}}$. [The complication in the definition of ψ_α for $\alpha < 1$ is due to the fact that the function $y = \exp(x^\alpha)$ is only convex for $x \geq \left(\frac{1-\alpha}{\alpha}\right)^{\frac{1}{\alpha}}$.] Note that for all $p > 0$ and all $\alpha > 0$ there is $c_{p,\alpha} < \infty$ such that

$$\|\xi\|_p \leq c_{p,\alpha}\|\xi\|_{\psi_\alpha}. \tag{4.3.3}$$

These norms extend naturally to vector-valued random variables as follows: if F is a normed vector space and X is an F-valued random variable such that $\|X\|$ is measurable, then $\|X\|_{\psi_\alpha}$ is defined as $\big\| \|X\| \big\|_{\psi_\alpha}$.

Our interest in exponential moduli stems from the fact that the hypercontractivity property of Rademacher chaos variables (Chapter 3, Section 2) implies, by virtue of Corollary 3.2.6, that for all $m \in \mathbb{N}$ there exists $c_m \in (0, \infty)$ such that, if X is an *F-valued homogeneous Rademacher chaos of order m* then

$$\|X\|_{\psi_{2/m}} \leq c_m \|X\|_2. \tag{4.3.4}$$

In particular, if $X = \sum_{(i_1,\ldots,i_m)\in I_n^m} \varepsilon_{i_1}\cdots\varepsilon_{i_m} a_{i_1,\ldots,i_m}$, $a_{\mathbf{i}} \in \mathbb{R}$, then

$$\|X\|_{\psi_{2/m}} \leq c_m \Big(\sum a_{i_1,\ldots,i_m}^2\Big)^{\frac{1}{2}}. \tag{4.3.4$'$}$$

Equations (4.3.4) and (4.3.4′) are obviously also valid for Gaussian chaos variables.

We can now state the aforementioned maximal inequality, valid for Young moduli slightly more general than ψ_α (but not for power moduli).

PROPOSITION 4.3.1. *Let ψ be a Young modulus such that*

$$\limsup_{x\wedge y\to\infty} \frac{\psi^{-1}(xy)}{\psi^{-1}(x)\psi^{-1}(y)} < \infty \tag{4.3.5}$$

and

$$\limsup_{x\to\infty} \frac{\psi^{-1}(x^2)}{\psi^{-1}(x)} < \infty. \tag{4.3.6}$$

Then, there exists a finite constant C_ψ such that, for every sequence of random variables $\{\xi_k : k \in \mathbf{N}\}$,

$$\left\| \sup_k \frac{|\xi_k|}{\psi^{-1}(k)} \right\|_\psi \leq C_\psi \sup_k \|\xi_k\|_\psi. \tag{4.3.7}$$

PROOF. For the proof, we can obviously delete a finite number of terms from the sup at the left of (4.3.7) as long as this number depends only on ψ. Also, by homogeneity, we can assume that $\sup_k \|\xi_k\|_\psi = 1$. Now let $M \geq 1/2$ and $c > 0$, $d > 0$ be finite constants such that $\psi^{-1}(xy) \leq c\psi^{-1}(x)\psi^{-1}(y)$ and $\psi^{-1}(x^2) \leq d\psi^{-1}(x)$ for all $x, y \geq M$, and let us set $k_0 = \max\{8, \psi\big(\frac{1}{d}\psi^{-1}(M)\big), M\}$, $\beta = \max\{\psi^{-1}(M^2)/\psi^{-1}(\frac{1}{2}), \psi^{-1}(M)/\psi^{-1}(\frac{1}{2}), d\}$ and $a = cd\beta$. In the computation that follows we use the properties of ψ together with the fact that, for $t \geq 1/2$, $\psi\big(\beta\psi^{-1}(t)\big) \geq t^2$. [Note that this inequality is obvious for $t \geq M$ because $\beta \geq d$, and that if $1/2 \le t < M$, then $\psi\big(\beta\psi^{-1}(t)\big) \ge \psi\big(\beta\psi^{-1}(\frac{1}{2})\big) \ge M^2 > t^2$ by the definition of β.] We then have, for $t \ge \frac{1}{2}$,

$$\begin{aligned}
\Pr\Big\{\psi\Big(\sup_{k\ge k_0}\frac{|\xi_k|}{a\psi^{-1}(k)}\Big) \ge t\Big\} &= \Pr\Big\{\sup_{k\ge k_0}\frac{|\xi_k|}{a\psi^{-1}(k)\psi^{-1}(t)} \ge 1\Big\}\\
&\le \sum_{k=k_0}^{\infty}\Pr\big\{|\xi_k| \ge a\psi^{-1}(k)\psi^{-1}(t)\big\}\\
&= \sum_{k=k_0}^{\infty}\Pr\big\{\psi\big(|\xi_k|\big) \ge \psi\big(a\psi^{-1}(k)\psi^{-1}(t)\big)\big\}\\
&\le \sum_{k=k_0}^{\infty}\frac{1}{\psi\big(a\psi^{-1}(k)\psi^{-1}(t)\big)}\\
&\le \sum_{k=k_0}^{\infty}\frac{1}{\psi\big(d\psi^{-1}(k)\big)\psi\big(\beta\psi^{-1}(t)\big)}\\
&\le \sum_{k\ge k_0}\frac{1}{k^2t^2} \le \frac{1}{4t^2}.
\end{aligned}$$

And this estimate gives

$$\begin{aligned}
\mathbb{E}\Big[\psi\Big(\sup_{k\ge k_0}\frac{|\xi_k|}{a\psi^{-1}(k)}\Big)\Big] &\le \frac{1}{2} + \int_{\frac{1}{2}}^{\infty}\Pr\Big\{\psi\Big(\sup_{k\ge k_0}\frac{|\xi_k|}{a\psi^{-1}(k)}\Big) \ge t\Big\}\,dt\\
&\le \frac{1}{2} + \frac{1}{4}\int_{\frac{1}{2}}^{\infty}\frac{dt}{t^2} = 1,
\end{aligned}$$

proving the proposition. □

It is interesting to point out that Proposition 4.3.1 does not hold for power-type Young functions x^p. For instance, there are random variables ξ with $\mathbb{E}|\xi| < \infty$ such that, if ξ_k are i.i.d. with the law of ξ, then $\mathbb{E}\sup_k |\xi_k/k| = \infty$; we leave it for the reader to verify that one such example is given by $\Pr\{\xi > t\} = et^{-1}(\log t)^{-2}$, $t \ge e$.

Inequality (4.3.7) is superior to inequality (4.3.1) whenever it applies, since $\max_{k\le N}|\xi_k|/\psi^{-1}(k) \ge \max_{k\le N}|\xi_k|/\psi^{-1}(N)$. For $N(0,1)$ variables ξ_k, (4.3.7) gives

$$\Big\|\sup_{k>2}\frac{|\xi_k|}{\sqrt{\log k}}\Big\|_{\psi_2} \le C < \infty,$$

which is of the right order if the variables are independent and cannot be deduced from inequality (4.3.1).

The previous maximal inequality will be combined with *blocking* to produce an inequality that will be the key to the LIL. Blocking, in turn, will require *Lévy's maximal inequality* for expectations,

$$\mathbb{E}\Big[\max_{k\le n}\Big\|\sum_{i=1}^{k}Y_i\Big\|^p\Big]\le 2\mathbb{E}\Big\|\sum_{i=1}^{n}Y_i\Big\|^p \tag{4.3.8}$$

for all $p<\infty$, where Y_i are independent symmetric random variables with values in a separable Banach space (Theorem 1.1.1).

LEMMA 4.3.2. *Let $b_n\nearrow\infty$ and let $x_{i_1,\dots,i_m}\in F$, F a normed linear space. Then,*

$$\mathbb{E}\sup_n\Big\|\frac{1}{b_n}\sum_{1\le i_1,\dots,i_m\le n}\varepsilon^1_{i_1}\cdots\varepsilon^m_{i_m}x_{i_1,\dots,i_m}\Big\|^p \le 2^m\mathbb{E}\sup_k\Big\|\frac{1}{b^*_k}\sum_{1\le i_1,\dots,i_m\le 2^{k+1}}\varepsilon^1_{i_1}\cdots\varepsilon^m_{i_m}x_{i_1,\dots,i_m}\Big\|^p$$

for all $p>0$, where $b^*_k=b_{2^k}$.

PROOF. Let us set $I(0)=\{1,2\}$ and $I(k)=(2^k,2^{k+1}]\cap\mathbb{N}$, $k\in\mathbb{N}$. It suffices to show that, for all $k_0<\infty$,

$$\mathbb{E}\max_{k\le k_0}\max_{n\in I(k)}\Big\|\frac{1}{b^*_k}\sum_{|\mathbf{i}|\le n}\boldsymbol{\epsilon}^{\mathrm{dec}}_{\mathbf{i}}x_{\mathbf{i}}\Big\|^p\le 2^m\mathbb{E}\max_{k\le k_0}\Big\|\frac{1}{b^*_k}\sum_{|\mathbf{i}|\le 2^{k+1}}\boldsymbol{\epsilon}^{\mathrm{dec}}_{\mathbf{i}}x_{\mathbf{i}}\Big\|^p, \tag{4.3.9}$$

where $|\mathbf{i}|=\max\{i_1,\dots,i_m\}$. Let us define new variables $\mathbf{x}_{\mathbf{i}}$ in $\ell^\infty_{k_0+1}(F)$ by $\mathbf{x}_{\mathbf{i}}=(0,\overset{r)}{\dots},0,x_{\mathbf{i}}/b^*_r,x_{\mathbf{i}}/b^*_{r+1},\dots,x_{\mathbf{i}}/b^*_{k_0})$ if $|\mathbf{i}|\in I(r)$, $r=0,\dots,k_0$. Here $\ell^\infty_{k_0+1}(F)$ denotes the space $F\times\overset{k_0+1)}{\cdots}\times F$ with norm equal to the maximum of the norms of the coordinates. Then we have

$$\max_{k\le k_0}\max_{n\in I(k)}\Big\|\frac{1}{b^*_k}\sum_{|\mathbf{i}|\le n}\boldsymbol{\epsilon}^{\mathrm{dec}}_{\mathbf{i}}x_{\mathbf{i}}\Big\|=\max_{n\le 2^{k_0+1}}\Big\|\sum_{|\mathbf{i}|\le n}\boldsymbol{\epsilon}^{\mathrm{dec}}_{\mathbf{i}}\mathbf{x}_{\mathbf{i}}\Big\|,$$

$$\max_{k\le k_0}\Big\|\frac{1}{b^*_k}\sum_{|\mathbf{i}|\le 2^{k+1}}\boldsymbol{\epsilon}^{\mathrm{dec}}_{\mathbf{i}}x_{\mathbf{i}}\Big\|=\Big\|\sum_{|\mathbf{i}|\le 2^{k_0+1}}\boldsymbol{\epsilon}^{\mathrm{dec}}_{\mathbf{i}}\mathbf{x}_{\mathbf{i}}\Big\|,$$

where the norm in the second terms is that of $\ell^\infty_{k_0+1}(F)$, that is, we have incorporated the max into the norm. Then, Lévy's inequality applied conditionally on all the variables except ε^1_i, $i\in\mathbb{N}$, after incorporating another maximum into a norm, gives

$$\begin{aligned}\mathbb{E}\max_{n\le 2^{k_0+1}}\Big\|\sum_{|\mathbf{i}|\le n}\boldsymbol{\epsilon}^{\mathrm{dec}}_{\mathbf{i}}\mathbf{x}_{\mathbf{i}}\Big\|^p&\le\mathbb{E}\max_{n_j\le 2^{k_0+1},1\le j\le m}\Big\|\sum_{i_j\le n_j,1\le j\le m}\boldsymbol{\epsilon}^{\mathrm{dec}}_{\mathbf{i}}\mathbf{x}_{\mathbf{i}}\Big\|^p\\&=\mathbb{E}\max_{n_1\le 2^{k_0+1}}\Big[\max_{n_j\le 2^{k_0+1},2\le j\le m}\Big\|\sum_{i_1\le n_1}\varepsilon^1_{i_1}\Big(\sum_{i_2\le n_2,\dots,i_m\le n_m}\varepsilon^2_{i_2}\cdots\varepsilon^m_{i_m}\mathbf{x}_{\mathbf{i}}\Big)\Big\|\Big]^p\\&\le 2\mathbb{E}\max_{n_j\le 2^{k_0+1},2\le j\le m}\Big\|\sum_{i_1\le 2^{k_0+1}}\varepsilon^1_{i_1}\Big(\sum_{i_2\le n_2,\dots,i_m\le n_m}\varepsilon^2_{i_2}\cdots\varepsilon^m_{i_m}\mathbf{x}_{\mathbf{i}}\Big)\Big\|^p.\end{aligned}$$

Concretely, Lévy's inequality applies to the third expectation above because the maximum over the n_js, $j \geq 2$, can be absorbed by the norm, thus becoming $\mathbb{E}\max_{n_1 \leq 2^{k_0+1}} \|\sum_{i \leq n_1} \varepsilon_i y_i\|^p$ for y_i in $\ell^\infty_{k_0}(\ell^\infty_{k_0+1}(F))$. Repeating this argument for $j = 2, \ldots, m$ yields the lemma. □

The "trick" of incorporating maxima into norms by enlarging the space, which is quite convenient, has been encountered before, in the proof of inequalities (1.2.9) and (3.1.17). It will be used more than once below.

Now we have at hand all the components of the proof of the basic LIL inequality, which we state and prove next. In what follows, with some abuse of notation, we write $\log x$ for $1 \vee \log x$.

THEOREM 4.3.3. *Let $(S, \mathcal{S}, P)$ be a probability space, let X_i, $i \in \mathbb{N}$, be i.i.d. random variables with values in S and law P, and let $h : S^m \to \mathbb{R}$ be a P-canonical kernel such that $\mathbb{E}h^2 < \infty$. Then, for every $0 < p < 2$, there exists a constant $C_{m,p} < \infty$ depending only on m and p such that*

$$\mathbb{E}\sup_{n\in\mathbb{N}}\left[\frac{1}{(2n\log\log n)^{m/2}}\left|\sum_{(i_1,\ldots,i_m)\in I_n^m} h(X_{i_1},\ldots,X_{i_m})\right|\right]^p \leq C_{m,p}\left(\mathbb{E}h^2(X_1,\ldots,X_m)\right)^{\frac{p}{2}}. \tag{4.3.10}$$

PROOF. To ease notation, we set

$$a_n = (2n\log\log n)^{m/2}$$

and $a_k^* := a_{2^k}$, and use throughout the abbreviated vector notation for multi-indices, products of Rademacher variables, vectors $(X_{i_1}, \ldots, X_{i_m})$, and their decoupled versions given in Section 3.5.

The first step is to decouple and randomize. Let K be a natural number. Letting $\mathbf{h}_r = (0, \overset{r-1}{\ldots}, 0, h/a_r, h/a_{r+1}, \ldots, h/a_{2^K}) \in \ell^\infty_{2^K}$, we have, as in the proof of (3.1.17),

$$\max_{n\leq 2^K}\frac{1}{a_n}\left|\sum_{\mathbf{i}\in I_n^m} h(\mathbf{X_i})\right| = \left\|\sum_{\mathbf{i}\in I_{2^K}^m} \mathbf{h}_{|\mathbf{i}|}(\mathbf{X_i})\right\|.$$

So, we can apply Theorem 3.5.3 [see Remark 3.5.4, 2)] for $r = m$ and obtain

$$\mathbb{E}\max_{n\leq 2^K}\left|\frac{1}{a_n}\sum_{\mathbf{i}\in I_n^m} h(\mathbf{X_i})\right|^p \leq C\,\mathbb{E}\max_{n\leq 2^K}\left|\frac{1}{a_n}\sum_{\mathbf{i}\in I_n^m} \boldsymbol{\epsilon}_{\mathbf{i}}^{\mathrm{dec}} h(\mathbf{X}_{\mathbf{i}}^{\mathrm{dec}})\right|^p \tag{4.3.11}$$

for some constant $C < \infty$ depending only on m and p.

Next we block: we can apply Lemma 4.3.2, more precisely inequality (4.3.9), conditionally on the X variables, with $b_n = a_n$ and $x_{\mathbf{i}} = h(\mathbf{X}_{\mathbf{i}}^{\mathrm{dec}}) \in \mathbb{R}$, to the effect that

$$\mathbb{E}\max_{n\leq 2^K}\left|\frac{1}{a_n}\sum_{\mathbf{i}\in I_n^m} \boldsymbol{\epsilon}_{\mathbf{i}}^{\mathrm{dec}} h(\mathbf{X}_{\mathbf{i}}^{\mathrm{dec}})\right|^p \leq 2^m\,\mathbb{E}\max_{k\leq K-1}\left|\frac{1}{a_k^*}\sum_{\mathbf{i}\in I_{2^{k+1}}^m} \boldsymbol{\epsilon}_{\mathbf{i}}^{\mathrm{dec}} h(\mathbf{X}_{\mathbf{i}}^{\mathrm{dec}})\right|^p. \tag{4.3.12}$$

Now we apply the maximal inequality (4.3.7), again conditionally on the X variables, with $\alpha = 2/m$, and in conjunction with inequality (4.3.4′) for Rademacher chaos, to basically reduce the LIL to a law of large numbers. These two inequalities, on account of inequality (4.3.3), yield the following:

$$\mathbb{E}_\varepsilon \max_{k\le K-1}\Big|\frac{1}{a_k^*}\sum_{\mathbf{i}\in I^m_{2^{k+1}}}\epsilon_{\mathbf{i}}^{\mathrm{dec}} h(\mathbf{X}_{\mathbf{i}}^{\mathrm{dec}})\Big|^p \le 3^{\frac{mp}{2}}\mathbb{E}_\varepsilon \max_{k\le K}\Big|\frac{1}{(\log k)^{\frac{m}{2}}}\frac{1}{2^{\frac{km}{2}}}\sum_{i\in I^m_{2^k}}\epsilon_{\mathbf{i}}^{\mathrm{dec}} h(\mathbf{X}_{\mathbf{i}}^{\mathrm{dec}})\Big|^p$$
$$\le C\max_{k\le K}\Big(\frac{1}{2^{km}}\sum_{\mathbf{i}\in I^m_{2^k}} h^2(\mathbf{X}_{\mathbf{i}}^{\mathrm{dec}})\Big)^{\frac{p}{2}}$$

for a constant $C<\infty$ depending only on p and m. So, integrating with respect to the X variables, we have

$$\mathbb{E}\max_{k\le K-1}\Big|\frac{1}{a_k^*}\sum_{\mathbf{i}\in I^m_{2^{k+1}}}\epsilon_{\mathbf{i}}^{\mathrm{dec}} h(\mathbf{X}_{\mathbf{i}}^{\mathrm{dec}})\Big|^p \le C\mathbb{E}\max_{k\le K}\Big(\frac{1}{2^{km}}\sum_{\mathbf{i}\in I^m_{2^k}} h^2(\mathbf{X}_{\mathbf{i}}^{\mathrm{dec}})\Big)^{\frac{p}{2}}. \tag{4.3.13}$$

It is classical and easy to check that the sequence of random variables $\Big\{\frac{(n-m)!}{n!}\sum_{\mathbf{i}\in I_n^m}|h(\mathbf{X}_{\mathbf{i}}^{\mathrm{dec}})|^2\Big\}_{n=1}^{\infty}$ is a reversed martingale since h is square integrable. Doob's maximal inequality for (reversed) martingales (e.g., Dudley, 1989, 10.4.2) then gives

$$\Pr\Big\{\sup_{n\in\mathbb{N}}\frac{(n-m)!}{n!}\sum_{\mathbf{i}\in I_n^m}\big|h(\mathbf{X}_{\mathbf{i}}^{\mathrm{dec}})\big|^2 > t\Big\} \le \frac{\mathbb{E}h^2(\mathbf{X})}{t}$$

and therefore, p being less than 2, we obtain

$$\mathbb{E}\max_{k\le K}\Big(\frac{1}{2^{km}}\sum_{\mathbf{i}\in I^m_{2^k}} h^2(\mathbf{X}_{\mathbf{i}}^{\mathrm{dec}})\Big)^{\frac{p}{2}} \le \mathbb{E}\sup_n\Big(\frac{1}{n^m}\sum_{\mathbf{i}\in I_n^m}|h(\mathbf{X}_{\mathbf{i}}^{\mathrm{dec}})|^2\Big)^{\frac{p}{2}}$$
$$\le \big(\mathbb{E}h^2(\mathbf{X})\big)^{\frac{p}{2}} + \frac{p}{2}\int_{\mathbb{E}h^2(\mathbf{X})}^{\infty} t^{\frac{p}{2}-1}\frac{\mathbb{E}h^2(\mathbf{X})}{t}dt$$
$$\le \frac{2}{2-p}\big(\mathbb{E}h^2(\mathbf{X})\big)^{\frac{p}{2}}. \tag{4.3.14}$$

Now, the theorem obtains by combining inequalities (4.3.11) to (4.3.14) and letting K tend to infinity (note that the constants in these inequalities do not depend on K, h or P). □

The proof of Theorem 4.3.3 shows how *decoupling* and *randomization* play an essential role in the LIL (Theorem 4.3.5 below).

REMARK 4.3.4. *(The bounded LIL).* Under the hypotheses of Theorem 4.3.3, we have

$$\limsup_{n\to\infty}\frac{1}{(2n\log\log n)^{m/2}}\Big|\sum_{(i_1,\dots,i_m)\in I_n^m} h(X_{i_1},\dots,X_{i_m})\Big|$$
$$\le C_{m,1}\big(\mathbb{E}h^2(X_1,\dots,X_m)\big)^{\frac{1}{2}} \quad \text{a.s.,} \tag{4.3.15}$$

that is, a bounded LIL. To see this note that, by Theorem 4.3.3, the random variable $\limsup_{n\to\infty} \frac{1}{a_n}\left|\sum_{\mathbf{i}\in I_n^m} h(\mathbf{X_i})\right|$ is almost surely finite and its expected value is dominated by the bound given in (4.3.10) for $p = 1$; since it is obviously invariant under permutations of the variables $\{X_i\}$, it is almost surely a constant by the Hewitt–Savage zero–one law, and this constant cannot exceed the bound for its expected value.

The bound given in the previous remark is not final: the exact a.s. value of the lim sup in (4.3.15) will shortly be obtained as a consequence of the more precise LIL that we will now prove, the *compact* LIL.

The proof of the compact LIL for canonical U-statistics has two main ingredients namely, the inequality in Theorem 4.3.3 and Strassen's LIL for sums of i.i.d. random vectors in $\mathbb{R}^d$. The following is the LIL in $\mathbb{R}^d$:

THEOREM 4.3.5. (STRASSEN) *Let* $\mathbf{Y}$, $\mathbf{Y}_i$, $i \in \mathbb{N}$, *be i.i.d. random vectors in* $\mathbb{R}^d$. *Then, the sequence*

$$\left\{\frac{1}{(2n\log\log n)^{\frac{1}{2}}}\sum_{i=1}^{n}\mathbf{Y}_i(\omega)\right\}_{n=1}^{\infty}$$

is relatively compact for almost every $\omega \in \Omega$ *if and only if* $\mathbb{E}\mathbf{Y} = 0$ *and* $\mathbb{E}|\mathbf{Y}|^2 < \infty$, *and then*

$$\lim\text{set}\left\{\frac{1}{(2n\log\log n)^{\frac{1}{2}}}\sum_{i=1}^{n}\mathbf{Y}_i(\omega)\right\} = K_{\mathbf{Y}} \quad \text{a.s.}$$

where $K_{\mathbf{Y}}$ *is the subset of* $\mathbb{R}^d$ *defined by*

$$K_{\mathbf{Y}} = \left\{\mathbb{E}\big[(g(\mathbf{Y}))\mathbf{Y}\big] : g \text{ real, measurable, and } \mathbb{E}g^2(\mathbf{Y}) \le 1\right\}.$$

Here and elsewhere, the notation $\lim\text{set}\{a_n\}$ refers to the set of all the subsequential limits of the sequence $\{a_n\}_{n=1}^{\infty}$.

The set $K_{\mathbf{Y}}$ is the unit ball of the reproducing kernel Hilbert space associated to the covariance of $\mathbf{Y}$ (see, e.g., Ledoux and Talagrand, 1991, page 208, for its definition). We will only use the "if" part of this statement, and a short proof of it can be found in pages 210 and 211 of the same reference (first part of the proof of Theorem 8.5 there).

To ease notation, given $h : S^m \mapsto \mathbb{R}$, that as usual we assume symmetric in its entries, we set

$$\alpha_n(h) := \frac{1}{(2n\log\log n)^{\frac{m}{2}}}\sum_{(i_1,\ldots,i_m)\in I_n^m} h(X_{i_1},\ldots,X_{i_m}). \tag{4.3.16}$$

The LIL for square integrable canonical U-statistics is then as follows:

THEOREM 4.3.6. *Let X, X_i, $i \in \mathbb{N}$, be i.i.d. random variables with values in a countably generated measurable space $(S, \mathcal{S})$ and common law P. Let $h_j : S^m \mapsto \mathbb{R}$ be P-canonical functions with $\mathbb{E}h_j^2 < \infty$, $j = 1, \ldots, d$. Then, with probability one, the sequence*

$$\left\{(\alpha_n(h_1), \ldots, \alpha_n(h_d))\right\}_{n=1}^{\infty} \tag{4.3.17}$$

is relatively compact in $\mathbb{R}^d$ and its limit set is

$$K := \Big\{\mathbb{E}\big[g(X_1)\cdots g(X_m)\big(h_1(X_1, \ldots, X_m), \ldots, h_d(X_1, \ldots, X_m)\big)\big] : \\ \mathbb{E}g^2(X) \leq 1\Big\}. \tag{4.3.18}$$

PROOF. Remark 4.3.4 (hence, Theorem 4.3.3) shows that the sequence in (4.3.17) is relatively compact for almost every $\omega \in \Omega$. So, we only have to determine its limit set. By (4.2.5) each function h_i, $1 \leq i \leq d$, is a series, convergent in the sense of $L_2(P^m)$, with partial sums of the form

$$\bar{h}_{i,R} = \sum_{r=1}^{R} c_r(h_i) h_m^{\psi_r},$$

where, we recall, the functions $\psi_r : S \mapsto \mathbb{R}$, which do not depend on h_i, are centered and square integrable for P,

$$h_m^{\psi_r}(x_1, \ldots, x_m) = \psi_r(x_1)\cdots\psi_r(x_m)$$

and $c_r(h_i)$ are real numbers. The first step consists in proving Theorem 4.3.6 for the simpler functions $h_m^{\psi_r}$ and $\bar{h}_{i,R}$. As we will see, Theorem 4.3.6 for these simpler functions is a consequence of Strassen's theorem in finite dimensions. The inequality in Theorem 4.3.3 will be used in order to extend the theorem from simple to general functions. Although the present proof is quite easy, its actual implementation is somewhat cumbersome.

By Strassen's LIL in $\mathbb{R}^R$, with probability one,

$$\operatorname{lim\,set}\left\{\frac{1}{(2n\log\log n)^{\frac{1}{2}}}\sum_{j=1}^{n}\big(\psi_1(X_j), \ldots, \psi_R(X_j)\big)\right\} \\ = \Big\{\mathbb{E}\big[g(X)\big(\psi_1(X), \ldots, \psi_R(X)\big)\big] : \mathbb{E}g^2(X) \leq 1\Big\} \tag{4.3.19}$$

(where we take this statement to mean that, moreover, the sequence in question is relatively compact with probability one). Also, for $k = 2, \ldots, m$,

$$\lim_{n\to\infty}\frac{1}{(2n\log\log n)^{\frac{k}{2}}}\sum_{j=1}^{n}\big((\psi_1(X_j))^k, \ldots, (\psi_R(X_j))^k\big) = 0 \quad \text{a.s.} \tag{4.3.20}$$

by the Kolmogorov law of large numbers for $k = 2$, and by the Marcinkiewicz law of large numbers for $2 < k \le m$. By Newton's identities, as in (4.2.6),

$$\alpha_n(h_m^{\psi_r}) = m!R_m\left(\frac{1}{(2n\log\log n)^{\frac{1}{2}}}\sum_{i=1}^n \psi_r(X_i), \dots, \frac{1}{(2n\log\log n)^{\frac{m}{2}}}\sum_{i=1}^n (\psi_r(X_i))^m\right) \tag{4.3.21}$$

for all r and n, where R_m are fixed homogeneous polynomials of degree m and $m!R_m(u, 0, \dots, 0) = u^m$.

The LIL for the functions $h_m^{\psi_r}$ will follow from the last three equations because of the following simple observation: if $\{z_n\}$ is a relatively compact sequence in a topological space with limit set Z and if λ is a continuous function (with values in not necessarily the same topological space), then the sequence $\{\lambda(z_n)\}$ is also relatively compact and its limit set is $\{\lambda(z) : z \in Z\}$. Since we will use this observation more than once, for lack of a better name, we will refer to it as "the continuous function principle."

Taking z_n to be the vector of $\mathbb{R}^{mR}$

$$z_n = \left(\frac{1}{(2n\log\log n)^{\frac{1}{2}}}\sum_{i=1}^n (\psi_1(X_i), \dots, \psi_R(X_i)), \dots, \frac{1}{(2n\log\log n)^{\frac{m}{2}}}\sum_{i=1}^n ((\psi_1(X_i))^m, \dots, (\psi_R(X_i))^m)\right)$$

and

$$\lambda(x_{1,1}, \dots, x_{1,R}, \dots, x_{m,1}, \dots, x_{m,R}) = \big(m!R_m(x_{1,1}, \dots, x_{m,1}), \dots, m!R_m(x_{1,R}, \dots, x_{m,R})\big),$$

in the continuous function principle, (4.3.19)–(4.3.21) give that, with probability one, the sequence $\{(\alpha_n(h_m^{\psi_1}), \dots, \alpha_n(h_m^{\psi_R}))\}$ is relatively compact and its limit set is

$$\Big\{\Big(m!R_m\big(\mathbb{E}[g(X)\psi_1(X)], 0, \dots, 0\big), \dots, m!R_m\big(\mathbb{E}[g(X)\psi_R(X)], 0, \dots, 0\big)\Big) : \mathbb{E}g^2(X) \le 1\Big\}.$$

But

$$\begin{aligned} m!R_m\big(\mathbb{E}[g(X)\psi_r(X)], 0, \dots, 0\big) &= \big(\mathbb{E}[g(X)\psi_r(X)]\big)^m \\ &= \mathbb{E}[g(X_1)\psi_r(X_1)]\cdots\mathbb{E}[g(X_m)\psi_r(X_m)] \\ &= \mathbb{E}[g(X_1)\cdots g(X_m)h_m^{\psi_r}(X_1, \dots, X_m)]. \end{aligned}$$

We thus have

$$\begin{aligned} &\lim\operatorname{set}\{(\alpha_n(h_m^{\psi_1}), \dots, \alpha_n(h_m^{\psi_R}))\} \\ &\quad = \Big\{\mathbb{E}\big[g(X_1)\cdots g(X_m)\big(h_m^{\psi_1}(\mathbf{X}), \dots, h_m^{\psi_R}(\mathbf{X})\big)\big] : \mathbb{E}g^2(X) \le 1\Big\} \quad \text{a.s.} \end{aligned}$$

Now, we can apply the continuous function principle to the last identity with

$$\lambda(x_1, \ldots, x_R) = \Big(\sum_{r=1}^{R} c_r(h_1)x_r, \ldots, \sum_{r=1}^{R} c_r(h_d)x_r\Big),$$

and obtain that, with probability one the sequence $\{(\alpha_n(\bar{h}_{1,R}), \ldots, \alpha_n(\bar{h}_{d,R}))\}_{n=1}^{\infty}$ is relatively compact and that

$$\begin{aligned}&\lim\operatorname{set}\{(\alpha_n(\bar{h}_{1,R}), \ldots, \alpha_n(\bar{h}_{d,R}))\}\\ &\quad= \Big\{\mathbb{E}\big[g(X_1)\cdots g(X_m)\big(\bar{h}_{1,R}(\mathbf{X}), \ldots, \bar{h}_{d,R}(\mathbf{X})\big)\big] : \mathbb{E}g^2(X) \le 1\Big\} \quad \text{a.s.,}\end{aligned} \tag{4.3.22}$$

which is Theorem 4.3.5 for the functions $\bar{h}_{i,R}$.

The next step in our proof is to show that

$$K \subseteq \lim\operatorname{set}\{(\alpha_n(h_1), \ldots, \alpha_n(h_d))\} \quad \text{a.s.} \tag{4.3.23}$$

If $\mathbf{x} \in K$, then there is g with $\mathbb{E}g^2(X) \le 1$ such that

$$\mathbf{x} = \mathbb{E}\big[g(X_1)\cdots g(X_m)\big(h_1(\mathbf{X}), \ldots, h_d(\mathbf{X})\big)\big].$$

As indicated at the beginning of the proof, given $\varepsilon > 0$, by (4.2.5), there exists $R < \infty$ such that

$$\Big|\mathbb{E}\big[g(X_1)\cdots g(X_m)\big(h_i(\mathbf{X})-\bar{h}_{i,R}(\mathbf{X})\big)\big]\Big| \le \big\|h_i(\mathbf{X})-\bar{h}_{i,R}(\mathbf{X})\big\|_2 < \varepsilon,\ i = 1, \ldots, d. \tag{4.3.24}$$

Then, by (4.3.15) (that is, again by Theorem 4.3.3), with probability one,

$$\limsup_n \big|\alpha_n(h_i - \bar{h}_{i,R})\big| < C_{m,1}\varepsilon, \qquad i = 1, \ldots, d. \tag{4.3.25}$$

By the first part of the theorem, concretely by (4.3.22), for almost every $\omega \in \Omega$ there is an infinite subset $J := J(\omega)$ of integers such that

$$\big|\alpha_n(\bar{h}_{i,R}) - \mathbb{E}\big[g(X_1)\cdots g(X_m)\bar{h}_{i,R}(\mathbf{X})\big]\big| < \varepsilon, \qquad i = 1, \ldots, d, \quad n \in J. \tag{4.3.26}$$

Then, (4.3.24)–(4.3.26) imply, by the triangle inequality, that for almost every $\omega \in \Omega$ there are infinitely many $n := n(\omega)$ such that

$$\big|\alpha_n(h_i) - \mathbb{E}\big[g(X_1)\cdots g(X_m)h_i(\mathbf{X})\big]\big| < (2+C_{m,1})\varepsilon, \qquad i = 1, \ldots, d. \tag{4.3.27}$$

Now taking $\varepsilon = 1/k$, $k \in \mathbb{N}$, we conclude from (4.3.27) that for almost every $\omega \in \Omega$ there is a subsequence $n_k = n_k(\omega)$ such that $\lim_{k\to\infty}\big(\alpha_{n_k}(h_1), \ldots, \alpha_{n_k}(h_d)\big) = \mathbf{x}$, proving the inclusion (4.3.23).

Finally, we conclude the proof of the theorem by showing that

$$\lim\operatorname{set}\{(\alpha_n(h_1), \ldots, \alpha_n(h_d))\} \subseteq K \quad \text{a.s.} \tag{4.3.28}$$

The previous paragraph shows in particular that there is a set $\Omega_1 \subset \Omega$ with $\Pr\Omega_1 = 1$ such that

$$\cap_{\omega\in\Omega_1}\Big[\lim\operatorname{set}\{(\alpha_n(h_1)(\omega), \ldots, \alpha_n(h_d)(\omega))\}\Big] \ne \emptyset.$$

We must prove that if $\mathbf{x}$ is in this set, then $\mathbf{x} \in K$. Let $\mathbf{x}$ be such a point and let $\varepsilon > 0$ be given. Then, by (4.3.24) and (4.3.25) there exists $R < \infty$ such that i) inequality (4.3.24) holds for the functions $\bar{h}_{i,R}$, $1 \le i \le d$, and ii) for every ω in a set of probability one there is a subsequence $n_k := n_k(\omega)$ for which

$$\lim_{k\to\infty} \left|\mathbf{x} - \left(\alpha_{n_k}(\bar{h}_{1,R}), \ldots, \alpha_{n_k}(\bar{h}_{d,R})\right)\right| < \varepsilon.$$

Since the limit set of the sequence $\left\{\left(\alpha_n(\bar{h}_{1,R}), \ldots, \alpha_n(\bar{h}_{d,R})\right)\right\}$ is almost surely equal to the right side of equation (4.3.22) (by the first part of the proof), there is a function g with $\mathbb{E}g^2(X) \le 1$ and a sub-subsequence $n_{k_\ell} := n_{k_\ell}(\omega)$ such that

$$\lim_{\ell\to\infty} \left(\alpha_{n_{k_\ell}}(\bar{h}_{1,R}) \ldots, \alpha_{n_{k_\ell}}(\bar{h}_{d,R})\right) = \mathbb{E}\left[g(X_1)\cdots g(X_m)\left(\bar{h}_{1,R}(\mathbf{X}), \ldots, \bar{h}_{d,R}(\mathbf{X})\right)\right]$$

for almost every ω. Taking limits in the inequality

$$\begin{aligned}
&\left|\mathbf{x} - \mathbb{E}\left[g(X_1)\cdots g(X_m)\left(h_1(\mathbf{X}), \ldots, h_d(\mathbf{X})\right)\right]\right| \\
&\quad \le \left|\mathbf{x} - \left(\alpha_{n_{k_\ell}}(\bar{h}_{1,R}), \ldots, \alpha_{n_{k_\ell}}(\bar{h}_{d,R})\right)\right| \\
&\quad + \left|\left(\alpha_{n_{k_\ell}}(\bar{h}_{1,R}), \ldots, \alpha_{n_{k_\ell}}(\bar{h}_{d,R})\right) - \mathbb{E}\left[g(X_1)\cdots g(X_m)\left(\bar{h}_{1,R}(\mathbf{X}), \ldots, \bar{h}_{d,R}(\mathbf{X})\right)\right]\right| \\
&\quad + \left|\mathbb{E}\left[g(X_1)\cdots g(X_m)\left(\bar{h}_{1,R}(\mathbf{X}) - h_1(\mathbf{X}), \ldots, \bar{h}_{d,R}(\mathbf{X}) - h_d(\mathbf{X})\right)\right]\right|,
\end{aligned}$$

we obtain

$$\left|\mathbf{x} - \mathbb{E}\left[g(X_1)\cdots g(X_m)\left(h_1(\mathbf{X}), \ldots, h_d(\mathbf{X})\right)\right]\right| < (\sqrt{d} + 1)\varepsilon.$$

Taking $\varepsilon = 1/k$, $k \in \mathbb{N}$, it then follows that there is a sequence of functions g_k with $\mathbb{E}g_k^2(X) \le 1$ such that

$$\lim_{k\to\infty} \mathbb{E}\left[g_k(X_1)\cdots g_k(X_m)\left(h_1(\mathbf{X}), \ldots, h_d(\mathbf{X})\right)\right] = \mathbf{x}. \tag{4.3.29}$$

Now, $L_2(P)$ being separable by the hypothesis on $(S, \mathcal{S})$, its unit ball is weakly sequentially compact, and a simple approximation argument shows that if $\{g_r\}$ converges weakly to f, then the sequence $\{g_r(x_1)\cdots g_r(x_m)\}$ converges weakly in $L_2(P^d)$ to $f(x_1)\cdots f(x_m)$. So, there exist a subsequence $\{g_{k_j}\}$ and a function f such that $\mathbb{E}f^2(X) \le 1$ and

$$\lim_{j\to\infty} \mathbb{E}[g_{k_j}(X_1)\cdots g_{k_j}(X_m)h_i(\mathbf{X})] = \mathbb{E}[f(X_1)\cdots f(X_m)h_i(\mathbf{X})], \qquad 1 \le i \le d.$$

It then follows from this limit and the limit (4.3.29) that

$$\mathbf{x} = \mathbb{E}\left[f(X_1)\cdots f(X_m)\left(h_1(\mathbf{X}), \ldots, h_d(\mathbf{X})\right)\right],$$

that is, $\mathbf{x} \in K$. □

REMARK 4.3.7. *The limit set for decoupled U-statistics.* In inequality (4.3.11) the left-hand side can obviously be replaced by $\max_{n\le 2^K} \left|\sum_{I_n^m} h(\mathbf{X}_\mathbf{i}^{\mathrm{dec}})/a_n\right|$. Hence, if

h is a P-canonical, square integrable kernel in m variables, then the two sequences whose generic terms are, respectively,

$$\alpha_n^{\mathrm{dec}}(h) := \frac{1}{a_n}\sum_{\mathbf{i}\in I_n^m} h(\mathbf{X}_{\mathbf{i}}^{\mathrm{dec}}), \quad \text{and} \quad \alpha_n^{\mathrm{sdec}}(h) := \frac{1}{a_n}\sum_{\mathbf{i}\in I_n^m} \boldsymbol{\epsilon}_{\mathbf{i}}^{\mathrm{dec}} h(\mathbf{X}_{\mathbf{i}}^{\mathrm{dec}}),$$

are both a.s. relatively compact in $\mathbb{R}^d$. Let now ψ_r, $1 \le r \le R$, be centered, square integrable functions of one variable. By Strassen's theorem, the limit set of the sequence

$$z_n = \frac{1}{2n\log\log n}\sum_{i=1}^n \Big(\psi_1(X_i^1),\dots,\psi_R(X_i^1),\dots,\psi_1(X_i^m),\dots,\psi_R(X_i^m)\Big)$$

is

$$\Big\{\mathbb{E}\Big[g(X^1,\dots,X^m)\Big(\psi_1(X^1),\dots,\psi_R(X^1),\dots,\psi_1(X^m),\dots,\psi_R(X^m)\Big)\Big] \\ : \mathbb{E}g^2(X^1,\dots,X^m) \le 1\Big\}.$$

It is easy to see (using that the functions ψ are centered) that this set coincides with the set

$$\Big\{\mathbb{E}\Big(g_1(X^1)\psi_1(X^1),\dots,g_1(X^1)\psi_R(X^1),\dots,g_m(X^m)\psi_1(X^m), \\ \dots,g_m(X^m)\psi_R(X^m)\Big) : \mathbb{E}g_j^2(X) \le 1,\ j=1,\dots,m\Big\}.$$

Moreover, if ψ is centered and square integrable, then, by the Marcinkiewicz law of large numbers (e.g., it is stated in Example 4.1.6), the "diagonals" in the sum $\sum_{i_1,\dots,i_m\le n}\psi(X_{i_1}^1)\cdots\psi(X_{i_m}^m)/a_n$ tend to zero a.s. and therefore, for the purpose of finding the limit set, we can replace $\alpha_n^{\mathrm{dec}}(h^{\psi_r})$, $r=1,\dots,R$, by $\prod_{k=1}^m\big[\sum_{i=1}^n\psi_r(X_i^{(k)})/(2n\log\log n)^{1/2}\big]$, which are continuous functions of z_n. These two observations, which apply as well to α_n^{sdec}, allow us to proceed as in Theorem 4.3.6 and conclude that, if $h_1,\dots,h_d$ are P-canonical square integrable kernels, then the limit sets of the two sequences

$$\big\{\big(\alpha_n^{\mathrm{dec}}(h_1),\dots,\alpha_n^{\mathrm{dec}}(h_d)\big)\big\}_{n=1}^\infty \quad \text{and} \quad \big\{\big(\alpha_n^{\mathrm{sdec}}(h_1),\dots,\alpha_n^{\mathrm{sdec}}(h_d)\big)\big\}_{n=1}^\infty \tag{4.3.30}$$

coincide and equal

$$K_{\mathrm{dec}} = \Big\{\mathbb{E}\Big[g_1(X_1)\cdots g_m(X_m)\big(h_1(\mathbf{X}),\dots,h_d(\mathbf{X})\big)\Big] \\ : \mathbb{E}g_j^2(X) < \infty,\ j=1,\dots,m\Big\}. \tag{4.3.31}$$

It is also useful to observe that if the functions h_r, $r = 1,\dots,d$, are square integrable, then, *even if they are not canonical*, the sequence $\big\{\big(\alpha_n^{\mathrm{sdec}}(h_1),\dots,\alpha_n^{\mathrm{sdec}}(h_d)\big)\big\}_{n=1}^\infty$ is a.s. relatively compact with limit set K_{dec} given by (4.3.31): one just applies the above to the kernels $f_r\big((X_1,\varepsilon_1),\dots,(X_m,\varepsilon_m)\big) = \varepsilon_1\cdots\varepsilon_m h(X_1,\dots,X_m)$, which are Q-canonical for $Q = \mathcal{L}(X_1,\varepsilon_1)$.

The limit set K of Theorem 4.3.6 for a single function h of two variables,

$$K = \left\{ \mathbb{E}\big[g(X_1)g(X_2)h(X_1, X_2)\big] : \mathbb{E}g^2(X) \le 1 \right\}$$

is the numerical range of the operator

$$T_h : L_2(S, P) \mapsto L_2(S, P)$$

defined as

$$T_h(g)(x) = \mathbb{E}\big[g(X)h(X, x)\big], \qquad g \in L_2(P), \quad x \in S.$$

This operator is Hilbert–Schmidt if and only if $\mathbb{E}h^2(X_1, X_2) < \infty$, but the set K is bounded whenever T_h is a bounded operator of $L_2(P)$. This indicates that finiteness of the second moment of h may not be a necessary condition for the LIL. It also indicates that this moment condition may be necessary for the class of functions h that we consider in the next proposition (functions for which T_h would have finite-dimensional range if it were defined).

PROPOSITION 4.3.8. *Suppose there exists a finite number R of measurable functions $\psi_r : S \to \mathbb{R}$ and real numbers λ_r such that $h = \sum_{r=1}^{R} \lambda_i h_m^{\psi_r}$. If*

$$\limsup_{n\to\infty} \frac{1}{(2n \log\log n)^{\frac{m}{2}}} \Big| \sum_{(i_1,\ldots,i_m)\in I_n^m} h(X_{i_1}, \ldots, X_{i_m}) \Big| < \infty \quad \text{a.s.} \tag{4.3.31}$$

then $Eh^2(X_1, \ldots, X_m) < \infty$ and h is P-canonical.

PROOF. If the bounded LIL (4.3.31) holds, then, by the Hewitt–Savage zero–one law, there is a finite constant c such that the limsup in (4.3.31) is almost surely bounded by c. Then, Theorem 3.5.7 gives, by Borel–Cantelli, that the decoupled and randomized sums also satisfy the bounded LIL, that is, there exists $c_0 < \infty$ such that

$$\limsup_n \frac{1}{a_n} \Big| \sum_{|\mathbf{i}|\le n} \epsilon_{\mathbf{i}}^{\text{dec}} h(\mathbf{X}_{\mathbf{i}}^{\text{dec}}) \Big| < c_0 \quad \text{a.s.} \tag{4.3.32}$$

The next step towards the proof of the theorem is *truncation* namely, showing that the LIL (4.3.32) implies

$$\sup_{\{A_1,\ldots,A_m\}\subset\mathcal{S}} \left[\limsup_n \frac{1}{a_n} \Big| \sum_{|\mathbf{i}|\le n} \epsilon_{\mathbf{i}}^{\text{dec}} h(\mathbf{X}_{\mathbf{i}}^{\text{dec}}) I_{\{X_{i_j}^j \in A_j : 1\le j\le m\}} \Big| \right] < c_0 \quad \text{a.s.} \tag{4.3.33}$$

To prove (4.3.33) we proceed as follows: for each $n \in \mathbb{N}$, we set $\tilde{\varepsilon}_n^r = \varepsilon_n^r$ for $1 \le r < m$, and define

$$\tilde{\varepsilon}_n^m = \begin{cases} \varepsilon_n^m & \text{if } X_n^m \in A_m \\ -\varepsilon_n^m & \text{if } X_n^m \notin A_m. \end{cases}$$

Then, the sequences $\{\tilde{\varepsilon}_n^r, X_n^r : n \in \mathbb{N}, 1 \le r \le m\}$ and $\{\varepsilon_n^r, X_n^r : n \in \mathbb{N}, 1 \le r \le m\}$

have the same finite dimensional distributions and therefore, (4.3.32) also holds with ε_i^r replaced by $\tilde{\varepsilon}_i^r$. Thus, it follows that, with probability one,

$$\limsup_{n\to\infty} \frac{1}{a_n} \sum_{|\mathbf{i}|\le n} \boldsymbol{\epsilon}_{\mathbf{i}}^{\mathrm{dec}} h(\mathbf{X}_{\mathbf{i}}^{\mathrm{dec}}) I_{X_{i_m}^m \in A_m} = \limsup_{n\to\infty} \frac{1}{2a_n} \sum_{|\mathbf{i}|\le n} \left[\boldsymbol{\epsilon}_{\mathbf{i}}^{\mathrm{dec}} + \tilde{\boldsymbol{\epsilon}}_{\mathbf{i}}^{\mathrm{dec}}\right] h(\mathbf{X}_{\mathbf{i}}^{\mathrm{dec}}) < c_0.$$

Repeating this argument m times yields (4.3.32).

Now, for each $t > 0$, we set $A(t) = \left\{x \in S : \max_{1\le r\le m} |\psi_r(x)| \le t\right\}$ and $h_t(x_1, \dots, x_m) = h(x_1, \dots, x_m) I_{\{x_j \in A(t): 1\le j\le m\}}$. h_t is bounded and therefore Remark 4.3.7 gives

$$\lim\mathrm{set}\{\alpha_n^{\mathrm{sdec}}(h_t)\} = \left\{\mathbb{E}\left[g_1(X_1)\cdots g_m(X_m) h_t(\mathbf{X})\right] : \mathbb{E}g_r^2(X) \le 1,\ 1\le r\le m\right\}.$$

We can apply (4.3.32) with $A_j = A(t)$ for all j, and conclude

$$\sup_{t>0}\left\{\left|\mathbb{E}\left[g_1(X_1)\cdots g_m(X_m) h_t(\mathbf{X})\right]\right| : \mathbb{E}g_r^2(X) \le 1,\ 1\le r\le m\right\} < c_0. \quad (4.3.34)$$

Let now $\{g_{i,t}(X) : i = 1, \dots, k\}$ be an orthonormal basis of the subspace of $L_2(P)$ generated by the functions $\{\psi_r(X) I_{X\in A(t)} : r = 1, \dots, R\}$. Then, $h_t(\mathbf{X})$ sits on the linear span of the orthonormal set $\{g_{i_1,t}(X_1)\cdots g_{i_m,t}(X_m) : 1 \le i_1, \dots, i_m \le k\}$ of $L_2(P^m)$. This gives, by (4.3.34),

$$\mathbb{E}h_t^2(\mathbf{X}) = \sum_{|\mathbf{i}|\le k} \left[\mathbb{E}\left(h_t(\mathbf{X}) g_{i_1,t}(X_1)\cdots g_{i_m,t}(X_m)\right)\right]^2 < k^m c_0^2.$$

Hence,

$$\mathbb{E}h^2(\mathbf{X}) = \lim_{t\nearrow\infty} \mathbb{E}h_t^2(\mathbf{X}) < k^m c_0^2 < \infty.$$

Proving that h is P-canonical involves Hoeffding's decomposition [(1.1)], the Marcinkiewicz law of large numbers for U-statistics (Proposition 5.2.8 below) and the central limit theorem for degenerate U-statistics, in a very straightforward way. As an illustration we consider the case $m = 2$. Since $\mathbb{E}h^2(X_1, X_2) < \infty$, the Hoeffding projections with respect to P, $\pi_k h$, which are P-canonical, are also square integrable (these projections are bounded operators in L_2). Then, in the decomposition

$$\frac{n}{\log\log n} U_n^{(2)}(h)$$
$$= \frac{n}{\log\log n}\mathbb{E}h(X_1, X_2) + \frac{2}{\log\log n}\sum_{i=1}^n \pi_1 h(X_i) + \frac{n}{\log\log n} U_n^{(2)}(\pi_2 h),$$

the term at the left is a.s. $O(1)$ by hypothesis, and so is the third summand at the right by the LIL for canonical U-statistics; the second summand is either zero a.s. or else is $O_P(n^{1/2}/\log\log n)$ by the CLT and $o(n^{1/(2-\delta)}/\log\log n)$ a.s. for all $0 < \delta < 2$ by the Marcinkiewicz LLN; and the first summand, is either zero or $O(n/\log\log n)$. Hence, the three summands at the right do not cancel each other and we thus conclude that the first and second must be zero a.s., that is, $\mathbb{E}h(X_1, X_2) = 0$ and $\mathbb{E}\pi_1 h(X) = 0$ a.s., showing that h is P-canonical. □

Finally, as another application of decoupling, we obtain a sharp necessary condition for the LIL in terms of integrability of the kernel. It will involve the version of Hoffmann-Jørgensen's inequality for U-statistics given in Theorem 4.1.1, the Paley–Zygmund argument (Theorem 3.3.6), and a convexity argument that turns the summability of tail probabilities of normalized decoupled U-statistics into stochastic boundedness of the same but with smaller normalizing constants. We single out this last part as a lemma.

LEMMA 4.3.9. *Let us set* $a(x) = (x \log\log x)^{m/2}$, $b(x) = a(x)/\sqrt{\log\log x}$, $n_j = 2^j$, $k_j = \left[(2t^{-2}\log j)^{1/m}\right]$, *and* $c_j = t2^{-1/2}k_j^m b(n_j)/a(n_jk_j)$ *(note that* $c_j \to 1$ *as* $j \to \infty$*). Let A be a convex set such that*

$$\sum_{j=1}^{\infty} \Pr\left\{\frac{1}{a(n_jk_j)} \sum_{|\mathbf{i}|\le n_jk_j} h(X_{i_1}^1, \dots, X_{i_m}^m) \in c_j\sqrt{2}A\right\} < \infty.$$

Then, for each $t > 0$, *the set* J_t *of integers* j *for which*

$$\Pr\left\{\frac{1}{b(n_j)} \sum_{|\mathbf{i}|\le n_j} h(X_{i_1}^1, \dots, X_{i_m}^m) \in tA\right\} \ge e^{-t^2/2}$$

is not too dense within the natural numbers in the sense that

$$\sum_{j\in J_t} \frac{1}{j} < \infty.$$

PROOF. Since the proof for general m is only formally different from the case $m = 2$, we assume $m = 2$ here. Letting $\mathbb{E}_i$, $\Pr_i$ denote expectation and probability with respect to the variables X^i only, by convexity of A we have

$$\begin{aligned}
&\Pr\left\{\frac{1}{a(n_jk_j)} \sum_{|\mathbf{i}|\le n_jk_j} h(X_{i_1}^1, X_{i_2}^2) \in c_j\sqrt{2}A\right\} \\
&= \mathbb{E}_2\Pr_1\left\{\frac{1}{k_j}\sum_{r=1}^{k_j}\Big[\sum_{i_1=n_j(r-1)+1}^{n_jr}\ \sum_{i_2=1}^{n_jk_j} h(X_{i_1}^1, X_{i_2}^2)\Big] \in \frac{a(n_jk_j)c_j\sqrt{2}}{k_j}A\right\} \\
&\ge \mathbb{E}_2\left[\Pr_1\left\{\sum_{i_1=1}^{n_j}\Big[\sum_{i_2=1}^{n_jk_j} h(X_{i_1}^1, X_{i_2}^2)\Big] \in \frac{a(n_jk_j)c_j\sqrt{2}}{k_j}A\right\}\right]^{k_j} \\
&\ge \left[\mathbb{E}_1\Pr_2\left\{\frac{1}{k_j}\sum_{r=1}^{k_j}\Big[\sum_{i_2=n_j(r-1)+1}^{n_jr}\ \sum_{i_1=1}^{n_j} h(X_{i_1}^1, X_{i_2}^2)\Big] \in \frac{a(n_jk_j)c_j\sqrt{2}}{k_j^2}A\right\}\right]^{k_j} \\
&\ge \left[\Pr\left\{\frac{1}{b(n_j)} \sum_{|\mathbf{i}|\le n_j} h(X_{i_1}^1, X_{i_2}^2) \in tA\right\}\right]^{k_j^2} \\
&\ge \left[\Pr\left\{\frac{1}{b(n_j)} \sum_{|\mathbf{i}|\le n_j} h(X_{i_1}^1, X_{i_2}^2) \in tA\right\}\right]^{2t^{-2}\log j}.
\end{aligned}$$

The conclusion of the lemma follows immediately from this inequality and the hypothesis. □

Before stating and proving the integrability theorem it is perhaps useful to explain how Hoffmann-Jørgensen's inequality (4.1.3) will be used (see also the proof of Proposition 4.1.7). The decoupled version of this inequality for non-negative symmetric kernels f in m variables, which has a similar but simpler proof than (4.1.3), asserts the existence of finite positive universal constants c_1, c_2 and $0 < c_3 < 1$ such that

$$\mathbb{E}\Big[\sum_{i_1,\dots,i_m\le n} f(X^1_{i_1},\dots,X^m_{i_m})\Big]^2 \le c_1\mathbb{E}\Big[\max_{i_1\le n}\sum_{i_2,\dots,i_m\le n} h(X^1_{i_1},\dots,X^m_{i_m})\Big]^2 + c_2t_0^2,$$

for any t_0 satisfying $\Pr\{\sum_{i_1,\dots,i_m\le n} f(X^1_{i_1},\dots,X^m_{i_m}) \ge t_0\} \le c_3$. So, if γ_n is such that $n^m\gamma_n \ge t_0$, and if we set $f_n = f\wedge(n\gamma_n)$, this inequality gives

$$\begin{aligned}(n^m\mathbb{E}f_n)^2 &\le c_1\Big\|\max_{i_1\le n}\sum_{i_2,\dots,i_m\le n} f_n(X^1_{i_1},\dots,X^m_{i_m})\Big\|_\infty^2 + c_2n^{2m}\gamma_n^2\\ &\le (c_1+c_2)n^{2m}\gamma_n^2.\end{aligned} \tag{4.3.35}$$

This upper bound for the L_1 norm of the truncated kernel f_n will be useful in the next theorem. In fact, it could also have been used in the proof of the converse CLT (Theorem 4.2.6) instead of the law of large numbers, just as it is used in the proof of Proposition 4.1.7.

THEOREM 4.3.10. *Suppose we have*

$$\limsup_{n\to\infty}\frac{1}{(2n\log\log n)^{\frac{m}{2}}}\Big|\sum_{(i_1,\dots,i_m)\in I_n^m} h(X_{i_1},\dots,X_{i_m})\Big| < \infty \quad \text{a.s.} \tag{4.3.31}$$

Then

$$\sup_{n\ge1}\frac{\mathbb{E}\min\{h^2(X_1,\dots,X_m),\, n(\log\log n)^{m-1}\}}{(\log\log n)^{m-1}} < \infty. \tag{4.3.36}$$

In particular, h is P-canonical and, for $m \ge 2$,

$$\mathbb{E}h^2(\log\log|h|)^{2-m}g(\log\log|h|) < \infty \tag{4.3.37}$$

for all bounded non-negative monotone decreasing integrable functions g on $\mathbb{R}^+$.

PROOF. Let $t > 0$ to be specified below, and let k_j, $c_j \to 1$ and $n_j = 2^j$ be as in Lemma 4.3.9. By Theorem 3.5.7, (4.3.31) implies

$$\sum_{j=1}^\infty \Pr\left\{\Big|\frac{1}{a(n_jk_j)}\sum_{|\mathbf{i}|\le n_jk_j}\epsilon_{\mathbf{i}}^{\mathrm{dec}}h(\mathbf{X}_{\mathbf{i}}^{\mathrm{dec}})\Big| > c_j\sqrt{2}c_0\right\} < \infty$$

for some $0 < c_0 < \infty$. Then, by Lemma 4.3.9 we have $\sum_{j\in J_\pm} 1/j < \infty$ with $J_\pm = J_\pm(t)$ being, respectively, the sets of integers j such that

$$\Pr\left\{\pm\frac{1}{b(n_j)}\sum_{|\mathbf{i}|\le n_j}\epsilon_{\mathbf{i}}^{\mathrm{dec}}h(\mathbf{X}_{\mathbf{i}}^{\mathrm{dec}}) > tc_0\right\} \ge e^{-t^2/2}.$$

Thus, letting $J(t) = J_+ \cup J_-$, we have

$$\Pr\left\{\left|\frac{1}{b(n_j)}\sum_{|\mathbf{i}|\le n_j}\epsilon_{\mathbf{i}}^{\mathrm{dec}}h(\mathbf{X}_{\mathbf{i}}^{\mathrm{dec}})\right| > tc_0\right\} \le e^{-t^2/2} \quad \text{for all } j \notin J(t)$$

and $\sum_{j\in J(t)} 1/j < \infty$. Since $\sum_{2^\ell\le j<2^{\ell+1}} 1/j \ge \frac{1}{2}$, the previous inequality implies the existence of $\ell_0 < \infty$ and of a sequence of integers $j_\ell < j_{\ell+1}$ such that $2^\ell \le j_\ell < 2^{\ell+1}$ and, for $\ell \ge \ell_0$,

$$\Pr\left\{\left|\frac{1}{b(n_{j_\ell})}\sum_{|\mathbf{i}|\le n_{j_\ell}}\epsilon_{\mathbf{i}}^{\mathrm{dec}}h(\mathbf{X}_{\mathbf{i}}^{\mathrm{dec}})\right| > tc_0\right\} \le 2e^{-t^2/2}.$$

Then, Theorem 3.3.6 [inequality (3.3.10)] implies that, for $n = n_{j_\ell}$, $\ell \ge \ell_0$,

$$\Pr\left\{\sum_{|\mathbf{i}|\le n} h^2(\mathbf{X}_{\mathbf{i}}^{\mathrm{dec}}) \ge 2n^m(\log\log n)^{m-1}t^2c_0^2\right\}$$
$$\le 2C_{m,2}\Pr\left\{\left|\frac{1}{b(n)}\sum_{|\mathbf{i}|\le n}\epsilon_{\mathbf{i}}^{dec}h(\mathbf{X}_{\mathbf{i}}^{dec})\right| \ge tc_0\right\} \le 2C_{m,2}e^{-t^2/2}.$$

If we choose t such that $2C_{m,2}e^{-t^2/2} < c_3$, where c_3 is one of the constants implicitly appearing in inequality (4.3.34), then the previous inequality shows that we can apply inequality (4.3.35) to $f = h^2$ with $\gamma_n = 2(\log\log n)^{m-1}t^2c_0^2$, and conclude

$$\sup_{\ell\ge1}\frac{\mathbb{E}\min\left\{h^2, n_{j_\ell}(\log\log n_{j_\ell})^{m-1}\right\}}{(\log\log n_{j_\ell})^{m-1}} < \infty,$$

which implies (4.3.36) due to the boundedness of $n_{j_{\ell+1}}/n_{j_\ell}$.

Condition (4.3.37) follows from (4.3.36): for $\tau_k = 2^{2^{2^k}}$, (4.3.36) gives that for all g monotone decrasing and integrable,

$$\mathbb{E}\Big[h^2(\log\log|h|)^{2-m}g(\log\log|h|)\Big]$$
$$= \sum_k \mathbb{E}\Big[h^2(\log\log|h|)^{2-m}g(\log\log|h|)I_{\tau_k(2^k)^{m-1}<h^2\le\tau_{k+1}(2^{k+1})^{m-1}}\Big]$$
$$\lesssim \sum_k 2^{k(2-m)}g(2^k)Eh^2I_{h^2\le\tau_{k+1}(2^{k+1})^{m-1}}$$
$$\lesssim \sum_k 2^kg(2^k) < \infty.$$

To prove that h is P-canonical, one proceeds as in the proof of Theorem 4.3.8, taking into account that (4.3.6) implies that $|h|^{2-\delta}$ is integrable for all $0 < \delta < 2$.

(Here the CLT is not available, since $\mathbb{E}(\pi_1 h)^2$ may be infinite, but if $\pi_1 h$ is not a.s. zero, then, by the converse CLT, the sequence $\{\sum_{i=1}^n \pi_1 h(X_i)/n^{1/2}\}$ is not tight, and this is all that is needed to complete the argument at the end of the proof of Theorem 4.3.8.) □

The integrability condition (4.3.36) for $m = 1$ reduces to $\mathbb{E}h^2 < \infty$, so that Theorem 4.3.10 recovers in particular Strassen's converse LIL. The following example shows that (4.3.36) and (4.3.37) cannot, in general, be improved.

EXAMPLE 4.3.11. Let I_r be the indicator function of the interval $\left(1 - \frac{1}{2^{r-1}}, 1 - \frac{1}{2^r}\right]$, for all $r \in \mathbb{N}$, and let

$$h\big((x,\alpha),(y,\beta)\big) = \sum_{r=1}^{\infty} \frac{2^r}{\sqrt{r}} I_r(x) I_r(y) \alpha\beta,$$

defined on S^2, $S = [0,1] \times \{-1, 1\}$. Let $P = \lambda \times \big(\delta_{-1} + \delta_1\big)/2$, where λ is Lebesgue measure. Then h is P-canonical and the opearor T_h is compact but not Hilbert–Schmidt (its eigenvalues are $1/r$, $r \in \mathbb{N}$). It can be proved, using truncation, binomial probabilities and the better exponential inequality (3.2.18) for Rademacher chaos, that this kernel does satisfy the bounded LIL, i.e., that

$$\limsup_{n\to\infty} \frac{1}{n \log\log n} \left| \sum_{(i,j)\in I_n^2} h\big((X_i,\varepsilon_i),(X_j,\varepsilon_j)\big) \right| < \infty \quad \text{a.s.}$$

However, not only $\mathbb{E}h^2(X_1, X_2) = \infty$, but in fact, $\mathbb{E}h^2 g(\log\log|h|) = \infty$ for all bounded, decreasing functions g such that $\int_0^\infty g(t)dt = \infty$. So, this example shows that for P-canonical U-statistics, (i) the condition $\mathbb{E}h^2 < \infty$ is not necessary for the LIL, and (ii) the necessary integrability conditions given in Theorem 4.3.10 are best possible, at least for $m = 2$. We refer to Giné and Zhang (1996) for details.

The problem of finding necessary conditions on a degenerate kernel h for the U-statistic $U_n(h)$ to satisfy the LIL, bounded or compact, is still open. See Goodman (1997) for a LLI in the case $m = 2$ under an integrability condition weaker than finiteness of the second moment of h but slightly stronger than (4.3.36).

4.4 References for Chapter 4

Hoffmann-Jørgensen's inequality for U-processes was obtained by Giné and Zinn (1992), but the present proof is partly based on an observation by M. Arcones. This inequality is farther from solving the problem of exact estimation of the moments of U-statistics than in the case of sums of independent random variables; this problem is open and we refer to Klass and Nowicki (1996) for significant progress. The law of large numbers is a clasical result of Hoeffding (1961), and the present proof is Berk's (1966). Proposition 4.1.5, the converse of the LLN for positive kernels, is implicitly contained in Giné and Zinn (1994), explicitly in Koltchinskii and Giné (1996); Example 4.1.6 is an adaptation of one of Zhang (1997) and Proposition 4.1.7 is new (the method of proof comes from Giné and Zhang, 1996). The problem of finding necessary and sufficient conditions for the law of large numbers for U-statistics was solved by Cuzick et al. (1995) and Zhang (1996) for the special kernels $x_1 \cdots x_m$, and it has recently been solved for randomized kernels by Zhang (1997) and Latała and Zinn (1998), and will appear. Sen (1974), Giné and Zinn (1992a), Teicher (1997), Zhang (1997) and Latała and Zinn (1998) consider Marcinkiewcz-type laws.

The proof of Bernstein's inequality given here is classsical and we learned it from Bennett (1962), which also contains inequality (4.1.19). The exponential inequalities in Theorem 4.1.12 were obtained by Arcones and Giné (1993), and inequality (4.1.31), with the constant c_1 completely specified, comes from Arcones (1995).

The construction of the Gaussian chaos given in Section 2 is inspired in Neveu (1968) and Bretagnolle (1983), and the proof of the central limit theorem (Theorem 4.2.4) is based on hints in Bretagnolle, loc. cit., as expanded in Arcones and Giné (1992). See Dynkin and Mandelbaum (1983) for a related but different account. Theorem 4.2.4 for $m = 2$ is given by Gregory (1977) and Serfling (1980), and Rubin and Vitale (1980) obtained the first proof of it for general m. Its converse, Theorem 4.2.6, was obtained by Giné and Zinn (1994). The decoupling inequality for homogeneous and tetrahedral Gaussian chaos vectors was obtained by Kwapień (1987), by Arcones and Giné (1992) for homogeneous Gaussian chaos vectors, and was observed in Giné (1998) in general, with the proof given here. Marcus (1997a) extends it to related processes.

The law of the iterated logarithm for nondegenerate U-statistics was early obtained by Serfling (1971). The law of the iterated logarithm for canonical square integrable U-statistics was obtained by Dehling (1989) in the case $m = 2$ and by Arcones and Giné (1995) for general m [Dehling and Utev (1996) contains a sketch of an alternate proof but no complete proof has followed]. The proof given here comes from the article of Arcones and Giné. Previously, Dehling (1989a) and Dehling et al. (1986) had obtained it for kernels with finite $2 + \delta$ absolute moments, and Teicher (1995) had proved the LIL for square integrable kernels of the form $x_1 \cdots x_m$. Giné and Zhang (1996) and particularly Goodman (1996) have proved the LIL under hypotheses that do not imply finiteness of the second moment of the kernel, however, the problem of finding necessary and sufficient conditions for the LIL is still unsolved.

The remaining results on the LIL presented here, namely, Proposition 4.3.8 on sharpness of the second moment condition, the integrability condition in Theorem 4.3.10 and Example 4.3.11 showing that $Eh^2 < \infty$ is not necessary for the LIL, all come from Giné and Zhang (1996). The simple but useful and elegant inequality (4.3.1) was noticed by Pisier (1983); Pollard (1990) replaced the L_1 norm at the left side of this inequality by the L_ψ norm. The bounds in Proposition 4.3.1 were obtained by Arcones and Giné (1995).

5

Limit Theorems for U-Processes

A U-process is a collection of U-statistics. Concretely, a U-process over a family $\mathcal{H}$ of kernels h of m variables, based on a probability measure P on $(S, \mathcal{S})$, is the collection $\{U_n^{(m)}(h, P) : h \in \mathcal{H}\}$ of U-statistics. This chapter is devoted to the asymptotic theory of U-processes: we are interested in finding conditions on $\mathcal{H}$ and P ensuring that the law of large numbers, or the central limit theorem, or the law of the iterated logarithm, hold for $U_n^{(m)}(h, P)$, *uniformly* in $h \in \mathcal{H}$. The theory of empirical processes deals with the same questions for the case $m = 1$, and U-process theory is patterned after it. This is a relatively new subject, at least in the generality presented here ($\mathcal{H}$ being an arbitrary collection of kernels which are defined on a general measurable space). There is therefore a need to indicate its usefulness. To this effect, a section on applications is added at the end of the chapter (Section 5.5); there we illustrate the use of each of the main theorems in this chapter by deriving properties of certain multidimensional generalizations of the median, and in general, of M-estimators, and by studying estimators of the cumulative hazard and distribution functions of a random variable based on truncated data.

In the first section of this chapter we give a self-contained account of the elements of the general theory that are needed in the development of the aspects of asymptotics of U-processes that we have chosen to present. It is divided into three parts: convergence in law (relative to the sup norm topology) of not necessarily separable processes, maximal inequalities based on metric entropy, and entropy bounds for Vapnik–Červonenkis classes of sets and functions. In the first, limit theorems are reduced to estimates of tail probabilities of suprema of random processes (in addition to finite-dimensional convergence, dealt with in the previous chapter). In the second, we describe a way to produce estimates for these probabilities in terms of integrals of certain metric entropies of the parameter space (the entropy of a metric space is a quantitative measure of its "total boundedness": see the definition in Section 5.1.2 below). The third part contains an introduction to classes of sets and functions whose $L_p(Q)$ metric entropies are well behaved (i.e., are small) unformly in $Q \in \mathcal{P}(S)$, the VC classes. Fortunately, most statistical applications involve these nice classes of set and functions. Our approach to these subjects is economical and expedient; the statements are tailored to our needs and are not always as general as possible.

The remaining three sections of the chapter are each devoted to one of the three types of limit theorems we consider (LLN, CLT, LIL). Regarding the law of large numbers, we present a necessary and sufficient condition for its validity in terms of "random entropies," in complete analogy with the celebrated results of Vapnink and Červonenkis on the LLN for empirical processes. The central limit theorem and the law of the iterated logarithm are developed only for VC classes of functions because general necessary and sufficient conditions do not exist at present. In passing, and with no extra effort, one also obtains the LLN for Banach valued U-statistics, for general separable Banach spaces B, and the CLT and the LIL for Banach spaces of type 2 (for the definition, see Section 5.3). We tend to view the separable Banach space case as a particular case of U-processes. An exponential inequality is also included in Section 5.3, as an important complement to the CLT.

The randomization results of Section 3.5 are used throughout. Randomization by single Rademacher variables ($r = 1$ in Theorem 3.5.3) plays a basic role in the proof of the law of large numbers: $U_n(h) - P^m h = U_n(h - P^m h)$ is the U-statistic of a centered variable and this allows only for comparison with $((n-m)!/n!) \sum_{I_n^m} \varepsilon_{i_1} h(X_{i_1}, \ldots, X_{i_m})$. The entropy bounds developed in the first section for Rademacher processes, based on their hypercontractivity, are then sufficient to solve the LLN problem for measurable classes of kernels admitting an integrable envelope. The $\sqrt{n}$-CLT for general U-processes reduces to the CLT for the empirical process in the first term of its Hoeffding expansion by means of randomization by products of two Rademacher variables ($r = 2$ in Theorem 3.5.3). The case of completely degenerate or canonical kernels is the crucial case for both the CLT and the LIL, and in this case we can apply complete randomization ($r = m$ in Theorem 3.5.3). Decoupling and randomization offer efficient ways of dealing with these problems and, in fact, it is not clear how most of the results in this chapter could be obtained without this tool.

5.1 Some Background on Asymptotics of Processes, Metric Entropy, and Vapnik–Červonenkis Classes of Functions: Maximal Inequalities

The three limit theorems for U-processes that we are concerned with in this chapter, namely, the law of large numbers, the central limit theorem, and the law of the iterated logarithm, all basically reduce to maximal inequalities, that is, to the control of quantities of the form

$$\Pr\Big\{\sup_{h\in\mathcal{H}} |U_n(h)| > c_n\Big\}, \tag{5.1.1}$$

where $\mathcal{H}$ is a finite or infinite family of kernels, and where Pr must be replaced by $\Pr^*$ when the variable within brackets is not measurable. This is clear for the law of large numbers and for the bounded law of the iterated logatithm, but it requires a nontrivial reduction for the central limit theorem, that we carry out below. There are now some standard powerful ways to obtain maximal inequalities for empirical and U-processes, and we will present the simplest such procedure, which is still probably the most useful: the metric entropy approach to maximal inequalities. Finally, we will introduce a type of families of kernels specially ubiquitous and easy to handle, namely, Vapnik–Červonenkis classes of functions; these are specially suited for the entropy bounds.

5.1.1 Convergence in Law of Sample Bounded Processes

Let T be a set and let $X_n(t)$, $t \in T$, be stochastic processes indexed by the set T. Assume these processes have versions with almost all their trajectories bounded (that is, the processes X_n are sample bounded), and let us continue denoting by X_n their sample bounded versions. Then $X_n(\cdot) \in \ell^\infty(T)$ almost surely, where $\ell^\infty(T)$ denotes the space of all bounded real functions on T. $\ell^\infty(T)$, equipped with the sup norm, $\|\cdot\|_T$, is a Banach space that is separable only if T is finite, and we do *not* assume that the finite-dimensional distributions of the processes $X_n(t)$, $\mathcal{L}\big(X_n(t_1), \ldots, X_n(t_k)\big)$, $k \in \mathbb{N}$, $t_i \in T$, correspond to the finite-dimensional projections $\mu_{n,t_1,\ldots,t_k}$ of individually tight Borel probability measures μ_n on $\ell^\infty(T)$. Now let $X(t)$, $t \in T$, be a sample bounded process whose finite-dimensional laws do correspond to the finite-dimensional laws of a tight Borel probability measure on $\ell^\infty(T)$. Then, we say that X_n *converges in law to* X *uniformly in* $t \in T$, or that

$$X_n \to_{\mathcal{L}} X \quad \text{in} \quad \ell^\infty(T) \tag{5.1.2}$$

if

$$\mathbb{E}^* H(X_n) \to \mathbb{E} H(X)$$

for all functions $H : \ell^\infty(T) \mapsto \mathbb{R}$ bounded and continuous. $\mathbb{E}^*$ denotes outer expectation. As with regular convergence in law, if F is a continuous function on $\ell^\infty(T)$ with values in another metric space and if $F(X_n)$ is measurable, then (5.1.2) implies that $F(X_n) \to_{\mathcal{L}} F(X)$ in the usual way, and this is what makes the concept of convergence in law in $\ell^\infty(T)$ so useful.

The outer expectation in the definition of convergence in law is necessary because even the most simple bounded process may fail to induce Borel probability laws on $\ell^\infty(T)$. For instance, let $T = [0, 1]$, let U be uniform on $[0, 1]$, and let $X(t) = I_{(U,1]}(t)$, $t \in T$. Let A be a nonmeasurable subset of $[0, 1]$ (which exists if we assume the axiom of choice) and let $F_A = \{I_{(s,1]} : s \in A\} \subset \ell^\infty(T)$.

Then, F_A is a discrete set for the sup norm, hence a closed set of $\ell^\infty(T)$. But $\{X \in F_A\} = \{U \in A\}$ is not measurable and therefore the law of X does not extend to a Borel probability measure on $\ell^\infty(T)$.

The following lemma clarifies the notion of process whose law is a tight Borel measure on $\ell^\infty(T)$.

LEMMA 5.1.1. *Let $X(t)$, $t \in T$, be a sample bounded stochastic process. Then the finite-dimensional distributions of X are those of a tight Borel probability measure on $\ell^\infty(T)$ if and only if there exists on T a pseudodistance d for which (T, d) is totally bounded and such that X has a version with almost all its sample paths uniformly continuous for d.*

PROOF. Let us assume the probability law of X is a tight Borel measure μ on $\ell^\infty(T)$, let K_n, $n \in \mathbb{N}$, be an increasing sequence of compact sets in $\ell^\infty(T)$ such that $\mu\big(\cup_{n=1}^\infty K_n\big) = 1$, and let $K = \cup_{n=1}^\infty K_n$. Then, we claim the pseudometric d defined on T by

$$d(s,t) = \sum_{n=1}^\infty 2^{-n}\big(1 \wedge d_n(s,t)\big),$$

with

$$d_n(s,t) = \sup\big\{|f(t) - f(s)| : f \in K_n\big\},$$

makes (T, d) totally bounded. To see this, given $\varepsilon > 0$, let m be such that $\sum_{n=m+1}^\infty 2^{-n} < \varepsilon/4$ and let $\{f_1, \dots, f_r\}$ be a finite subset of $\cup_{n=1}^m K_n$, $\varepsilon/4$-dense in $\cup_{n=1}^m K_n$ for the sup norm, that is, for each $f \in \cup_{n=1}^m K_n$ there is $i \le r$ such that $\|f - f_i\|_\infty \le \varepsilon/4$ (such a set of functions exists by compactness). The subset A of $\mathbb{R}^r$ defined by $A = \big\{\big(f_1(t), \dots, f_r(t)\big) : t \in T\big\}$ is bounded: the subset $\cup_{n=1}^m K_n$ of $\ell^\infty(T)$ being compact, is *a fortiori* bounded. Hence, A is totally bounded and therefore there exists a finite set $T_\varepsilon = \{t_i : 1 \le i \le N\}$ such that, for each $t \in T$, there is $i \le N$ for which we have $\max_{1\le s\le r} |f_s(t) - f_s(t_i)| \le \varepsilon/4$. Then, T_ε is ε-dense in T for the pseudometric d: if t and t_i are as above, we have for $n \le m$ that

$$d_n(t, t_i) = \sup_{f \in K_n} |f(t) - f(t_i)| \le \max_{s \le r} |f_s(t) - f_s(t_i)| + \varepsilon/2 \le \frac{3\varepsilon}{4},$$

so that

$$d(t, t_i) \le \frac{\varepsilon}{4} + \sum_{n=1}^m 2^{-n} d_n(t, t_i) \le \varepsilon.$$

This proves (T, d) is totally bounded. Moreover, the functions $f \in K$ are uniformly d-continuous, since, if $f \in K_n$, then $|f(s) - f(t)| \le d_n(s,t) \le 2^n d(s,t)$ for all $s, t \in T$ with $d(s,t) \le 1$. Now, since $\mu(K) = 1$, the identity map of $(\ell^\infty(T), \mathcal{B}, \mu)$ is a version of X with almost all of its trajectories in K, hence in $C_u(T, d)$, the space of bounded uniformly d-continuous functions on T. This proves the first part of the lemma.

Conversely, let $X(t)$, $t \in T$, be a process with a version whose sample paths are almost all in $C_u(T, d)$ for a distance or pseudodistance d on T for which (T, d) is totally bounded, and let us continue denoting X such a version. We can assume all the trajectories of X are uniformly continuous. If Ω is the probability space where X is defined, then the map $X : \Omega \mapsto C_u(T)$ is Borel measurable because the random vectors $(X(t_1), \dots, X(t_k))$, $t_i \in T$, $k \in \mathbb{N}$, are measurable and the Borel σ-algebra of $C_u(T)$ is generated by the "finite-dimensional sets" $\{f \in C_u : (f(t_1), \dots, f(t_k)) \in A\}$, for all Borel sets A of $\mathbb{R}^k$, $t_i \in T$, $k \in \mathbb{N}$. Hence, the probability law of X is a Borel measure in $C_u(T, d)$, and therefore, (T, d) being precompact, the law of X is in fact a tight Borel measure by a well-known theorem of Ulam (e.g., Araujo and Giné, 1980, Theorem 1.1.2, or Dudley, 1989, Theorem 7.1.4). But a tight Borel probability measure on $C_u(T)$ is a tight Borel measure on $\ell^\infty(T)$ since the inclusion of C_u into ℓ^∞ is continuous. □

The following theorem reduces convergence in law in $\ell^\infty(T)$ to maximal inequalities. We will often refer to this theorem as *the asymptotic equicontinuity criterion* for convergence in law in $\ell^\infty(T)$.

THEOREM 5.1.2. *The following statements are equivalent:*

i) *The finite-dimensional distributions of the sample bounded processes X_n converge in law and there exists a pseudometric d on T such that both, (T, d) is totally bounded and*

$$\lim_{\delta \to 0} \limsup_{n \to \infty} \Pr{}^* \Big\{ \sup_{d(s,t) \le \delta} \big|X_n(t) - X_n(s)\big| > \varepsilon \Big\} = 0 \tag{5.1.3}$$

for all $\varepsilon > 0$.

ii) *There exists a process X whose law is a tight Borel probability measure on $\ell^\infty(T)$ and such that*

$$X_n \to_{\mathcal{L}} X \quad \text{in} \quad \ell^\infty(T).$$

If i) holds, then the process X in ii), which is completely determined by the limit laws of the finite-dimensional distributions of X_n, has a version with bounded uniformly continous paths for d. If X in ii) has a version with almost all of its trajectories in $C_u(T, \rho)$ for a pseudodistance ρ for which (T, ρ) is totally bounded, then the distance d in i) can be taken to be $d = \rho$.

PROOF. Let us assume i) holds. Let T_0 be a countable d-dense subset of T, and let T_k, $k \in \mathbb{N}$, be finite sets increasing to T_0. The limit laws of the finite-dimensional distributions of the processes X_n are consistent and thus define a stochastic process X on T, and, moreover,

$$\begin{aligned}\Pr\Big\{\max_{d(s,t)\le\delta,\ s,t\in T_k}|X(t)-X(s)|>\varepsilon\Big\}&\\ \le \liminf_{n\to\infty}\Pr\Big\{&\max_{d(s,t)\le\delta,\ s,t\in T_k}|X_n(t)-X_n(s)|>\varepsilon\Big\}\\ \le \liminf_{n\to\infty}\Pr\Big\{&\max_{d(s,t)\le\delta,\ s,t\in T_0}|X_n(t)-X_n(s)|>\varepsilon\Big\}.\end{aligned}$$

Hence, taking limits as $k\to\infty$ and using condition (5.1.3), we obtain that there exists a sequence $\delta_r\searrow 0$, $\delta_r>0$, such that

$$\Pr\Big\{\sup_{d(s,t)\le\delta_r,\ s,t\in T_0}|X(t)-X(s)|>2^{-r}\Big\}\le 2^{-r}.$$

Then, by Borel–Cantelli, there exists $r(\omega)<\infty$ a.s. such that

$$\sup_{d(s,t)\le\delta_r,\ s,t\in T_0}|X(t,\omega)-X(s,\omega)|\le 2^{-r}$$

for all $r>r(\omega)$. Hence, $X(t,\omega)$ is a d-uniformly continuous function of $t\in T_0$ for almost every ω; T being totally bounded, the restriction to T_0 of $X(t,\omega)$ is also bounded. The extension to T by uniform continuity of the restriction of X to T_0 (only the ω set where X is uniformly continuous needs be considered) produces a version of X whose trajectories are all in $C_u(T,d)$ and, in particular, the law of X admits a tight extension to the Borel σ-algebra of $\ell^\infty(T)$ (Lemma 5.1.1).

Before proving convergence, we recall a useful fact (whose simple proof we omit): if $f:\ell^\infty(T)\mapsto\mathbb{R}$ is bounded and continuous, and if $K\subset\ell^\infty(T)$ is compact, then for every $\varepsilon>0$ there exists $\delta>0$ such that

$$\|u-v\|_T<\delta,\quad u\in K,\quad v\in\ell^\infty(T)\Longrightarrow|f(u)-f(v)|<\varepsilon. \tag{5.1.4}$$

Since (T,d) is totally bounded, for every $\tau>0$ there exists a finite set of points $t_1,\dots,t_{N(\tau)}$ that is τ-dense in (T,d) in the sense that $T\subseteq\cup_{i=1}^{N(\tau)}B(t_i,\tau)$, where $B(t,\tau)$ denotes the open ball of center t and radius τ. Then, for each $t\in T$ we can choose $\pi_\tau(t)\in\{t_1,\dots,t_{N(\tau)}\}$ so that $d(\pi_\tau(t),t)<\tau$. We then define processes $X_{n,\tau}$, $n\in\mathbb{N}$, and X_τ as

$$\begin{aligned}X_{n,\tau}(t)&=X_n(\pi_\tau(t)),\\ X_\tau(t)&=X(\pi_\tau(t)),\qquad t\in T.\end{aligned}$$

These are approximations of X_n and X that take at most a finite number $N(\tau)$ of values. Convergence of the finite-dimensional distributions of X_n to those of X implies that

$$X_{n,\tau}\to_{\mathcal{L}}X_\tau\quad\text{in}\quad\ell^\infty(T). \tag{5.1.5}$$

Moreover, the uniform continuity of the sample paths of X implies

$$\lim_{\tau\to 0}\|X-X_\tau\|=0\quad\text{a.s.} \tag{5.1.6}$$

Let now $f : \ell^\infty(T) \mapsto \mathbb{R}$ be a bounded continuous function. We have

$$\begin{aligned}\left|\mathbb{E}^* f(X_n) - \mathbb{E} f(X)\right| &\le \left|\mathbb{E}^* f(X_n) - \mathbb{E} f(X_{n,\tau})\right| \\ &\quad + \left|\mathbb{E} f(X_{n,\tau}) - \mathbb{E} f(X_\tau)\right| + \left|\mathbb{E} f(X_\tau) - \mathbb{E} f(X)\right| \\ &:= I_{n,\tau} + II_{n,\tau} + III_\tau.\end{aligned}$$

In order to prove that ii) holds we must show that the limits $\lim_{\tau\to 0}\limsup_{n\to\infty}$ of each of these three quantities is 0. This is true for $II_{n,\tau}$ by (5.1.5). Next we show it for III_τ. Given $\varepsilon > 0$ let $K \subset \ell^\infty(T)$ be a compact set such that $\Pr\{X \in K^c\} < \varepsilon/(6\|f\|_\infty)$, let $\delta > 0$ be such that (5.1.4) works for K and $\varepsilon/6$, and let $\tau_1 > 0$ be such that $\Pr\big\{\|X_\tau - X\|_T \ge \delta\big\} < \varepsilon/(6\|f\|_\infty)$ for all $\tau < \tau_1$ [possible by (5.1.6)]. Then,

$$\begin{aligned}\left|\mathbb{E} f(X_\tau) - \mathbb{E} f(X)\right| &\le 2\|f\|_\infty \Pr\big\{X \in K^c \quad\text{or}\quad \|X_\tau - X\|_T \ge \delta\big\} \\ &\quad + \sup\big\{|f(u) - f(v)| : u \in K,\ \|u - v\|_T < \delta\big\} \\ &\le 2\|f\|_\infty\Big(\frac{\varepsilon}{6\|f\|_\infty} + \frac{\varepsilon}{6\|f\|_\infty}\Big) + \frac{\varepsilon}{6} < \varepsilon,\end{aligned}$$

proving $\lim_{\tau\to 0} III_\tau = 0$. For the same ε, δ and K, we have

$$\begin{aligned}&\left|\mathbb{E}^* f(X_n) - \mathbb{E} f(X_{n,\tau})\right| \\ &\quad \le 2\|f\|_\infty\Big[\Pr^*\big\{\|X_n - X_{n,\tau}\|_T \ge \frac{\delta}{2}\big\} + \Pr\big\{X_{n,\tau} \in \big(K_{\delta/2}\big)^c\big\}\Big] \\ &\quad\quad + 2\sup\big\{|f(u) - f(v)| : u \in K,\ \|u - v\|_T < \delta\big\}, \qquad (5.1.7)\end{aligned}$$

where $K_{\delta/2}$ is the $\delta/2$ open neighborhood of the set K for the sup norm. [To verify inequality (5.1.7) note that if $X_{n,\tau} \in K_{\delta/2}$ and $\|X_n - X_{n,\tau}\|_T < \delta/2$ then there exists $u \in K$ such that $\|u - X_{n,\tau}\| < \delta/2$ and $\|u - X_n\|_T < \delta$.] Since the hypothesis (5.1.3) implies that there is $\tau_2 > 0$ such that

$$\limsup_{n\to\infty} \Pr^*\Big\{\|X_{n,\tau} - X_n\|_T \ge \frac{\delta}{2}\Big\} < \frac{\varepsilon}{6\|f\|_\infty}$$

for all $\tau < \tau_2$, and finite-dimensional convergence gives

$$\limsup_{n\to\infty} \Pr\Big\{X_{n,\tau} \in \big(K_{\delta/2}\big)^c\Big\} \le \Pr\Big\{X_\tau \in \big(K_{\delta/2}\big)^c\Big\} \le \frac{\varepsilon}{6\|f\|_\infty},$$

we obtain from (5.1.7), as above, that for $\tau < \tau_1 \wedge \tau_2$,

$$\limsup_{n\to\infty} \left|\mathbb{E}^* f(X_n) - \mathbb{E} f(X_{n,\tau})\right| < \varepsilon,$$

showing that $\lim_{\tau\to 0}\limsup_{n\to\infty} I_{n,\tau} = 0$. Hence, i) implies ii).

For the converse, we first observe that if $X_n \to_{\mathcal{L}} X$ in $\ell^\infty(T)$, then, as for regular convergence in law, $\limsup_{n\to\infty}\Pr^*\{X_n \in F\} \le \Pr\{X \in F\}$ for every closed set of $\ell^\infty(T)$. The proof is the same as for regular convergence in law and is omitted (it can be found in many texts under the heading "portmanteau's lemma," e.g. Araujo and Giné, 1980, Theorem 1.2.3, or Dudley, 1989, Theorem 11.1.1). Suppose now that ii) holds. Then, by Lemma 5.1.1, there exists a pseudodistance d on T for which (T, d) is totally bounded and such that X has a version (that we still denote by X) with all its sample paths in $C_u(T, d)$. Take $F_{\delta,\varepsilon} = \{u \in \ell^\infty(T) : \sup_{d(s,t)\le\delta} |u(s) - u(t)| \ge \varepsilon\}$. Then, applying the previous observation to $F_{\delta,\varepsilon}$, we obtain, by the convergence hypothesis, that

$$\lim_{\delta\to 0} \limsup_{n\to\infty} \Pr{}^*\Big\{\sup_{d(s,t)\le\delta} \big|X_n(t) - X_n(s)\big| \ge \varepsilon\Big\} \le \lim_{\delta\to 0} \Pr\Big\{\sup_{d(s,t)\le\delta} \big|X(t) - X(s)\big| \ge \varepsilon\Big\} = 0$$

for all $\varepsilon > 0$. □

Let $BL_1\big(\ell^\infty(T)\big)$ denote the set of real functionals f on $\ell^\infty(T)$ such that

$$\sup_{x\in\ell^\infty(T)} |f(x)| \le 1 \quad \text{and} \quad \sup_{x\ne y, x,y\in\ell^\infty(T)} |f(y) - f(x)|/\|y - x\|_T \le 1$$

(BL_1 stands for the unit ball of the space of bounded Lipschitz functions). It is well known that if B is a separable metric space, then the distance between probability measures on B $d_{BL}(\mu, \nu) := \sup\big\{|\int f d(\mu - \nu)| : f \in BL_1(B)\big\}$ metrizes weak convergence (e.g., Dudley, 1989, page 310, or Araujo and Giné, 1980, pages 10–11). If Y is a process on T with almost all its trajectories bounded and X a process whose law is a tight Borel measure on $\ell^\infty(T)$, and we also denote by X one of its versions almost all of whose sample paths are in $\ell^\infty(T)$, we define

$$d_{BL}\big(Y, X\big) := \sup\Big\{\big|\mathbb{E}^* f(Y) - \mathbb{E} f(X)\big| : f \in BL_1(\ell^\infty(T))\Big\}, \tag{5.1.8}$$

the *(dual) bounded Lipschitz distance* between X and Y. With this definition we have the following corollary on metrizability of convergence in law (it does not add a lot to Theorem 5.1.2 but it is sometimes useful).

COROLLARY 5.1.3. *If the law of X is defined by a tight Borel mesure on $\ell^\infty(T)$ and almost all the trajectories of the processes $X_n(t)$, $t \in T$, are bounded, then*

$$X_n \to_{\mathcal{L}} X \quad \text{in} \quad \ell^\infty(T)$$

if and only if

$$d_{BL}(X_n, X) \to 0.$$

PROOF. Let $X_n \to_{\mathcal{L}} X$. We use the same letter X to denote a version of X whose paths are all in $C_u(T, d)$ for some d for which (T, d) is totally bounded, and keep the definitions and notation from the proof of Theorem 5.1.2. Consider the decomposition of the previous proof,

$$\big|\mathbb{E}^* f(X_n) - \mathbb{E} f(X)\big| \le I_{n,\tau} + II_{n,\tau} + III_\tau,$$

but now for f bounded Lipschitz. Then, since, for τ fixed, $X_{n,\tau} \to_{\mathcal{L}} X_\tau$ as random vectors in $\mathbb{R}^{N(\tau)}$, a convergence metrized by d_{BL}, it follows that

$$\lim_{n\to\infty} \sup_{f\in BL_1(\ell^\infty(T))} |II_{n,\tau}| = 0$$

for all $\tau > 0$. Since $\|X_\tau - X\|_T \to 0$ a.s.,

$$\lim_{\tau\to 0} \sup_{f\in BL_1(\ell^\infty(T))} |III_\tau| \le \lim_{\tau\to 0} \mathbb{E}\big(\|X_\tau - X\|_T \wedge 2\big) = 0$$

by bounded convergence. Finally,

$$\begin{aligned}\lim_{\tau\to 0}\limsup_{n\to\infty} \sup_{f\in BL_1(\ell^\infty(T))} |I_{n,\tau}| &\le \limsup_{n\to\infty} \mathbb{E}^*\Big[\sup_{d(s,t)\le\tau} \big|X_n(t) - X_n(s)\big| \wedge 2\Big] \\ &\le 2\lim_{\tau\to 0}\limsup_{n\to\infty} \Pr{}^*\Big\{\sup_{d(s,t)\le\tau} \big|X_n(t) - X_n(s)\big| > \varepsilon\Big\} + \varepsilon \\ &= \varepsilon\end{aligned}$$

for all $\varepsilon > 0$, by Theorem 5.1.2 [(5.1.3)]. Thus, $d_{BL}(X_n, X) \to 0$.

Conversely, let us assume $d_{BL}(X_n, X) \to 0$. For $\delta > 0$ fixed, and all $\varepsilon > 0$, let

$$A_\varepsilon = \Big\{x \in \ell^\infty(T) : \sup_{d(s,t)\le\delta} |x(t) - x(s)| \ge \varepsilon\Big\}.$$

If $x \in A_\varepsilon$ and $y \in A^c_{\varepsilon/2}$, then $\|x - y\|_T \ge \varepsilon/5$: by the triangle inequality, if $|x(t) - y(t)| < \varepsilon/5$, then there is s close to t by at most δ such that $|x(s) - y(s)| \ge 3\varepsilon/10 > \varepsilon/5$. Therefore, the restriction to the set $A_\varepsilon \cap A^c_{\varepsilon/2}$ of the indicator function I_{A_ε} is Lipschitz with constant bounded by $5/\varepsilon$. Hence, by the Kirzbraun–McShane extension theorem (e.g., Dudley, 1989, page 141 or Araujo and Giné, 1980, pages 2–3), there exists a bounded Lipschitz function f on $\ell^\infty(T)$ that is 0 on $A^c_{\varepsilon/2}$ and 1 on A_ε, which is non-negative and bounded by 1, and whose Lipschitz constant is bounded by $5/\varepsilon$. Then, the assumption implies

$$\begin{aligned}\limsup_{n\to\infty} \Pr{}^*\{X_n \in A_\varepsilon\} &\le \limsup_{n\to\infty} \mathbb{E}^* f(X_n) = Ef(X) \\ &\le \Pr\Big\{\sup_{d(s,t)\le\delta} \big|X(t) - X(s)\big| \ge \frac{\varepsilon}{2}\Big\}.\end{aligned}$$

Now, the asymptotic equicontinuity condition (5.1.3) follows from the uniform continuity with respect to d of the sample paths of X, which implies that the last probability tends to zero as $\delta \to 0$ by bounded convergence. Then, Theorem 5.1.2 gives $X_n \to_{\mathcal{L}} X$. □

5.1.2 Maximal Inequalities Based on Metric Entropy

If a process $X(t)$ on a metric space (T, d) satisfies good integrability properties, then inequality (4.3.7) can be used, in combination with a measure of the size of (T, d), to estimate the size of $X(t)$ and of its increments, concretely, to obtain upper bounds for exponential moments of $\sup_{t\in T} |X(t)|$ and $\sup_{d(s,t)\le\delta} |X(t) - X(s)|$. This will be applied in later sections to Gaussian and Rademacher chaos processes, and this is the reason why only exponential moments will be considered. In fact, we do not require the full force of inequality (4.3.7), but only its immediate consequence

$$\left\| \sup_{k \le N} \xi_k \right\|_\psi \le C_\psi \psi^{-1}(N) \max_{k \le N} \left\| \xi_k \right\|_\psi \tag{5.1.9}$$

for Young moduli satisfying conditions (4.3.5) and (4.3.6). Inequality (5.1.9) holds, in fact, for moduli satisfying (4.3.5) only, as can be easily verified with a proof similar to that of (4.3.1).

The estimates in question will be in terms of the measure of the size of the index set T of the process that we now define. The *covering number* $N(T, d, \varepsilon)$, $\varepsilon > 0$, of a metric or pseudometric space (T, d) is the smallest number of open balls of radius at most ε and centers in T required to cover T, that is,

$$N(T, d, \varepsilon) := \min\{n : \text{there exist } t_1, \dots, t_n \in T \text{ such that } T \subseteq \cup_{i=1}^n B(t_i, \varepsilon)\}.$$

The *packing number* $D(T, d, \varepsilon)$, $\varepsilon > 0$, of (T, d) is the largest n for which there are n points $t_1, \dots t_n$ in T at least ε apart for d [i.e., satisfying $d(t_i, t_j) \ge \varepsilon$ if $i \ne j$]. The *metric ε-entropy* or ε-entropy of (T, d) is $\log N(T, d, \varepsilon)$, and its *$\varepsilon$-capacity*, $\log D(T, d, \varepsilon)$. N and D are equivalent in the sense that

$$D(T, d, 2\varepsilon) \le N(T, d, \varepsilon) \le D(T, d, \varepsilon), \tag{5.1.10}$$

as the reader can easily verify. The following theorem is the simplest version of a series of results of the same kind, but is sufficient for our purposes. Its proof is known as a *chaining argument* for reasons that will become apparent. What we present here is an instance of a chaining argument that is valid only for Young moduli of exponential type (it requires nontrivial modifications for more general moduli).

THEOREM 5.1.4. *Let (T, d) be a pseudometric space, let $X(t)$, $t \in T$, be a stochastic process indexed by T, and let ψ be a Young modulus satisfying conditions (4.3.5) and (4.3.6), such that*

$$\|X(t) - X(s)\|_\psi \le d(s, t), \ s, t \in T. \tag{5.1.11}$$

Then, there exists a constant $K < \infty$ depending only on ψ such that, for all finite subsets $S \subseteq T$, $t_0 \in T$ and $\delta > 0$, the following inequalities hold:

$$\left\| \max_{t \in S} |X(t)| \right\|_\psi \le \|X(t_0)\|_\psi + K \int_0^D \psi^{-1}\big(N(T, d, \varepsilon)\big) d\varepsilon, \tag{5.1.12}$$

where D is the diameter of (T, d), and

$$\left\| \max_{\substack{d(s,t) \le \delta \\ s,t \in S}} |X(t) - X(s)| \right\|_\psi \le K \int_0^\delta \psi^{-1}\big(N(T, d, \varepsilon)\big) d\varepsilon. \tag{5.1.13}$$

PROOF. If (T, d) is not totally bounded, then the right sides of (5.1.12) and (5.1.13) are infinite. Then, without loss of generality, we assume T totally bounded and with diameter less than 1. Given $S \subset T$ finite and $t_0 \in T$, we can take $S \cup \{t_0\}$ instead, so that $t_0 \in S$. We can also assume $X(t_0) = 0$. For each $k = 0, 1, \dots$ let

$\{s_1^k, \ldots, s_{N_k}^k\} = S_k$ be the centers of $N_k := N(S, d, 2^{-k})$ open balls of radius at most 2^{-k} and centers in S covering S. Note that S_0 consists of one point, which we may take to be t_0. For each $k \geq 0$ let $\pi_k : S \mapsto S_k$ be a function satisfying $d(s, \pi_k(s)) < 2^{-k}$ for all $s \in S$, which obviously exists. Moreover, S being finite, there is k_S such that, if $k \geq k_S$ and $s \in S$, then $d(\pi_k(s), s) = 0$; this implies, by (5.1.11), that $X(s) = X(\pi_k(s))$ a.s. Then, for $s \in S$ we have:

$$X(s) = \sum_{r=1}^{k_S} \big(X(\pi_k(s)) - X(\pi_{k-1}(s)) \big) \quad \text{a.s.}$$

Then, since

$$d(\pi_k(s), \pi_{k-1}(s)) \leq d(\pi_k(s), s) + d(s, \pi_{k-1}(s)) < 3 \cdot 2^{-k},$$

inequality (5.1.9) and the hypothesis (5.1.11) give, using (4.3.6) in the last step,

$$\begin{aligned} \Big\| \max_{s \in S} |X(s)| \Big\|_\psi &\leq \sum_{k=1}^{k_S} \Big\| \max_{\substack{t \in S_k, s \in S_{k-1} \\ d(s,t) < 3 \cdot 2^{-k}}} |X(t) - X(s)| \Big\|_\psi \\ &\leq 3C_\psi \sum_{k=1}^{k_S} 2^{-k} \psi^{-1}(N_k N_{k-1}) \\ &\leq K \sum_{k=1}^{k_S} 2^{-k} \psi^{-1}(N_k), \end{aligned} \tag{5.1.14}$$

giving inequality (5.1.12) on account of the fact that $N(S, d, 2\varepsilon) \leq N(T, d, \varepsilon)$ (if an ε-ball with center in T intersects S, it is contained in a 2ε ball with center in S).

Consider now, for given $\delta > 0$, the set $U = \big\{(s, t) : s, t \in T, d(s, t) \leq \delta\big\}$ and on it the process $Y(u) = X(t_u) - X(s_u)$ where $u = (s_u, t_u)$. Take on U the pseudodistance $\rho(u, v) = \|Y(v) - Y(u)\|_\psi$. We can assume $\delta \leq \operatorname{diam} T$. The ρ-diameter of U is dominated by $2 \max_{u \in U} \|Y(u)\|_\psi \leq 2\delta$. We also have

$$\rho(u, v) \leq \big\| X(t_v) - X(t_u) \big\|_\psi + \big\| X(s_v) - X(s_u) \big\|_\psi \leq d(t_v, t_u) + d(s_v, s_u).$$

So, if $t_1, \ldots, t_N$ are the centers of a covering of T by $N = N(T, d, \varepsilon)$ open balls of radius at most ε, then the set of open balls with centers in $\{(t_i, t_j) : i, j \leq N\}$ and ρ-radius 2ε cover U; some of the (t_i, t_j) may not be in U, but if the 2ε ball about (t_i, t_j) has a nonvoid intersection with U, then it is contained in a ball of radius 4ε, centered at a point in U. Hence,

$$N(U, \rho, 4\varepsilon) \leq N^2(T, d, \varepsilon).$$

By definition, the process $Y(u)$, $u \in U$, satisfies inequality (5.1.11) for ρ. Therefore we can apply inequality (5.1.12) to it with $u_0 = (s, s)$ for any $s \in S$ (so that $Y(u_0) = 0$), thus obtaining

$$\begin{aligned}\Big\| \max_{\substack{d(s,t)\le\delta \\ s,t\in S}} |X(t) - X(s)| \Big\|_\psi &\le K \int_0^{2\delta} \psi^{-1}(N(U, \rho, \tau))d\tau \\ &\le K \int_0^{2\delta} \psi^{-1}(N^2(T, d, \tau/4))d\tau \\ &\le K' \int_0^{\delta/2} \psi^{-1}\big(N(T, d, \varepsilon)\big)d\varepsilon.\end{aligned}$$

[Note that (4.3.6) is also used here in the last step.] □

A process $X(t)$, $t \in T$, (T, d) a metric space, is *separable* if there exist a countable set $T_0 \subseteq T$ and a set $\Omega_0 \subseteq \Omega$ with $\Pr \Omega_0 = 0$ such that for all ω not in $\Omega_0, t \in T$ and $\varepsilon > 0$, $X(t, \omega)$ is in the closure of the set $\{X(s, \omega) : s \in T_0 \cap B(t, \varepsilon)\}$. Obviously, if X is separable,

$$\Big\|\sup_{t\in T} |X(t)|\Big\|_\psi = \sup_{\substack{S\subseteq T \\ S \text{ finite}}} \Big\|\max_{t\in S} |X(t)|\Big\|_\psi$$

and likewise for $\big\|\sup_{d(s,t)\le\delta,\, s,t\in T} |X(t) - X(s)|\big\|_\psi$. It is well known and easy to see that if (T, d) is a separable metric or pseudometric space and X is uniformly continuous in probability for d, then X admits a separable version. Since $N(T, d, \varepsilon) < \infty$ for all $\varepsilon > 0$ means that (T, d) is totally bounded, and condition (5.1.11) implies that X is uniformly continuous in probability, we have

COROLLARY 5.1.5. *Let (T, d) be a pseudometric space of diameter D, let ψ be a Young modulus satisfying conditions (4.3.5) and (4.3.6), such that*

$$\int_0^D \psi^{-1}\big(N(T, d, \varepsilon)\big)d\varepsilon < \infty, \tag{5.1.15}$$

and let $X(t)$, $t \in T$, be a stochastic process satisfying condition (5.1.11). Then, there is a version $\tilde{X}$ of X with almost all its sample paths in $C_u(T, d)$ and such that, moreover,

$$\Big\|\sup_{t\in T} |\tilde{X}(t)|\Big\|_\psi \le \|\tilde{X}(t_0)\|_\psi + K \int_0^D \psi^{-1}\big(N(T, d, \varepsilon)\big)d\varepsilon, \tag{5.1.16}$$

and

$$\Big\| \sup_{\substack{d(s,t)\le\delta \\ s,t\in T}} |\tilde{X}(t) - \tilde{X}(s)| \Big\|_\psi \le K \int_0^\delta \psi^{-1}\big(N(T, d, \varepsilon)\big)d\varepsilon \tag{5.1.17}$$

for all $0 < \delta \le D$ and a finite constant K that depends only on ψ. In fact, any separable version $\tilde{X}$ of X satisfies inequalities (5.1.16) and (5.1.17).

If a process X admits a version $\tilde{X}$ with almost all its sample paths in $C_u(T, d)$, then we say that X is *sample continuous* with respect to d. So, Corollary 5.1.5 gives conditions for a process X to be sample continuous.

This corollary, in particular, applies to Gaussian and Rademacher processes, and to Gaussian and Rademacher chaos processes. We give the details for Gaussian processes, and then indicate the results for the other types of processes. We recall that a Gaussian process $X(t)$, $t \in T$, is a stochastic process whose finite-dimensional distributions $\mathcal{L}\big(X(t_1), \ldots, X(t_r)\big)$, for any $r \in \mathbb{N}$ and $t_1, \ldots, t_r$ in T, are all multivariate normal. Also, see (4.3.2) for the defintion of the exponential Young modulus ψ_2.

COROLLARY 5.1.6. (DUDLEY'S THEOREM) *Let $X(t), t \in T$, be a Gaussian process and let $d_X(s,t) := \Big[\mathbb{E}\big(X(t) - X(s)\big)^2\Big]^{1/2}$, $s, t \in T$. If*

$$\int_0^D \big(\log N(T, d_X, \varepsilon)\big)^{1/2} d\varepsilon < \infty, \tag{5.1.18}$$

then there is a version of X, which we keep denoting as X, with almost all of its sample paths in $C_u(T, d_X)$ and such that

$$\Big\|\sup_{t\in T} |X(t)|\Big\|_{\psi_2} \le \|X(t_0)\|_{\psi_2} + K \int_0^D \Big(\log\big(N(T, d_X, \varepsilon)\big)\Big)^{1/2} d\varepsilon, \tag{5.1.19}$$

and

$$\Big\| \sup_{\substack{d_X(s,t)\le\delta \\ s,t\in T}} |X(t) - X(s)|\Big\|_{\psi_2} \le K \int_0^\delta \Big(\log\big(N(T, d_X, \varepsilon)\big)\Big)^{1/2} d\varepsilon \tag{5.1.20}$$

for all $0 < \delta \le D$, where K is a universal constant and D is the diameter of T for the pseudodistance d_X. In fact, every separable version of X satisfies these properties.

PROOF. As indicated at the beginning of Section 4.3, if ξ is standard normal, then $\mathbb{E}e^{\xi^2/c^2} = 1/\sqrt{1 - 2/c^2} < \infty$ for $c^2 > 2$, and $\mathbb{E}e^{\xi^2/c^2} = 2$ for $c^2 = 8/3$. So, setting $\psi(x) = \psi_2(x) = e^{x^2} - 1$ [see (4.3.2)], we have $\|\xi\|_\psi = \sqrt{8/3}$ and, by homogeneity, $\|\sigma\xi\|_\psi = \sqrt{8/3}\sigma$. Therefore,

$$\|X(t) - X(s)\|_{\psi_2} = \sqrt{8/3}\big[\mathbb{E}(X(t) - X(s))^2\big]^{1/2},$$

and we can take $\psi = \psi_2$ and $d = \sqrt{8/3}d_X$ in Corollary 5.1.6. Since the inverse function of ψ_2 is $[\log(1+x)]^{1/2}$, inequalties (5.1.16) and (5.1.17) do not translate exactly into (5.1.19) and (5.1.20), but they do, for different constants K, if the following two trivial observations are taken into account. 1) Using that there is $c < \infty$ such that $[\log(1+x)]^{1/2} \le c(\log x)^{1/2}$ for $x \ge 2$, and that N is monotone decreasing, $N(T, d, D/2) \geq 2$ and $N(T, d, D) = 1$ we easily conclude that if $0 < \delta \leq D/2$, then

$$\int_0^\delta \left[\log\bigl(1 + N(T, d, \varepsilon)\bigr)\right]^{1/2} d\varepsilon \leq C \int_0^\delta \left[\log N(T, d, \varepsilon)\right]^{1/2} d\varepsilon,$$

and that, if $D/2 < \delta \leq D$, then

$$\begin{aligned}\int_0^\delta \left[\log\bigl(1 + N(T, d, \varepsilon)\bigr)\right]^{1/2} d\varepsilon &\leq 2 \int_0^{D/2} \left[\log\bigl(1 + N(T, d, \varepsilon)\bigr)\right]^{1/2} d\varepsilon \\ &\leq 2C \int_0^{D/2} \left[\log N(T, d, \varepsilon)\right]^{1/2} d\varepsilon \\ &\leq 2C \int_0^{\delta} \left[\log N(T, d, \varepsilon)\right]^{1/2} d\varepsilon.\end{aligned}$$

And 2), if $\alpha > 0$, then $\int_0^\delta \left[\log N(T, \alpha d, \varepsilon)\right]^{1/2} d\varepsilon = \alpha \int_0^{\delta/\alpha} \left[\log N(T, d, \varepsilon)\right]^{1/2} d\varepsilon$ by change of variables. □

For instance, the bounds (5.1.19) and (5.1.20) apply to Brownian motion on $T = [a, b]$, with $d_X(s, t) = |t - s|^{1/2}$.

REMARK 5.1.7. Corollary 5.1.6 applies verbatim to *subgaussian processes*. A process $X(t)$, $t \in T$, is subgaussian if

$$\|X(t) - X(s)\|_{\psi_2} \leq c\left[\mathbb{E}(X(t) - X(s))^2\right]^{1/2}$$

for some $c < \infty$ and all $s, t \in T$. For example, if $X(t) = \sum_{i=1}^\infty x_i(t)\varepsilon_i$, where $x_i(t)$ are functions and $\{\varepsilon_i\}$ is a Rademacher sequence, then the process X is subgaussian by virtue of Corollary 3.2.6 for $d = 1$.

Now let X be a process on T such that, for all $s, t \in T$, the joint law of $(X(s), X(t))$ coincides with the joint law of two Gaussian or Rademacher chaos variables of order m. If X satisfies this property, we say that X is a (weak) *Gaussian* or *Rademacher chaos process of degree m*. Then, Theorem 3.2.10 and Corollary 3.2.6 give that there exists a finite constant c such that

$$\|X(t) - X(s)\|_{\psi_{2/m}} \leq c\left[\mathbb{E}(X(t) - X(s))^2\right]^{1/2}$$

for all $s, t \in T$, where $\psi_{2/m}$ is the Young modulus of exponential type defined by equations (4.3.2), which satisfies conditions (4.3.5) and (4.3.6). Obviously, then, the previous proof applies to these processes and gives the following:

COROLLARY 5.1.8. *Let $X(t)$, $t \in T$, be a (weak) Gaussian or Rademacher chaos process of degree m and let $d_X(s, t) := \left[\mathbb{E}\bigl(X(t) - X(s)\bigr)^2\right]^{1/2}$, $s, t \in T$. If*

$$\int_0^D \bigl(\log N(T, d_X, \varepsilon)\bigr)^{m/2} d\varepsilon < \infty, \tag{5.1.21}$$

then there is a version of X, which we keep denoting as X, with almost all of its sample paths in $C_u(T, d_X)$ and such that

$$\left\|\sup_{t\in T}|X(t)|\right\|_{\psi_{2/m}} \leq \|X(t_0)\|_{\psi_2} + K\int_0^D \Big(\log\big(N(T, d_X, \varepsilon)\big)\Big)^{m/2} d\varepsilon, \qquad (5.1.22)$$

and

$$\Big\|\sup_{\substack{d_X(s,t)\leq\delta \\ s,t\in T}} |X(t) - X(s)|\Big\|_{\psi_{2/m}} \leq K\int_0^\delta \Big(\log\big(N(T, d_X, \varepsilon)\big)\Big)^{m/2} d\varepsilon \qquad (5.1.23)$$

for all $0 < \delta \leq D$, where K is a universal constant and D is the diameter of T for the pseudodistance d_X. In fact, every separable version of X satisfies these properties.

These results are best possible only for Gaussian processes: condition (5.1.18) is necessary for sample continuity of stationary Gaussian processes (Fernique, 1975). Whereas there are exact necessary and sufficient conditions on the metric space (T, d_X) for a general Gaussian process X to be sample bounded or sample continuous (Talagrand, 1987), and they are in some sense close to condition (5.1.19), no such result exists -or even can exist- for chaos processes (see, e.g., Ledoux and Talagrand, 1991).

5.1.3 Vapnik–Červonenkis Classes of Functions

In this section we identify a type of collections of kernels for which the results in the previous subsections can be directly applied to obtain limit theorems for the U-processes based on them. We begin with classes of sets.

Let $\mathcal{C}$ be a class of subsets of a set S. Let $A \subseteq S$ be a finite set. $\Delta^{\mathcal{C}}(A)$ is defined as the number of subsets of A obtained by intersection of A with sets in the class $\mathcal{C}$, and we say $\mathcal{C}$ *shatters* A if $\Delta^{\mathcal{C}}(A) = 2^{\mathrm{Card}(A)}$, that is, if $\mathcal{C}$ picks out all possible subsets of A. Let

$$m^{\mathcal{C}}(k) = \sup_{\substack{A\subseteq S \\ \mathrm{Card}(A)=k}} \Delta^{\mathcal{C}}(A).$$

DEFINITION 5.1.9. *$\mathcal{C}$ is a Vapnik-Červonenkis class ($\mathcal{C}$ is VC) if the quantity*

$$v(\mathcal{C}) := \begin{cases} \min\{k : m^{\mathcal{C}}(k) < 2^k\} \\ \infty \quad \textit{if no such } k \textit{ exists} \end{cases}$$

is finite, that is, if there exists $k < \infty$ such that $\mathcal{C}$ does not shatter any subsets of S of cardinality k. The VC index of the class $\mathcal{C}$ is defined as $v(\mathcal{C})$.

For instance, $\mathcal{C} = \{(-\infty, t] : t \in \mathbb{R}\}$ is VC and $v(\mathcal{C}) = 2$ since $\{x_1, x_2\}$ is not shattered by $\mathcal{C}$ as $\Delta^{\mathcal{C}}(x_1, x_2) = 3 < 2^2$ ($\mathcal{C}$ cannot pick out $\{x_1 \vee x_2\}$.)

It is easy to see that $v(\mathcal{C}) = 0$ if and only if $\mathcal{C}$ is empty and $v(\mathcal{C}) = 1$ if and only if $\mathcal{C}$ consists of only one set. Thus, we can assume in the sequel that $v(\mathcal{C}) \geq 2$.

The following important combinatorial lemma asserts that *either* $m^{\mathcal{C}}(k) = 2^k$ *for all* k *or* $m^{\mathcal{C}}(k)$ *grows only polynomially in* k.

THEOREM 5.1.10. *Let $\mathcal{C}$ be a VC class and let $v = v(\mathcal{C})$. Then for any finite set $A \subseteq S$,*

$$\Delta^{\mathcal{C}}(A) \leq \text{Card}\{B \subseteq A : \text{Card } B < v\} = \sum_{j=0}^{v-1} \binom{\text{Card } A}{j}$$

and therefore

$$m^{\mathcal{C}}(n) \leq \sum_{j=0}^{v-1} \binom{n}{j},$$

a polynomial in n of degree $v - 1$. In particular there is a constant $B(v) < \infty$, depending only on v, such that $m^{\mathcal{C}}(n) \leq B(v)n^{v-1}$ for all $n \geq v - 1$.

The result follows from the following proposition by letting

$$\mathcal{U} = \{C \cap A : C \in \mathcal{C}\}.$$

PROPOSITION 5.1.11. *Let A be a finite set and let $\mathcal{U}$ be a class of subsets of A. Then*

$$\text{Card } \mathcal{U} \leq \text{Card}\{B \subseteq A : B \textit{ is shattered by } \mathcal{U}\}.$$

PROOF. Say that $\mathcal{U}'$ is hereditary if $B \in \mathcal{U}'$ implies all subsets of B are also in $\mathcal{U}'$. If $\mathcal{U}'$ is hereditary, then $\mathcal{U}'$ shatters each set it contains so that

$$\text{Card } \mathcal{U}' \leq \text{Card}\{B \subseteq A : B \text{ is shattered by } \mathcal{U}'\}.$$

Hence, the proposition will be proved if we construct a collection $\mathcal{U}'$ from $\mathcal{U}$ such that

i) $\mathcal{U}'$ is hereditary,

ii) Card $\mathcal{U} =$ Card $\mathcal{U}'$,

iii) $\mathcal{U}$ shatters at least as many sets as $\mathcal{U}'$ shatters.

We will obtain $\mathcal{U}'$ by repeated application of a simple operation T_x. Given $x \in A$, let $T_x(\mathcal{U}) = \{T_x(U) : U \in \mathcal{U}\}$ where for $U \in \mathcal{U}$, $T_x(U) = U \backslash \{x\}$ if $x \in U$ and $U \backslash \{x\} \notin \mathcal{U}$, and $T_x(U) = U$ otherwise. So, all the sets in $T_x(\mathcal{U})$ satisfy that if $V \in T_x(\mathcal{U})$ and $x \in V$, then $V \backslash \{x\} \in T_x(\mathcal{U})$, which is a step towards transforming $\mathcal{U}$ into a hereditary set. Another property of $T_x(\mathcal{U})$ to be used below, that follows directly from the definition, is that if $V \in T_x(\mathcal{U})$ and $x \in V$ then $V \in \mathcal{U}$, $V \backslash \{x\} \in \mathcal{U}$ and $T_x(V) = V$.

Let $w(\mathcal{U}) := \sum_{U \in \mathcal{U}} \text{Card}(U)$. Then, since Card$(T_x(U))$ is either Card(U) or Card$(U) - 1$, it follows that $w \circ T_x \leq w$, and A being finite, this implies that the minimum of w among all the collections $\mathcal{U}'$ obtained from $\mathcal{U}$ by repeated applications of T_x, $x \in A$, is attained. Let $\mathcal{U}'$ be one of the possibly more than one such collections on which w is minimal. Then, $w(\mathcal{U}') = w(T_x(\mathcal{U}'))$ for all $x \in A$. But this implies that $\mathcal{U}'$ is hereditary: if $\mathcal{U}'$ were not hereditary then there would exist a set $B \in \mathcal{U}'$ and a point $x \in A$ such that $B\backslash\{x\} \notin \mathcal{U}'$, implying $\text{Card}(T_x(B)) < \text{Card}(B)$, hence, also $w(\mathcal{U}') < w(T_x(\mathcal{U}'))$.

Since $\mathcal{U}'$ is obtained from $\mathcal{U}$ by a finite number of operations T_x, in order to prove that $\mathcal{U}'$ satisfies properties ii) and iii) above, it suffices to see that they are satisfied by $T_x(\mathcal{U})$. So, we must show that Card $T_x(\mathcal{U}) =$ Card $(\mathcal{U})$ and that $\mathcal{U}$ shatters at least as many sets as $T_x(\mathcal{U})$ does. To prove the first, we observe that T_x is one-to-one: The operation T_x keeps unchanged those sets $U \in \mathcal{U}$ which contain x and are such that $U\backslash\{x\}$ is in $\mathcal{U}$ and also those sets $U \in \mathcal{U}$ which do not contain x; and T_x replaces U by $U\backslash\{x\}$ if U contains x but $U\backslash\{x\}$ is not in $\mathcal{U}$. This is obviously a one-to-one replacement. To see the second, we just note that if $T_x(\mathcal{U})$ shatters B then also $\mathcal{U}$ shatters B: Suppose $T_x(\mathcal{U})$ shatters B. If $x \notin B$ then $\mathcal{U}$ shatters B because $U \in \mathcal{U}$ differs from $T_x(U) \in T_x(\mathcal{U})$ at most by x. If $x \in B$, then for every $B' \subseteq B\backslash\{x\}$ there exists $V \in T_x(\mathcal{U})$ such that $V \cap B = B' \cup \{x\}$ since $T_x(\mathcal{U})$ shatters B. This implies $x \in V$ and therefore V is also in $\mathcal{U}$, thus showing that $U \cap B = B' \cup \{x\}$ for some $U \in \mathcal{U}$; moreover, $V\backslash\{x\} \in \mathcal{U}$ and $V\backslash\{x\} \cap B = B'$. Hence, $\mathcal{U}$ shatters B. □

It is not always easy to prove that a class of sets is VC. The following two propositions can be very helpful.

PROPOSITION 5.1.12. *If $\mathcal{G}$ is a finite-dimensional vector space of real functions on S, then the class of sets $\mathcal{C} := [\{g \geq 0\} : g \in \mathcal{G}]$ is VC with $v(\mathcal{C}) \leq \dim \mathcal{G} + 1$.*

PROOF. Let $v - 1 = \dim \mathcal{G}$ and let $\{s_1, \ldots, s_v\}$ be v distinct points of S. (There is nothing to prove if S contains fewer than v points.) Let $L : \mathcal{G} \mapsto \mathbb{R}^v$ be given by $L(g) = (g(s_1), \ldots, g(s_v))$. Then $\dim L(\mathcal{G}) \leq v - 1$. Let then $w = (w_1, \ldots, w_v)$ be a nonzero vector orthogonal to $L(\mathcal{G})$. Then,

$$\sum w_i I(w_i \geq 0) g(s_i) = -\sum w_i I(w_i < 0) g(s_i)$$

and we can assume that the set of i with $w_i < 0$ is nonempty (otherwise we can replace w by $-w$). If there existed g for which $\{g \geq 0\} \cap \{s_1, \ldots, s_v\} = \{s_i : w_i \geq 0\}$ then the left side in the above equation for this g would be larger than or equal to zero whereas the right side would be negative, which is impossible. So, there is a subset of $\{s_1, \ldots, s_v\}$ that is not the intersection of $\{s_1, \ldots, s_v\}$ with any set $\{g \geq 0\}$. Hence, the class $\mathcal{C}$ is VC and $v(\mathcal{C}) \leq \dim \mathcal{G} + 1$. □

For example, this proposition immediately gives that the class of all closed halfspaces of $\mathbb{R}^d$ is VC.

The proof of the following proposition is straightforward and we leave it to the reader.

PROPOSITION 5.1.13. i) *If $\mathcal{C}$ is VC, then $\mathcal{C}^c : \{C^c : C \in \mathcal{C}\}$ is VC.*

ii) *If $\mathcal{C}$ and $\mathcal{D}$ are VC, then $\mathcal{C} \cup \mathcal{D}$ and $\mathcal{C} \cap \mathcal{D}$ are VC. Here $\mathcal{C} \cup \mathcal{D} \equiv \{C \cup D : C \in \mathcal{C}, D \in \mathcal{D}\}$ and $\mathcal{C} \cap \mathcal{D} \equiv \{C \cap D : C \in \mathcal{C}, D \in \mathcal{D}\}$.*

iii) *If $\mathcal{C}$ and $\mathcal{D}$ are VC, then* $\mathcal{C} \times \mathcal{D} \equiv \{C \times D : C \in \mathcal{C}, D \in \mathcal{D}\}$ *is VC.*

iv) *If $\mathcal{C} \subset \mathcal{D}$ and $\mathcal{D}$ is VC, then $\mathcal{C}$ is VC.*

Next we consider classes of functions.

DEFINITION 5.1.14. *The subgraph of a real function f on S is the set*

$$G_f = \{(s,t) : s \in S,\ t \in \mathbb{R},\ 0 \le t \le f(s) \quad \text{or} \quad f(s) \le t \le 0\}.$$

A class of functions $\mathcal{F}$ is VC-subgraph if the class of sets $\mathcal{C} = \{G_f : f \in \mathcal{F}\}$ is VC.

The family of indicator functions of the sets in a VC class is a VC subgraph class of functions. More generally, if f is a non-negative function on S and $\mathcal{C}$ is a VC class of sets, then the class of functions $\mathcal{F} = \{f I_C : C \in \mathcal{C}\}$ is VC-subgraph. To see this, we just note that, in order that a subset $\{(s_1, t_1), \ldots, (s_n, t_n)\}$ of $S \times \mathbb{R}$ be shattered by the subgraphs of functions in $\mathcal{F}$, it is necessary that the S-coordinates $s_1, \ldots, s_n$ be all different and that the set $\{s_1, \ldots, s_n\}$ be shattered by $\mathcal{C}$.

Hence, the following key proposition for VC-subgraph classes of functions applies to VC classes of sets and beyond. We must first set up some notation. Let $\mathcal{F}$ be a class of functions in $L_p(S, \mathcal{S}, P)$, $0 < p < \infty$. Then, letting d_p denote the $L_p(P)$ pseudodistance, $d_p(f, g) = \|f - g\|_p = \int |f - g|^p dP$ for $0 < p < 1$, $d_p(f, g) = \|f - g\|_p = \left[\int |f - g|^p dP\right]^{1/p}$ for $p \ge 1$, we set

$$D_p(\mathcal{F}, P, \varepsilon) := D(\mathcal{F}, d_p, \varepsilon) \tag{5.1.24}$$

and N_p is defined in an analoguous way. We recall that, given a class of functions $\mathcal{F}$, an *envelope* F of $\mathcal{F}$ is any *everywhere finite, measurable* function F such that

$$\sup_{f \in \mathcal{F}} |f(s)| \le F(s), \qquad s \in S. \tag{5.1.25}$$

Of course, a class $\mathcal{F}$ may not admit any envelopes. We recall also the notation Pf for $\int f dP$. Finally, we should emphasize that, although $\mathcal{F}$ is always a subset of measurable functions on a probability space, we do *not* identify functions which are a.s. equal.

THEOREM 5.1.15. *Let $\mathcal{F}$ be a VC-subgraph class of real functions admitting an envelope $F \in L_p(S, \mathcal{S}, P)$ for some $0 < p < \infty$. Suppose the class of subgraphs has index v. Then, there exist finite constants $A(v, p)$ depending only on v and p (and not on $\mathcal{F}$ or P) such that, for all $0 < \varepsilon \le 1$,*

$$D_p(\mathcal{F}, P, \varepsilon\|F\|_p) \le A(v, p)\varepsilon^{-pv} \quad \text{if } p \ge 1, \tag{5.1.26}$$

and

$$D_p(\mathcal{F}, P, \varepsilon\|F\|_p) \le A(v, p)\varepsilon^{-v/p} \quad \text{if } p < 1. \tag{5.1.27}$$

Hence, by (5.1.10), the same bounds apply to $N(\mathcal{F}, d_p, \varepsilon\|F\|_p)$.

PROOF. Assume $p \geq 1$. Let $f_1, \ldots, f_m$ be a maximal collection of functions in $\mathcal{F}$ for which

$$P|f_i - f_j|^p \geq \varepsilon^p P F^p, \quad i \neq j.$$

Then $m = D_p(\mathcal{F}, P, \varepsilon\|F\|_p)$. Choose points in $S \times \mathbb{R}$ as follows: sample s_i independently from the distribution defined by $P_F(A) = P(I_A F^p)/PF^p$,, $A \in \mathcal{S}$; then, independently for each i, given s_i, sample t_i from the uniform distribution on $[-F(s_i), F(s_i)]$, $i = 1, \ldots, k$, with k to be specified below. In particular, the two-dimensional vectors (s_i, t_i), $1 \leq i \leq k$, are independent. Let G_i denote the subgraph of f_i.

The probability that at least one pair of graphs picks out the same set of points from the sample $\{(s_i, t_i),\ i \leq k\}$, is at most

$$\begin{aligned}
&\binom{m}{2} \max_{i \neq j} \Pr\{G_i \text{ and } G_j \text{ pick out the same sets of points}\} \\
&= \binom{m}{2} \max_{i \neq j} \prod_{r=1}^{k} \Pr\{(s_r, t_r) \notin G_i \Delta G_j\} \\
&= \binom{m}{2} \max_{i \neq j} \prod_{r=1}^{k} \Big[1 - \Pr\{(s_r, t_r) \in G_i \Delta G_j\}\Big] \\
&= \binom{m}{2} \max_{i \neq j} \prod_{r=1}^{k} \Big[1 - \Pr\big\{t_r \in \text{ the interval with end points } \{f_i(s_r), f_j(s_r)\}\big\}\Big] \\
&= \binom{m}{2} \max_{i \neq j} \left[1 - \frac{1}{\|F\|_p^p} \int \frac{|f_i - f_j|}{2F} F^p dP\right]^k \\
&\leq \binom{m}{2} \max_{i \neq j} \left[1 - \frac{1}{\|F\|_p^p} \int \frac{|f_i - f_j|^p}{2^p} dP\right]^k \\
&\leq \binom{m}{2} \left[1 - \frac{\varepsilon^p}{2^p}\right]^k \\
&\leq \binom{m}{2} \exp\left(-\frac{\varepsilon^p k}{2^p}\right) \leq \frac{1}{2} \exp\left(2 \log m - \frac{\varepsilon^p k}{2^p}\right).
\end{aligned}$$

This last number is strictly smaller than 1 if $k = \left\lceil \frac{1+2^{p+1}\log m}{\varepsilon^p} \right\rceil$ and $0 < \varepsilon \leq 1$. So, in this case, *with positive probability*, the graphs pick out different subsets from the k-sample, and we have m graphs. But, by Theorem 5.1.10, the sets G_i, $i \leq m$, can pick out at most Bk^{v-1} subsets of any sample of size k, where $B = B(v)$ is a fixed number. By the choice of k, $Bk^{v-1} \leq B\left(\frac{1+2^{p+1}\log m}{\varepsilon^p}\right)^{v-1}$, and we have thus proved that

$$m \leq B\left(\frac{1 + 2^{p+1}\log m}{\varepsilon^p}\right)^{v-1}$$

So, if n_0 is such that $(1 + 2^{p+1} \log n)^{v-1} \le n^{\frac{1}{v}}$ for all $n \ge n_0$, then either $m \le n_0$ or $m \le Bm^{\frac{1}{v}}\varepsilon^{p(1-v)}$. That is, $m \le n_0 \vee B^{\frac{v}{v-1}}\varepsilon^{-vp}$. Thus, inequality (5.1.26) is proved with $A(v, p) = n_0 \vee B^{v/(v-1)}$. The proof for $p < 1$ is similar. Here are the changes: The functions f_i satisfy $P|f_i - f_j|^p \ge \varepsilon P F^p$ and one uses the following estimate of $\Pr\{(s_r, t_r) \in G_i \Delta G_j\}$ in the above string of inequalities:

$$\begin{aligned}\Pr\{(s_r, t_r) \in G_i \Delta G_j\} &= P_F\Big(\frac{|f_i(s_r) - f_j(s_r)|}{2F(s_r)}\Big)\\ &\ge \Big[P_F\Big(\frac{|f_i(s_r) - f_j(s_r)|}{2F(s_r)}\Big)^p\Big]^{\frac{1}{p}}\\ &= \Big(\frac{P|f_i - f_j|^p}{2^p P F^p}\Big)^{\frac{1}{p}} \ge \frac{\varepsilon^{1/p}}{2}. \qquad \square\end{aligned}$$

This theorem, in combination with the main theorems in the previous two subsections, will allow us to prove several limit theorems for U-processes indexed by VC classes of kernels in the sections that follow.

5.2 The Law of Large Numbers for U-Processes

Let $\mathcal{H}$ be a family of measurable real-valued functions on $(S^m, \mathcal{S}^m, P^m)$ admitting an envelope H in $L_1(S^m, \mathcal{S}^m, P^m)$, and let, for each $h \in \mathcal{H}$,

$$U_n(h) := \frac{(n-m)!}{n!} \sum_{I_n^m} h(X_{i_1}, \dots, X_{i_m})$$

be the U-statistic based on the *coordinate functions* X_i of $S^{\mathbb{N}}$. As seen in Chapter 4, Section 4.1, the law of large numbers for U-statistics asserts that $|U_n(h) - P^m h| \to 0$ a.s. for each $h \in \mathcal{H}$. The next step, which is the problem we consider in this section, should be to determine whether, or under what conditions, does *this limit hold uniformly in* $h \in \mathcal{H}$, that is, when do we have

$$\sup_{h\in\mathcal{H}} |U_n(h) - P^m h| := \big\|U_n(h) - P^m h\big\|_{\mathcal{H}} \to 0 \quad \text{a.s.}$$

These variables may not be measurable and therefore we should consider their measurable envelope $\big\|U_n(h) - P^m h\big\|_{\mathcal{H}}^*$—however, since we will restrict ourselves to measurable classes $\mathcal{H}$, in the end this will not be necessary.

The reversed submartingale convergence theorem already provides a first reduction of the problem, as follows. For each $n \in \mathbb{N}$, let $\bar{\mathcal{G}}_n$ be the σ-algebra of $S^{\mathbb{N}}$ generated by the sets in $\mathcal{S}^{\mathbb{N}}$ which are invariant under permutations of the first n coordinates and by the sets of $P^{\mathbb{N}}$ measure zero. Then, it can be safely left to the reader to verify that the sequence of random variables

$$\Big\{\big\|U_n(h) - P^m h\big\|_{\mathcal{H}}^*\Big\}_{n=m}^{\infty}$$

is a reversed submartingale with respect to the σ-algebras $\bar{\mathcal{G}}_n$ (Strobl, 1995, observes this for $m = 1$). Thus, it follows from reversed submartingale convergence that, provided an integrable envelope of the class $\mathcal{H}$ exists, the above sequence always converges a.s. and in L_1. So, the only remaining problem is that of determination of the limit. Therefore the strong law of large numbers for U-processes is equivalent to the weak law, and even more, concretely,

$$\big\| U_n(h) - P^m h \big\|_{\mathcal{H}}^* \to 0 \quad \text{a.s.} \tag{5.2.1}$$

if and only if

$$\big\| U_n(h) - P^m h \big\|_{\mathcal{H}}^* \to 0 \quad \text{in pr.,} \tag{5.2.2}$$

if and only if

$$\mathbb{E}\big\| U_n(h) - P^m h \big\|_{\mathcal{H}}^* \to 0. \tag{5.2.3}$$

In this way, a.s. convergence becomes a problem of estimation of expected values of suprema of families of U-statistics. By the randomization theorem for expected values in Section 3.5 (Theorem 3.5.3, with $r = 1$), the last condition is equivalent to

$$\mathbb{E}^* \left\| \frac{(n-m)!}{n!} \sum_{I_n^m} \varepsilon_{i_1} \big(h(X_{i_1}, \ldots, X_{i_m}) - P^m h \big) \right\|_{\mathcal{H}} \to 0.$$

Since the norm of the centering term is just $\| P^m h \|_{\mathcal{H}} \mathbb{E} \big| \sum_{i_1=1}^n \varepsilon_{i_1}/n \big| < P^m H/\sqrt{n}$, which tends to zero by integrability of H, the last limit can be simplified and becomes

$$\mathbb{E}^* \left\| \frac{(n-m)!}{n!} \sum_{I_n^m} \varepsilon_{i_1} h(X_{i_1}, \ldots, X_{i_m}) \right\|_{\mathcal{H}} \to 0. \tag{5.2.4}$$

Note that we can randomize by *only* one Rademacher variable, as opposed to a product of several such variables, because we are not assuming any degeneracy on the kernels h. We should record these observations in a lemma:

LEMMA 5.2.1. *Let $\mathcal{H}$ be a collection of measurable real functions of m variables from S admitting a P^m-integrable envelope H. Then, if one of the limits (5.2.1), (5.2.2), (5.2.3) or (5.2.4) holds, so do the others.*

Even the very limited randomization present in the limit (5.2.4) will allow us to solve the problem at hand by just applying some of the good properties of Rademacher variables to these processes, once we condition on the sample. This leads to consideration of certain *random distances*, in particular the L_p distances with respect to certain empirical measures which, in the case of $p = 2$, coincide with the conditional L_2 norm, given the sample, of the randomized U-statistics.

Next we define the random distances that will appear in the LLN for U-processes, in its proof and in its corollaries. If X_i, $i \in \mathbb{N}$, are the coordinate functions of $S^{\mathbb{N}}$, we set, for $n \geq m$, $0 < p < \infty$ and $f, g \in \mathcal{H}$,

$$\tilde{e}_{n,p}(f,g) := \left[\frac{1}{n}\sum_{i_1=1}^{n}\left|\frac{(n-m)!}{(n-1)!}\sum_{(i_2,\dots,i_m):\mathbf{i}\in I_n^m}(f-g)(X_{i_1},\dots,X_{i_m})\right|^p\right]^{1/(p\vee 1)}, \tag{5.2.5}$$

and

$$\begin{aligned} e_{n,p}(f,g) &:= \left[U_n^{(m)}(|f-g|^p)\right]^{1/(p\vee 1)} \\ &= \left[\frac{(n-m)!}{n!}\sum_{I_n^m}|f-g|^p(X_{i_1},\dots,X_{i_m})\right]^{1/(p\vee 1)}. \end{aligned} \tag{5.2.6}$$

We note that

$$\tilde{e}_{n,2}(f,g) = \left[\mathbb{E}_\varepsilon\Big(\frac{n^{1/2}(n-m)!}{n!}\sum_{I_n^m}\varepsilon_{i_1}(f-g)(X_{i_1},\dots,X_{i_m})\Big)^2\right]^{1/2}, \tag{5.2.7}$$

and that $e_{n,p}$ is the L_p distance corresponding to the random measure that assigns mass $(n-m)!/n!$ to each of the points $(X_{i_1},\dots,X_{i_m}) \in S^m$, $\mathbf{i} \in I_n^m$.

We should next recall that, when dealing with U-processes, we redefine the X_i variables and the Rademacher variables ε_i as the coordiantes of the product probability space

$$(\Omega, \Sigma, \Pr) := \left(S^{\mathbb{N}} \times \{-1,1\}^{\mathbb{N}}, \mathcal{S}^{\mathbb{N}} \otimes \left(2^{\{-1,1\}}\right)^{\mathbb{N}}, P^{\mathbb{N}} \times \nu^{\mathbb{N}}\right),$$

with $\nu = (\delta_{-1} + \delta_1)/2$, so that, in particular, the X and the ε variables are all independent. We adhere to this setup for the rest of this chapter.

The basic argument in the proof of the main result in this section consists in using Rademacher randomization and then a maximal inequality for linear combinations of Rademacher variables, conditionally on the X variables. The inequality in question is the simple inequality (4.3.1) in Chapter 4, and we comment on it now. If ξ_i are linear combinations of Rademacher variables, then, by Corollary 3.2.6, there exist $\lambda < \infty$ such that

$$\mathbb{E}\exp\Big(\frac{|\xi|^2}{\lambda^2\mathbb{E}\xi^2}\Big) \leq 2.$$

So, if ξ_i are N Rademacher linear combinations, inequality (4.3.1) with $\Phi(x) = \exp x^2$ gives

$$\mathbb{E}\max_{i\leq N}|\xi_i| \leq \lambda\big(\log(2N)\big)^{1/2}\max_{i\leq N}\big(\mathbb{E}\xi_i^2\big)^{1/2}. \tag{5.2.8}$$

Now we can state (and prove part of) the main theorem in this section:

THEOREM 5.2.2. *Let $\mathcal{H}$ be a measurable (:=image-admissible Suslin) class of real symmetric kernels on $(S^m, \mathcal{S}^m)$ with everywhere finite P^m-integrable envelope H. Then the following statements are equivalent:*

i) $\|U_n - P^m\|_{\mathcal{H}} \to 0$ a.s.,

ii) $\mathbb{E}\Big\|\frac{1}{n^m}\sum_{I_n^m} \varepsilon_{i_1} h(X_{i_1}, \dots, X_{i_m})\Big\|_{\mathcal{H}} \to 0,$

iii) $\frac{1}{n}\log N\big(\mathcal{H}, \tilde{e}_{n,1}, \varepsilon\big) \to 0$ *in pr* for all* $\varepsilon > 0$.

Moreover, the condition

iv) $\frac{1}{n}\log N\big(\mathcal{H}_M, \tilde{e}_{n,1}, \varepsilon\big) \to 0$ *in pr* for all* $\varepsilon > 0$ *and* $0 < M < \infty$, *where* $\mathcal{H}_M := \{hI_{H\le M} : h \in \mathcal{H}\}$, *is sufficient for the law of large numbers i).*

PROOF (PARTIAL). Lemma 5.2.1 implies i) $\Leftrightarrow$ ii). The implication ii) $\Rightarrow$ iii) follows from a Sudakov-type minorization inequality for suprema of linear combinations of Rademacher variables due to Carl and Pajor (1988) (see Ledoux and Talagrand, 1991, Corollary 4.14). We do not reproduce the proof of the Carl–Pajor inequality because we are primarily interested in the implication iii) $\Rightarrow$ i), but we indicate how to use it. This inequality is as follows: if $T \subset \mathbb{R}^N$ and $r(T) = \mathbb{E}\sup_{\mathbf{t}\in T}\big|\sum_{i=1}^N \varepsilon_i t_i\big|$, $\mathbf{t} = (t_1, \dots, t_N)$, then, for all $\tau > 0$,

$$\tau\big(\log N(T, d_2, \tau)\big)^{1/2} \le Kr(T)\Big(\log\big(2 + N^{1/2}/r(T)\big)\Big)^{1/2}, \tag{5.2.9}$$

where d_2 is Euclidean distance and K is a universal constant. iii) will follow from ii) by using this inequality on the processes

$$\frac{n^{1/2}(n-m)!}{n!}\sum_{I_n^m} \varepsilon_{i_1} h(X_{i_1}, \dots, X_{i_m}), \qquad h \in \mathcal{H},$$

conditionally on the X variables. So, we take for $T = T_n$ the subset of $\mathbb{R}^n$ defined by

$$T_n = \Big\{\Big(n^{1/2}\frac{(n-m)!}{n!}\sum_{(i_2,\dots,i_m):\mathbf{i}\in I_n^m} h(X_{i_1}, \dots, X_{i_m}) : i_1 = 1, \dots, n\Big) : h \in \mathcal{H}\Big\},$$

for $X_1, \dots, X_n$ fixed. Then $d_2 = \tilde{e}_{n,2}$ and

$$r(T_n) = n^{1/2}\mathbb{E}_\varepsilon\Big\|\frac{(n-m)!}{n!}\sum_{I_n^m} \varepsilon_{i_1} h(X_{i_1}, \dots, X_{i_m})\Big\|_{\mathcal{H}}.$$

Statement ii) implies that $r(T_n)/n^{1/2} \to 0$ in probability, and, since $x_n \to 0$ implies $x_n\big(\log(2 + 1/x_n)\big)^{1/2} \to 0$, it follows that

$$n^{-1/2}r(T_n)\big(\log(2 + n^{1/2}/r(T_n))\big)^{1/2} \to 0$$

in probability. (Note that these variables are measurable by the measurability hypothesis on $\mathcal{H}$.) Then, inequality (5.2.9) gives

$$\frac{1}{n}\big[\log N(\mathcal{H}, \tilde{e}_{n,2}, \tau)\big]^* \to 0 \tag{5.2.10}$$

in probability for all $\tau > 0$. This is even a stronger statement than iii) is because, by Hölder, $\tilde{e}_{n,1} \le \tilde{e}_{n,2}$, which gives $N(\mathcal{H}, \tilde{e}_{n,1}, \tau) \le N(\mathcal{H}, \tilde{e}_{n,2}, \tau)$ for all $\tau > 0$.

To prove that iii) implies ii) we begin with truncation:

$$\mathbb{E}\Big\|\frac{(n-m)!}{n!}\sum_{I_n^m}\varepsilon_{i_1}h(X_{i_1},\dots,X_{i_m})\Big\|_{\mathcal{H}} \le \mathbb{E}\Big\|\frac{(n-m)!}{n!}\sum_{I_n^m}\varepsilon_{i_1}(hI_{H\le M})(X_{i_1},\dots,X_{i_m})\Big\|_{\mathcal{H}} + P^m\big(HI_{H>M}\big).$$

Since, by integrability of H, $P^m\big(HI_{H>M}\big) \to 0$ as $M \to \infty$, it follows that, to prove ii), it suffices to show

$$\lim_{M\to\infty}\limsup_{n\to\infty}\mathbb{E}\Big\|\frac{(n-m)!}{n!}\sum_{I_n^m}\varepsilon_{i_1}h(X_{i_1},\dots,X_{i_m})\Big\|_{\mathcal{H}_M} = 0. \tag{5.2.11}$$

To prove (5.2.11), we will apply the maximal inequality (5.2.8) conditionally on the X variables, as follows. For $\{X_i\}$ *fixed*, let $\mathcal{H}_{\tau,M}$ be a subset of $\mathcal{H}_M$ of cardinality $N(\mathcal{H}_M, \tilde{e}_{n,1}, \tau)$, τ-dense in $\mathcal{H}_M$ for the distance $\tilde{e}_{n,1}$ (which depends on the variables X_i). Let $\mathbb{E}_\varepsilon$ denote integration with respect to the Rademacher variables only, as usual. Then, by the triangle inequality,

$$\mathbb{E}_\varepsilon\Big\|\frac{(n-m)!}{n!}\sum_{I_n^m}\varepsilon_{i_1}h(X_{i_1},\dots,X_{i_m})\Big\|_{\mathcal{H}_M} \le \tau + \mathbb{E}_\varepsilon\Big\|\frac{(n-m)!}{n!}\sum_{I_n^m}\varepsilon_{i_1}h(X_{i_1},\dots,X_{i_m})\Big\|_{\mathcal{H}_{\tau,M}},$$

and, by inequality (5.2.8),

$$\begin{aligned}
&\mathbb{E}_\varepsilon\Big\|\frac{(n-m)!}{n!}\sum_{I_n^m}\varepsilon_{i_1}h(X_{i_1},\dots,X_{i_m})\Big\|_{\mathcal{H}_{\tau,M}}\\
&\quad\le \lambda\frac{(n-m)!}{n!}\big[\log 2 + \log N(\mathcal{H}_M, \tilde{e}_{n,1}, \tau)\big]^{1/2}\\
&\quad\quad\times \sup_{h\in\mathcal{H}_{\tau,M}}\Big[\sum_{i_1=1}^{n}\Big(\sum_{(i_2,\dots,i_m):\mathbf{i}\in I_n^m} h(X_{i_1},\dots,X_{i_m})\Big)^2\Big]^{1/2}\\
&\quad\le \lambda M\frac{(n-m)!}{n!}\Big[\frac{n(n!)^2}{((n-m+1)!)^2}\Big]^{1/2}\big[\log 2 + \log N(\mathcal{H}_M, \tilde{e}_{n,1}, \tau)\big]^{1/2}\\
&\quad\le \lambda M\frac{n^{1/2}}{n-m+1}\big[\log 2 + \log N(\mathcal{H}_M, \tilde{e}_{n,1}, \tau)\big]^{1/2}.
\end{aligned}$$

Hence, in order to prove that the limit (5.2.11) holds, it remains only to show that for all $\tau > 0$ there exists $M_\tau < \infty$ such that

$$\limsup_{n\to\infty}\mathbb{E}\Big[\Big(\frac{1}{n}\log N(\mathcal{H}_M, \tilde{e}_{n,1}, \tau)\Big)^*\Big]^{1/2} = 0 \tag{5.2.12}$$

whenever $M \geq M_\tau$. In order to prove that (5.2.12) follows from condition iii) we must relate the entropy numbers of $\mathcal{H}$ and $\mathcal{H}_M$. Since

$$\tilde{e}_{n,1}(fI_{H\leq M}, gI_{H\leq M}) \leq \tilde{e}_{n,1}(f,g) + 2U_n(HI_{H>M}),$$

any covering of $\mathcal{H}$ by $\tilde{e}_{n,1}$-balls of radius $\tau/2$ induces a covering of $\mathcal{H}_M$ by $\tilde{e}_{n,1}$-balls of radius $\tau/2 + 2U_n(HI_{H>M})$ of the same cardinality. Therefore, for all $\varepsilon > 0$,

$$\Pr{}^*\left\{\frac{1}{n}\log N(\mathcal{H}_M, \tilde{e}_{n,1}, \tau) > \varepsilon\right\} \leq \Pr{}^*\left\{\frac{1}{n}\log N(\mathcal{H}, \tilde{e}_{n,1}, \tau/2) > \varepsilon\right\} + \Pr\left\{U_n(HI_{H>M}) > \frac{\tau}{4}\right\}.$$

Condition iii) and the law of large numbers for U-statistics then show that

$$\frac{1}{n}\log N(\mathcal{H}_M, \tilde{e}_{n,1}, \tau) \to 0 \quad \text{in} \quad \text{pr}^* \quad \text{as } n \to \infty$$

for all M such that $\mathbb{E}HI_{H>M} < \tau/4$, hence, for all $M \geq M_\tau$ for some $M_\tau < \infty$. If we set $|(a_1, \ldots, a_n)| = \max |a_i|$, it follows immdiately that

$$N(\mathcal{H}_M, \tilde{e}_{n,1}, \tau) \leq N([-M, M]^n, |\cdot|, \tau) \leq \left(\frac{3M}{\tau}\right)^n$$

for all $\tau \leq 1$ and $n > m$, so that we can apply bounded convergence to the previous limit and obtain (5.2.12). This completes the proof of iii) $\Rightarrow$ ii), and also that of iv) $\Rightarrow$ ii). □

As in several proofs in Chapter 3, the above proof is another instance of how decoupling and randomization allows for use of poweful properties of Rademacher averages to reach the desired conclusions.

We should mention that the previous proof shows that the random distance $\tilde{e}_{n,1}$ can be replaced by $\tilde{e}_{n,p}$, for any $0 < p \leq 2$ in condition iii) of the last theorem. In the case $\|H\|_\infty < \infty$, these distances can even be replaced by any $\tilde{e}_{n,p}$, $0 < p \leq \infty$, where

$$\tilde{e}_{n,\infty}(f,g) = \max_{1\leq i_1\leq n} \frac{1}{n}\left|\frac{(n-m)!}{(n-1)!}\sum_{(i_2,\ldots,i_m):\mathbf{i}\in I_n^m}(f-g)(X_{i_1}, \ldots, X_{i_m})\right|.$$

This is trivial for sufficiency [iii) $\Rightarrow$ i)] but requires a simple extra argument that we will not give (Talagrand, 1987, page 863) for necessity. So, Theorem 5.2.2 is the complete analog for U-processes of the celebrated Vapnik–Červonenkis law of large numbers for empirical processes (Vapnik and Červonenkis, 1971, 1981).

An advantage of conditions in terms of *random entropies*, such as iii) above, is that they do not depend on the cancellation inherent in the average that defines the U-process, but a disadvantage is that they still depend on the sample. Next, as corollaries to Theorem 5.2.2, we give sufficient conditions for the LLN that do not depend on the sample but only on the class $\mathcal{H}$ and on a single integral with respect to P^m, and are therefore more user-friendly than Theorem 5.2.2.

The random distance $e_{n,1}$ between not necesarily symmetric functions f and g of m variables, as mentioned, is the L_1 distance between f and g for the random probability measure on S^m that gives equal mass to each of the points $(X_{i_1}, \ldots, X_{i_m})$, $\mathbf{i} \in I_n^m$. Theorem 5.1.15 thus implies that if $\mathcal{H}$ is a VC-subgraph class of functions on S^m and $\mathbb{E}H < \infty$, then

$$\frac{1}{n} \log N(\mathcal{H}, e_{n,1}, \varepsilon) \leq \frac{c(1 + \log \varepsilon^{-1})}{n} \tag{5.2.13}$$

for some $c = c(\mathcal{H}) < \infty$ and all $0 < \varepsilon < 1$ and $n \in \mathbb{N}$, independently of $X_1, \ldots, X_n$. [See Definitions 5.1.9, 5.1.14 for VC classes.] On the other hand, if

$$s_m f(x_1, \ldots, x_m) := \sum f(x_{i_1}, \ldots, x_{i_m})/m!,$$

with the sum extended to all the permutations $(i_1, \ldots, i_m)$ of $(1, \ldots, m)$, denotes the symmetrization of f, it is clear that

$$\tilde{e}_{n,1}(s_m f, s_m g) \leq e_{n,1}(s_m f, s_m g) = e_{n,1}(f, g).$$

This inequality is inherited by the corresponding entropies. So, inequality (5.2.13) and the previous theorem give the following:

COROLLARY 5.2.3. *Let $\mathcal{H}$ be a measurable class of real-valued measurable functions on S^m with an envelope H in $L_1(P^m)$. Then, if $\mathcal{H}$ is a VC-subgraph class,*

$$\|U_n(h) - P^m h\|_{\mathcal{H}} \to 0 \quad \text{a.s.}$$

Corollary 5.2.3 is sufficent for most applications in Statistics since most classes of functions of interest are VC-subgraph or are combinations of VC-subgraph classes.

EXAMPLE 5.2.4. As a concrete example, we will obtain the law of large numbers for the *simplicial depth process* corresponding to a sample in Euclidean space, an object of some interest in multivariate statistics. For simplicity we will consider it only in the plane $\mathbb{R}^2$. Let X_i, $i \in \mathbb{N}$, be i.i.d. random vectors in the plane, with common law P. For each $\theta \in \mathbb{R}^2$ let $C_\theta \subset \mathbb{R}^2 \times \mathbb{R}^2 \times \mathbb{R}^2$ be the set of all triplets $(x_1, x_2, x_3) \in (\mathbb{R}^2)^3$ such that $\theta \in S(x_1, x_2, x_3)$, where $S(x_1, x_2, x_3)$ is the (open) triangle determined by the points x_1, x_2, x_3. Then, the simplicial depth process corresponding to the given sample is defined as

$$D_n(\theta) := U_n^{(3)}(C_\theta) = \frac{1}{\binom{n}{3}} \sum_{1 \leq i < j < k \leq n} I_{C_\theta}(X_i, X_j, X_k), \qquad \theta \in \mathbb{R}^2, \tag{5.2.14}$$

that is, $D_n(\theta)$ is the proportion of sample triangles that contain the point θ. The *depth function* associated to the probability measure P on $\mathbb{R}^2$ is defined as $P^3 C_\theta$, $\theta \in \mathbb{R}^2$. These notions are used in the definition of a multidimensional location parameter, the simplicial median (an argmax of the depth function) and its estimator, the empirical simplicial median (an argmax of the depth process), as well as multidimensional order statistics. The class of sets $\mathcal{C} = \{C_\theta : \theta \in \mathbb{R}^2\}$

is image-admissible Suslin as it is parametrized by $\mathbb{R}^2$ and the evaluation map is jointly measurable in θ and x_1, x_2, x_3. Next we see that this class of subsets $\mathcal{C}$ of $(\mathbb{R}^2)^3$ is VC. Given n triplets of points $S_1, \ldots, S_n \in (\mathbb{R}^2)^3$ we have, e.g.,

$$C_\theta \cap \{S_1, \ldots, S_n\} = \{S_1, S_2\} \quad \text{iff} \quad \theta \in S_1 \cap S_2 \cap S_3^c \cap \cdots \cap S_n^c,$$

where we identify a triplet of points with the triangle with vertices these points. So, $\Delta^{\mathcal{C}}(S_1, \ldots, S_n)$ is less than or equal to the largest possible number of regions that $3n$ lines determine on the plane. A single line determines two regions, a second line breaks each of these two regions into two (or less, if parallel to the first), a third line crosses at most three of the regions obtained at the second stage and breaks each into two, and so on, so that this number is easily seen by recurrence to be

$$2 + 2 + 3 + \cdots + 3n = 1 + \frac{3n(3n+1)}{2},$$

a polynomial of degree 2. Hence $\mathcal{C}$ is a measurable VC class of sets and Corollary 5.2.3 immediately gives

$$\|D_n(\theta) - P^3 C_\theta\|_\infty \to 0 \quad \text{a.s.},$$

the uniform strong law of large numbers for the simplicial depth process, with no conditions on the probability law P of the data.

As another application we obtain for U-processes the analog of another well-known LLN for empirical processes, the Blum–DeHardt law of large numbers. It is in terms of the *bracketing numbers* $N_{[\,]}^{(1)}(\mathcal{H}, P^m, \varepsilon)$, $\varepsilon > 0$. The bracketing number $N_{[\,]}^{(1)}(\mathcal{H}, P^m, \varepsilon)$ is defined as the minimum number r such that there exist functions $f_1, \ldots, f_r$ and $\Delta_1, \ldots, \Delta_r$, all in $L_1(P^m)$ but not necessarily in $\mathcal{H}$, with $\Delta_i \geq 0$ and $P^m \Delta_i < \varepsilon$, such that, for all $h \in \mathcal{H}$, there exists $i \leq r$ with $|f_i - h| < \Delta_i$. The set of fs such that $|f_i - f| < \Delta_i$ is the *ith bracket A_i*.

COROLLARY 5.2.5. *Let $\mathcal{H}$ be a measurable class of functions on S^m such that $N_{[\,]}^{(1)}(\mathcal{H}, P^m, \varepsilon) < \infty$ for all $\varepsilon > 0$. Then $\|U_n(h) - P^m h\|_{\mathcal{H}} \to 0$ a.s.*

PROOF (SKETCH). Given $\varepsilon > 0$, let $\{f_j\}$ and $\{\Delta_j\}$ be a set of functions as in the definition of $N_{[\,]}^{(1)}(\mathcal{H}, P^m, \varepsilon)$. Then, since $P^m \Delta_j < \varepsilon$, the law of large numbers for U-statistics implies that, for f in the bracket A_j, $1 \leq j \leq N$,

$$\frac{(n-m)!}{n!} \sum_{I_n^m} |f - f_j|(X_{i_1}, \ldots, X_{i_m}) \leq \frac{(n-m)!}{n!} \sum_{I_n^m} \Delta_j(X_{i_1}, \ldots, X_{i_m}) \leq \varepsilon$$

for all $n > n_0(\omega)$, for some $n_0(\omega) < \infty$ a.s. Hence, for each $n > n_0(\omega)$, $N(\mathcal{H}, e_{n,1}, \varepsilon) \leq N_{[\,]}^{(1)}(\mathcal{H}, P^m, \varepsilon)$. Since, as observed above, $\tilde{e}_{n,1}(s_m f, s_m g) \leq e_{n,1}(f, g)$, the result follows directly from Theorem 5.2.2. [Here we are being slopy about measurablity by considering N instead of N^*: see the proof of Theorem 5.2.6 below for a careful treatment of an identical situation.] $\square$

More interesting perhaps is the following *law of large numbers for B-valued U-statistics*, B a separable Banach space, which is the analog of the Maurier law of large numbers for sums of i.i.d. B-valued random variables (e.g. Ledoux and Talagrand, 1991, Corollary 7.10). It is also a direct corollary of Theorem 5.2.2.

THEOREM 5.2.6. *Let B be a separable Banach space and let $H : S^m \mapsto B$ be a measurable B-valued function of m variables such that $P^m\|H\| < \infty$. Then,*

$$\|U_n(H) - P^m H\| \to 0 \quad \text{a.s.}$$

PROOF. Let D be a countable subset of the unit ball of the dual of B such that $\|x\| = \sup_{f\in D} f(x)$ for all $x \in B$, and let $\mathcal{H} := \{f \circ H : f \in D\}$. Obviously, proving Theorem 5.2.6 is equivalent to showing that the U-processes $\{U_n(h) : h \in \mathcal{H}\}$ satisfy the law of large numbers. $\mathcal{H}$ being a countable class, it is measurable, and by Theorem 5.2.2 and a previous observation it therefore suffices to show that

$$\frac{1}{n}\log N(\mathcal{H}, e_{n,1}, \varepsilon) \to 0 \quad \text{in} \quad \text{pr}^*. \tag{5.2.15}$$

Since $P^m H < \infty$, the map $\mu : A \to \mathbb{E}\big(\|H\| I_A(H)\big)$, $A \in \mathcal{B}$, is a finite Borel measure on B, hence it is a tight Borel measure by separability of B. Therefore, given $\varepsilon > 0$ there exists a compact set $K \subset B$ such that $\mu\big(K^c\big) < \varepsilon/3$, that is, $\mathbb{E}\big(\|H\| I_{K^c}(H)\big) < \varepsilon/3$. Let $\mathcal{F} = \{f|_K : f \in D\}$. Then, $\mathcal{F}$ is an equibounded and equicontinuous set of functions on the compact set K, thus it is relatively compact for the sup norm over K, that we will denote by $\|\cdot\|_K$. Let $f_1, \ldots, f_r$ then be a finite subset of D $(\varepsilon/3)$-dense in $\mathcal{F}$ for the sup norm over K. For $f \in D$ we obviously have

$$\begin{aligned} e_{n,1}(f \circ H, f_i \circ H) &= U_n(|(f - f_i) \circ H|) \\ &\le U_n(|(f - f_i) \circ H| I_{K^c}(H)) + \|f - f_i\|_K. \end{aligned}$$

Hence, if $\|f - f_i\|_K < \varepsilon/3$, then

$$e_{n,1}(f \circ H, f_i \circ H) \le 2U_n\big(\|H\| I_{K^c}(H)\big) + \frac{\varepsilon}{3}.$$

This shows

$$N\Big(\mathcal{H}, e_{n,1}, 2U_n(\|H\| I_{K^c}(H)) + \frac{\varepsilon}{3}\Big) \le r.$$

Since the counting numbers $N(\varepsilon)$ are nonincreasing in ε, it follows that

$$\begin{aligned} \Pr{}^*\big\{N(\mathcal{H}, e_{n,1}, \varepsilon) > r\big\} &\le \Pr{}^*\Big\{N(\mathcal{H}, e_{n,1}, \varepsilon) > r,\, U_n(\|H\| I_{K^c}(H)) \le \frac{\varepsilon}{3}\Big\} \\ &\quad + \Pr\Big\{U_n(\|H\| I_{K^c}(H)) > \frac{\varepsilon}{3}\Big\} \\ &= \Pr\Big\{U_n(\|H\| I_{K^c}(H)) > \frac{\varepsilon}{3}\Big\}. \end{aligned}$$

But the law of large numbers for U-statistics implies

$$U_n(\|H\|I_{K^c}(H)) \to_{\text{a.s.}} \mathbb{E}\big(\|H\|I_{K^c}(H)\big) < \varepsilon/3,$$

so that $\Pr\big\{U_n(\|H\|I_{K^c}(H)) > \varepsilon/3\big\} \to 0$. We then conclude

$$\Pr^*\big\{N(\mathcal{H}, e_{n,1}, \varepsilon) > r\big\} \to 0,$$

which is a stronger statement than the limit (5.2.15). □

REMARK 5.2.7. *Decoupled U-statistics.* We should remark that the laws of large numbers just obtained hold true as well for decoupled U-statistics and processes. This is due to the fact that the a.s. limits in these theorems are all equivalent to limits in L_1, and these can be decoupled (Section 4.5). Not even decoupling needs to be invoked for this because the proofs work exactly in the same way for regular and for decoupled U-statistics and processes; in particular we can add to Theorem 5.2.2 several extra equivalent conditions, such as 1) $\|U_n^{\text{dec}} - P^m\|_{\mathcal{H}} \to 0$ a.s., where $U_n^{\text{dec}}(h) := \big((n-m)!/n!\big)\sum_{I_n^m} h(X_{i_1}^1, \ldots, X_{i_m}^m)$, and 2) the decoupled versions of conditions ii)–iv). Another variation: letting $e_{n,1}^{\text{dec}}(f, g) = U_n^{\text{dec}}(|f-g|)$, which is the decoupled version of the distance defined by (5.2.6), a sufficient condition for the law of large numbers for a U-process over a measurable class $\mathcal{H}$ with an integrable envelope, as well as for its decoupled version, is that $n^{-1}\log N(\mathcal{H}_M, e_{n,1}^{\text{dec}}, \varepsilon) \to 0$ in pr* for all $\varepsilon > 0$ and $M < \infty$. This follows from the decoupled version of Theorem 5.2.2 because $e_{n,1}^{\text{dec}}$ dominates $\tilde{e}_{n,1}^{\text{dec}}$. We will use this criterion in Section 5.4.

A V-statistic or a *von Mises* statistic is a U-statistic plus the "diagonal" terms, more precisely,

$$V_n(h) := P_n^m h = \frac{1}{n^m} \sum_{1 \le i_1, \ldots, i_m \le n} h(X_{i_1}, \ldots, X_{i_m}). \tag{5.2.16}$$

These statistics are important because they appear in the Taylor–von Mises development of smooth statistics. They can offen be reduced to the corresponding U-statistic, and we will see next an instance regarding the law of large numbers. The following very simple proposition, a Marcinkiewicz-type law of large numbers for U-statistics, is relevant for this reduction.

PROPOSITION 5.2.8. *If $\mathbb{E}|h|^p < \infty$ with $0 < p < 1$, then*

$$\frac{1}{n^{m/p}} \sum_{1 \le i_1 < \cdots < i_m \le n} \big|h(X_{i_1}, \ldots, X_{i_m})\big| \to 0 \quad \text{a.s.}$$

PROOF. We assume $h \ge 0$ and set $s := m/p > m$. By Kronecker's lemma, it suffices to show that

$$\sum_{j=m}^{\infty} \frac{1}{j^s} \sum_{1 \le i_1 < \cdots < i_{m-1} < j} h(X_{i_1}, \ldots, X_{i_{m-1}}, X_j) < \infty \quad \text{a.s.}$$

Since

$$\sum_{j=m}^{\infty}(j-1)^{m-1}\Pr\{h > j^s\} \le \int_{m-1}^{\infty} x^{m-1}\Pr\{h > x^s\}dx \le m^{-1}\mathbb{E}h^{m/s} < \infty,$$

we can truncate at the level j^s and therefore it suffices to show

$$\sum_{j=m}^{\infty}\frac{1}{j^s}\sum_{1\le i_1<\cdots<i_{m-1}<j}(hI_{h\le j^s})(X_{i_1},\ldots,X_{i_{m-1}},X_j) < \infty \quad \text{a.s.}$$

But this follows by B. Levi's lemma and the following estimate of the series of expected values:

$$\begin{aligned}
\sum_{j=m}^{\infty}\frac{1}{(j-1)^{s-m+1}}\mathbb{E}hI_{h\le j^s} &\le \int_{m-1}^{\infty} x^{m-s-1}\mathbb{E}hI_{h\le x^s}dx \\
&\le \int_{m-1}^{\infty}\int_0^{x^s} x^{m-s-1}\Pr\{h > t\}\,dtdx \\
&= \frac{(m-1)^{m-s}}{s-m}\int_0^{(m-1)^s}\Pr\{h > t\}\,dt \\
&\quad + \int_{(m-1)^s}^{\infty}\left(\int_{t^{1/s}}^{\infty} x^{m-s-1}dx\right)\Pr\{h > t\}\,dt \\
&\le \frac{(m-1)^m}{s-m} + \frac{1}{s-m}\int_{(m-1)^s}^{\infty} t^{\frac{m}{s}-1}\Pr\{h > t\}dt < \infty.
\end{aligned}$$

□

There are better laws of large numbers for completely as well as partially degenerate U-statistics (Giné and Zinn, 1992; Teicher, 1997; and, particularly, Zhang, 1997), but they are not relevant for our purposes here.

Here is a useful law of large numbers for V-processes. We denote $\#\{i_1,\ldots,i_m\}$ the number of distinct subindices among $i_1,\ldots,i_m$ (e.g. $\#\{2,2\}=1$).

THEOREM 5.2.9. *If $\mathcal{H}$ is measurable class of real-valued functions of m variables with envelope H satisfying the integrability conditions*

$$E|H(X_{i_1},\ldots,X_{i_m})|^{\#\{i_1,\ldots,i_m\}/m} < \infty$$

for all $1\le i_1,\ldots,i_m\le m$, then the V-processes $\{V_n^n(h): h\in\mathcal{H}\}$ satisfy the strong law of large numbers, that is

$$\|V_n(h) - P^m h\|_{\mathcal{H}} \to 0 \quad \text{a.s.},$$

if and only if the U-processes $\{U_n(h): h\in\mathcal{H}\}$ do.

PROOF. For notational convenience we only consider $m = 2$: the general case is similar. In this case,

$$\Big\|\frac{1}{n^2}\sum_{i,j=1}^{n}(h(X_i, X_j) - P^2h) - \frac{n-1}{n}(U_n(h) - P^2h)\Big\|_{\mathcal{H}}$$

$$\leq \frac{1}{n^2}\sum_{i=1}^{n} H(X_i, X_i) + \frac{1}{n}\|P^2h\|_{\mathcal{H}}.$$

But by the Marcinkiewicz law of large numbers (Proposition 5.2.8 above with $m = 1$ and $p = 1/2$, as $\mathbb{E}h^{1/2} < \infty$) $\sum_{i=1}^{n} H(X_i, X_i)/n^2 \to 0$ a.s. (Note that the proof for $m > 2$ requires use of Proposition 5.2.8 for U-statistics of order higher than 1.) □

5.3 The Central Limit Theorem for U-Processes

A complete result such as Theorem 5.2.2 is not available for the central limit theorem. There are however useful results, particularly for VC-subgraph and related classes of kernels. Here, as in the case of the law of large numbers, randomization by Rademacher variables (Section 3.5) is basic. Two main results on the CLT will be presented, one for general U-processes and one for canonical U-processes over VC-subgraph classes of kernels. Banach space valued kernels will also be briefly considered. The section ends with an exponential bound for degenerate U-processes.

A class of measurable square P-integrable functions $\mathcal{F}$ on a probability space $(S, \mathcal{S}, P)$ is *P-Donsker* if the centered Gaussian process G_P indexed by $\mathcal{F}$ and with covariance

$$\mathbb{E}\big[G_P(f)G_P(g)\big] = P(fg) - (Pf)(Pg), \quad f, g \in \mathcal{F},$$

defines a tight Borel probability law in $\ell^\infty(\mathcal{F})$ and the empirical process indexed by $\mathcal{F}$ and constructed from the coordinate functions X_i of $S^{\mathbb{N}}$,

$$\nu_n(f) := n^{1/2}(P_n f - Pf) := \frac{1}{\sqrt{n}}\sum_{i=1}^{n}(f(X_i) - Pf), \quad f \in \mathcal{F},$$

converges in law in $\ell^\infty(\mathcal{F})$ to G_P. [See (5.1.2) for the definition of convergence in law in ℓ^∞.] That is, $\mathcal{F}$ is P-Donsker iff the empirical process based on P and indexed by $\mathcal{F}$ satisfies the CLT, and we alternatively write $\mathcal{F} \in CLT(P)$. There is a vast literature about conditions on $\mathcal{F}$ and P for $\mathcal{F}$ to be P-Donsker [to name just a few references, Dudley (1984), Pollard (1984), Giné and Zinn (1984, 1986), Alexander (1987), Ossiander (1987) and van der Vaart and Wellner (1996)]. In particular, measurable VC-subgraph classes of functions are P-Donsker for all P (this result, due to Dudley and Pollard, is contained in most of the above references). The second main result in this section will extend this theorem to U-processes.

The first result consists of a reduction of the $\sqrt{n}$-CLT for not necessarily degenerate U-processes to the CLT for empirical processes.

THEOREM 5.3.1. *Let $\mathcal{H}$ be a measurable class of real symmetric square integrable functions of m variables on $(S^m, \mathcal{S}^m, P^m)$. If the class of functions $P^{m-1}\mathcal{H} := \{P^{m-1}h : h \in H\}$ is P-Donsker and if*

$$\lim_{n\to\infty} \mathbb{E}\Big\| \frac{1}{n^{m-1/2}} \sum_{I_n^m} \varepsilon_{i_1}\varepsilon_{i_2} h(X_{i_1}, \ldots, X_{i_m}) \Big\|_{\mathcal{H}} = 0, \tag{5.3.1}$$

then the U-processes based on P and indexed by $\mathcal{H}$ satisfy the central limit theorem in the sense that

$$\big\{n^{1/2}\big(U_n(h) - P^m h\big) : h \in \mathcal{H}\big\} \to_{\mathcal{L}} \{mG_P(P^{m-1}h) : h \in \mathcal{H}\} \quad \text{in } \ell^\infty(\mathcal{H}). \tag{5.3.2}$$

The converse of this theorem is also true assuming $\mathcal{H}$ admits a measurable envelope such that $t^2 \Pr\{H > t\} \to 0$ as $t \to \infty$ (Arcones and Giné, 1993), but this will not be proved.

PROOF. For $h \in \mathcal{H}$, we define

$$\tilde{h}(x_1, \ldots, x_m) := h(x_1, \ldots, x_m) - P^m h - \sum_{i=1}^m (\pi_1 h)(x_i), \qquad x_i \in S,$$

and recall from Section 3.5 that $\pi_1 h = P^{m-1}h - P^m h$. Then, $\tilde{h}$ is centered and P-degenerate of order 1 since

$$\int \tilde{h}(x_1, \ldots, x_m) dP(x_1) \cdots dP(x_{i-1}) dP(x_{i+1}) \cdots dP(x_m) = 0$$

for $i = 1, \ldots, m$. Therefore, by the randomization theorem (Theorem 3.5.3) with $r = 2$,

$$n^{-m+1/2}\mathbb{E}\Big\| \sum_{I_n^m} \tilde{h}(X_{i_1}, \ldots, X_{i_m}) \Big\|_{\mathcal{H}} \simeq n^{-m+1/2}\mathbb{E}\Big\| \sum_{I_n^m} \varepsilon_{i_1}\varepsilon_{i_2}\tilde{h}(X_{i_1}, \ldots, X_{i_m}) \Big\|_{\mathcal{H}}.$$

Jensen's inequality implies that, for $k \le m$,

$$\mathbb{E}\Big\| \sum_{I_n^m} \varepsilon_{i_1}\varepsilon_{i_2} P^{m-k} h(X_{i_1}, \ldots, X_{i_k}) \Big\|_{\mathcal{H}} \le \mathbb{E}\Big\| \sum_{I_n^m} \varepsilon_{i_1}\varepsilon_{i_2} h(X_{i_1}, \ldots, X_{i_m}) \Big\|_{\mathcal{H}}$$

and this, applied with $k = 0, 1$, together with the previous inequality and hypothesis (5.3.1), gives

$$\lim_{n\to\infty} \mathbb{E}\Big\| \frac{1}{n^{m-1/2}} \sum_{I_n^m} \tilde{h}(X_{i_1}, \ldots, X_{i_m}) \Big\|_{\mathcal{H}} = 0, \tag{5.3.3}$$

that is, $n^{1/2}\mathbb{E}\|U_n(\tilde{h})\|_{\mathcal{H}} \to 0$. Now, the result follows from the assumption that $P^{m-1}\mathcal{H}$ is P-Donsker because

$$n^{1/2}(U_n(h) - P^m h) = mn^{1/2}\frac{1}{n}\sum_{i=1}^{n}\big((P^{m-1}h)(X_i) - P^m h\big) + n^{1/2}U_n(\tilde{h})$$

so that

$$\Big\|n^{1/2}(U_n(h) - P^m h) - \frac{m}{\sqrt{n}}\sum_{i=1}^{n}\big((P^{m-1}h)(X_i) - P^m h\big)\Big\|_{\mathcal{H}} \to 0 \quad \text{in pr}$$

by (5.3.3). (Regarding measurability, recall from Section 3.5 that if $\mathcal{H}$ is measurable, so is $P^{m-1}\mathcal{H}$.) □

If we estimate the expected value in (5.3.1) by means of entropies with respect to the random distances $e_{n,1}$ introduced in the previous section, we will obtain a result in the same vein as Theorem 5.2.2. Here is the required maximal inequality: Corollary 3.2.6 gives the existence of a universal constant $\lambda \in (0, \infty)$ such that, if $\xi_i, i \le N < \infty$, are (not necessarily independent) quadratic forms in (independent) Rademacher variables, then

$$\mathbb{E}\exp\Big(\frac{|\xi_i|}{\lambda\|\xi_i\|_2}\Big) \le 2$$

and therefore, the maximal inequality (4.3.1) becomes

$$\mathbb{E}\max_{i\le N}|\xi_i| \le \lambda\big(\log 2N\big)\max_{i\le N}\big(\mathbb{E}\xi_i^2\big)^{1/2}, \tag{5.3.4}$$

which is what we use in the proof that follows.

COROLLARY 5.3.2. *Let $\mathcal{H}$ be a measurable class of real symmetric functions of m variables on S^m with an envelope H square integrable for P^m, P a probability measure on $(S, \mathcal{S})$. If the class of functions $P^{m-1}\mathcal{H} := \{P^{m-1}h : h \in H\}$ is P-Donsker and if*

$$\lim_{n\to\infty}\mathbb{E}^*\Big[\frac{1}{n^{1/2}}\log N\big(\mathcal{H}, e_{n,1}, \delta/n^{1/2}\big)\Big]^2 = 0 \tag{5.3.5}$$

for all $\delta > 0$, then

$$n^{1/2}(U_n - P^m) \to_{\mathcal{L}} mG_P \circ P^{m-1} \quad \text{in } \ell^\infty(\mathcal{H}).$$

PROOF. For each $n \ge m$, let $\mathcal{H}_n$ be a minimal $(\delta/n^{1/2})$-dense set of $(\mathcal{H}, e_{n,1})$. Then, inequality (5.3.4) above gives

$$E\Big\|\frac{n^{1/2}(n-m)!}{n!}\sum_{I_n^m}\varepsilon_{i_1}\varepsilon_{i_2}h(X_{i_1},\dots,X_{i_m})\Big\|_{\mathcal{H}}$$
$$\le \delta + \mathbb{E}\Big\|\frac{n^{1/2}(n-m)!}{n!}\sum_{I_n^m}\varepsilon_{i_1}\varepsilon_{i_2}h(X_{i_1},\dots,X_{i_m})\Big\|_{\mathcal{H}_n}$$

$$\le \delta + \lambda \mathbb{E}^*\Bigg[\Big(\log 2 + \log N\big(\mathcal{H}, e_{n,1}, \delta n^{-1/2}\big)\Big) \times \max_{h\in\mathcal{H}_n}\Big(E_\varepsilon\Big|\frac{n^{1/2}(n-m)!}{n!}\sum_{I_n^m}\varepsilon_{i_1}\varepsilon_{i_2}h(X_{i_1},\dots,X_{i_m})\Big|^2\Big)^{1/2}\Bigg] \tag{5.3.6}$$

for all $n \ge m$. Since

$$\begin{aligned}\mathbb{E}\Big(\max_{h\in\mathcal{H}_n} E_\varepsilon\Big|\sum_{I_n^m}\varepsilon_{i_1}\varepsilon_{i_2}h(X_{i_1},\dots,X_{i_m})\Big|^2\Big) &\le \mathbb{E}\Bigg[4\sum_{1\le i_1\le i_2\le n}\Big(\sum_{(i_3,\dots,i_m):\mathbf{i}\in I_n^m} H(X_{i_1},\dots,X_{i_m})\Big)^2\Bigg]\\ &= 4n(n-1)\Big(\frac{n!}{(n-m+2)!}\Big)^2\mathbb{E}H^2(X_1,\dots,X_m)\\ &\simeq 4n^{2(m-1)}\mathbb{E}H^2(X_1,\dots,X_m),\end{aligned}$$

it follows from (5.3.6), by Holder, that

$$\begin{aligned}&E\Big\|\frac{n^{1/2}(n-m)!}{n!}\sum_{I_n^m}\varepsilon_{i_1}\varepsilon_{i_2}h(X_{i_1},\dots,X_{i_m})\Big\|_{\mathcal{H}}\\ &\le \delta + \lambda c n^{-1/2}\Big[\mathbb{E}^*\big(\log 2 + \log N\big(\mathcal{H}, e_{n,1}, \delta n^{-1/2}\big)\big)^2\Big]^{1/2}\big[\mathbb{E}H^2(X_1,\dots,X_m)\big]^{1/2}\end{aligned}$$

for some $c < \infty$ and all $n \ge m$. So,

$$\lim_{n\to\infty}\mathbb{E}\Big\|\frac{n^{1/2}(n-m)!}{n!}\sum_{I_n^m}\varepsilon_{i_1}\varepsilon_{i_2}h(X_{i_1},\dots,X_{i_m})\Big\|_{\mathcal{H}} = 0$$

by hypothesis (5.3.5). Now, the result follows directly from Theorem 5.3.1. □

THEOREM 5.3.3. *Let $\mathcal{H}$ be a measurable VC-subgraph class of real symmetric functions on S^m with an envelope H square integrable for P^m, P a probability measure on $(S, \mathcal{S})$. Then, the central limit theorem in (5.3.2) holds for $\mathcal{H}$ and P.*

PROOF. By Theorem 5.1.15 there exist $A, v < \infty$ such that

$$N(\mathcal{H}, e_{n,1}, \varepsilon) \le A\Big(\frac{U_n(2H)}{\varepsilon}\Big)^v$$

for all $0 < \varepsilon \le U_n(2H)$ [and, by definition, $N(\mathcal{H}, e_{n,1}, \varepsilon) = 1$ for $\varepsilon \ge U_n(2H)$]. Then,

$$\frac{1}{n^{1/2}}\log N\big(\mathcal{H}, e_{n,1}, \delta/n^{1/2}\big) \le \frac{\log A}{n^{1/2}} + \frac{v\log\big(1\vee U_n(2H)\big)}{n^{1/2}} + \frac{v\log(n^{1/2}/\delta)}{n^{1/2}},$$

which tends to zero in $L_2(P^{\mathbb{N}})$ because, since $U_n(2H) \to 2P^mH$ a.s. and in L_1 (LLN: Theorem 4.1.4), we have, by uniform integrability, that

$$\mathbb{E}\big(\log 1 \vee U_n(2H)\big)^2 \to \big(\log 1 \vee P^m(2H)\big)^2.$$

Condition (5.3.5) is therefore satisfied by $\mathcal{H}$ and P. We postpone proving that the class $P^{m-1}\mathcal{H}$ is P-Donsker to the proof of the next theorem. Then, modulo this postponement, the conclusion follows from Corollary 5.3.2. □

Next we turn to degenerate U-processes, but will only consider VC-subgraph classes. The proof of the theorem will require two easy lemmas on metric entropy.

LEMMA 5.3.4. *Let $\mathcal{H}$ be a set and let $d, d_1, \ldots, d_k$ be pseudodistances on $\mathcal{H}$. If $d^2 \leq \sum_{i=1}^k c_i^2 d_i^2$ (with $0 < c_i < \infty$), then*

$$N(\mathcal{H}, d, \tau) \leq \prod_{i=1}^{k} N\Big(\mathcal{H}, d_i, \frac{\tau}{2k^{1/2}c_i}\Big)$$

for all $\tau > 0$.

PROOF. Taking $k = 2$ does not change the proof. So, we assume $d^2 \leq c_1^2 d_1^2 + c_2^2 d_2^2$. For $i = 1, 2$, let $N_i := N\big(\mathcal{H}, d_i, \tau/(2\sqrt{2}c_i)\big)$ and let $\{f_{\ell,i}\}_{\ell=1}^{N_i}$ be the centers of minimal coverings of $\mathcal{H}$ by d_i-balls of radius at most $\tau/(2\sqrt{2}c_i)$. Then, a routine verification shows that any set of elements of $\mathcal{H}$, $h_{i,j}$, $0 < i \leq N_1$, $0 < j \leq N_2$, satisfying

$$h_{ij} \in B_{d_1}\big(f_{i,1}, \tau/(2\sqrt{2}c_1)\big) \cap B_{d_2}\big(f_{j,2}, \tau/(2\sqrt{2}c_2)\big)$$

are the centers of a covering of $\mathcal{H}$ by d-balls of radius at most τ. (Here, $B_{d_i}(f, \varepsilon)$ stands for the d_i ball of center f and radius ε.) □

For the next lemma, we should recall the definition of the Hoeffding projections $\pi_{k,m}^P$ from Chapter 3, Section 2. Since m and P are fixed, we write π_k for $\pi_{k,m}^P$, and they are defined as

$$\pi_k h(x_1, \ldots, x_k) = \big(\delta_{x_1} - P\big) \times \cdots \times \big(\delta_{x_k} - P\big) \times P^{m-k}h.$$

The functions $\pi_k h$ are P-canonical: their P-integral with respect to any one of the variables is zero.

LEMMA 5.3.5. *Let $\mathcal{H}$ be a class of real symmetric measurable functions on S^m and, as usual, let P be a probability measure on $(S, \mathcal{S})$ and X_i the coordinates of $S^{\mathbb{N}}$. For $k = 1, \ldots, m$ and for $n \geq m$, we define pseudodistances $e_{n,k,2}$ on $\mathcal{H}$, depending on P and on $\{X_i\}$, as follows:*

$$e_{n,k,2}^2(f, g) := \frac{n^k}{\binom{n}{k}} U_n^{(k)}\big((\pi_k(f-g))^2\big).$$

Then, there exist positive finite constants $c_{k,r}$ independent of P and $\{X_i\}$ such that, for all $\tau > 0$,

$$N\big(\mathcal{H}, e_{n,k,2}, \tau\big) \leq \prod_{r=0}^{k} N\Big(\mathcal{H}, \|\cdot\|_{L_2(U_n^{(r)} \times P^{m-r})}, \frac{\tau}{2(k+1)^{1/2}c_{k,r}}\Big),$$

where, for $r \neq 0$, $U_n^{(r)} \times P^{m-r}$ denotes the random probability measure

$$U_n^{(r)} \times P^{m-r} = \Big(\frac{(n-r)!}{n!} \sum_{I_n^r} \delta_{(X_{i_1},\dots,X_{i_r})}\Big) \times P^{m-r}$$

defined on $(S^m, \mathcal{S}^m)$ and, for $r = 0$, $L_2(U_n^{(0)} \times P^m)$ just means $|P^m|$ (with some abuse of notation).

PROOF. Developing $\pi_k(f-g)$ gives that, for suitable constants $0 < d_{r,k} < \infty$,

$$\begin{aligned}\big[\pi_k(f-g)\big]^2 &= \Big[\sum_{r=0}^{k} \frac{1}{r!} \sum_{I_k^r} (-1)^{k-r} P^{m-r}(f-g)(x_{i_1},\dots,x_{i_r})\Big]^2 \\ &\leq \sum_{r=0}^{k} \sum_{I_k^r} d_{k,r}\big[P^{m-r}(f-g)(x_{i_1},\dots,x_{i_r})\big]^2.\end{aligned}$$

Then, applying the operator $U_n^{(k)}$, which is monotone, we obtain that, for suitable constants $0 < c_{k,r}^2 < \infty$,

$$\begin{aligned}e_{n,k,2}^2(f,g) &\leq \sum_{r=0}^{k} c_{k,r}^2 U_n^{(r)}\big([P^{m-r}(f-g)]^2\big) \\ &\leq c_{k,1}^2 |P^m(f-g)|^2 + \sum_{r=1}^{k} c_{k,r}^2 U_n^{(r)}\big([P^{m-r}(f-g)^2]\big),\end{aligned}$$

and Lemma 5.3.4 gives the result. □

The CLT for canonical U-processes over VC-classes will be a consequence of the randomization inequality, the entropy maximal inequality for Rademacher chaos variables (inequalities (5.1.22) and (5.1.23)), the good entropy properties of VC-classes (Theorem 5.1.15) and the CLT and LLN for U-statistics, all combined in a proof whose pattern comes from empirical process theory and is by now almost classical. The previous lemma will be of technical help. Before proving the CLT we will isolate as another lemma a simple application of the LLN for U-processes which will allow us to interchange deterministic and random distances in the asymptotic equicontinuity condition that we must prove in the next theorem.

LEMMA 5.3.6. *Let $\mathcal{H}$ be a measurable VC-subgraph collection of symmetric functions on S^m with a P^m-square integrable envelope H. Then,*

$$\sup_{(f,g)\in\mathcal{H}\times\mathcal{H}} \Big| U_n^{(k)}\big([\pi_k(f-g)]^2\big) - P^k[\pi_k(f-g)]^2 \Big| \to 0 \quad \text{a.s.} \tag{5.3.7}$$

and

$$\sup_{(f,g)\in\mathcal{H}\times\mathcal{H}} \Big| U_n\big((f-g)^2\big) - P^m(f-g)^2 \Big| \to 0 \quad \text{a.s.} \tag{5.3.8}$$

as well as

$$\sup_{(f,g)\in\mathcal{H}\times\mathcal{H}}\left|U_n^{\mathrm{dec}}\big((f-g)^2\big)-P^m(f-g)^2\right|\to 0 \quad \text{a.s.} \tag{5.3.8$'$}$$

PROOF. The proofs of (5.3.7), (5.3.8) and (5.3.8′) are very similar and the proof of (5.3.7), which requires computations similar to those in the previous lemma, is more cumbersome; so we only prove (5.3.8) [see Remark 5.2.7 on decoupled U-statistics for (5.3.8′)]. The class $\mathcal{H}$ being measurable, it follows from the definition that the class $\mathcal{G} := \{(f-g)^2 : f, g \in \mathcal{H}\}$ is also measurable. Moreover, $\mathcal{G}$ admits the integrable envelope $G := 4H^2$. Since $\tilde{e}_{n,1}$ is dominated by the pseudodistance $e_{n,1}$ [see (5.2.5) and (5.2.6) for the definitions], the law of large numbers [the sufficiency of condition iv) in Theorem 5.2.2] will give (5.3.8) if we prove

$$\lim_{n\to\infty}\frac{\log N(\mathcal{G}_M, e_{n,1}, \varepsilon)}{n}\to 0 \quad \text{in} \quad \mathrm{pr}^* \tag{5.3.9}$$

for all $\varepsilon > 0$ and $M < \infty$. This we will do by comparing the covering numbers of $\mathcal{G}_M$ and $\mathcal{H}$. We obviously have, for $f, g, \bar{f}, \bar{g}$ in $\mathcal{H}$,

$$\begin{aligned}
e_{n,1}\Big((f-g)^2 I_{G\le M}, (\bar{f}-\bar{g})^2 I_{G\le M}\Big) &= U_n^{(k)}\Big(\big|(f-g)^2 I_{G\le M}-(\bar{f}-\bar{g})^2 I_{G\le M}\big|\Big)\\
&\le 4MU_n^{(k)}\Big(|f-g-\bar{f}+\bar{g}|\Big)\\
&\le 4M\Big[e_{n,1}(f,\bar{f})+e_{n,1}(g,\bar{g})\Big].
\end{aligned}$$

So, if $\{f_i\}$ are the centers of a covering of $\mathcal{H}$ by $e_{n,1}$ balls of radius at most $\varepsilon/8M$, then the $e_{n,1}$ balls of radius ε and centers in the set $\{(f_i - f_j)^2 I_{G\le M}\}$ cover $\mathcal{G}_M$. This implies that

$$N(\mathcal{G}_M, e_{n,1}, \varepsilon) \le N^2(\mathcal{H}, e_{n,1}, \varepsilon/8M)$$

and the limit (5.3.9) follows from (5.2.13) (that is, from Theorem 5.1.15). □

The following theorem, giving the CLT for each term in the Hoeffding expansion of a U-process indexed by a VC class, is the main result of this section.

THEOREM 5.3.7. *If $\mathcal{H}$ is a measurable VC-subgraph class of real symmetric functions on S^m with a P^m-square integrable envelope H, where P is a probability measure on $(S, \mathcal{S})$, then the classes of functions*

$$\pi_{k,m}^P\mathcal{H} = \pi_k(\mathcal{H}) := \{\pi_k h : h \in \mathcal{H}\}, \qquad k = 1, \ldots, m,$$

satisfy the central limit theorem, that is,

$$\Big\{n^{k/2}U_n^{(k)}(\pi_k h) : h\in\mathcal{H}\Big\} \to_{\mathcal{L}} \Big\{(k!)^{1/2}K_P(\pi_k h) : h \in \mathcal{H}\Big\} \quad \text{in } \ell^\infty(\mathcal{H}), \tag{5.3.10}$$

for $1 \le k \le m$, where K_P is the chaos process defined in Section 4.2. In particular, the restriction of K_P to $\pi_k\mathcal{H}$ is then sample continuous with respect to the L_2 distance for all $1 \le k \le m$.

PROOF. The central limit theorem for U-statistics, Theorem 4.2.4, gives convergence of the finite-dimensional distributions in (5.3.10). We fix $1 \le k \le m$ and define the pseudodistance

$$d^2(\pi_k f, \pi_k g) = P^m(f-g)^2, \qquad f, g \in \mathcal{H}.$$

Since $\mathcal{H}$ is VC-subgraph and $P^m H^2 < \infty$, Theorem 5.1.15 shows that $N(\mathcal{H}, d, \varepsilon)$ is finite for all $\varepsilon > 0$ and, therefore, the pseudometric space $(\pi_k \mathcal{H}, d)$ is totally bounded. These two observations together with Theorem 5.1.2 reduce the proof of the limit (5.3.10) to showing

$$\lim_{\delta \to 0} \limsup_{n \to \infty} \Pr\Big\{ \sup_{f,g \in \mathcal{H}: d(f,g) \le \delta} n^{k/2} \big| U_n^{(k)}\big(\pi_k(f-g)\big)\big| > \varepsilon \Big\} = 0 \tag{5.3.11}$$

for all $\varepsilon > 0$. In fact, we will prove a little more, namely

$$\lim_{\delta \to 0} \limsup_{n \to \infty} \mathbb{E}\Big[\sup_{f,g \in \mathcal{H}: d(f,g) \le \delta} n^{k/2} \big| U_n^{(k)}\big(\pi_k(f-g)\big)\big| \Big] = 0. \tag{5.3.11$'$}$$

Since $\pi_k h$ is a P-canonical kernel for all h, we can apply the randomization theorem (Theorem 3.5.3) and randomize $\pi_k f$ by a product of k independent Rademacher variables as follows:

$$\begin{aligned} &\mathbb{E} \sup_{f,g \in \mathcal{H}: d(f,g) \le \delta} n^{k/2} \big| U_n^{(k)}\big(\pi_k(f-g)\big)\big| \qquad (5.3.12) \\ &\le K \mathbb{E} \sup_{f,g \in \mathcal{H}: d(f,g) \le \delta} n^{k/2} \bigg| \binom{n}{k}^{-1} \sum_{1 \le i_1 < \cdots < i_k \le n} \varepsilon_{i_1} \cdots \varepsilon_{i_k} \pi_k(f-g)\big(X_{i_1}, \ldots, X_{i_k}\big) \bigg| \end{aligned}$$

where K is a constant independent of $\{X_i\}$. The L_2 distance associated to the conditional Rademacher chaos process in (5.3.12) is the pseudodistance $e_{n,k,2}$ already defined in Lemma 5.3.5,

$$e_{n,k,2}^2(f,g) = \frac{n^k}{\binom{n}{k}^2} \sum_{1 \le i_1 < \cdots < i_k \le n} \big[\pi_k(f-g)(X_{i_1}, \ldots, X_{i_k})\big]^2.$$

For $n \ge 2k$, let

$$D_{n,k}(\delta) := \Big\{ \Big\| e_{n,k,2}^2(f,g) - \frac{n^k}{\binom{n}{k}} P^m[\pi_k(f-g)]^2 \Big\|_{\mathcal{H} \times \mathcal{H}} > 2^k k! \delta^2 \Big\}.$$

The law of large numbers in Lemma 5.3.6 implies

$$\Pr\big(D_{n,k}(\delta)\big) \to 0 \tag{5.3.13}$$

as $n \to \infty$ for all $\delta > 0$. Since π_k is a projection in L_2, $\{f, g \in \mathcal{H} : d(f,g) < \delta\} \subseteq \{f, g \in \mathcal{H} : P^m[\pi_k(f-g)]^2 < \delta^2\}$ and therefore,

$$\{f, g \in \mathcal{H} : d(f,g) < \delta\} \subseteq \{f, g \in \mathcal{H} : e_{n,k,2}^2(f,g) < 2^{k+1} k! \delta^2\}$$

on the set $D_{n,k}(\delta)^c$. The entropy maximal inequality for Rademacher chaos processes (Corollary 5.1.8) then gives

$$\begin{aligned}\mathbb{E}_\varepsilon \sup_{f,g\in\mathcal{H}:d(f,g)\le\delta} & n^{k/2}\left|\binom{n}{k}^{-1}\sum_{1\le i_1<\cdots<i_k\le n}\varepsilon_{i_1}\cdots\varepsilon_{i_k}\pi_k(f-g)\big(X_{i_1},\ldots,X_{i_k}\big)\right| \\ &\le K\int_0^{\sqrt{k!}2^k\delta}\Big(\log N\big(\mathcal{H},e_{n,k,2},\varepsilon\big)\Big)^{k/2}d\varepsilon \\ &\quad + K\left[\int_0^\infty\Big(\log N\big(\mathcal{H},e_{n,k,2},\varepsilon\big)\Big)^{k/2}d\varepsilon\right]I_{D_{n,k}(\delta)}. \qquad (5.3.14)\end{aligned}$$

By Lemma 5.3.5, the right-hand side of (5.3.14) is dominated by a constant times the sum from $r=0$ to $r=k$ of

$$\begin{aligned}&\int_0^{\delta'}\Big(\log N\big(\mathcal{H},\|\cdot\|_{L_2(U_n^{(r)}\times P^{m-r})},\varepsilon\big)\Big)^{k/2}d\varepsilon \\ &\qquad + \left[\int_0^\infty\Big(\log N\big(\mathcal{H},\|\cdot\|_{L_2(U_n^{(r)}\times P^{m-r})},\varepsilon\big)\Big)^{k/2}d\varepsilon\right]I_{D_{n,k}(\delta)} \\ &\qquad := I_r(n,\delta), \qquad (5.3.15)\end{aligned}$$

where δ' is a function of δ that tends to 0 as $\delta\to 0$. So, by (5.3.12), (5.3.14), and (5.3.15), the limit (5.3.11′), hence the theorem, will be proved if we show

$$\lim_{\delta\to 0}\limsup_{n\to\infty}\mathbb{E}^* I_r(n,\delta)=0. \qquad (5.3.16)$$

For $r=0$, $\|f-g\|_{L_2(U_n^{(0)}\times P^m)} := |P^m f - P^m g|$, as mentioned in Lemma 5.3.5, and therefore, $N\big(\mathcal{H},\|\cdot\|_{L_2(U_n^{(0)}\times P^m)},\varepsilon\big)\simeq 2P^mH/\varepsilon$ if $\varepsilon\le 2P^mH$ and equal to 1 otherwise. Hence

$$I_0(n,\delta)\le c(\delta)+CI_{D_{n,k}(\delta)}$$

for nonrandom $c(\delta)\to 0$ as $\delta\to 0$ and $C<\infty$. Taking expectations, we obtain (5.3.16) for $r=0$ because of the limit (5.3.13). For $1\le r\le k$ we use the bound (5.1.26) given in Theorem 5.1.15 for the packing numbers [hence, for the entropy numbers: see (5.1.10)] of VC-subgraph classes, which asserts the existence of positive finite constants A, v depending only on $\mathcal{H}$, such that

$$N\Big(\mathcal{H},\|\cdot\|_{L_2(U_n^{(r)}\times P^{m-r})},2\varepsilon\big[U_n^{(r)}(P^{m-r}H^2)\big]^{1/2}\Big)\le\Big(\frac{A}{\varepsilon}\Big)^v$$

for $0 < \varepsilon \le 1$; this covering number is clearly dominated by 1 for $\varepsilon > 1$. The function N being monotone decreasing, there is no loss of generality in assuming $H(x_1, \dots, x_m) \ge 1$ for all $x_i \in S$, and we do so. Then, for $1 \le r \le k$,

$$\int_0^\delta \Big(\log N\big(\mathcal{H}, \|\cdot\|_{L_2(U_n^{(r)} \times P^{m-r})}, \varepsilon\big)\Big)^{k/2} d\varepsilon$$
$$= \big[U_n^{(r)}(P^{m-r}H^2)\big]^{1/2}$$
$$\times \int_0^{\delta/[U_n^{(r)}(P^{m-r}H^2)]^{1/2}} \Big(\log N\big(\mathcal{H}, \|\cdot\|_{L_2(U_n^{(r)} \times P^{m-r})}, \varepsilon\big[U_n^{(r)}(P^{m-r}H^2)\big]^{1/2}\big)\Big)^{k/2} d\varepsilon$$
$$\le \big[U_n^{(r)}(P^{m-r}H^2)\big]^{1/2} \int_0^\delta \Big[\log\Big(\frac{2A}{\varepsilon}\Big)^v\Big]^{k/2} d\varepsilon$$
$$:= \big[U_n^{(r)}(P^{m-r}H^2)\big]^{1/2} c(\delta),$$

where $c(\delta)$ depends only on $\mathcal{H}$ and tends to zero as $\delta \to 0$. By the law of large numbers (Theorem 4.1.4) if $P^r|h| < \infty$, then $U_n^{(r)}(h)$ converges to $P^r h$ a.s. and in L_1; hence, $\mathbb{E}\big[U_n^{(r)}(P^{m-r}H^2)\big]^{1/2} \to \big(P^m H^2\big)^{1/2}$ and we have

$$\lim_{\delta\to 0} \limsup_{n\to\infty} \mathbb{E}^* \int_0^\delta \Big(\log N\big(\mathcal{H}, \|\cdot\|_{L_2(U_n^{(r)} \times P^{m-r})}, \varepsilon\big)\Big)^{k/2} d\varepsilon$$
$$\le \lim_{\delta\to 0} \limsup_{n\to\infty} c(\delta)\mathbb{E}\big[U_n^{(r)}(P^{m-r}H^2)\big]^{1/2} = \big(P^m H^2\big)^{1/2} \lim_{\delta\to 0} c(\delta) = 0.$$

The previous computations also show

$$\int_0^\infty \Big(\log N\big(\mathcal{H}, \|\cdot\|_{L_2(U_n^{(r)} \times P^{m-r})}, \varepsilon\big)\Big)^{k/2} d\varepsilon$$
$$\le \big[U_n^{(r)}(P^{m-r}H^2)\big]^{1/2} \int_0^\infty 0 \vee \Big[\log\Big(\frac{2A}{\varepsilon}\Big)^v\Big]^{k/2} d\varepsilon$$
$$= C\big[U_n^{(r)}(P^{m-r}H^2)\big]^{1/2}$$

for some finite constant C. The uniform integrability of the sequence $\big[U_n^{(r)}(P^{m-r}H^2)\big]^{1/2}$ and the limit (5.3.13) then give

$$\lim_{n\to\infty} \mathbb{E}^*\Big[\Big(\int_0^\infty \Big(\log N\big(\mathcal{H}, \|\cdot\|_{L_2(U_n^{(r)} \times P^{m-r})}, \varepsilon\big)\Big)^{k/2} d\varepsilon\Big) I_{D_{n,k}(\delta)}\Big] = 0$$

for all $\delta > 0$. The last two limits give (5.3.16) for $0 < r \le k$, concluding the proof of the theorem. □

It should be emphasized that the proof of the CLT in Theorem 5.3.7 consists of nothing but gaining control of the size of the oscillations of the processes involved, that is, it consists in proving the limits (5.3.11) or (5.3.11′). Sometimes, these limits, or the estimates on which they are based (maximal inequalities), are even more useful than the limit theorems themselves: see, for instance, the proof of the CLT for the simplicial median in Section 5.5.

Theorem 5.3.7 for $k = 1$ simply asserts that the class of functions $P^{m-1}\mathcal{H}$ is P-Donsker, thus completing the proof of Theorem 5.3.3.

Theorem 5.3.7 for $m = 1$ is nothing but the CLT for empirical processes indexed by VC-subgraph classes of functions (assuming the envelope is square integrable: it needs not be, as shown by Alexander, 1987).

By Hoeffding's decomposition, Theorem 5.3.7 contains the central limit theorem for VC-subgraph classes of any degree of degeneracy (see Definition 3.5.1):

COROLLARY 5.3.8. *Let $\mathcal{H}$ be a measurable VC-subgraph class with a square P^m-integrable envelope consisting of real symmetric functions h of m variables which are P-degenerate of order at least $r-1$, $1 < r \le m$. Then,*

$$\left\{n^{r/2}(U_n(h) - P^m h) : h \in \mathcal{H}\right\} \to_{\mathcal{L}} \left\{(r!)^{1/2}\binom{m}{r} K_P(\pi_r h) : h \in \mathcal{H}\right\} \quad (5.3.17)$$

in $\ell^\infty(\mathcal{H})$. In particular, if h is $m-1$ degenerate and $P^m h = 0$ for all $h \in \mathcal{H}$, that is, if $\mathcal{H}$ consists of P-canonical kernels, then

$$\left\{n^{m/2}(U_n(h) - P^m h) : h \in \mathcal{H}\right\} \to_{\mathcal{L}} \left\{(m!)^{1/2} K_P(h) : h \in \mathcal{H}\right\} \quad (5.3.18)$$

in $\ell^\infty(\mathcal{H})$.

PROOF. Theorem 5.3.7 shows that the first term in the Hoeffding decomposition (3.5.2) of $U_n(h) - P^m h$ is $O_p(n^{-r/2})$ and the remaining terms are of smaller orders so that only the first term of the expansion contributes to the limit if we normalize by $n^{r/2}$, and its contribution is just (5.3.17). (5.3.18) is just (5.3.17) for $r = m$ and $P^m h = 0$. □

REMARK 5.3.9. It is important to realize that in the last theorems (5.3.3, 5.3.7, 5.3.8) the class $\mathcal{H}$ of kernels needs not be VC-subgraph in order for it to satisfy the corresponding central limit theorems. Let us consider, for instance, Theorem 5.3.7. Simple inspection of the proof shows that if a measurable class $\mathcal{H}$ with a P^m-square integrable envelope H satifies that

$$\left[\sup_Q \log N\left(\mathcal{H}, \|\cdot\|_{L_2(Q)}, \varepsilon(QH^2)^{1/2}\right)\right]^{m/2} \in L_1([0,\infty), \lambda), \quad (5.3.19)$$

where the sup is taken over all the probability measures Q on S^m such that $QH^2 < \infty$, and where λ is Lebesgue measure, then the central limit theorems (5.3.10) hold for $\mathcal{H}$. It is this property of VC-subgraph classes (Theorem 5.1.15) what makes the proof work. This applies, for instance to *sums of VC-subgraph classes*, which do not need to be VC-subgraph themselves, but which still satisfy the entropy bounds in Theorem 5.1.15, and therefore (5.3.17). Since we will use this observartion in several examples below, we should sketch a proof: Let $\mathcal{H}_1$ and $\mathcal{H}_2$ be VC-subgraph with measurable envelopes H_1 and H_2, and let $\mathcal{H} = \{h_1 + h_2 : h_i \in \mathcal{H}_i, i = 1, 2\}$.

Since if h_i is in the neighborhood of radius r_i from f_i, then $d(h_1+h_2, f_1+f_2) \leq r_1+r_2$ for L_p distances, it follows that

$$N\big(\mathcal{H}, \|\cdot\|_{L_p(Q)}, \varepsilon(Q(H_1+H_2)^p)^{1/p}\big) \leq N\big(\mathcal{H}_1, \|\cdot\|_{L_p(Q)}, \varepsilon(QH_1^p)^{1/p}/2\big) \times N\big(\mathcal{H}_2, \|\cdot\|_{L_p(Q)}, \varepsilon(QH_2^p)^{1/p}/2\big).$$

This shows that the bounds of Theorem 5.1.15 are also enjoyed (perhaps with different constants) by $\mathcal{H}$, and therefore, $\mathcal{H}$ satisfies the central limit theorems above (both, Theorem 5.3.7 and Theorem 5.3.3). This also applies to subsets of $\mathcal{H}$. Finally, the functions in the class $\mathcal{H}$ *do not have to be symmetric* as long as we use the symmetrized definition of a U-statistic, with the sum extended to all of I_n^m (not just to the lower triangle); this is due to the easily verifiable fact that the bounds on the covering numbers of $\mathcal{H}$ translate into bounds of the same order, perhaps with different constants, for the symmetrized class $s\mathcal{H} := \{sh : h \in \mathcal{H}\}$, where $sh(x_1, \ldots, x_m) = \sum_\sigma h(x_{\sigma(1)}, \ldots, x_{\sigma(m)})/m!$, with the sum extended to all the permutations σ of m elements.

The two observations in the previous remark are trivial but convenient: they do save work sometimes, as in some examples in this and the next section.

EXAMPLE 5.3.10. Theorem 5.3.3 immediately gives the CLT for the simplicial depth process of Example 5.2.4 with no conditions on P. So, we have

$$\big\{n^{1/2}(D_n(\theta) - P^3C_\theta) : \theta \in \mathbb{R}^2\big\} \to_{\mathcal{L}} \big\{G_\theta : \theta \in \mathbb{R}^2\big\}$$

in $\ell^\infty(\mathbb{R}^2)$, where G_θ, $\theta \in \mathbb{R}^2$, is a sample continuous centered Gaussian process (whose covariance, depending on P, can in principle be computed). In particular,

$$n^{1/2}\|D_n(\theta) - P^3C_\theta\|_\infty \to \|G_\theta\|_\infty$$

in distribution. These results hold as well for $\mathbb{R}^d$, for any d.

EXAMPLE 5.3.11. Theorem 5.3.7 finds most of its uses in proving that the remainder terms in the Hoeffding decomposition of a U-process are of lower order. However, there are instances of direct application. Silverman (1978) proposes using the following U-process to test uniformity on the circle. Let $\widehat{xy}$ denote the unoriented angle, in radians, between $x, y \in T$, T the circle (e.g., of radius 1 and center 0 in $\mathbb{R}^2$); if X_i are i.i.d., with common law the uniform distribution on the circle, the U-process in question is

$$\frac{1}{n(n-1)} \sum_{I_n^2} \big(I_{\widehat{X_iX_j} \leq t} - t/\pi\big), \qquad t \in [0, \pi).$$

Now, by a simple quotient of lengths, it is obvious that $\Pr\{\widehat{X_1X_2} \leq t | X_1\} = \Pr\{\widehat{X_1X_2} \leq t\} = t/\pi$, so that the kernel $I_{\widehat{x,y} \leq t}$ is degenerate, that is, canonical when centered. Since the class of kernels $\mathcal{H} = \{I_{\widehat{x,y} \leq t} - t/\pi : t \in [0, \pi)\}$ is parametrized by $[0, \pi)$ and these kernels are jointly measurable in x, y, t, this class is measurable. Also, this class of kernels is a subset of the sum of two bounded VC-subgraph classes $\mathcal{H}_1 = [0, \pi)$ and $\mathcal{H}_2 = \{I_{\widehat{xy} \le t} : t \in [0, \pi)\}$ (the constant functions are a VC-subgraph class since their graphs cannot shatter any set of two or more points, and $\mathcal{H}_2$ is VC-subgraph because the graphs of the indicators $I_{\widehat{xy} \le t}$ are ordered by inclusion and therefore they cannot shatter any set of two or more points). Then, Theorem 5.3.7, with the help of Remark 5.3.9, shows that there exist a sample continuous process K_P, which is a Gaussian chaos process of order two, such that

$$\left\{\frac{1}{n} \sum_{i \ne j \le n} \left(I_{\widehat{X_i X_j} \le t} - t/\pi\right) : t \in [0, \pi)\right\} \to_{\mathcal{L}} \left\{K_P(t) : t \in [0, \pi)\right\}$$

in $\ell^\infty[0, \pi)$. It is not difficult to obtain a representation of the limit K_P by developing the kernels into Fourier series and applying the theory from Section 4.2. This is developed e.g. in Silverman, loc. cit., and will not be pursued.

Let us now consider U-statistics whose kernels are B-valued, B a separable Banach space. U-processes are also Banach valued U-statistics, but the Banach space in this case is $\ell^\infty(\mathcal{H})$, which is not separable if $\mathcal{H}$ is infinite, and special methods, such as the ones developed so far, are required. Not even the central limit theorem for sums of independent B-valued random variables has a true solution unless the space satisfies certain geometric properties (e.g., Araujo and Giné, 1980, and Ledoux and Talagrand, 1991). It turns out that if the space is of *type 2*, then a neat result similar to those for VC-classes is available with almost no effort. Here is a brief description.

A Banach space B is of type 2 if it satisfies the following one-sided parallelogram inequality: there exist $c < \infty$ such that, for all $x_1, \dots, x_n \in B$, $n \in \mathbb{N}$,

$$\mathbb{E}\Big\| \sum_{i=1}^n \varepsilon_i x_i \Big\|^2 \le c \sum_{i=1}^n \|x_i\|^2, \tag{5.3.20}$$

where $\{\varepsilon_i\}$ is, as usual, a sequence of independent Rademacher variables. For instance, the L_p spaces, with $p \ge 2$, are type 2. We will deal in the proof below with distances to subspaces, hence it is convenient to mention that if B is type 2 with constant c in (5.3.20), and if F is a closed subspace of B, then the quotient Banach space B/F is also type 2 and admits the same constant: let $\|x\|_F$ denote the B/F norm, that is, the inf of the norms $\|x + y\|$, $y \in F$, or, what is the same, the distance of x to the subspace F; if given $\delta > 0$, y_i are vectors in F such that $\|x_i + y_i\|^2 \le \|x_i\|_F^2 + \delta/n$, then

$$\begin{aligned}
\mathbb{E}\Big\| \sum_{i=1}^n \varepsilon_i x_i \Big\|_F^2 &\le \mathbb{E}\Big\| \sum_{i=1}^n \varepsilon_i x_i + \sum_{i=1}^n \varepsilon_i y_i \Big\|^2 \\
&\le c \sum_{i=1}^n \|x_i + y_i\|^2 \\
&\le c \sum_{i=1}^n \|x_i\|_F^2 + c\delta,
\end{aligned}$$

so that inequality (5.3.20) holds for $\|\cdot\|_F$.

In the case of U-processes, we used Theorem 5.1.2 for tightness. Here we use the following simpler device. Total boundedness, hence compactness, in a Banach space can obviously be described in terms of boundedness plus arbitrary closeness to finite-dimensional spaces; this leads to the following version of Prohorov's theorem:

PROPOSITION 5.3.12. *A family Γ of Borel probability measures on a separable Banach space B is relatively compact for the topology of weak convergence if and only if*

i) $\lim_{M\to\infty}\sup_{\mu\in\Gamma}\mu\{\|x\|\geq M\}=0$, *and*

ii) *there exists a sequence of finite-dimensional subspaces $F_n\subset B$ such that*

$$\lim_{n\to\infty}\sup_{\mu\in\Gamma}\mu\{\|x\|_{F_n}>\varepsilon\}=0$$

for all $\varepsilon>0$.

Given the well known Prohorov's theorem, this proposition has a very simple proof that we leave as an exercise [alternatively, see, e.g., Araujo and Giné (1980), Theorem 1.4.5]. The proposition itself is not too different from Theorem 5.1.2 above on convergence in law of processes: the key in both criteria is finite-dimensional approximation in uniform μ-probability, $\mu\in\Gamma$. We should also recall that, as in the case of processes, weak converge is equivalent to relative compactness together with weak convergence of the finite-dimensional distributions, that is, $Z_n\to_{\mathcal{L}} Z$ in B if and only if the sequence of probability laws $\mathcal{L}(Z_n)$ is relatively compact (for weak convergence) and

$$(f_1(Z_n),\dots,f_k(Z_n))\to_{\mathcal{L}}(f_1(Z),\dots,f_k(Z)) \tag{5.3.21}$$

(as $\mathbb{R}^k$ valued random variables) for all $f_i\in B'$, the dual of B, and all $k\in\mathbb{N}$. We refer to the convergence in (5.3.21) as "convergence of the finite-dimensional distributions."

THEOREM 5.3.13. *Let B be a separable Banach space of type 2 and let $H:S^m\mapsto B$ be a measurable symmetric function on $(S^m,\mathcal{S}^m,P^m)$ such that $P^m\|H\|^2<\infty$. Then, for all $1\leq k\leq m$ there exist B-valued Gaussian chaos variables of order k, $K_{P,k}$, such that*

$$n^{k/2}U_n^{(k)}(\pi_k H)\to_{\mathcal{L}} K_{P,k} \tag{5.3.22}$$

in B.

PROOF. Given k, the finite-dimensional distributions in (5.3.22) converge, by Theorem 4.2.4, the CLT for canonical U-statistics in several dimensions. So, we must prove only that the sequence of laws of the variables $n^{k/2}U_n^{(k)}(\pi_k H)$, $n\geq k$, is relatively compact. By the randomization theorem (Theorem 3.5.3), since $\pi_k H$ is P-canonical,

$$\mathbb{E}\Big\|\sum_{I_n^k}(\pi_k H)(X_{i_1},\ldots,X_{i_k})\Big\|^2 \le K\mathbb{E}\Big\|\sum_{I_n^k}\varepsilon_{i_1}^1\cdots\varepsilon_{i_k}^k(\pi_k H)(X_{i_1},\ldots,X_{i_k})\Big\|^2,$$

where ε_i^r are all independent Rademacher variables and K is a constant that depends only on k. Moreover, by applying the type 2 inequality (5.3.20) conditionally on the X variables, iteratively, we obtain

$$\mathbb{E}\Big\|\sum_{I_n^k}\varepsilon_{i_1}^1\cdots\varepsilon_{i_k}^k(\pi_k H)(X_{i_1},\ldots,X_{i_k})\Big\|^2 \le c^k\frac{n!}{(n-k)!}\mathbb{E}\|\pi_k H\|^2.$$

By using iteratively the fact that, by Jensen, if X' is an independent copy of X, then

$$\mathbb{E}\|X-\mathbb{E}X\|^2 = \mathbb{E}\|X-\mathbb{E}X'\|^2 \le \mathbb{E}\|X-X'\|^2 \le 4\mathbb{E}\|X\|^2,$$

it follows that

$$\mathbb{E}\|\pi_k H\|^2 \le 4^k\mathbb{E}\|H\|^2.$$

These observations hold as well for the pseudonorm $\|\cdot\|_F$, with F any closed subspace of B. Applied just to the B norm, they give

$$\begin{aligned}\Pr\Big\{\big\|n^{k/2}U_n(\pi_k H)\big\| > M\Big\} &\le \frac{1}{M^2}\mathbb{E}\big\|n^{k/2}U_n(\pi_k H)\big\|^2\\ &\le \frac{K4^kc^k}{M^2}\frac{n^k(n-k)!}{n!}\mathbb{E}\|H\|^2.\end{aligned}$$

Hence,

$$\lim_{M\to\infty}\sup_{n\ge k}\Pr\Big\{\big\|n^{k/2}U_n(\pi_k H)\big\| > M\Big\} = 0$$

for all $\varepsilon > 0$, proving the stochastic boundedness condition in Proposition 5.3.12 for the laws of the variables $n^{k/2}U_n(\pi_k H)$, $n \ge k$. Let now F_n be an incresing sequence of finite-dimensional subspaces of B such that their union is dense in B (such a sequence exists by separability). Then, for all $x \in B$, $\|x\|_{F_n} \to 0$ and, therefore, since $\|x\|_{F_n} \le \|x\|$, it follows by dominated convergence that $\mathbb{E}\|H\|_{F_n}^2 \to 0$. The same computations above, but now for the $\|\cdot\|_{F_n}$ norms, which are also type 2, thus give

$$\begin{aligned}&\lim_{n\to\infty}\sup_{n\ge k}\Pr\Big\{\big\|n^{k/2}U_n(\pi_k H)\big\|_{F_n} > \varepsilon\Big\}\\ &\qquad\le \frac{K4^kc^k}{\varepsilon^2}\lim_{n\to\infty}\sup_{n\ge k}\frac{n^k(n-k)!}{n!}\mathbb{E}\|H\|_{F_n}^2 = 0.\end{aligned}$$

This is just the finite-dimensional approximation condition (condition ii)) in Proposition 5.3.12 for the laws of the variables $n^{k/2}U_n(\pi_k H)$, $n \ge k$. These laws are therefore relatively compact, proving the theorem. □

Back to U-processes, we conclude this section with an exponential bound for canonical U-processes. Exponential bounds are important complements of central limit theorems in that they apply for each fixed n and often produce tail bounds of the same order as the limit. Although best possible bounds for degenerate (canonical) U-processes seem out of reach at present, the power of t in the following analog of the bound in Theorem 4.1.12 (b) for U-statistics is of the right order, at least for large t. A similar statement for B-valued U-statistics, B of type 2, can be safely left to the reader.

THEOREM 5.3.14. *Let $\mathcal{H}$ be a uniformly bounded measurable VC-subgraph class of real-valued kernels h on $(S^m, \mathcal{S}^m)$, symmetric in their entries. Then, for each $1 \le k \le m$ there exist constants $c_k, d_k \in (0, \infty)$ such that, for all P on $(S, \mathcal{S})$, $n \ge m$ and $t > 0$,*

$$\Pr\Big\{ \big\| n^{k/2} U_n^{(k)}(\pi_{k,m}^P h, P) \big\|_{\mathcal{H}} > t \Big\} \le c_k e^{-d_k t^{2/k}}. \tag{5.3.23}$$

PROOF. We can assume, without loss of generality, that $\|h\|_\infty \le 1$ for all $h \in \mathcal{H}$. For each $X_1, \dots, X_n$ fixed, where $\{X_i\}$ are the coordinate functions on $S^{\mathbb{N}}$, we consider the Rademacher process

$$Z(h) = \frac{n^{k/2}}{\binom{n}{k}} \sum_{1 \le i_1 < \cdots < i_k \le n} \varepsilon_{i_1} \cdots \varepsilon_{i_k} (\pi_{k,m}^P h)(X_{i_1}, \dots, X_{i_k}), \qquad h \in \mathcal{H},$$

which is the randomized version of the normalized U-process $n^{k/2} U_n^{(k)}(\pi_{k,m}^P h)$, $h \in \mathcal{H}$, and which we view as a Rademacher process by conditioning on the X variables. The L_2 pseudodistance associated to the Rademacher process Z is the square root of

$$\mathbb{E}_\varepsilon \big(Z(f) - Z(g)\big)^2 = e_{n,k,2}^2(f, g) = \frac{n^k}{\binom{n}{k}} U_n^{(k)}\big((\pi_k(f - g))^2\big),$$

already encountered in Lemma 5.3.5 above. Here, as usual, we write π_k for $\pi_{k,m}^P$. By Lemma 5.2.5, the covering number of $(\mathcal{H}, e_{n,k,2})$ is a finite product of covering numbers of $\mathcal{H}$ by L_2 distances of probability measures on S^m, and therefore, $\mathcal{H}$ being VC-subgraph, Theorem 5.1.15 gives

$$N\big(\mathcal{H}, e_{n,k,2}, \tau\big) \le \Big(\frac{A}{\tau}\Big)^v, \qquad 0 < \tau \le 1,$$

for finite constants A and v that depend only on k and the VC index of $\mathcal{H}$. Hence, since

$$\int_0^1 \Big[\log\Big(\frac{A}{\varepsilon}\Big)^v\Big]^{k/2} d\varepsilon < \infty,$$

it follows from Corollary 5.1.5 (entropy bounds) and Corollary 3.2.6 (integrability of Rademacher chaos), in complete analogy with Corollary 5.1.8, that

$$\big\| \|Z\|_{\mathcal{H}} \big\|_{\psi_{2/k}} \le K$$

for some constant K that depends only on k and the VC index of $\mathcal{H}$, and where the function $\psi_{2/k}$ is the exponential Young modulus defined by equation (4.3.2). In other words,

$$\mathbb{E}_\varepsilon \psi_{2/k}\big(\|Z\|_{\mathcal{H}}/K\big) \le 1.$$

Hence, by Fubini,

$$\mathbb{E}\psi_{2/k}\big(\|Z\|_{\mathcal{H}}/K\big) \le 1.$$

Now, since $\pi_k(h)$ is canonical for all h, the randomization theorem 3.5.3, as extended by Remark 3.5.4, implies the existence of a constant C depending only on k such that

$$\mathbb{E}\psi_{2/k}\big(\|n^{k/2}U_n^{(k)}\big(\pi_k^P(h), P)\|_{\mathcal{H}}/(CK)\big) \le C.$$

Finally, Markov's inequality completes the proof of the theorem:

$$\begin{aligned}
\Pr\Big\{ \|n^{k/2}U_n^{(k)}(\pi_{k,m}^P h, P)\|_{\mathcal{H}} > t\Big\} & \\
&\le \Pr\Big\{\psi_{2/k}\big(\|n^{k/2}U_n^{(k)}(\pi_{k,m}^P h, P)\|_{\mathcal{H}}/(CK)\big) > \psi_{2/k}\big(t/(CK)\big)\Big\} \\
&\le \frac{C}{\psi_{2/k}\big(t/(CK)\big)} \wedge 1 = \frac{C}{\psi_{2/k}\big(t/(CK)\big) \vee C} \le c_k e^{-t^{2/k}/(CK)^{2/k}},
\end{aligned}$$

where the last inequality follows from the definition of $\psi_{2/k}$. □

REMARK 5.3.15. As with the CLT, the exponential bound in Theorem 5.3.14 also holds for measurable classes $\mathcal{H}$ of nonsymmetric kernels satisfying the entropy condition (5.3.19), and the proof is the same: see Remark 5.3.9.

5.4 The Law of the Iterated Logarithm for Canonical U-Processes

Again, only U-processes over VC-type classes of functions and U-statistics with kernels in type 2 Banach spaces will be considered. The proofs do not formally differ from those in the real case, if one allows for new ingredients to appear at a few specific places. This is best seen in connection with the bounded LIL.

5.4.1 *The Bounded LIL*

Our object here is to prove the analog of Theorem 4.3.3. The maximal inequality (4.3.7) applied to randomized (and decoupled) U-statistics is crucial in the proof of this theorem, and what makes it possible is inequality (4.3.4′), a specific property of *real* Rademacher chaos, that is,

$$\Big\|\sum_{I_n^m} \varepsilon_{i_1}\cdots\varepsilon_{i_m} a_{i_1,\dots,i_m}\Big\|_{\psi_{2/m}} \le c_m\Big(\sum_{I_n^m} a_{i_1,\dots,i_m}^2\Big)^{1/2}. \qquad (5.4.1)$$

(This inequality is only needed for decoupled chaos, which is in fact a particular case of regular chaos.) Given the analogs of this inequality that we are going to derive now, the proof of the bounded LIL in type 2 spaces or for VC classes is the same as the proof of Theorem 4.3.3, with the formal change of replacing absolute values by norms. If $H : S^m \mapsto B$ is a B-valued kernel, B being a separable Banach space of type 2 [see equation (5.3.20) for the definition], and $\{x_i^j : i \le n, j \le m\} \subset S$, $m \le n < \infty$, then, as indicated in the proof of Theorem 5.3.13, iterating the type 2 inequality conditionally on the X variables gives

$$\mathbb{E}\Big\|\sum_{I_n^m} \varepsilon_{i_1}^1 \cdots \varepsilon_{i_m}^m H(x_{i_1}^1, \ldots, x_{i_m}^m)\Big\|^2 \le c^m \sum_{I_n^m} \big\|H(x_{i_1}^1, \ldots, x_{i_m}^m)\big\|^2,$$

where c is the type 2 constant of B. Since, on the other hand, Corollary 3.2.6 on integrability of F-valued Rademacher chaos variables implies that

$$\Big\|\sum_{I_n^m} \varepsilon_{i_1}^1 \cdots \varepsilon_{i_m}^m H(x_{i_1}^1, \ldots, x_{i_m}^m)\Big\|_{\psi_{2/m}} \le c_m \Big\|\sum_{I_n^m} \varepsilon_{i_1}^1 \cdots \varepsilon_{i_m}^m H(x_{i_1}^1, \ldots, x_{i_m}^m)\Big\|_2$$

for a universal constant $c_m < \infty$, we obtain

$$\Big\|\sum_{I_n^m} \varepsilon_{i_1}^1 \cdots \varepsilon_{i_m}^m H(x_{i_1}^1, \ldots, x_{i_m}^m)\Big\|_{\psi_{2/m}} \le c_m c^{m/2} \Big(\sum_{I_n^m} \big\|H(x_{i_1}^1, \ldots, x_{i_m}^m)\big\|^2\Big)^{1/2}. \tag{5.4.2}$$

This is the inequality that plays the role of inequality (5.4.1) in the proof of the LIL when the kernel H is B-valued and B is type 2. This inequality can be undecoupled, of course, but this is not required.

Let now $\mathcal{H}$ be a measurable VC-subgraph class of functions with a square integrable envelope H. We can and do assume that $0 \in \mathcal{H}$. For each $n \in \mathbb{N}$ we set, as in (5.2.6),

$$\begin{aligned} e_{n,2}^2(f,g) &:= \mathbb{E}\Big|\sqrt{\frac{(n-m)!}{n!}} \sum_{I_n^m} \varepsilon_{i_1}^1 \cdots \varepsilon_{i_m}^m (f-g)(x_{i_1}^1, \ldots, x_{i_m}^m)\Big|^2 \\ &= \frac{(n-m)!}{n!} \sum_{I_n^m} (f-g)^2(x_{i_1}^1, \ldots, x_{i_m}^m), \end{aligned}$$

which is the square of the L_2 distance with respect to a probability measure on S^m [namely, the uniform probability measure on the set $\{(x_{i_1}, \ldots, x_{i_m}) : \mathbf{i} \in I_n^m\}$]. Then, the covering number bound given in Theorem 5.1.15 for VC-subgraph classes applies to $(\mathcal{H}, e_{n,2})$. Since the diameter of $\mathcal{H}$ for this norm is dominated by

$$H_{n,2} := 2\Big[\frac{(n-m)!}{n!} \sum_{I_n^m} H^2(x_{i_1}^1, \ldots, x_{i_m}^m)\Big]^{1/2},$$

combining the covering number bound from Theorem 5.1.15 with the entropy maximal inequality for Rademacher chaos processes, Corollary 5.1.8, we obtain

$$\Big\|\Big\|\sqrt{\frac{(n-m)!}{n!}}\sum_{I_n^m}\varepsilon_{i_1}^1\cdots\varepsilon_{i_m}^m h(x_{i_1}^1,\ldots,x_{i_m}^m)\Big\|_{\mathcal{H}}\Big\|_{\psi_{2/m}}$$
$$\le K\int_0^{H_{n,2}}\Big(\log\frac{AH_{n,2}^{2v}}{\varepsilon^{2v}}\Big)^{m/2}d\varepsilon$$
$$= KH_{n,2}\int_0^1\Big(\log\frac{A}{\tau^{2v}}\Big)^{m/2}d\tau$$
$$= C\Big[\frac{(n-m)!}{n!}\sum_{I_n^m}H^2(x_{i_1}^1,\ldots,x_{i_m}^m)\Big]^{1/2}$$

for some finite constants A, v, C, K depending only on $\mathcal{H}$ (recall $0 \in \mathcal{H}$). Or,

$$\Big\|\Big\|\sum_{I_n^m}\varepsilon_{i_1}^1\cdots\varepsilon_{i_m}^m h(x_{i_1}^1,\ldots,x_{i_m}^m)\Big\|_{\mathcal{H}}\Big\|_{\psi_{2/m}} \le C\Big[\sum_{I_n^m}H^2(x_{i_1}^1,\ldots,x_{i_m}^m)\Big]^{1/2}, \quad (5.4.3)$$

which is the equation to use in place of (5.4.1).

Using inequalities (5.4.2) and (5.4.3) with $x_{i_j}^j = X_{i_j}^j$ in the proof of Theorem 4.3.3, concretely in the step following (4.3.12) (taking into account that the ψ-norm dominates the L_2 norm), we obtain the following two inequalities, which constitute the bounded LIL for canonical U-processes over VC classes and for canonical B-valued U-statistics, B of type 2.

THEOREM 5.4.1. *Let $\mathcal{H}$ is a measurable VC-subgraph class of symmetric P-canonical kernels on S^m with a P^m-square integrable envelope H, where P is a probability measure on $(S, \mathcal{S})$, and let X_i, $i \in \mathbb{N}$, denote the coordinates of $S^{\mathbb{N}}$. Then, for every $p \in (0, 2)$ there exists a constant $C := C(\mathcal{H}, p)$ such that*

$$\mathbb{E}\sup_{n\in\mathbb{N}}\Big[\frac{1}{(2n\log\log n)^{m/2}}\Big\|\sum_{(i_1,\ldots,i_m)\in I_n^m}h(X_{i_1},\ldots,X_{i_m})\Big\|_{\mathcal{H}}\Big]^p$$
$$\le C\big(\mathbb{E}H^2(X_1,\ldots,X_m)\big)^{p/2}. \quad (5.4.4)$$

Remark 5.3.9 on extension to not necessarily VC-subgraph classes of not necessasrily symmetric functions also applies to Theorem 5.4.1.

THEOREM 5.4.2. *Let B be a separable Banach space of type 2, let $(S, \mathcal{S}, P)$ be a probability space, let X_i, $i \in \mathbb{N}$, be i.i.d. S valued random variables with law P, and let $H : S^m \mapsto B$ be a P-canonical symmetric function such that $P^m\|H\|^2 < \infty$. Then, for every $p \in (0, 2)$ there exists a constant $C := C(B, p)$ such that*

$$\mathbb{E}\sup_{n\in\mathbb{N}}\Big[\frac{1}{(2n\log\log n)^{m/2}}\Big\|\sum_{(i_1,\ldots,i_m)\in I_n^m}H(X_{i_1},\ldots,X_{i_m})\Big\|\Big]^p$$
$$\le C\big(\mathbb{E}\|H(X_1,\ldots,X_m)\|^2\big)^{p/2}. \quad (5.4.5)$$

As in Remark 4.3.4, inequalities (5.4.4) and (5.4.5) imply the following: Under the conditions of Theorem 5.4.1, we have

$$\limsup_{n\to\infty} \frac{1}{(2n\log\log n)^{m/2}} \Big\| \sum_{(i_1,\ldots,i_m)\in I_n^m} h(X_{i_1},\ldots,X_{i_m}) \Big\|_{\mathcal{H}} \le C\big(\mathbb{E}H^2(X_1,\ldots,X_m)\big)^{1/2} \quad \text{a.s.,} \tag{5.4.6}$$

and, under the conditions of Theorem 5.4.2,

$$\limsup_{n\to\infty} \frac{1}{(2n\log\log n)^{m/2}} \Big\| \sum_{(i_1,\ldots,i_m)\in I_n^m} H(X_{i_1},\ldots,X_{i_m}) \Big\| \le C\big(\mathbb{E}\|H(X_1,\ldots,X_m)\|^2\big)^{1/2} \quad \text{a.s.,} \tag{5.4.7}$$

bounded laws of the iterated logarithm that, as in the real case, do hold with better constants as a consequence of the compact LIL (that we consider next).

We should finally note that *inequalities (5.4.4) to (5.4.7) also hold for decoupled U-processes and U-statistics*: this can either be seen directly because the same proofs with the obvious formal changes work for the decoupled statistics, or by applying the decoupling inequalities of Section 3.1 [see (3.1.17′) for instance] to (5.4.4)–(5.4.7).

5.4.2 *The Compact LIL*

To prove the compact LIL for processes, we need first to reduce it to an asymptotic equicontinuity condition, in the spirit of Theorem 5.1.2. The following general theorem will accomplish this reduction. Its proof will be given in full, except that, as in previous chapters, we refer the reader to Dudley (1984) for an argument on measurability (as usual, if the class of functions $\mathcal{H}$ is assumed to be countable or stochastically separable instead of just measurable, the issue of measurability disappears). We recall from the notation set up in Chapter 4 that

$$\alpha_n(h) := \frac{1}{(2n\log\log n)^{m/2}} \sum_{(i_1,\ldots,i_m)\in I_n^m} h(X_{i_1},\ldots,X_{i_m}), \tag{5.4.8}$$

where, as usual when dealing with processes, X_i are the coordinate maps $X_i : S^{\mathbb{N}} \mapsto S$.

THEOREM 5.4.3. *Let $\mathcal{H}$ be a measurable class of symmetric P-canonical kernels $h : S^m \mapsto \mathbb{R}$ square integrable with respect to P, a probability measure on $(S, \mathcal{S})$, and let X_i, $i \in \mathbb{N}$, denote the coordinate functions on $S^{\mathbb{N}}$. We assume that $\mathcal{H}$ and the sequence of processes $\{\alpha_n(h) : h \in \mathcal{H}\}$ satisfy the following two conditions:*

i) *the class of functions $\mathcal{H}$ is totally bounded for the $L_2(P^m)$ norm;*

ii)

$$\lim_{\delta\to 0}\limsup_{n\to\infty}\sup_{\substack{f,g\in\mathcal{H}\\ \|f-g\|_{L_2(P^m)}\le\delta}}|\alpha_n(f-g)| = 0 \quad \text{a.s.} \tag{5.4.9}$$

Then, the sequence of processes $\{\alpha_n(h) : h \in \mathcal{H}\}_{n=1}^\infty$ *is relatively compact in* $\ell^\infty(\mathcal{H})$ *for almost every* ω *and*

$$\lim \operatorname{set}\{\alpha_n(h) : h \in \mathcal{H}\}_{n=1}^\infty = K \quad \text{a.s.}, \tag{5.4.10}$$

where K *is the (deterministic) subset of* $\ell^\infty(\mathcal{H})$ *defined by*

$$K = \Big\{\big\{\mathbb{E}\big[h(X_1,\dots,X_m)g(X_1),\dots,g(X_m)\big] : h \in \mathcal{H}\big\} : \mathbb{E}g^2(X_1) \le 1\Big\}, \tag{5.4.11}$$

which is compact.

PROOF. We divide the proof into several steps.

CLAIM 1. *For almost every* ω*, both, the sequence* $\{\alpha_n(h,\omega) : h \in \mathcal{H}\}_{n=1}^\infty$ *is relatively compact in* $\ell^\infty(\mathcal{H})$ *and any subsequential limit* $\alpha(\omega)$ *of this sequence is in* $C_u(\mathcal{H}, \|\cdot\|_{L_2(P^m)})$*, the space of functions on* $\mathcal{H}$ *which are uniformly continuous for the* $L_2(P^m)$ *norm.*

The proof of this claim is nothing but the standard Arzelà-Ascoli argument. Let $\bar{\mathcal{H}}$ be a countable dense subset of $\mathcal{H}$ for the $L_2(P^m)$ distance, and let Ω_1 be a set of probability one on which the sequences $\{\alpha_n(h,\omega)\}$, $h \in \bar{\mathcal{H}}$, are all relatively compact (Theorem 4.3.6) and on which the limit (5.4.9) holds. Given $\omega \in \Omega_1$ and given $\varepsilon > 0$, let $\delta > 0$ and $N_1 < \infty$ be such that

$$\sup_{\substack{f,g\in\mathcal{H}\\ \|f-g\|_{L_2(P^m)}\le\delta}}|\alpha_n(f-g,\omega)| < \varepsilon/3$$

for all $n \ge N_1$, and let $h_1,\dots,h_k$ be a subset of $\bar{\mathcal{H}}$ δ-dense in $\mathcal{H}$ for the $L_2(P^m)$ norm. For any given subsequence of $\mathbb{N}$, let $\{n'\}$ be a further subsequence for which the sequences $\{\alpha_{n'}(h_i,\omega)\}$, $i \le k$, converge, and let N_2 be such that

$$\max_{i\le k}\big|\alpha_{n'}(h_i,\omega) - \alpha_{N_2}(h_i,\omega)\big| < \varepsilon/6$$

for all $n' \ge N_2$. Then, if $N' = \min\{n' : n' \ge N_1 \vee N_2\}$, for all $n' \ge N'$, $n' \in \{n'\}$, we have

$$\begin{aligned}&\big\|\alpha_{n'}(h,\omega) - \alpha_{N'}(h,\omega)\big\|_{\mathcal{H}}\\ &\quad\le \max_{i\le k}\big|\alpha_{n'}(h_i,\omega) - \alpha_{N'}(h_i,\omega)\big|\\ &\qquad + 2\sup_{n'\in\{n'\}:n'\ge N'}\sup_{h\in\mathcal{H}}\min_{i\le k}\big|\alpha_{n'}(h,\omega) - \alpha_{n'}(h_i,\omega)\big| < \varepsilon.\end{aligned}$$

So, the sequence $\{\alpha_{n'}(h,\omega) : h \in \mathcal{H}\}$ is convergent in $\ell^\infty(\mathcal{H})$. This proves that the sequence $\{\alpha_n(h,\omega) : h \in \mathcal{H}\}_{n=1}^\infty$ is relatively compact in $\ell^\infty(\mathcal{H})$ for each $\omega \in \Omega_1$. Taking limits along a subsequence in (5.4.9) shows that all subsequential limits $\{\alpha(h, \omega)\}$ of this sequence are uniformly continuous in h for the $L_2(P^m)$ distance [and also bounded by hypothesis i)].

CLAIM 2. *Let $L(\omega) \subset \mathcal{H}$ denote the limit set of the sequence $\{\alpha_n(h, \omega) : h \in \mathcal{H}\}$. Then, there exists a nonrandom compact set $L \subset \mathcal{H}$ such that $L(\omega) = L$ a.s.*

$\mathcal{H}$ being measurable implies, by Theorem 10.2.2 and the proof of Theorem 10.3.2 in Dudley (1984), that the sets $\{\|\alpha_n - x\|_{\mathcal{H}} \leq t\}$ are measurable for all $t \geq 0$ and $x \in C_u(\mathcal{H}, \|\cdot\|_{L_2(P^m)}) := C_u$. So, for any such x, the set

$$\Omega_x := \{\omega : x \in L(\omega)\} = \{\omega : x \text{ is a limit point of } \{\alpha_n(\omega)\}\}$$
$$= \cap_{m\geq 1} \limsup_n \{\omega : \|\alpha_n(\omega) - x\|_{\mathcal{H}} \leq 1/m\}$$

is measurable. It is also invariant under permutations of the independent sequence X_i, and therefore, by the Hewitt-Savage zero–one law, it has probability 0 or 1. We then set

$$L := \{x \in C_u : \Pr(\Omega_x) = 1\}.$$

Since $(\mathcal{H}, \|\cdot\|_{L_2(P^m)})$ is totally bounded, the space C_u is separable; so, there exists a countable set $\{x_k\} \subset L$ dense in L. Now, $\omega \in \cap\Omega_{x_k}$ if and only if $\{x_k\} \subseteq L(\omega)$ by definition, and $\{x_k\} \subset L(\omega)$ if and only if $L \subseteq L(\omega)$ since $L(\omega)$ is closed. Hence,

$$\{\omega : L \subseteq L(\omega)\} = \cap_{k=1}^{\infty} \Omega_{x_k},$$

proving

$$\Pr\{\omega : L \subseteq L(\omega)\} = 1. \tag{5.4.12}$$

Next we prove

$$\Pr^*\{\omega : L(\omega) \not\subseteq L\} = 0. \tag{5.4.13}$$

The set L is closed: if $x_k \in L$ and $x_k \to x$ uniformly on $\mathcal{H}$, then $\cap\Omega_{x_k} \subset \Omega_x$ and $\Pr(\cap\Omega_{x_k}) = 1$, hence $\Pr \Omega_x = 1$, so that $x \in L$. So, the set $C_u \setminus L$ is open in C_u, and it is separable since C_u is separable. Hence $C_u \setminus L = \cup_n B_n$ where B_n are open sets of C_u whose diameters are smaller than 1/2 and such that $\bar{B}_n \cap L = \emptyset$. Since $\{L(\omega) \not\subseteq L\} = \cup\{L(\omega) \cap B_n \neq \emptyset\}$ and $\{L(\omega) \cap B_n \neq \emptyset\} \subseteq \limsup_k \{\alpha_k \in B_n\}$, it follows that if (5.4.13) does not hold, that is, if $\Pr^*\{L(\omega) \not\subseteq L\} \neq 0$, then there exists $N < \infty$ such that $\Pr\{\limsup_k \{\alpha_k \in B_N\}\} > 0$. But then, by the zero–one law, this probability must be one. Proceeding in the same way, for every $r \in \mathbb{N}$ we can find nested subsets $D_r \subseteq \bar{B}_N$ of diameter smaller than $1/2^r$ such that $\Pr\{\limsup_k \{\alpha_k \in D_r\}\} = 1$. C_u being complete, the intersection of the sets D_r is not empty. Let $x \in \cap D_r \subseteq \bar{B}_N$. Since $\Omega_x \supseteq \cap_r \{\limsup_k \{\alpha_k \in D_r\}\}$, it follows that $\Pr(\Omega_x) = 1$, that is, $x \in \bar{B}_N \cap L$, contradiction. (5.4.13) is thus proved and, together with (5.4.12), it shows that

$$\lim \text{set}\{\alpha_n\}_{n=1}^{\infty} = L \quad \text{a.s.} \tag{5.4.14}$$

Since then, by (5.4.14) and Claim 1, L is the set of limit points of a relatively compact sequence ($\{\alpha_n(\omega)\}$ for some ω), it follows that L is a compact subset of $C_u\big(\mathcal{H}, \|\cdot\|_{L_2(P^m)}\big)$. Claim 2 is proved.

CLAIM 3. *The set K of Claim 2 satisfies equation (5.4.11).*

Let K be as defined by equation (5.4.11). K is obviously a subset of $C_u\big(\mathcal{H}, \|\cdot\|_{L_2(P^m)}\big)$. Theorem 4.3.6 shows that for all $k < \infty$ and $h_1, \ldots, h_k$, with probability one, the sequence $\big\{\big(\alpha_n(h_1), \ldots, \alpha_n(h_k)\big)\big\}$ is relatively compact and the set of its limit points is precisely $\big\{\big(x(h_1), \ldots, x(h_k)\big) : x \in K\big\}$. This implies, in particular, that K is a closed subset of $\ell^\infty(\mathcal{H})$ (if $x_r \to x$ and $\big(x_r(h_1), \ldots, x_r(h_k)\big)$ are limit points of $\big\{\big(\alpha_n(h_1, \omega), \ldots, \alpha_n(h_k, \omega)\big)\big\}$, so is x). A simple modification of the proof of Claim 1 shows that $K \subseteq L$, where L is defined in Claim 2: for $x \in K$ fixed, choose δ in that proof to satisfy also that

$$\sup_{\substack{f,g \in \mathcal{H} \\ \|f-g\|_{L_2(P^m)} \leq \delta}} |x(f) - x(g)| < \varepsilon/3,$$

(any $\delta \leq \varepsilon/3$ does), and choose $\{n'\}$ and N_2 such that

$$\max_{i \leq k} \big|\alpha_{n'}(h_i, \omega) - x(h_i)\big| < \varepsilon/3$$

for all $n' > N_2$. These choices give that $\|\alpha_{n'} - x\|_{\mathcal{H}} < \varepsilon$ for all $n' \geq N'$, $n' \in \{n'\}$, and this implies that x is a limit point of $\{\alpha_n(\omega)\}$ for all $\omega \in \Omega_1$, that is, $x \in L$ by Claim 2. In particular, L being compact, so is K. Suppose $x \in L$, that is, x is a limit point of almost every sequence $\{\alpha_n(\omega)\}$, and let $\{h_i\}$ be a dense subset of $\mathcal{H}$. Then, for each $k < \infty$, $\big(x(h_1), \ldots, x(h_k)\big)$ is a limit point of almost every sequence $\{\big(\alpha_n(h_1, \omega), \ldots, \alpha_n(h_k, \omega)\big)\}$, and in particular there is $x_k \in K$ such that $\big(x_k(h_1), \ldots, x_k(h_k)\big) = \big(x(h_1), \ldots, x(h_k)\big)$. K being compact, it contains a limit point of the sequence $\{x_k\}$, say y. Then, x and y coincide on $\{h_k\}$, which is dense in $\mathcal{H}$ [for the $L_2(P^m)$ norm]. Since both are continuous functions on $\mathcal{H}$ for the $L_2(P^m)$ norm (as $K \subseteq L \subseteq C_u$), it follows that $x = y$ and therefore, $x \in K$, that is, $L \subseteq K$. Claim 3 is proved and, with it, the theorem. □

In order to prove the limit (5.4.9) we will require an entropy based maximal inequality for the increments of the processes $\alpha_n(h)$. Let $\mathcal{H}$ be a VC-subgraph class of symmetric kernels on S^m with a measurable envelope H square integrable for P^m. Let $e_{n,2}$ and $H_{n,2}$ be as in the derivation of inequality (5.4.3) above. Without loss of generality we may assume $H_{n,2}(x_1, \ldots, x_m) \geq 1$ just by taking the envelope $H \vee 1$ instead of H, which is also square integrable. Then, if in the proof of inequality (5.4.3) we replace the maximal inequality (5.1.22) for the sup of a process by inequality (5.1.23) for its modulus of continuity, we obtain

$$\Big\| \sup_{\substack{f,g\in\mathcal{H}\\ e_{n,2}(f,g)\le\delta}} \sqrt{\frac{(n-m)!}{n!}} \Big| \sum_{I_n^m} \varepsilon_{i_1}^1 \cdots \varepsilon_{i_m}^m (f-g)(x_{i_1}^1,\ldots,x_{i_m}^m)\Big| \Big\|_{\psi_{2/m}}$$

$$\le K\int_0^\delta \Big(\log\frac{AH_{n,2}^v}{\varepsilon^v}\Big)^{m/2} d\varepsilon$$

$$\le KH_{n,2}\int_0^\delta \Big(\log\frac{A}{\varepsilon^v}\Big)^{m/2} d\varepsilon$$

$$= \kappa(\delta)\Big[\frac{(n-m)!}{n!}\sum_{I_n^m} H^2(x_{i_1}^1,\ldots,x_{i_m}^m)\Big]^{1/2},$$

where $\kappa(\delta)\searrow 0$ as $\delta\searrow 0$. That is,

$$\Big\| \sup_{\substack{f,g\in\mathcal{H}\\ e_{n,2}(f,g)\le\delta}} \Big| \sum_{I_n^m} \varepsilon_{i_1}^1 \cdots \varepsilon_{i_m}^m (f-g)(x_{i_1}^1,\ldots,x_{i_m}^m)\Big| \Big\|_{\psi_{2/m}}$$

$$\le \kappa(\delta)\Big[\sum_{I_n^m} H^2(x_{i_1}^1,\ldots,x_{i_m}^m)\Big]^{1/2}. \tag{5.4.15}$$

Theorem 5.4.3 and inequality (5.4.15) is all we need to set in motion a fast proof of the compact LIL for degenerate U-processes:

THEOREM 5.4.4. *Let $\mathcal{H}$ is a measurable VC-subgraph class of symmetric P-canonical functions on S^m with a P^m-square integrable envelope H, where P is a probability measure on $(S,\mathcal{S})$, and let X_i, $i\in\mathbb{N}$, denote the coordinates of $S^{\mathbb{N}}$. Then, almost surely, the sequence of processes*

$$\Big\{\alpha_n(h) := \frac{1}{(2n\log\log n)^{\frac{m}{2}}} \sum_{(i_1,\ldots,i_m)\in I_n^m} h(X_{i_1},\ldots,X_{i_m}) : h\in\mathcal{H}\Big\}$$

is relatively compact in $\ell^\infty(\mathcal{H})$ and its limit set is

$$K = \Big\{\big\{\mathbb{E}\big[h(X_1,\ldots,X_m)g(X_1),\ldots,g(X_m)\big] : h\in\mathcal{H}\big\} : \mathbb{E}g^2(X_1)\le 1\Big\},$$

which is compact.

PROOF. By Theorem 5.4.3, since the VC property implies that $(\mathcal{H},\|\cdot\|_{L_2(P^m)})$ is totally bounded, it suffices to prove the asymptotic equicontinuity condition (5.4.9). By monotonicity in δ, this is equivalent to showing that the random variables $\limsup_{n\to\infty}\|\alpha_n\|_{\mathcal{H}'_\delta}$ converge to zero in probability as $\delta\searrow 0$, where

$$\mathcal{H}'_\delta := \{f-g : f,g\in\mathcal{H},\ \|f-g\|_{L_2(P^m)}<\delta\}.$$

Hence, Theorem 5.4.4 will be proved if we show

$$\lim_{\delta\to 0}\lim_{r\to\infty}\mathbb{E}\sup_{n\ge r}\|\alpha_n\|_{\mathcal{H}'_\delta} = 0. \tag{5.4.16}$$

Let $s = [\log_2 r]$. Using decoupling and randomization (Theorem 3.5.3 via Remark 3.5.4, 2), as in (4.3.11)) and blocking (Lemma 4.3.2), we obtain

$$\mathbb{E} \sup_{n \geq r} \|\alpha_n\|_{\mathcal{H}'_\delta} \leq C \mathbb{E} \sup_{k \geq s} \frac{1}{a_k^*} \Big\| \sum_{\mathbf{i} \in I_{2^k}^m} \epsilon_i^{\text{dec}} h(\mathbf{X}_{\mathbf{i}}^{\text{dec}}) \Big\|_{\mathcal{H}'_\delta}. \tag{5.4.17}$$

(Here we do not worry about having $n \leq 2^K$ as in the proof of Theorem 4.3.3 because, by Theorem 5.4.1, decoupled version, these infinite sups are integrable.) As in the proof of the CLT for degenerate U-processes, we will split the integrals in (5.4.17) into two parts so that we can apply the bound (5.4.15). To this end we define the analog of the distance $e_{n,2}$ in (5.4.15), that we denote by $e_{k,2}^{\text{dec}}$ for obvious reasons:

$$\begin{aligned}\left(e_{k,2}^{\text{dec}}(f, g)\right)^2 &:= \frac{(2^k - m)!}{2^k!} \mathbb{E}_\varepsilon \Big(\sum_{\mathbf{i} \in I_{2^k}^m} \epsilon_{\mathbf{i}}^{\text{dec}} (f - g)(\mathbf{X}_{\mathbf{i}}^{\text{dec}}) \Big)^2 \\ &= \frac{(2^k - m)!}{2^k!} \sum_{\mathbf{i} \in I_{2^k}^m} (f - g)^2 (\mathbf{X}_{\mathbf{i}}^{\text{dec}}).\end{aligned}$$

The law of large numbers (5.3.8′) in Lemma 5.3.6 yields

$$\left\| \left(e_{k,2}^{\text{dec}}(f, g)\right)^2 - P^m (f - g)^2 \right\|_{\mathcal{H} \times \mathcal{H}} \to 0 \quad \text{a.s.}$$

Hence, the probability of the set

$$D_s(\delta) := \Big\{ \sup_{k \geq s} \left\| \left(e_{k,2}^{\text{dec}}(f, g)\right)^2 - P^m (f - g)^2 \right\|_{\mathcal{H} \times \mathcal{H}} > \delta^2 \Big\}$$

tends to zero as $s \to \infty$ for all $\delta > 0$. This, together with the bounded LIL for the decoupled process (the decoupled version of Theorem 5.4.1) implies, by uniform integrability,

$$\begin{aligned}&\mathbb{E}\Big[\sup_{k \geq s} \frac{1}{a_k^*} \Big\| \sum_{\mathbf{i} \in I_{2^k}^m} \epsilon_i^{\text{dec}} h(\mathbf{X}_{\mathbf{i}}^{\text{dec}}) \Big\|_{\mathcal{H}'_\delta} I_{D_s(\delta)} \Big] \\ &\qquad \leq 2\mathbb{E}\Big[\sup_{k} \frac{1}{a_k^*} \Big\| \sum_{\mathbf{i} \in I_{2^k}^m} \epsilon_i^{\text{dec}} h(\mathbf{X}_{\mathbf{i}}^{\text{dec}}) \Big\|_{\mathcal{H}} I_{D_s(\delta)} \Big] \to 0\end{aligned}$$

as $s \to \infty$ (hence, as $r \to \infty$), for all $\delta > 0$. Since on the set $D_s(\delta)^c$, and for $k \geq s$, we have the inclusion

$$\left\{ f, g \in \mathcal{H} : P^m (f - g)^2 \leq \delta^2 \right\} \subseteq \left\{ f, g \in \mathcal{H} : e_{k,2}^{\text{dec}}(f, g) \leq \sqrt{2}\delta \right\},$$

the last limit and (5.4.17) reduce the proof of (5.4.16) to showing

$$\lim_{\delta \to 0} \lim_{s \to \infty} \mathbb{E} \sup_{k \geq s} \frac{1}{a_k^*} \Big\| \sum_{\mathbf{i} \in I_{2^k}^m} \epsilon_i^{\text{dec}} h(\mathbf{X}_{\mathbf{i}}^{\text{dec}}) \Big\|_{\mathcal{H}'(\delta, e_{k,2}^{\text{dec}})} = 0 \tag{5.4.18}$$

where

$$\mathcal{H}'(\delta, e_{k,2}^{\text{dec}}) := \{f - g : f, g \in \mathcal{H}, e_{n,2}^{\text{dec}}(f - g) < \delta\}.$$

To prove (5.4.18), we begin by proceeding as in the proof of (4.3.13) in Theorem 4.3.3, but applying the maximal inequality (5.4.15) instead of (4.3.4′) [together, of course, with Proposition 4.3.1 and inequality (4.3.3)], to obtain

$$\mathbb{E} \sup_{k \geq s} \frac{1}{a_k^*} \Big\| \sum_{\mathbf{i} \in I_{2^k}^m} \epsilon_i^{\text{dec}} h(\mathbf{X}_{\mathbf{i}}^{\text{dec}}) \Big\|_{\mathcal{H}'(\delta, e_{k,2}^{\text{dec}})} \leq C\kappa(\delta) \mathbb{E} \sup_{k \geq s} \Big(\frac{1}{2^{km}} \sum_{I_{2^k}^m} H^2(\mathbf{X}_{\mathbf{i}}^{\text{dec}}) \Big)^{1/2},$$

where $\kappa(\delta) \to 0$ as $\delta \to 0$ and C is a universal constant. (We jumped two steps here, but these are just as in the proof of Theorem 4.3.3.) Now, also as in the proof of Theorem 4.3.3, the reversed martingale inequality bounds the above expectation by $2\big(\mathbb{E}H^2\big)^{1/2}$ and (5.4.18) follows since $\kappa(\delta) \to 0$. □

We could have chosen to state Theorem 5.4.4 in the same style as Theorem 5.3.7, that is, by stating a different LIL for each family of kernels $\{\pi_k h : h \in \mathcal{H}\}$, $k = 1, \ldots, m$.

We do not need to assume the symmetry of $h \in \mathcal{H}$, or the strict VC subgraph property for $\mathcal{H}$: *if $\mathcal{H}$ is a measurable class of not necessarily symmetric kernels satisfying the entropy condition (5.3.19), then the LIL holds for each of the classes $\{\pi_k(sh) : h \in \mathcal{H}\}$, $k = 1, \ldots, m$, where sh is the symmetrization of h, defined in Remark 5.3.9.* Statement and details of proof are straightforward.

We conclude this section by sketching the proof of the LIL for degenerate B-valued kernels when B is of type 2.

THEOREM 5.4.5. *Let B be a separable Banach space of type 2, let $(S, \mathcal{S}, P)$ be a probability space, let X_i, $i \in \mathbb{N}$, be i.i.d. S valued random variables with law P, and let $H : S^m \mapsto B$ be a P-canonical symmetric function such that $P^m \|H\|^2 < \infty$. Then, almost surely, the sequence*

$$\Big\{\alpha_n(H) := \frac{1}{(2n \log\log n)^{\frac{m}{2}}} \sum_{(i_1, \ldots, i_m) \in I_n^m} H(X_{i_1}, \ldots, X_{i_m})\Big\}$$

is relatively compact in B and its limit set is

$$K = \Big\{\mathbb{E}\big[H(X_1, \ldots, X_m) g(X_1), \ldots, g(X_m)\big] : \mathbb{E}g^2(X_1) \leq 1\Big\}, \tag{5.4.19}$$

which is compact.

PROOF. We divide it into three claims with simple proofs.

CLAIM 1. *For every $\varepsilon > 0$ there exists a simple (= measurable, finitely valued) symmetric canonical kernel $\tilde{H}$ such that $\mathbb{E}\|H - \tilde{H}\|^2 < \varepsilon$.*

Since the map $A \mapsto \int_{H \in A} \|H\|^2 dP$, $A \in \mathcal{B}$, is a finite Borel measure on B, it is a tight Borel measure and therefore there is a compact set $K \subset B$ such that

$\mathbb{E}H^2 I_{H\in K^c} < \varepsilon/2^{2m+1}$. Now we partition K into a finite number of sets $A_1, \ldots, A_k$ of diameter not exceeding $\sqrt{\varepsilon/2^{2m+1}}$, select arbitrary points $x_i \in A_i$ and define

$$\bar{H}(x) = \sum_{i=1}^{k} x_i I_{H\in A_i}.$$

If we let $\tilde{H} := \pi_m(s\bar{H})$, we have, by Jensen

$$\begin{aligned}\mathbb{E}\|H - \tilde{H}\|^2 &= \mathbb{E}\|\pi_m s(H - \bar{H})\|^2 = \mathbb{E}\Big\|\Big[\prod_{i=1}^{m}(\delta_{X_i} - P)\Big]\big(s(H - \bar{H})\big)\Big\|^2 \\ &\le 2^{2m}\mathbb{E}\|s(H - \bar{H})\|^2 \le 2^{2m}\mathbb{E}\|H - \bar{H}\|^2 \\ &\le 2^{2m}\big(\mathbb{E}\|H\|^2 I_{H\in K^c} + \mathbb{E}\|H - \bar{H}\|^2 I_{H\in K}\big) \\ &\le 2^{2m}\big(\varepsilon/2^{2m+1} + \varepsilon/2^{2m+1}\big) = \varepsilon.\end{aligned}$$

CLAIM 2. *The sequence $\{\alpha_n(H,\omega)\}_{n=1}^{\infty}$ is relatively compact for almost every ω.*

A sequence $\{x_n\} \subset B$ is totally bounded (hence relatively compact) if and only if eventually all the points in it are close enough to the first N, that is, if and only if

$$\lim_{N\to\infty} \sup_{n\in\mathbb{N}} \inf_{1\le i\le N} \|x_n - x_i\| = 0.$$

We will apply this to the sequence $x_n = \alpha_n(H,\omega)$. If we set

$$\alpha_H(\omega) := \lim_{N\to\infty} \sup_{n\in\mathbb{N}} \inf_{1\le i\le N} \|\alpha_n(H,\omega) - \alpha_i(H,\omega)\|,$$

and if $\tilde{H}$ is as in Claim 1 for a given $\varepsilon > 0$, we have

$$\alpha_H(\omega) \le \alpha_{\tilde{H}}(\omega) + 2\sup_n \|\alpha_n(H - \tilde{H},\omega)\|.$$

Now, $\alpha_{\tilde{H}}(\omega) = 0$ a.s. by the LIL in finite dimensions given in the previous chapter (as $\tilde{H}$ is finitely valued). Moreover, the bounded LIL for B-valued kernels, B of type 2 (Theorem 5.4.2), gives that

$$\mathbb{E}\sup_n \|\alpha_n(H - \tilde{H},\omega)\| \le C\big(\mathbb{E}\|H - \tilde{H}\|^2\big)^{1/2} \le C\varepsilon.$$

Since $\varepsilon > 0$ is arbitrary, it follows that $\alpha_H = 0$ a.s., proving the claim.

CLAIM 3. *The limit set of $\{\alpha_n(H)\}$ coincides a.s. with the set K defined by (5.4.19)*

This follows from the arguments in the proof of Claims 2 and 3 in Theorem 5.4.3, with only formal, obvious changes. □

5.5 Statistical Applications

Whereas there are many instances of applications of U-statistics in the literature [e.g., the books of Serfling (1980), Lee (1990), Borovskikh and Korolyuk (1994)], the theory of U-processes as developped in the previous sections is a newer subject and applications have not yet appeared in books. In this section we develop typical instances of application of the LLN, the CLT and the LIL for U-processes, illustrative of how this theory can be put to use. We first present an application of the law of large numbers for U-processes to obtain the law of large numbers for the empirical simplicial median encountered in Example 5.2.4. The procedure can be both sharpened and generalized to M-estimators based on criterion functions of several variables, but this is not pursued here. Then, we use the central limit theorem for degenerate U-processes (actually, the associated asymptotic equicontinuity condition) in order to obtain the CLT for the empirical simplicial median. In this case, the central limit theorem for canonical kernels is used in order to dispose of the higher-order terms in the Hoeffding decomposition of a U-process. The methods of proof for both the LLN and the CLT for the simplicial median generalize to M-estimators whose criterion function is multivariate. The third application consists of an analysis of the remainder term in the linearization of a "smooth" statistic, in our case, the Woodroofe estimator of the hazard function and the Lynden–Bell estimator of the cumulative distribution function for truncated data (analogous to the Nelson–Aalen and Kaplan–Meier estimators for censored data). Here, the CLT and the LIL for degenerate U-statistics determine the order of the remainder.

5.5.1 *The Law of Large Numbers for the Simplicial Median*

It is shown in Example 5.2.4 that the *empirical simplicial depth process* $D_n(\theta)$ satisfies the uniform law of large numbers. For notational simplicity, and because the loss of generality is only formal, we work in $\mathbb{R}^2$ instead of $\mathbb{R}^d$, $d \geq 1$. We should recall the definition of $D_n(\theta)$ from (5.2.14):

$$D_n(\theta) := U_n^{(3)}(C_\theta) = \frac{1}{\binom{n}{3}} \sum_{1 \leq i < j < k \leq n} I_{C_\theta}(X_i, X_j, X_k), \qquad \theta \in \mathbb{R}^2, \quad (5.2.14)$$

where X_i are the coordinates in $((\mathbb{R}^2)^{\mathbb{N}}, \mathcal{B}_2^{\mathbb{N}}, P^{\mathbb{N}})$,

$$C_\theta := \left\{(x_1, x_2, x_3) \in (\mathbb{R}^2)^3 : \theta \in S(x_1, x_2, x_3)\right\}$$

and $S(x_1, x_2, x_3)$ is the open triangle determined by the three points x_i, $1 \leq i \leq 3$. $D_n(\theta)$ is thus the proportion of sample triangles that contain the point θ, and it estimates the *depth function* $D(\theta)$ associated to the probability measure P on $\mathbb{R}^2$ (P is the probability law of the variables X), which is defined as the probability that the triangle determined by three points independently sampled from P contains θ, that is,

$$D(\theta) = P^3 C_\theta, \qquad \theta \in \mathbb{R}^2. \quad (5.5.1)$$

The law of large numbers in Section 5.2, as shown in Example 5.2.4, implies

$$\|D_n(\theta) - D(\theta)\|_\infty \to 0 \quad \text{a.s.} \tag{5.5.2}$$

Any point in the set

$$A := \{\theta : D(\theta) \text{ is maximum}\} := \text{argmax}\, D(\theta) \tag{5.5.3}$$

is a *simplicial median* for P. This is an interesting multidimensional location parameter, an analog of the median. The object of this subsection is to show that the *empirical simplicial median*, which is defined as any (measurable) choice from the random set

$$A_n := \text{argmax}\, D_n(\theta), \tag{5.5.4}$$

is a consistent estimator of the simplicial median. The empirical simplicial median is an M-estimator in that it is obtained by maximizing a statistic. Typically, M-estimators are obtained by maximizing (in θ) statistics of the form $\sum_{i=1}^n \phi(X_i, \theta)/n$, and the function $\phi(x, \theta)$ is the "criterion function" on which the estimator is based. In our case, the criterion function is $I_{C_\theta}(x_1, x_2, x_3)$, a function of several variables in the sample, and the average is then over the multiple sum without ties, that is, a U-process indexed by θ. The method of obtaining the law of large numbers for the empirical simplicial median from the law of large numbers (5.5.2) for the U-process that we give now is typical and can be applied elsewhere. We begin with a simple lemma.

LEMMA 5.5.1. *Let P be a probability measure on $\mathbb{R}^2$ such that $P(L) = 0$ for all straight lines $L \subset \mathbb{R}^2$. Then,*

i) *the depth function $D(\theta)$ is continuous;*

ii) *the set A of simplicial medians is compact and nonempty, and*

iii) *the set A is* identifiable *in the sense that, for all $\varepsilon > 0$,*

$$\delta_\varepsilon := \sup_{\theta \in \mathbb{R}^2} D(\theta) - \sup_{\theta : d(\theta, A) > \varepsilon} D(\theta) > 0,$$

where $d(\theta, A) = \inf_{y \in A} |\theta - y|$ is the distance from the point θ to the set A.

PROOF. By Fatou's lemma and the fact that the simplices S are open, if $\theta_n \to \theta$ we have

$$\begin{aligned} D(\theta) &= P^3\{\theta \in S(x_1, x_2, x_3)\} \le P^3\big\{\theta_n \in S(x_1, x_2, x_3) \quad \text{from some } n \text{ on}\big\} \\ &= P^3\big(\liminf_n \{\theta_n \in S(x_1, x_2, x_3)\}\big) \le \liminf_n P^3\{\theta_n \in S(x_1, x_2, x_3)\} \\ &= \liminf_n D(\theta_n). \end{aligned}$$

Hence, D is lower semicontinuous.

Let $\tilde{C}_\theta = \{(x_1, x_2, x_3) : \theta \in \overline{S(x_1, x_2, x_3)}\}$. We will prove that the hypothesis on P implies that $P^3(C_\theta) = P^3(\tilde{C}_\theta)$. We note first that the probability that X_1 and X_2 are aligned with a fixed point θ is zero as this is the integral with respect to $dP(x)$ of the probability of X_1 being in the line determined by the two points x and θ, which is zero; so, if $B = \{(x_1, x_2) : x_1$ and x_2 are not aligned with $\theta\}$ then $P^2B = 1$, and therefore, letting $C_\theta(x_1, x_2) = \{x : \theta \in S(x_1, x_2, x)\}$ and $\tilde{C}_\theta(x_1, x_2) = \{x : \theta \in \overline{S(x_1, x_2, x)}\}$, we have

$$P^3(C_\theta) = \iint_B P(C_\theta(x_1, x_2))dP(x_1)dP(x_2),$$

$$P^3(\tilde{C}_\theta) = \iint_B P(\tilde{C}_\theta(x_1, x_2))dP(x_1)dP(x_2).$$

Now, $C_\theta(x_1, x_2)$, for x_1, x_2 not aligned with θ, is the open cone of $\mathbb{R}^2$ with vertex at θ, boundaries $\theta + \lambda(\theta - x_1)$, $\theta + \mu(\theta - x_2)$, $\lambda, \mu > 0$, and angle $\alpha < \pi$, and $\tilde{C}_\theta(x_1, x_2)$ is the closure of this cone. Hence, the difference between $C_\theta(x_1, x_2)$ and $\tilde{C}_\theta(x_1, x_2)$ consists of the union of two half lines, which has probability zero, and therefore, the two probabilities coincide and so do the integrals.

Let now $\theta_n \to \theta$. Then, the fact that $P^3(C_\theta) = P^3(\tilde{C}_\theta)$ gives, by Fatou's lemma, that

$$\begin{aligned}\limsup_n D(\theta_n) &= \limsup_n P^3(C_{\theta_n}) = \limsup_n P^3(\tilde{C}_{\theta_n}) \\ &\le P^3\left(\limsup_n \tilde{C}_{\theta_n}\right) = P^3\left(\limsup_n \{\theta_n \in \overline{S(x_1, x_2, x_3)}\}\right) \\ &\le P^3\{\theta \in \overline{S(x_1, x_2, x_3)}\} = P^3(\tilde{C}_\theta) = P^3(C_\theta) = D(\theta).\end{aligned}$$

Hence, D is upper semicontinuous and therefore, continuous. As an immediate consequence, the set A of simplicial medians is closed (if $\theta_n \in A$ converge to θ, then $D(\theta)$ is also maximum and therefore $\theta \in A$).

By tightness of P^3 in $\left(\mathbb{R}^2\right)^3$, $P^3\{|x_i| \ge M : 1 \le i \le 3\} \to 0$ as $M \to \infty$, and therefore, $D(\theta) \to 0$ as $|\theta| \to \infty$. On the other hand, since P gives mass zero to lines,

$$a := \sup_\theta D(\theta) > 0$$

[otherwise $S(X_1, X_2, X_3)$ contains no rational points with probability one, hence it has a.s. empty interior, and then, by Fubini, $\Pr\{X_3 \in L(x_1, x_2)\} = 1$ P^2-a.s., where $L(x_1, x_2)$ is the line determined by x_1 and x_2]. These two facts imply that A is bounded. Hence, A is compact.

Finally, if $a := \sup_\theta D(\theta) = \sup_{\theta : d(\theta, A) \ge \varepsilon} D(\theta)$ for some $\varepsilon > 0$, then there is a finite number M such that $a = \sup_{\theta : d(\theta, A) > \varepsilon, |\theta| \le M} D(\theta)$ [it suffices to choose $M < \infty$ so that $D(\theta)$ is smaller than $a/2$ for $|\theta| > M$, which exists by the argument in the previous paragraph]. So, by compactness of $\{|\theta| \le M\}$, there is a convergent sequence $\theta_n \to \theta$ satisfying $d(\theta_n, A) \ge \varepsilon$ and $D(\theta_n) \to a$; but then, by upper semicontinuity of D, $D(\theta) = a$, that is $\theta \in A$, contradiction. □

Note that if we take the simplices closed in the definition of C_θ, then the previous proof works equally well: in this case, upper semicontinuity is automatic, and lower semicontinuity requires $P(L) = 0$ for all L. Taking the simplices open or closed is mostly a matter of convenience but there are some advantages at taking them open (e.g., the corresponding estimator is smoother—a fact that helps in the next subsection—and is also more robust (Chen, 1995)).

Here is the law of large numbers, valid even for nonmeasurable simplicial medians.

THEOREM 5.5.2. *Let P be a probability measure on $\mathbb{R}^2$ giving mass zero to every straight line, let θ_n be any point in the random set $A_n = \operatorname{argmax} D_n$, and let A be the simplicial median set for P, $A = \operatorname{argmax} D$. Then,*

$$\left[d(\theta_n, A)\right]^* \to 0 \quad \text{a.s.}, \tag{5.5.5}$$

where * denotes outer expectation.

PROOF. Let $\theta \in A$ and, given $\varepsilon > 0$, let $\delta = \delta(\varepsilon)$ be as prescribed by the identifiability condition in Lemma 5.5.1. Then, since $D_n(\theta) - D_n(\theta_n) \leq 0$ for all n, we have

$$\begin{aligned}
\Pr{}^*\left\{\sup_{n\geq\ell} d(\theta_n, A) > \varepsilon\right\} &\leq \Pr{}^*\left\{\sup_{n\geq\ell}[D(\theta) - D(\theta_n)] \geq \delta\right\} \\
&= \Pr{}^*\left\{\sup_{n\geq\ell}\left[(D(\theta) - D_n(\theta)) + (D_n(\theta) - D_n(\theta_n)) + (D_n(\theta_n) - D(\theta_n))\right] \geq \delta\right\} \\
&\leq \Pr{}^*\left\{\sup_{n\geq\ell}\left[(D(\theta) - D_n(\theta)) + (D_n(\theta_n) - D(\theta_n))\right] \geq \delta\right\} \\
&\leq \Pr\left\{\sup_{n\geq\ell}(D(\theta) - D_n(\theta)) \geq \delta/2\right\} + \Pr{}^*\left\{\sup_{n\geq\ell}(D(\theta_n) - D_n(\theta_n)) \geq \delta/2\right\} \\
&\leq 2\Pr\left\{\sup_{n\geq\ell}\|D(\theta) - D_n(\theta)\|_\infty \geq \delta/2\right\} \to 0
\end{aligned}$$

as $\ell \to \infty$, by the law of large numbers (5.5.2) from Example 5.2.4. □

It is interesting to note that Theorem 5.5.2 requires only a *one-sided* uniform law of large numbers, $\sup_\theta\left(D(\theta) - D_n(\theta)\right) \to 0$ a.s.

We may ask whether the law of large numbers is still true if we look at the maximizer of D_n *restricted to the sample points* for the definition of the simplicial median. Under some extra regularity we can still obtain the LLN for this modified empirical simplicial median. This observation is important because such a parameter is much easier to compute than the one considered above. Concretely, we define the modified empirical simplicial median to be any of the deepest data points, that is, any point in the set

$$\tilde{A}_n := \{X_i : \max_{1\leq j\leq n} D_n(X_j) = D_n(X_i)\}. \tag{5.5.6}$$

The previous proposition then has the following corollary:

COROLLARY 5.5.3. *If P gives zero mass to straight lines and if there exists $\theta_0 \in A$ such that $P(B(\theta_0, \delta)) > 0$ for all $\delta > 0$, where $B(\theta, \delta)$ is the open circle with center θ and radius δ, then*

$$\max_{1 \le i \le n} D_n(X_i) \to \sup_\theta D(\theta) \quad \text{a.s.} \tag{5.5.7}$$

and

$$\Big[\max_{\theta \in \tilde{A}_n} d(\theta, A)\Big]^* \to 0 \quad \text{a.s.} \tag{5.5.8}$$

as $n \to \infty$. If in addition A is convex, in particular, if A consists of a single point, and if we define $\tilde{\theta}_n$ to be the average of all the points in $\tilde{A}_n$, then

$$d(\tilde{\theta}_n, A) \to 0 \quad \text{a.s.} \tag{5.5.9}$$

PROOF. By the law of large numbers (5.5.2) for $D_n - D$,

$$\max_{1 \le i \le n} D_n(X_i) \le \sup_\theta D_n(\theta) \le \sup_\theta D(\theta) + \|D_n - D\|_\infty \to \sup_\theta D(\theta) \quad \text{a.s.}$$

To prove the reverse inequality, let X_n^* denote the (random) point closest to θ_0 among $X_1, \ldots, X_n$ (we can choose the one with lowest subindex in case of ties). Then X_n^* converges to θ_0 a.s. by the Borel–Cantelli lemma: for all $\delta > 0$,

$$\sum \Pr\big\{X_n^* \notin B(\theta_0, \delta)\big\} = \sum (1 - P\{B(\theta_0, \delta)\})^n < \infty$$

since $P\{B(\theta_0, \delta)\} > 0$ by hypothesis. Hence, D being continuous by Lemma 5.5.1, $D(X_n^*) \to D(\theta_0)$ a.s. and we have

$$\max_{1 \le i \le n} D_n(X_i) \ge D_n(X_n^*) \ge D(X_n^*) - \|D_n - D\|_\infty \to D(\theta_0) = \sup_\theta D(\theta) \quad \text{a.s.}$$

This completes the proof of (5.5.7). As for (5.5.8), given $\varepsilon > 0$, we take $\delta = \delta_\varepsilon > 0$ from the identifiability condition in Lemma 5.5.1, and set $a = \sup_\theta D(\theta) > 0$ as above, to obtain

$$\begin{aligned}
\Pr^*\Big\{\sup_{n \ge \ell} \max_{\theta \in \tilde{A}_n} d(\theta, A) > \varepsilon\Big\} &\le \Pr^*\Big\{\sup_{n \ge \ell} \max_{\theta \in \tilde{A}_n}\big[a - D(\theta)\big] \ge \delta\Big\} \\
&\le \Pr\Big\{\sup_{n \ge \ell}\big[a - \min_{\theta \in \tilde{A}_n} D_n(\theta)\big] \ge \delta/2\Big\} + \Pr\Big\{\sup_{n \ge \ell} \max_{\theta \in \tilde{A}_n}\big[D_n(\theta) - D(\theta)\big] > \delta/2\Big\} \\
&\le \Pr\Big\{\sup_{n \ge \ell}\big[\sup_\theta D(\theta) - \max_{1 \le i \le n} D_n(X_i)\big] \ge \delta/2\Big\} + \Pr\Big\{\sup_{n \ge \ell} \|D_n - D\|_\infty \ge \delta/2\Big\}.
\end{aligned}$$

Now, the first summand tends to zero because of (5.5.7), and so does the second, by the law of large numbers (5.5.2). The law of large numbers (5.5.8) is thus proved. (5.5.9) follows trivially from (5.5.8). □

In connection with the limit (5.5.9), it is interesting to mention that, as far as we know, there is only one general class of probability laws in $\mathbb{R}^d$ for which the simplicial median is unique. These are the angularly symmetric probability laws that assign positive mass to any neighborhood of their center of symmetry. A random vector X on $\mathbb{R}^d$, or its probability law P, is *angularly symmetric about the point* θ if the random variable $(X-\theta)/|X-\theta|$, which takes values in the unit sphere of $\mathbb{R}^d$, is symmetric. The next lemma holds in $\mathbb{R}^d$ with obvious changes ("straight lines" by "hyperplanes," "4" by "2^d") but it is only stated and proved for the plane.

LEMMA 5.5.4. *Let P be a probability law on the plane giving zero mass to straight lines. If P is angularly symmetric about a point θ, then*

$$D(x) \leq D(\theta) = 1/4 \tag{5.5.10}$$

and

$$\Pr\{\theta \in S(X_1, X_2, x)\} = 1/4 \tag{5.5.11}$$

for all $x \in \mathbb{R}^2 \setminus \{\theta\}$. If, in addition, $P\big(B(\theta,\delta)\big) > 0$ for all $\delta > 0$, then

$$D(\theta) > D(x) \tag{5.5.12}$$

for all $x \in \mathbb{R}^2 \setminus \{\theta\}$, in particular, θ is the only simplicial median for P.

PROOF. We prove equation (5.5.11) first. We can assume $\theta = 0$. Recall the definition of $C_0(x_2, x_3)$ given in the proof of Lemma 5.5.1: $C_0(x_2, x_3)$ is the set of $x \in \mathbb{R}^2$ such that $0 \in S(x, x_2, x_3)$ and it coincides with the (planar) cone of vertex 0, angle $\alpha < \pi$ and boundary the half lines $-\lambda x_2$, $-\mu x_3$, λ, $\mu > 0$. Then, the cones $C_0(x_2, x_3)$ and $C_0(-x_2, x_3)$ are disjoint and their union is a half plane (the one determined by the line through 0 and x_2 that contains $-x_3$) minus a half line (the one through 0 and $-x_3$). Hence, angular symmetry of P about 0 and $P(L) = 0$ for all lines L, imply

$$P\{C_0(x_2,x_3)\} + P\{C_0(-x_2,x_3)\} = P\big[C_0(x_2,x_3) \cup C_0(-x_2,x_3)\big] = 1/2.$$

On the other hand, since, again by angular symmetry, if X_1 and X_2 are independent with law P, then the pairs of random vectors $(X_1/|X_1|, X_2/|X_2|)$ and $(X_1/|X_1|, -X_2/|X_2|)$ are equally distributed, it follows that, for all $x_3 \in \mathbb{R}^2 \setminus \{0\}$,

$$\begin{aligned}\Pr\{0 \in S(X_1, X_2, x_3)\} &= \Pr\{0 \in S(X_1/|X_1|, X_2/|X_2|, x_3)\}\\ &= \Pr\{0 \in S(X_1/|X_1|, -X_2/|X_2|, x_3)\}\\ &= \Pr\{0 \in S(X_1, -X_2, x_3)\}.\end{aligned}$$

Then, the two previous observations give

$$\begin{aligned}\Pr\{0 \in S(X_1, X_2, x_3)\} &= \frac{1}{2}\Big[\Pr\{0 \in S(X_1, X_2, x_3)\} + \Pr\{0 \in S(X_1, -X_2, x_3)\}\Big]\\ &= \frac{1}{2}\int \big[P\{C_0(x_2,x_3)\} + P\{C_0(-x_2,x_3)\}\big] dP(x_2) = \frac{1}{4},\end{aligned}$$

proving (5.5.11).

To prove (5.5.10) and (5.5.12), we need some extra notation. We define $A_{\text{in}}(x_1, x_2, x_3)$ as the set of points $(x_1, x_2, x_3) \in (\mathbb{R}^2)^3$ for which the vector $\overrightarrow{0x}$ "enters" the triangle $S(x_1, x_2, x_3)$, that is, as the collection of the triplets of vertices corresponding to triangles that intersect the segment $\overline{0x}$ but that do not contain the point 0; likewise, $A_{\text{out}}(x_1, x_2, x_3)$ is the set of points $(x_1, x_2, x_3) \in (\mathbb{R}^2)^3$ such that the triangle $S(x_1, x_2, x_3)$ intersects the segment $\overline{0x}$ but does not contain the point x [the vector $\overrightarrow{0x}$ "leaves" the triangle $S(x_1, x_2, x_3)$]. Then,

$$D(0) - D(x) = \Pr\{A_{\text{out}} \setminus A_{\text{in}}\} - \Pr\{A_{\text{in}} \setminus A_{\text{out}}\}$$

(here we are discounting the probability that a triangle contain both 0 and x from the definitions of $D(0)$ and $D(x)$); further adding and subtracting $\Pr\{A_{\text{in}} \cap A_{\text{out}}\}$, we obtain

$$D(0) - D(x) = \Pr\{A_{\text{out}}\} - \Pr\{A_{\text{in}}\}.$$

If $S(x_1, x_2, x_3)$ is in A_{in} or A_{out} then the segment $\overline{0x}$ cuts the boundary ∂S only once, hence, up to a null set, it cuts only one of the three sides $\overline{x_{i_1} x_{i_2}}$, $1 \le i_1 < i_2 \le 3$. Hence, if H_{x_1,x_2} denotes the half plane that contains 0 and whose boundary is the line joining x_1 and x_2, and if $C := \{(a, b) \in (\mathbb{R}^2)^2 : \overline{ab} \cap \overline{0x} \neq \emptyset\}$ is the set of end points of segments that intersect $\overline{0x}$, we have

$$\begin{aligned}
\Pr(A_{\text{out}}) - \Pr(A_{\text{in}}) &= 3\Pr\{X_3 \in H_{X_1,X_2}, (X_1, X_2) \in C\} \\
&\quad - 3\Pr\{X_3 \notin H_{X_1,X_2}, (X_1, X_2) \in C\} \\
&= 3\int_C \big[\Pr\{X_3 \in H_{x_1,x_2}\} - \Pr\{X_3 \notin H_{X_1,X_2}\}\big] dP(x_1) dP(x_2).
\end{aligned}$$

But, since the half plane H contains zero, angular symmetry implies that the difference of probabilities under the integral sign is non-negative, and it is positive if every neigborhood of zero has positive mass; since the set C has also positive probability in this last case, (5.5.10) and (5.5.12) are both proved. □

As a consequence of Proposition 5.5.2, Corollary 5.5.3 and Lemma 5.5.4, we have the following: Let P be an angularly symmetric probability measure on $\mathbb{R}^2$ (or $\mathbb{R}^d$) with a nonvanishing density, and let θ be its center of symmetry. Let θ_n denote any selection (measurable or not) from the random set A_n of empirical simplicial medians, or even the baricenter of A_n and, likewise, let $\tilde{\theta}_n$ denote any selection from the random set $\tilde{A}_n$ of modified empirical simplicial medians, or even the baricenter of $\tilde{A}_n$. Then

$$|\theta_n - \theta|^* \to 0 \quad \text{a.s.} \quad \text{and} \quad |\tilde{\theta}_n - \theta|^* \to 0 \quad \text{a.s.,} \tag{5.5.13}$$

that is, *both* θ_n *and* $\tilde{\theta}_n$ *are consistent estimators of the center of angular symmetry of* P.

The baricenter of A_n (resp. $\tilde{A}_n$) is measurable, but it may not belong to A_n (resp. $\tilde{A}_n$). It should be noted that measurable selections θ_n from A_n (or $\tilde{\theta}_n$ from $\tilde{A}_n$) do exist: this is a consequence of selection theorems, e.g. Corollary 8.5.4 in Cohn (1980), page 286. In other words, θ_n and $\tilde{\theta}_n$ may be taken to be true random variables. We will make this assumption in the next subsection, where we prove the central limit theorem for θ_n.

5.5.2 The Central Limit Theorem for the Simplicial Median

We will show that, if we assume some extra smoothness on P, the estimator θ_n of the center of angular symmetry θ of P in (5.5.13) is $\sqrt{n}$ *consistent*, in fact, asymptotically normal. We assume, without loss of generality, that $\theta = 0$, that is, that P *is angularly symmetric about zero*. We also assume, to begin, that

(P.1) P has a bounded density f such that $f(0) \neq 0$.

(P.2) f is twice differentiable with continuous second derivatives, its partial derivatives f_i are Lebesgue integrable and there exists $\delta > 0$ such that the functions $\sup_{|\theta|<\delta} |f_{ij}(x+\theta)|$ are also Lebesgue integrable.

We also need some notation. For $\theta \in \mathbb{R}^2$ and $\mathbf{x} = (x_1, \ldots, x_k) \in \left(\mathbb{R}^2\right)^k$, $k = 2, 3$, we define the joint density

$$\phi_k(\mathbf{x}, \theta) = f(x_1 + \theta) \cdots f(x_k + \theta)$$

and let $\phi_k'(\mathbf{x}, 0)$ and $\phi_k''(\mathbf{x}, 0)$ denote respectively the first and second derivatives of ϕ_k with respect to θ at $\theta = 0$, that is

$$\phi_k'(\mathbf{x}, 0) = \Big(\frac{\partial \phi_k}{\partial \theta_j}(\mathbf{x}, 0) : j = 1, 2\Big), \ \phi_k''(\mathbf{x}, 0) = \Big(\frac{\partial^2 \phi_k}{\partial \theta_i \partial \theta_j}(\mathbf{x}, 0) : i, j = 1, 2\Big).$$

The following symmetric matrix will be important below:

$$B(= B_f) := -\int_{\left(\mathbb{R}^2\right)^3} I_{C_0}(\mathbf{x}) \phi_3''(\mathbf{x}, 0) d\mathbf{x}, \tag{5.5.14}$$

where, as above, $C_\theta := \{\mathbf{x} \in \left(\mathbb{R}^2\right)^3 : \theta \in S(\mathbf{x})\}$. We need to make another assumption related to the matrix B:

(P.3) $\det(B) \neq 0$.

We begin the proof of the central limit theorem for θ_n by showing that $D(\theta) = P^3 I_{C_\theta}$ has the "right" development at 0:

CLAIM 1. *The matrix B is positive definite and*

$$D(\theta) - D(0) = -\frac{1}{2}\theta \cdot B \cdot \theta' + o(|\theta|^2). \tag{5.5.15}$$

PROOF. Using obvious changes of coordinates, using $(P.2)$ to justify differentiation and passage to the limit under the integral sign, and letting $\mathbf{x} = (x_1, x_2, x_3) \in (\mathbb{R}^2)^3$, we have

$$\begin{aligned} D(\theta) - D(0) &= \int I_{C_0}(\mathbf{x})\big[\phi_3(\mathbf{x}, \theta) - \phi_3(\mathbf{x}, 0)\big]d\mathbf{x} \\ &= \langle \theta, \int I_{C_0}(\mathbf{x})\phi_3'(\mathbf{x}, 0)d\mathbf{x}\rangle - \frac{1}{2}\theta \cdot B \cdot \theta^t \\ &\quad + \frac{1}{2}\theta \cdot \Big[\int I_{C_0}(\mathbf{x})\big[\phi_3''(\mathbf{x}, \eta\theta) - \phi_3''(\mathbf{x}, 0)\big]d\mathbf{x}\Big] \cdot \theta^t \\ &= \langle \theta, \int I_{C_0}(\mathbf{x})\phi_3'(\mathbf{x}, 0)d\mathbf{x}\rangle - \frac{1}{2}\theta \cdot B \cdot \theta^t + o(|\theta|^2), \end{aligned}$$

where $0 \le \eta \le 1$ is a number that depends on θ and $\mathbf{x}$. Now, by Lemma 5.5.4, $-D$ has gradient zero at zero and the matrix of its second partial derivatives at $\theta = 0$ is non-negative definite. Hence, it follows from the previous calculations and $(P.3)$ that $\int I_{C_0}(\mathbf{x})\phi_3'(\mathbf{x}, 0)d\mathbf{x} = 0$ and that B is positive definite. □

Let now θ_n be a measurable empirical simplicial median of P, that is, a random variable such that $\theta_n \in A_n$ a.s. By (5.5.13),

$$\theta_n \to 0 \quad \text{a.s.} \tag{5.5.13$'$}$$

This and Claim 1 allow us to bound $n|\theta_n|^2$ in terms of $n\big(D(\theta_n) - D(0)\big)$, taking us closer to U-processes:

CLAIM 2. *There exists $c > 0$ such that*

$$cn|\theta_n|^2 \le n\big(D(0) - D(\theta_n)\big) + o_P(1). \tag{5.5.16}$$

PROOF. Since B is positive definite there is $c > 0$ such that $2c|\theta|^2 \le \theta \cdot B \cdot \theta^t/2$ for all $\theta \in \mathbb{R}^2$. By Claim 1 and the law of large numbers (5.5.13$'$) for θ_n, we have

$$\Pr\Big\{\frac{1}{2}\theta_n \cdot B \cdot \theta_n^t - D(0) + D(\theta_n) \le c|\theta_n|^2\Big\} \to 1$$

as $n \to \infty$, so that $Z_n I_{\frac{1}{2}\theta_n \cdot B \cdot \theta_n^t - D(0) + D(\theta_n) > c|\theta_n|^2} \to 0$ in probability for any random variables Z_n. Hence,

$$\begin{aligned} cn|\theta_n|^2 &\le \frac{n}{2}\theta_n \cdot B \cdot \theta_n^t - cn|\theta_n|^2 \\ &\le n\big(D(0) - D(\theta_n)\big) + \big(\frac{n}{2}\theta_n \cdot B \cdot \theta_n^t - cn|\theta_n|^2\big) I_{\frac{1}{2}\theta_n \cdot B \cdot \theta_n^t - D(0) + D(\theta_n) > c|\theta_n|^2} \\ &= n\big(D(0) - D(\theta_n)\big) + o_P(1). \end{aligned}$$

□

The next step relates $n|\theta_n|^2$ to empirical processes via Hoeffding's decomposition and the central limit theorem for degenerate U-processes over VC classes of sets (Theorem 5.3.7).

CLAIM 3.

$$cn|\theta_n|^2 \le 3n(P_n - P)(P^2(C_{\theta_n}) - P^2(C_0)) + o_P(1). \tag{5.5.17}$$

PROOF. First we add and subtract the appropriate variables in (5.5.16). We should recall that $P_n = \sum_{i=1}^n \delta_{X_i}/n$ is the empirical measure corresponding to the observations X_i, $i = 1, \ldots, n$, that $D(\theta) = P^3(C_\theta)$, $D_n(\theta) = U_n^{(3)}(I_{C_\theta})$ and that, by definition, $D_n(\theta_n) \ge D_n(\theta)$ for all $\theta \in \mathbb{R}^2$. Then, (5.5.16) gives

$$\begin{aligned} cn|\theta_n|^2 &\le n\big(U_n^{(3)} - P^3\big)(I_{C_{\theta_n}} - I_{C_0}) + n\big(D_n(0) - D_n(\theta_n)\big) + o_P(1) \\ &\le n\big(U_n^{(3)} - P^3\big)(I_{C_{\theta_n}} - I_{C_0}) + o_P(1). \end{aligned} \tag{5.5.18}$$

By Hoeffding's decomposition,

$$\begin{aligned} n\big(U_n^{(3)} - P^3\big)(I_{C_{\theta_n}} - I_{C_0}) &= 3n(P_n - P)(P^2 I_{C_{\theta_n}} - P^2 I_{C_0}) \\ &\quad + n\sum_{k=2}^{3}\binom{3}{k} U_n^{(k)}\big(\pi_k(I_{C_{\theta_n}} - I_{C_0})\big). \end{aligned} \tag{5.5.19}$$

We want to show that the higher-order terms in (5.5.19) are negligible by applying Theorem 5.3.7. As shown in Example 5.2.4, the class of sets $\{C_\theta : \theta \in \mathbb{R}^2\}$ is measurable VC, and therefore Theorem 5.3.7 applies to it. Actually, we will apply the limit (5.3.11′), to the effect that

$$\lim_{\delta\to 0}\sup_n n^{k/2}\mathbb{E}\Big[\sup_{\theta,\theta' : P^3(I_{C_\theta} - I_{C_{\theta'}})^2 \le \delta^2} \big|U_n^{(k)}\big(\pi_k(I_{C_\theta} - I_{C_{\theta'}})\big)\big|\Big] = 0. \tag{5.5.20}$$

Since the planar cones $C_\theta(x_1, x_2) = \{x : (x_1, x_2, x) \in C_\theta\}$ satisfy that

$$I_{C_\theta(x_1,x_2)}(x) \to I_{C_0(x_1,x_2)}(x)$$

as $\theta \to 0$, except for $x \in \partial(C_0(x_1, x_2))$, which is a continuity set for P, repeated application of the bounded convergence theorem gives

$$P^3(I_{C_\theta} - I_{C_0})^2 \to 0 \quad \text{as} \quad \theta \to 0, \tag{5.5.21}$$

and this, combined with (5.5.20), yields

$$\lim_{\delta\to 0}\sup_n n^{k/2}\mathbb{E}\Big[\sup_{|\theta|\le\delta} \big|U_n^{(k)}\big(\pi_k(I_{C_\theta} - I_{C_0})\big)\big|\Big] = 0. \tag{5.5.20′}$$

Replacing this limit in (5.5.19) (recall $\theta_n \to 0$ a.s.) and then in (5.5.18), we obtain (5.5.17). □

Now, we must study the class of functions $\{P^2 I_{C_\theta} : |\theta| < \delta\}$, for $\delta > 0$. These functions should be well behaved because they are smoothed out by two integrations. Since, changing coordinates,

$$(P^2 I_{C_\theta})(x) = \int_{(\mathbb{R}^2)^2} I_{C_0}(x_1, x_2, x - \theta)\phi_2((x_1, x_2), \theta)\,dx_1 dx_2,$$

and the derivative of $I_{C_0}(x_1, x_2, x-\theta)$ with respect to θ_i at $\theta = 0$ is zero except if x_1 or x_2 are in the half lines determined by $-x$ and 0, formally the vector of derivatives at zero of $(P^2 I_{C_\theta})(x)$ is

$$\Delta(x) := \int_{(\mathbb{R}^2)^2} I_{C_0}(\mathbf{x}, x)\phi_2'(\mathbf{x}, 0)d\mathbf{x}. \tag{5.5.22}$$

Now, by the identity (5.5.11) in Lemma 5.5.4, if X has law P, letting Δ_i denote the ith coordinate of the vector Δ, we have, for $i = 1, 2$,

$$\mathbb{E}\Delta_i(X) = 2\int I_{C_0}(x_1, x_2, x_3) f_i(x_1) f(x_2) f(x_3) dx_1 dx_2 dx_3 = \frac{1}{2}\int f_i(x)dx;$$

it is easy to see (mean value theorem) that condition (P.2) implies that the functions $\sup_{|\eta|<\delta}|f_i(x+\eta)|$, $i = 1, 2$, are integrable, and that this gives, by dominated convergence, that partial differentiation and integration can be exchanged in the last integral, thus obtaining $\mathbb{E}\Delta_i(X) = 0$. Also,

$$\begin{aligned}
&\mathbb{E}\big(\Delta_i(X)\big)^2 \\
&\quad = \int\Big(\int I_{C_0}(x_1, x_2, x_3)\big[f_i(x_1)f(x_2) + f(x_1)f_i(x_2)\big]dx_1dx_2\Big)^2 f(x_3)dx_3 \\
&\quad \le 4\int\Big(\int I_{C_0}(x_1, x_2, x_3) f_i(x_1) f(x_2) dx_1 dx_2\Big)^2 f(x_3)dx_3 \\
&\quad \le \int |f_i(x_1)| f(x_2) |f_i(y_1)| f(y_2) f(x_3) dx_1 dx_2 dy_1 dy_2 dx_3 \\
&\quad = 4\Big(\int |f_i| dx\Big)^2 < \infty
\end{aligned}$$

again by (P.2). Hence, $\mathbb{E}\Delta(X) = 0$ and $\mathbb{E}|\Delta(X)|^2 < \infty$, implying that $\Delta(X)$ satisfies the central limit theorem, that is

$$\mathcal{L}\big(n^{1/2}P_n\Delta\big) \to_{\mathcal{L}} N\big(0, \operatorname{cov}(\Delta(X))\big). \tag{5.5.23}$$

The stochastic boundedness of $\{n^{1/2}P_n\Delta\}$ allows us to prove

CLAIM 4. *Let $r(x,\theta)$ be defined for $x \in \mathbb{R}^2$ and $\theta \in \mathbb{R}^2 \setminus \{0\}$ by the relation*

$$(P^2C_\theta)(x) = (P^2C_0)(x) + \theta\cdot\Delta(x) + |\theta| r(x,\theta), \tag{5.5.24}$$

and let $r(x,0) = 0$. Suppose that

$$n^{1/2}(P_n - P)(r(\cdot, \theta_n)) = O_P(1). \tag{5.5.25}$$

Then, the sequence $\{n^{1/2}|\theta_n|\}$ is stochastically bounded, that is,

$$n^{1/2}|\theta_n| = O_P(1). \tag{5.5.26}$$

Note that if $\Delta(x)$ were a true derivative of $P^2(C_\theta)(x)$ at $\theta = 0$, then $r(x,\theta_n)$ would tend to zero a.s. However, since $\Delta(x)$ is only a derivative in a weak sense, likewise, $r(x,\theta_n)$ must tend to zero in a weak sense. Condition (5.5.25), with $O_P(1)$ replaced by $o_P(1)$, together with condition (5.5.25″) below, is the kind of *stochastic differentiablity* we need to impose on $P^2(C_\theta)(x)$ in order to obtain the CLT for θ_n.

PROOF OF CLAIM 4. By Claim 3 and (5.5.24), we have

$$\frac{c}{3}n|\theta_n|^2 \le n^{1/2}|\theta_n|\big[|n^{1/2}(P_n - P)(\Delta(\cdot))| + |n^{1/2}(P_n - P)(r(\cdot, \theta_n))|\big] + o_P(1).$$

Then, the central limit theorem (5.5.26) for Δ and condition (5.5.25) for r give

$$\frac{c}{3}\big(n^{1/2}|\theta_n|\big)^2 \le \xi_n\big(n^{1/2}|\theta_n|\big) + \eta_n$$

where $\xi_n = \big(O_P(1)\big)$ and $\eta_n = o_P(1)$. Solving the last relation yields (5.5.26). □

Modulo proving (5.5.25), we have thus obtained the *bounded CLT for the empirical simplicial median.* It turns out that with very little extra work we can also prove that $n^{1/2}\theta_n$ is asymptotically normal.

CLAIM 5. *If (5.5.25) holds and moreover there exists a sequence of random variables δ_n such that $n^{1/2}\delta_n \to 0$ in probability for which, both*

$$n^{1/2}(P_n - P)(r(\cdot, \theta_n)) = O_P(1) \tag{5.5.25$'$}$$

and

$$n^{1/2}(P_n - P)\big(r(\cdot, 3(P_n\Delta)B^{-1} + \delta_n)\big) = o_P(1), \tag{5.5.25$''$}$$

then

$$\mathcal{L}\big(n^{1/2}\theta_n\big) \to_w N(0, \Gamma) \tag{5.5.27}$$

where $\Gamma = 9B^{-1}(\mathrm{cov}_P\,\Delta)B^{-1}$.

PROOF. To avoid technicalities in an already too long proof, we will assume that (5.5.25″) holds for $\delta_n = 0$, and leave it to the reader to convince himself that everything goes through if (5.5.25″) only holds for a $o_P(n^{-1/2})$ perturbation of $3(P_n\Delta)B^{-1}$. To ease notation, we set

$$\Delta_n := 3P_n(\Delta).$$

We begin by estimating $n(D_n(\Delta_n B^{-1}) - D_n(0))$ using the developments (5.5.15) and (5.5.19), together with (5.5.20′), the definition of r in (5.5.24), and the fact that $\Delta_n B^{-1} \to 0$ in probability [by (5.5.23)]:

$$\begin{aligned}
&n(D_n(\Delta_n B^{-1}) - D_n(0))\\
&\quad= n(D(\Delta_n B^{-1}) - D(0)) + n\big((D_n - D)(\Delta_n B^{-1}) - (D_n - D)(0)\big)\\
&\quad= -\frac{1}{2}n\Delta_n B^{-1}\Delta_n^t + 3n(P_n - P)\big[\Delta_n B^{-1}\Delta^t(\cdot) + |\Delta_n B^{-1}|r(\cdot, \Delta_n B^{-1})\big] + o_P(1)\\
&\quad= \frac{1}{2}n\Delta_n B^{-1}\Delta_n^t + 3\big[n^{1/2}|\Delta_n B^{-1}|\big]\big[n^{1/2}(P_n - P)\big(r(\cdot, \Delta_n B^{-1})\big)\big] + o_P(1).
\end{aligned}$$

Then, condition (5.5.25″) together with the stochastic boundedness of $\{n^{1/2}\Delta_n\}$ give

$$n(D_n(\Delta_n B^{-1}) - D_n(0)) = \frac{1}{2}n\Delta_n B^{-1}\Delta_n^t + o_P(1).$$

The same arguments, but using (5.5.25′) instead of (5.5.25″), and stochastic boundedness of $\{n^{1/2}\theta_n\}$ (Claim 4) instead of (5.5.23), give

$$n(D_n(\theta_n) - D_n(0)) = -\frac{1}{2}n\theta_n \cdot B \cdot \theta_n^t + n\theta_n\Delta_n^t + o_P(1).$$

Since, by definition of θ_n, $D_n(\Delta_n B^{-1}) \leq D_n(\theta_n)$, the two previous identities imply

$$\frac{1}{2}n\Delta_n B^{-1}\Delta_n^t + \frac{1}{2}n\theta_n \cdot B \cdot \theta^t - n\theta_n\Delta_n^t \leq o_P(1).$$

B being positive definite and symmetric, there is a nondegenerate matrix C such that $CBC^t = I$. Then, we can rewrite the last identity as

$$n|\Delta_n C^t - \theta_n C^{-1}|^2 = o_P(1).$$

Hence the sequence $\{n^{1/2}\theta_n\}$ has the same limit law as $\{n^{1/2}\Delta_n B^{-1}\}$, which is $N(0, \Gamma)$. □

We are then left with the task of proving the limits (5.5.25′) and (5.5.25″), and this is an exercise on empirical processes, that we will not complete in full detail.

CLAIM 6. *The limits (5.5.25′) and (5.5.25″) hold true.*

PROOF (PARTIAL). Only a complete proof of (5.5.25), hence of the bounded CLT (5.5.26), will be given since (5.5.25′) and (5.5.25″) require a little more empirical process theory than that so far developed in this book. We will also indicate how to complete the proof of the CLT (5.5.27) if we assume that, in addition to (P.2), the functions $f(x_1)f_i(x_2)$, $i = 1, 2$, are Riemann integrable. The function r in (5.5.24) does not satisfy

$$\lim_{\delta\to 0}\limsup_n \Pr\Big\{\sup_{|\theta|<\delta}\big|n^{1/2}(P_n - P)(r(\cdot,\theta))\big| > \varepsilon\Big\} = 0,$$

which is what empirical process theory is taylored to give us; so, we will have to modify r, leaving it unchanged at the values (X_i, θ_n) and $(X_i, \Delta_n B^{-1} + \delta_n)$ of its arguments. The facts that P is angularly symmetric and that the simplices are taken to be open in the definitions of D and D_n help. By (5.5.24),

$$r(x,\theta) = \frac{1}{|\theta|}\big[(P^2C_\theta)(x) - (P^2C_0)(x) - \theta\cdot\Delta(x)\big].$$

By Lemma 5.5.4, $(P^2C_0)(X_i) = (P^2C_0)(X_i - \theta_n) = 1/4$ as long as neither X_i nor $X_i - \theta_n$ are zero. This is a.s. true for X_i as P is continuous, and it is also true for $X_i - \theta_n$ because $\theta_n \in \operatorname{argmax} D_n$ and $\operatorname{argmax} D_n$ does not contain any of $X_1, \ldots, X_n$ (as can be shown by a simple geometric argument based in part on the fact that we take our simplices open). If $\Delta_n B^{-1} \neq X_i$ for $i = 1, \ldots, n$ then we take $\delta_n = 0$ in (5.5.25″), and otherwise, we take $\delta_n = \min\{n^{-1} \wedge |X_i - X_j| : 1 \leq i \neq j \leq n\}$; this ensures $\Delta_n B^{-1} + \delta_n \neq X_i$ for $i = 1, \ldots n$, and therefore, also $(P^2 C_0)(X_i) = (P^2 C_0)(X_i - \Delta_n B^{-1} - \delta_n)$. It is also clear that $(P^2 C_0)(x) = (P^2 C_0)(x - \theta)$ for P—almost all x and $\theta \neq 0$. Hence, (5.5.25) and (5.5.25″) need only be verified for the following modification $\tilde{r}$ of r:

$$\begin{aligned}
\tilde{r}(x,\theta) &= \frac{1}{|\theta|}\big[(P^2 C_\theta)(x) - (P^2 C_0)(x-\theta) - \theta \cdot \Delta(x-\theta)\big] \\
&\quad + \frac{\theta}{|\theta|} \cdot \big[\Delta(x-\theta) - \Delta(x)\big] \\
&= \frac{1}{|\theta|} \int_{(\mathbb{R}^2)^2} I_{C_0}(\mathbf{x}, x-\theta)\big[\phi_2(\mathbf{x},\theta) - \phi_2(\mathbf{x},0) - \theta \cdot \phi_2'(\mathbf{x},0)\big] d\mathbf{x} \\
&\quad + \frac{\theta}{|\theta|} \cdot \big[\Delta(x-\theta) - \Delta(x)\big] \\
&:= \tilde{r}_1(x,\theta) + \tilde{r}_2(x,\theta),
\end{aligned}$$

and the last term, $\tilde{r}_2(x,\theta)$ is a (finite) sum of integrals of the form

$$\frac{\theta_i}{|\theta|} \int_{(\mathbb{R}^2)^2} \big[I_{C_0}(x_1, x_2, x-\theta) - I_{C_0}(x_1, x_2, x)\big] f_i(x_1) f(x_2) dx_1 dx_2.$$

Now, for x_1, x_2, θ fixed, the set of xs such that $(x_1, x_2, x - \theta) \in C_0$ is the translation by θ of the cone $C_0(x_1, x_2)$ encountered in the proof of Lemma 5.5.1; we denote it as $C_0^\theta(x_1, x_2)$. Note that $C_0^\theta(x_1, x_1) = \emptyset$ and $C_0^\theta(x_1, 0) = \emptyset$. Since a cone is the intersection of two half planes, and the family of half planes is VC by Proposition 5.1.12, Proposition 5.1.13 gives that the class of sets $\{C_0^\theta(x_1, x_2) : \theta, x_1, x_2 \in \mathbb{R}^2\}$ is VC, and it is measurable because the map $(\theta, x_1, x_2, x) \mapsto I_{C_0^\theta(x_1,x_2)}(x)$ is measurable (see, e.g. the comments following equation (3.5.2)). Hence, the central limit theorem for measurable VC classes (Theorem 5.2.7 with $m = 1$) implies

$$\lim_{M\to\infty} \sup_n \Pr\Big\{ \sup_{x_1,x_2,\theta} \big|n^{1/2}(P_n - P)(C_0^\theta(x_1, x_2))\big| > M \Big\} = 0. \tag{5.5.28}$$

On the other hand, condition (P.2) gives, by Taylor's formula, that there exist $K < \infty$ such that, for all $\delta > 0$ small enough,

$$\int_{(\mathbb{R}^2)^2} \sup_{|\theta|<\delta} \left| \frac{\phi_2(\mathbf{x},\theta) - \phi_2(\mathbf{x},0) - \theta \cdot \phi_2'(\mathbf{x},0)}{|\theta|} \right| d\mathbf{x} \leq K\delta.$$

This and (5.5.28) yield

$$\lim_{\delta\to 0} \sup_n \Pr\Big\{ \sup_{|\theta|<\delta} \big|n^{1/2}(P_n - P)(\tilde{r}_1(\cdot,\theta))\big| > \varepsilon \Big\} = 0 \tag{5.5.29}$$

for all $\varepsilon > 0$. As for $\tilde{r}_2$, the two estimates

$$\int |f_i(x_1)| f(x_2) dx_1 dx_2 < \infty,$$

which follows from (P.2), and

$$\lim_{M\to\infty} \sup_n \Pr\Big\{ \sup_{\theta, x_1, x_2} \big| n^{1/2}(P_n - P)(I_{C_0^\theta(x_1,x_2)} - I_{C_0(x_1,x_2)})\big| > M \Big\}$$
$$\leq 2 \lim_{M\to\infty} \sup_n \Pr\Big\{ \sup_{x_1,x_2,\theta} \big| n^{1/2}(P_n - P)(C_0^\theta(x_1, x_2))\big| > M/2 \Big\} = 0,$$

which follows from (5.5.28), give

$$\lim_{M\to\infty} \sup_n \Pr\Big\{ \sup_\theta \big| n^{1/2}(P_n - P)(\tilde{r}_2(\cdot, \theta))\big| > M \Big\} = 0. \qquad (5.5.30)$$

(5.5.29) and (5.5.30) prove (5.5.25). This concludes the proof of stochastic boundedness of the sequence $n^{1/2}\theta_n$.

It is $\tilde{r}_2$ that poses more problems in order to prove (5.5.25$'$) and (5.5.25$''$), hence, the complete CLT for the empirical simplicial median. We will prove it here under the additional assumption mentioned above,

(P.2$'$) Besides (P.2), the functions of $\mathbb{R}^4$, $f_i(x_1)f(x_2)$, $i = 1, 2$, are Riemann integrable.

In this case, the class of functions

$$\Big\{ \int I_{C_0^\theta(x_1,x_2)}(\cdot) f_i(x_1) f(x_2) dx_1 dx_2 : \theta \in \mathbb{R}^2 \Big\}$$

is VC–hull, in the notation of Dudley (1987), and, according to Corollary 5.8 and Figure 1 in the same article (or Corollary 2.6.12 in van der Vaart and Wellner, 1996), this class is P-Donsker for all P on $\mathbb{R}^2$. Then, the asymptotic equicontinuity condition in Theorem 5.1.2 implies that, if

$$\mathbb{E}\Big[\int \big(I_{C_0^\theta(x_1,x_2)}(X) - I_{C_0(x_1,x_2)}(X)\big) f_i(x_1) f(x_2) dx_1 dx_2 \Big]^2 \to 0 \quad \text{as } \theta \to 0, \qquad (5.5.31)$$

then

$$\lim_{\delta\to 0} \sup_n \Pr\Big\{ \sup_{|\theta|<\delta} \Big| n^{1/2}(P_n - P) \int \big(I_{C_0^\theta(x_1,x_2)} - I_{C_0(x_1,x_2)}\big) f_i(x_1) f(x_2)\, dx_1 dx_2 \Big| > \varepsilon \Big\}$$
$$= 0. \qquad (5.5.32)$$

(5.5.32) and (5.5.29) immediately give (5.5.25$'$) and (5.5.25$''$), concluding the proof of the central limit theorem for the simplicial median under (P.2$'$), modulo the limit (5.5.31), which we now prove.

Since

$$|I_{C_0^\theta(x_1,x_2)} - I_{C_0(x_1,x_2)}| \leq I_{\partial C_0(x_1,x_2)_{|\theta|}},$$

where $A_{\varepsilon]}$ is the set of points at distance at most ε from A, for any set A, and since $\int |f_1(x_1)|f(x_2)dx_1dx_2 = \int |f_1(x_1)|dx_1 := K < \infty$, we have

$$\mathbb{E}\Big[\int \big(I_{C_0^\theta(x_1,x_2)}(X) - I_{C_0(x_1,x_2)}(X)\big) f_i(x_1)f(x_2)dx_1dx_2\Big]^2$$
$$\leq K\mathbb{E}\Big[\int \big(I_{C_0^\theta(x_1,x_2)}(X) - I_{C_0(x_1,x_2)}(X)\big) f_i(x_1)f(x_2)dx_1dx_2\Big]$$
$$\leq K \int P\{\partial C_0(x_1,x_2)_{|\theta|]}\}|f_1(x_1)|f(x_2)dx_1dx_2 \to 0$$

as $\theta \to 0$ by dominated convergence, given that P has a density. □

Condition (P.2′) is not required, but to prove the last claim under only (P.2) necessitates the bracketing CLT of Ossiander (1987).

Collecting all the above claims, we have proved the following:

THEOREM 5.5.5. *Let P be a probability measure in $\mathbb{R}^2$ angularly symmetric about θ_0 and satisfying hypotheses (P.1), (P.2) and (P.3). Let θ_n be measurable empirical simplical medians corresponding to a random sample from P. Then,*

$$\mathcal{L}\big[n^{1/2}(\theta_n - \theta_0)\big] \to_w N(0, \Gamma),$$

where $\Gamma = 9B^{-1}(\mathrm{cov}_P\,\Delta)B^{-1}$, with B and Δ as defined respectively by equations (5.5.14) and (5.5.22).

EXAMPLE 5.5.6. If $P = N(0, I)$ is the normal probability law in the plane with center θ_0 and covariance equal to the identity, then it can be shown with relatively straightforward computations that we omit, that $\Delta(x) = (2/\pi^3)^{1/2}x/|x|$ for $x \neq 0$, and that $B = 3I/2\pi$. It then follows that $\mathrm{cov}_P\,\Delta = \pi^{-3}I$ and, therefore,

$$\mathcal{L}\big(n^{1/2}(\theta_n - \theta_0)\big) \to_w N\Big(0, \frac{4}{\pi}I\Big).$$

The empirical simplicial median θ_n is an example of an M-estimator with a criterion function of several variables. Imbedded in the proof of Claims 2 to 5 is a general result on such M-estimators, that we now describe without proof.

Let $(S, \mathcal{S}, P)$ be a probability space and let X_i be the coordinate functions from the infinite product probability space $\big(S^{\mathbb{N}}, \mathcal{S}^{\mathbb{N}}, P^{\mathbb{N}}\big)$. Let $\mathcal{F} = \{f_\theta : \theta \in \Theta\}$ be a measurable class of functions on S^m indexed by a set Θ of $\mathbb{R}^d$ containing a neighborhood of the origin, and assume each f_θ is P^m-integrable and symmetric in its arguments.

THEOREM 5.5.7. *Assume that the following holds:*

i) *$U(\theta) := P^m f_\theta$, $\theta \in \Theta$, has a unique maximum, attained at $\theta = 0$, and admits the development*

$$U(\theta) = U(0) - \frac{1}{2}\theta A\theta^t + o(|\theta|^2)$$

near zero, where A is a symmetric, (strictly) positive definite $d \times d$ matrix.

ii) *For* $2 \le k \le m$ *and for all* $\varepsilon > 0$

$$\lim_{\delta\to 0}\limsup_{n\to\infty}\Pr\Big\{\sup_{|\theta|\le\delta}|nU_n(\pi_k(f_\theta - f_0))| > \varepsilon\Big\} = 0;$$

iii) *there exists a measurable function* $\Delta : S \to \mathbb{R}^d$ *satisfying* $E\Delta(X) = 0$ *and* $E|\Delta|^2 < \infty$, *and such that the functions defined by* $r(x, 0) = 0$ *for* $x \in \mathbb{R}^d$ *and*

$$r(x,\theta) = \frac{\pi_1 f_\theta(x) - \pi_1 f_0(x) - \theta\cdot\Delta(x)}{|\theta|}, \qquad \theta \in \Theta\setminus\{0\}, \quad x \in \mathbb{R}^d,$$

satisfy

$$\lim_{\delta\to 0}\limsup_{n\to\infty}\Pr\Big\{\sup_{|\theta|\le\delta}|n^{1/2}P_n r(\cdot,\theta)| > \varepsilon\Big\} = 0$$

for all $\varepsilon > 0$;

iv) *Let* $\{\theta_n\}$ *be a sequence of random variables such that*

$$\theta_n \to 0 \; in \; pr.$$

and

$$n\big(\sup_{\theta\in\Theta} U_n(\theta) - U_n(\theta_n)\big) \to 0 \; in \; pr.$$

Then

$$n^{\frac{1}{2}}\theta_n \to_{\mathcal{L}} Z$$

where Z *is* $N(0,\Gamma)$ *and* $\Gamma = m^2A^{-1}(\operatorname{cov}\Delta(X))A^{-1}$. *(In (iii),* r *can be replaced by* r_n *obtained from* r *by setting the denominator equal to* $|\theta| \vee n^{-\frac{1}{2}}$.*)*

The meaning of conditions (i) and (ii) is obvious, condition (iii) is a *stochastic differentiability* condition at zero for the function $\pi_1 f_\theta(x)$, and the conditions on θ_n indicate that θ_n is consistent and is close to maximizing the function $U_n(f_\theta)$, that is, is an *approximate M-estimator with criterion function* $f_\theta(x_1,\dots,x_m)$. The asymptotic equicontinuity condition for r can be relaxed to conditions analogous to (5.5.25′) and (5.5.25″), and this is important in the case of the empirical simplicial median.

5.5.3 *Truncated Data*

In this subsection we will apply the LIL for U-processes to obtain an exact a.s. error bound for the linearization of an estimator. The estimator we consider is the Woodroofe estimator of the cumulative hazard function for left truncated data, and, through it, the Lynden–Bell product limit estimator of the cumulative distribution function for left truncated data. Given (X, Y), (X_i, Y_i), $1 \le i \le N$, i.i.d. random vectors in $\mathbb{R}^2$ with X independent of Y, let

$$F(x) = \Pr\{X \le x\}\,, \; G(y) = \Pr\{Y \le y\}$$

be the corresponding distribution functions, and let

$$a_F = \inf\{x : F(x) > 0\}, \qquad b_F = \sup\{x : F(x) < 1\}$$
$$a_G = \inf\{x : G(x) > 0\}, \qquad b_G = \sup\{x : G(x) < 1\}$$

be the limits of their natural domains. The problem is to estimate the distribution function F of X if we only observe the pairs (X_i, Y_i), $i \leq N$ (with N unknown), when $X_i \geq Y_i$. Situations like this occur in Astronomy, Geology, etc.: an object is observed only if it is large enough, the truncating constant changing with each observation. A first natural assumption is

$$a_F \geq a_G, \tag{5.5.33}$$

otherwise the observations give little information about the part of F below a_G. We assume that (5.5.33) holds in all that follows. We also assume, for simplicity, that F and G are continuous and that $a_G \geq 0$. Although N is unknown, if $n = \#\{i \leq N : Y_i \leq X_i\}$ is the number of actual observations, we have $n/N \to \Pr\{Y \leq X\}$ a.s. by the law of large numbers, and this probability is strictly positive by (5.5.33). For F continuous, the cumulative hazard function is defined as

$$\Lambda(x) = \int_{a_F}^{x} \frac{dF(t)}{1 - F(t)}, \quad x \geq a_F.$$

Since $G(t) > 0$ on (a_F, ∞) [again by (5.5.33)], we can multiply numerator and denominator of the integrand by $G(t)$. Setting

$$F^*(x) = \Pr\{Y \leq X\} = \int_{a_F}^{x} G\,dF$$

and

$$C(x) = \Pr\{Y \leq x \leq X\} = G(x)\big(1 - F(x)\big),$$

this gives

$$\Lambda(x) = \int_{a_F}^{x} \frac{dF^*(t)}{C(t)}, \quad x \geq a_F.$$

This is very interesting because the empirical counterparts of F^* and C depend only on N and the observed data, and the unknown N cancels by quotient; therefore, using these empirical counterparts produces an estimator of Λ that involves observed quantities only. It is as follows: setting

$$C_N(x) = \frac{1}{N}\sum_{i=1}^{N} I_{Y_i \leq x \leq X_i} \quad \text{and} \quad F_N^*(x) = \frac{1}{N}\sum_{i=1}^{N} I_{Y_i \leq X_i \leq x}, \ x \geq a_F,$$

the Woodroofe estimator of the cumulative hazard function is defined to be

$$\Lambda_N(x) = \int_{a_F}^{x} \frac{dF_N^*(t)}{C_N(t)} = \sum_{i \leq N : Y_i \leq X_i \leq x} \frac{1}{NC_N(X_i)}, \quad x \geq a_F. \tag{5.5.34}$$

This estimator is analogous to the Nelson–Aalen estimator for right censored data. Λ_N is well defined because, for each $i \le N$, either $Y_i \le X_i$ and then $NC_N(X_i) \ge 1$ or, otherwise, the ith term is missing from the sum. The identity

$$1 - F(x) = \exp\big(-\Lambda(x)\big), \; a_F \le x \le b < b_F \tag{5.5.35}$$

follows by elementary calculus for F differentiable (both sides are 1 at $x = a_F$ and their logarithmic derivatives coincide), and by approximation for F continuous (e.g., by convolving F with an approximate identity). Then, it is natural to estimate $F(x)$ by $\bar{F}_N(x)$ as defined by the equation

$$1 - \bar{F}_N(x) = \exp\big(-\Lambda_N(x)\big), \qquad a_F \le x \le b < b_F. \tag{5.5.35$'$}$$

It turns out that the maximum likelihood estimator $\hat{F}_N$ of F for this problem, the Lynden–Bell estimator, is obtained by replacing e^{-a} by $1 - a$ for $a = 1/(NC_N(X_i))$, $i = 1, \dots, N$, in the definition of $\bar{F}_N(x)$, that is

$$1 - \hat{F}_N(x) = \prod_{i \le N : Y_i \le X_i \le x} \Big(\frac{NC_N(X_i) - 1}{NC_N(X_i)}\Big), \qquad a_F \le x \le b < b_F. \tag{5.5.36}$$

It can be proved using empirical process theory that, assuming $\int_{a_F}^{\infty} dF(t)/G^2(t) < \infty$,

$$\sup_{a_F \le x \le b} \big|\hat{F}_N(x) - \bar{F}_N(x)\big| = O\Big(\frac{1}{N}\Big) \quad \text{a.s.} \tag{5.5.37}$$

for $a_F \le b < b_F$. The proof of (5.5.37) will be indicated without much detail at the end of the section; it does not use U-processes, and it relies on the same arguments we will make in the proof of the main result which, in order to avoid repetition, we will only prove in full for the estimator $\bar{F}_N$ (at any rate, $\bar{F}_N$ is comparable to $\hat{F}_N$, at least for large samples).

In the analysis of $\Lambda_N - \Lambda$ and $\bar{F}_N - F$ we will encounter expressions with the quantities $C_N(X_i)$ in the denominator. The following lemma, which is a nice application of Doob's maximal inequalities, will allow us to replace $C_N(X_i)$ by the simpler variable $C(X_i)$ at a negligible cost.

LEMMA 5.5.8. *For $b < b_F$,*

$$\sup_{i \le N : Y_i \le X_i \le b} \frac{C(X_i)}{C_N(X_i)} = O(\log N) \quad \text{a.s.} \tag{5.5.38}$$

PROOF. First we reduce the problem to one on classical empirical processes in the plane and then use Doob's maximal inequalities. Since the set $\{i \le N : Y_i \le X_i \le b\}$ is eventually a.s. nonvoid, it suffices to prove (5.5.38) for B_N defined as the sup in (5.5.38) when $\{i \le N : Y_i \le X_i \le b\} \ne \emptyset$, and as zero otherwise. By definition, $Y_i \le X_i$ implies $C_N(X_i) \ge 1/N$; hence, if moreover $C(X_i)/C_N(X_i) \ge \lambda > 0$, then $C(X_i) \ge \lambda/N$. Therefore,

$$\big\{B_N \ge \lambda\big\} \subseteq \Big\{ \sup_{t : C(t) \ge \lambda/N, t \le b} -\frac{C_N(t)}{C(t)} \ge -\frac{1}{\lambda}\Big\}.$$

Let U_i, V_i, $i = 1, \ldots, N$, be i.i.d. random variables uniform on [0, 1] and let G^{-1} and L^{-1}, respectively, be the left continuous inverses of the cdf G of Y and the cdf $1 - F(-x)$ of $-X$. Then we can assume, without loss of generality, that $Y_i = G^{-1}(V_i)$ and $X_i = -L^{-1}(U_i)$, $i = 1, \ldots, N$. Since $G(t)\big(1 - F(t)\big) = C(t) \geq \lambda/N$ implies $G(t) \geq \lambda/N$, and $1 - F(t) \geq 1 - F(b)$ if $t \leq b$, setting $u = 1 - F(t)$ and $v = G(t)$, we have

$$\sup_{t:C(t)\geq\lambda/N, t\leq b} -\frac{C_N(t)}{C(t)} = \sup_{t:C(t)\geq\lambda/N, t\leq b} -\frac{1}{N}\frac{1}{G(t)\big(1 - F(t)\big)}\sum_{i=1}^{N} I_{G^{-1}(V_i)\leq t\leq -L^{-1}(U_i)}$$
$$\leq \sup_{\substack{1-F(b)\leq u\leq 1\\ \lambda/N\leq v\leq 1}} -\frac{1}{N}\frac{1}{uv}\sum_{i=1}^{N} I_{U_i\leq u, V_i\leq v}.$$

That is,

$$\{B_n \geq \lambda\} \subseteq \left\{ \sup_{\substack{1-F(b)\leq u\leq 1\\ \lambda/N\leq v\leq 1}} -\frac{1}{N}\frac{1}{uv}\sum_{i=1}^{N} I_{U_i\leq u, V_i\leq v} \geq -\frac{1}{\lambda}\right\} \tag{5.5.39}$$

For ease of notation, we set $u_0 := 1 - F(b)$, $v_0 := \lambda/N$ and

$$H_N(u, v) := -\frac{1}{N}\frac{1}{uv}\sum_{i=1}^{N} I_{U_i\leq u, V_i\leq v}.$$

For u fixed, the random variables $H_N(u, v)$, $v_0 \leq v \leq 1$, form a reverse martingale with respect to the σ-algebras $\mathcal{F}_v = \sigma\big\{I_{V_i\leq t}, U_i : 1 \leq i \leq N, v \leq t \leq 1\big\}$. The proof of this assertion reduces to checking that if $v \leq v'$ then $\mathbb{E}\big(I_{V_i\leq v}\big|I_{V_i\leq v'}\big) = (v/v')I_{V_i\leq v'}$, which is straightforward. So, for each $r > 0$, the process $\sup_{u_0\leq u\leq 1}\exp\big(rH_n(u, v)\big)$, $v_0 \leq v \leq 1$, is a reverse submartingale, and Doob's maximal inequality then gives

$$\Pr\Big\{\sup_{\substack{u_0\leq u\leq 1\\ v_0\leq v\leq 1}} H_N(u, v) \geq -\eta\Big\} \leq e^{\eta r}\mathbb{E}\Big[\sup_{u_0\leq u\leq 1}\exp\big(rH_N(u, v_0)\big)\Big].$$

Similarly, the process $\exp\big(p^{-1}rH_N(u, v_0)\big)$, $u_0 \leq u \leq 1$, is also a reverse submartingale, and we can apply to it Doob's inequality for pth moments (see, e.g., Chow and Teicher, 1988, page 247, for Doob's inequalities) to obtain

$$\Pr\Big\{\sup_{\substack{u_0\leq u\leq 1\\ v_0\leq v\leq 1}} H_N(u, v) \geq -\eta\Big\} \leq e^{\eta r}\Big(\frac{p}{p-1}\Big)^p\mathbb{E}\big[\exp\big(rH_N(u_0, v_0)\big)\big].$$

Letting $p \to \infty$ and taking into account that

$$\mathbb{E}\big[\exp\big(rH_N(u_0, v_0)\big)\big] = \Big[\mathbb{E}\big[\exp(-(r/N)I_{U_1\leq u_0, V_1\leq v_0})\big]\Big]^N$$
$$= \big[1 - u_0v_0 + u_0v_0e^{-r/(Nu_0v_0)}\big]^N,$$

we obtain

$$\Pr\Big\{\sup_{\substack{u_0\le u\le 1\\ v_0\le v\le 1}} H_N(u,v)\ge -\eta\Big\} \le e^{\eta r+1}\big[1-u_0v_0+u_0v_0e^{-r/(Nu_0v_0)}\big]^N.$$

Now, replacing η by $1/\lambda$ for $\lambda>1$, taking $r=Nu_0v_0\log\lambda$, and using the estimate $1-(1-\lambda^{-1})u_0v_0\le e^{-(1-\lambda^{-1})u_0v_0}$, give

$$\Pr\Big\{\sup_{\substack{u_0\le u\le 1\\ v_0\le v\le 1}} H_N(u,v)\ge -\frac{1}{\lambda}\Big\} \le \exp\Big[1+Nu_0v_0\Big(\frac{\log\lambda}{\lambda}+\frac{1}{\lambda}-1\Big)\Big].$$

Replacing u_0 and v_0 by their values, the last exponent becomes $1+\big(1-F(b)\big)\times(-\lambda+1+\log\lambda)$, finally giving [by (5.5.39)] that, for $\lambda>1$,

$$\Pr\{B_n\ge\lambda\}\le\lambda e^2\exp\big(-(1-F(b))\lambda\big),$$

so that

$$\sum_N \Pr\Big\{B_n\ge\frac{2}{1-F(b)}\log N\Big\}<\infty.$$

Now, the lemma follows by Borel–Cantelli. □

The following theorem decomposes $\Lambda_N-\Lambda$ and $\bar F_N-F$ into a "linear' part, that is, a sum of independent random functions (an empirical process), and a remainder term, and gives the exact a.s. order of the remainder term uniformly over an interval starting at a_F. Hence, by (5.5.37), it does the same for $\hat F_N-F$. It applies, for instance, to density estimation, but we will not treat this here.

THEOREM 5.5.9. *Let F and G be continuous cdf's and let $a_F\le b<b_F$. Let*

$$L_N(x):=\int_{a_F}^x\frac{1}{C(t)}d(F_N^*(t)-F^*(t))-\int_{a_F}^x\frac{C_N(t)-C(t)}{C^2(t)}dF^*(t).$$

Then, if

$$a_G\le a_F \quad\text{and}\quad \int_{a_F}^\infty\frac{dF(t)}{G^3(t)}<\infty \tag{5.5.40}$$

we have both

$$\Lambda_N(x)-\Lambda(x)=L_N(x)+R_{N,1}(x) \tag{5.5.41}$$

and

$$\bar F_N(x)-F(x)=\big(1-F(x)\big)L_N(x)+R_{N,2}(x) \tag{5.5.42}$$

with

$$\sup_{a_F\le x\le b}|R_{N,i}(x)|=O\Big(\frac{\log\log N}{N}\Big)\quad\text{a.s.,}\qquad i=1,2. \tag{5.5.43}$$

PROOF. We first consider (5.5.41). Simple algebra shows

$$\begin{aligned} R_{N,1}(x) &= \int_{a_F}^{x} \frac{C(t)-C_N(t)}{C^2(t)} d(F_N^*(t)-F^*(t)) \\ &+ \int_{a_F}^{x} \frac{(C(t)-C_N(t))^3}{C^3(t)C_N(t)} dF_N^*(t) + \int_{a_F}^{x} \frac{(C(t)-C_N(t))^2}{C^3(t)} dF_N^*(t) \\ &:= (I)+(II)+(III). \end{aligned}$$

We look at each term separately. Denoting just by π_2 the Hoeffding projection $\pi_{2,2}^{\mathcal{L}(X,Y)}$, it is clear that

$$(I) = \frac{1}{N^2} \sum_{i,j=1}^{N} \pi_2 f_x(\,(X_i,Y_i),(X_j,Y_j)\,)$$

where

$$f_x\big((x_1,y_1),(x_2,y_2)\big) = -\frac{1}{C^2(x_2)} I_{y_1 \le x_2 \le x_1,\, y_2 \le x_2 \le x}.$$

The class of subsets of $\mathbb{R}^2 \times \mathbb{R}^2$ $\{A_x = I_{y_1 \le x_2 \le x_1, y_2 \le x_2 \le x} : x \in \mathbb{R}\}$ is a measurable VC class of sets since each set is the intersection of at most four halfspaces of $\mathbb{R}^4$ (Propositions 5.1.12, 5.1.13). Hence, the class of functions

$$\{\frac{1}{C^2(x_2)} I_{A_x} : a_F \le x \le b\}$$

is measurable VC subgraph by the comment following Definition 5.1.14. This class admits the envelope $H(x_1,y_1,x_2,y_2) = I_{A_b}/C^2(x_2)$ which is square integrable for the law of (X_1,Y_1,X_2,Y_2):

$$EH^2 = \int_{a_F}^{b} \frac{dF(x_2)}{G^2(x_2)(1-F(x_2))^3},$$

finite by condition (5.5.40). Hence, Theorem 5.4.4 (see the comment immediately below its proof; Theorem 5.4.2 suffices) implies

$$\sup_{a_F \le x \le b} \Big| \frac{1}{N^2} \sum_{1 \le i \ne j \le N} \pi_2 f_x(\,(X_i,Y_i),(X_j,Y_j)\,) \Big| = O\Big(\frac{\log\log N}{N}\Big),$$

and this order is exact (that is, it is attained). The sum of the diagonal terms in the V-statistic (I) is

$$\begin{aligned} -\frac{1}{N^2} \sum_{i=1}^{N} \Big[\Big(\frac{1}{C^2(X_i)} + \frac{1}{C(X_i)} \Big) I_{Y_i \le X_i \le x} \\ + \int_{a_F}^{x} \frac{1}{G(t)(1-F(t))} I_{Y_i \le t \le X_i} dF^*(t) - \Lambda(x) \Big] \end{aligned}$$

which is a biased empirical process over a VC class of functions. The law of large numbers (Theorem 5.2.3 for $m = 1$) implies that its sup norm is a.s. $O\left(\frac{1}{N}\right)$. Thus, we conclude

$$\sup_{a_F \le x \le b} |(I)| = O\left(\frac{\log\log N}{N}\right).$$

As for (II) we have

$$(II) \le \sup_{a_F \le x \le b} |C(x) - C_N(x)|^3 \int_{a_F}^{b} \frac{1}{C^3(t)C_N(t)} dF_N^*(t).$$

By the LIL for the empirical process over VC classes of functions (Theorem 5.4.2 for $m = 1$),

$$\sup_{a_F \le x \le b} |C(x) - C_N(x)|^3 = O\left(\left(\frac{\log\log N}{N}\right)^{3/2}\right) \quad \text{a.s.}$$

By Lemma 5.5.8 and the law of large numbers,

$$\int_{a_F}^{b} \frac{1}{C^3(t)C_N(t)} dF_N^*(t) \le O(\log N) \frac{1}{N} \sum_{i=1}^{N} \frac{I_{Y_i \le X_i \le b}}{C^4(X_i)} = O(\log N) \quad \text{a.s.}$$

(Note that the law of large numbers applies since

$$E[I_{Y \le X \le b}/C^4(X)] = \int_{a_F}^{b} dF(x)/\big(G^3(x)(1 - F(x))^4\big)$$

is finite by (5.5.40) and the condition on b). Therefore,

$$\sup_{a_F \le x \le b} |(II)| \le O\left(\frac{(\log\log N)^{3/2} \log N}{N^{3/2}}\right) \quad \text{a.s.}$$

(III) can be bounded similarly: The LIL and the law of large numbers for empirical processes over VC classes (hence, these theorems for U-processes with $m = 1$) give

$$\sup_{a_F \le x \le b} |C(x) - C_N(x)|^2 = O\left(\frac{\log\log N}{N}\right) \quad \text{a.s.}$$

and

$$\int_{a_F}^{b} \frac{1}{C^3(t)} dF_N^*(t) = \frac{1}{N} \sum_{i=1}^{N} \frac{I_{Y_i \le X_i \le b}}{C^3(X_i)} = O(1) \quad \text{a.s.}$$

so that

$$\sup_{a_F \le x \le b} |(III)| = O\left(\frac{\log\log N}{N}\right) \quad \text{a.s.}$$

We have thus proved that Λ_N admits the representation (5.5.41) with remainder term bounded by (5.5.43).

The process $L_N(x)$ is the sum of two empirical processes over VC classes of functions with square integrable envelope, hence it satisfies the LIL (Theorem 5.4.2 with $m = 1$), that is, its sup over $a_F \le x \le b$ is a.s. of the order of $\big((\log\log N)/N\big)^{1/2}$. So, by the previous part of the proof, the sup of $|\Lambda_N(x) - \Lambda(x)|$ is of the same order. Therefore, by Taylor,

$$\begin{aligned}\sup_{a_F \le x \le b} \left| e^{-\Lambda(x)} - e^{-\Lambda_N(x)} - (\Lambda_N(x) - \Lambda(x))e^{-\Lambda(x)} \right| \\ &= O\Big(\sup_{a_F \le x \le b} \left|\Lambda_N(x) - \Lambda(x)\right|^2\Big) \\ &= O\Big(\frac{\log\log N}{N}\Big) \quad \text{a.s.} \qquad (5.5.44)\end{aligned}$$

Now, the expansion (5.5.42) for $\bar{F}_N$ with the bound on the remainder given by (5.5.43) follows from (5.5.41) and (5.5.44) on account of (5.5.35) and (5.5.35′). □

The proof of the bound (5.5.37) uses estimates not unlike those in the previous proof. First one shows that, introducing the modification $\tilde{F}_N$ of $\hat{F}_N$ given by

$$1 - \tilde{F}_N = \prod_{i \le N: Y_i \le X_i \le x} \Big(\frac{NC_N(X_i)}{NC_N(X_i) + 1}\Big), \qquad a_F \le x \le b < b_F,$$

one has

$$\sup_{a_F \le x \le b} \left|\hat{F}_N(x) - \tilde{F}_N(x)\right| = O\Big(\frac{1}{N}\Big) \quad \text{a.s.}$$

Then, the fact that the factors in the definition of $1 - \tilde{F}_N$ are uniformly bounded away from 0 and smaller than one allows us to develop the logarithm and compare to $\Lambda_N = -\log(1 - \bar{F}_N)$, concretely,

$$\left|\log(1 - \tilde{F}_N(x)) - \Lambda_N(x)\right| \le \frac{1}{N}\int_{a_F}^{b} \frac{dF_N^*(x)}{C_N^2(x)},$$

which is a.s. uniformly $O(1/N)$ by previous arguments. Hence

$$\sup_{a_F \le x \le b} \left|\tilde{F}_N(x) - \bar{F}_N(x)\right| = \sup_{a_F \le x \le b} \left|1 - \tilde{F}_N(x) - e^{-\Lambda_N(x)}\right| \le O\Big(\frac{1}{N}\Big)$$

(since $|e^{-a} - e^{-b}| \le |b - a|$ if $a, b \ge 0$). Then, putting these bounds together, we obtain (5.5.37).

The bound (5.5.37), combined with Theorem 5.5.9, shows that, *under the hypotheses of Theorem 5.5.9,*

$$\hat{F}_N(x) - F(x) = \big(1 - F(x)\big)L_N(x) + R_N(x)$$

with

$$\sup_{a_F \le x \le b} |R_N(x)| = O\Big(\frac{\log\log N}{N}\Big) \quad \text{a.s.,}$$

thus decomposing the Lynden–Bell product limit estimator $\hat{F}_N(x)$ into a linear term and an a.s. uniformly small remainder.

We should remark that the order $(\log\log N)/N$ of the remainders $R_{N,i}$, R_N above is exact, that is, the limsup of $N/\log\log N$ times the sup of R_N or $R_{N,i}$ over $[a_F, b]$ is a.s. a constant different from zero. This follows from the proof of Theorem 5.5.9, applying the compact law of the iterated logarithm instead of the bounded one.

On the other hand, if we are interested in the uniform order of the remainders *in probability* instead of a.s., then we can also proceed as in the proof of Theorem 5.5.9, but now invoking the CLT for degenerate U-processes and for empirical processes instead of the LIL, and easily obtain an order of $1/N$.

For a somewhat similar study of the product limit estimator of F in the case of truncated data, the Kaplan–Meyer estimator, we refer to Stute (1994); the U-process bounds used in this work are (a factor of) a power of a logarithm away from being exact.

5.6 References for Chapter 5

The definition of weak convergence of processes in its present form is due to Hoffmann-Jørgensen (1991, in private circulation since 1984); it modifies a similar one by Dudley (1966, 1967). The asymptotic equicontinuity criterion, Theorem 5.1.2, has a history starting with Prohorov (1956), where he proves a version of the theorem for $C(S)$-valued random variables by combining his tightness theorem with Arzelà-Ascoli's. In the present generality, it is ascribed by Andersen and Dobrić (1987) to Hoffmann-Jørgensen, with important previous contributions by Dudley (1978) and Dudley and Philipp (1983). The present simple proof, comes from Giné and Zinn (1986, Theorem 1.1.3). Lemma 5.1.1 on Radon measures and sample continuity is due to Hoffmann-Jørgensen. The observation about metrizability of weak convergence in this general context was made, as needed, by several authors at the same time (Dudley, 1990; Giné and Zinn, 1990; van der Vaart and Wellner, 1989, unpublished): it is a trivial corollary of the Giné–Zinn proof of Theorem 5.1.2.

Metric entropy was introduced into the study of the regularity of Gaussian processes by Sudakov (1966, 1969). Dudley (1967a) proved that finiteness of the entropy integral (5.1.18) implies sample continuity of the Gaussian process, and in fact his result contains Corollary 5.1.6 (he also proved by examples that there is no necessary and sufficient condition for sample continuity of Gaussian processes in terms of metric entropy). Dudley's proof has been elaborated upon by several authors, particularly Pisier (1983) and Fernique (1983). It was soon realized (e.g., Marcus and Pisier, 1981) that Dudley's proof extends to non-Gaussian processes. Pisier (1983) then simplified the proof by directly computing with expected values instead of probabilities, using his inequality (4.3.1), and he also subtly modified the chaining in order to handle power moduli. Fernique (1983) further modified Pisier's proof and made it possible, in particular, to obtain exponential moment bounds in general, directly from chaining arguments (exponential moment bounds come for free from expectation bounds when adequate hypercontractivity is present, as is the case for Gaussian and Rademacher chaos processes). Ledoux and Talagrand, (1991), pages 299–308, render their own account of the Pisier–Fernique modifications of Dudley's proof, including explicit, direct entropy bounds of exponential Orlicz norms as in Theorem 5.1.4 above. Pollard (1990) obtains entropy exponential bounds in general in a different way, by simply replacing, in Pisier's proof, inequality (4.3.1) by his inequality (5.1.9). Theorem 5.1.4 and its corollaries, statements and proofs, are based on Pisier's proof, on the Ledoux–Talagrand exposition and on the proof of an exponential bound for empirical processes in Giné and Zinn (1991) (correction: Giné, 1997a). [We were unaware of Pollard's (1990) proof until we researched these references: his proof is similar to ours and precedes it by several years.] The main differences with these sources are that we simplify the chaining at the cost of restricting ourselves to only exponential moduli and that we present Pollard's inequality (5.1.9) as a consequence of the sharper bound (4.3.7) from Arcones and Giné (1995), which is needed elsewhere in the book (LIL).

Vapnik–Červonenkis classes of functions were introduced by Vapnik and Červonenkis (1971, 1981) in a landmark study of the law of large numbers for empirical processes, and they also proved the basic Theorem 5.1.10 (independently obtained also by Sauer, 1972, and Shelah, 1972). The present proof, along with Proposition 5.1.11, is from Frankl (1983) and our exposition follows Ledoux and Talagrand (1991). Dudley (1978) showed that VC classes of sets have power type covering numbers for the $L_1(P)$ pseudodistance (hence any $L_p(P)$ pseudodistance) uniformly in P, and this allowed him to prove that empirical proceses over measurable VC classes of sets satisfy the CLT for all P. Pollard (1982, 1984) extended Dudley's lemma to classes of functions (Theorem 5.1.15). Proposition 5.1.12 and 5.1.13 are in Dudley (1978, 1984, 1997).

The law of large numbers for U-processes was proved by Helmers, Janssen, and Serfling (1988) for "empirical processes of U-statistic structure," that is, for the special kernels $f_t(x_1, \ldots, x_m) = I_{h(x_1,\ldots,x_m)\le t}$, and it was proved for VC-subgraph classes and $m = 2$ by Nolan and Pollard (1987). The general LLN, Theorem 5.2.2, as well as its application to VC-subgraph classes (Corollary 5.2.3) and to B-valued U-statistics (Theorem 5.2.6) were obtained by Arcones and Giné (1993). Example 5.2.4, the LLN for the empirical simplicial depth process, comes from Liu (1990), who introduced these processes, from Dümbgen (1990) and from Arcones and Giné (1993). The law of large numbers for B-valued U-statistics was also published in Borovskikh and Korolyuk (1994), and it seems to have existed in preprint since 1988. The Marcinkiewicz law we present here is due to Sen (1974), with the proof taken from Giné and Zinn (1992a).

The central limit theorems for U-processes given in Section 5.3 were obtained by Arcones and Giné (1993) in the present generality. Nolan and Pollard (1988) contains analogs of Theorems 5.3.3 and 5.3.7, the CLT for VC-subgraph classes, for the case $m = 2$; also for $m = 2$, they proved a theorem under random entropy conditions. Sherman (1994) proved maximal inequalities for degenerate U-processes and showed as well that the sequence $\{n^{k/2}\|U_n(h)\|_{\mathcal{H}}\}$ is stochastically bounded if $\mathcal{H}$ is measurable VC-subgraph (or has similar entropy bounds) and $P^m H^2 < \infty$; this is close to Theorem 5.3.7. The CLT for type 2 Banach space valued U-statistics is also given in Borovskikh (1996), with a different proof (we obtained it independently and approximately at the same time). The exponential inequality in Theorem 5.3.14 comes from Arcones and Giné (1994). Example 5.3.10, the CLT for the empirical simplicial depth process, was given in Arcones and Giné (1993), and Dümbgen (1990) contains a slightly weaker version. Example 5.3.11, from statistics of directions, comes from Silverman (1978) by way of Nolan and Pollard (1988).

The law of the iterated logarithm for degenerate U-processes over VC classes and for degenerate B-valued U-statistics presented in Section 5.4 was proved by Arcones and Giné (1995), and the general structure of their proof as well as some of the arguments in it are adapted from Pisier (1975/76, 1976). Dehling, Denker and Philipp (1986) proved the LIL for Hilbert space valued canonical U-statistics assuming finiteness of the $(2+\delta)$th moment of the norm. Serfling (1971) and Arcones (1993) consider the LIL for U-processes in the nondegenerate case. See Giné (1997) for a short account on the applications of decoupling to the limit theory of U-statistics and processes.

The law of large numbers for the empirical simplicial median, in the generality given here (Section 5.5.1), comes from Arcones and Giné (1993) and from Chen (1995). Liu (1990), who introduced it, obtained it under additional hypotheses. Pollard (1985) put forward a general framework for proving central limit theorems for M-estimators from the CLT for the empirical process; Arcones, Chen and Giné (1994) adapted it to criterion functions of several variables by means of U-processes in order to obtain the CLT for the empirical simplicial median. Independently, Sherman (1994) did a similar adaptation of Pollard's method to U-processes and applied it to parameter estimation in a semiparametric regression model. [For other approaches to the theory of M-estimators see Ibrahimov and Has'minski, 1981, and Hoffmann-Jørgensen, 1992].

The Woodroofe (1985) estimator of the cumulative hazard function as well as the Lynden–Bell (1971) estimator of the cdf for truncated data, were first analized up to second order terms by Stute (1993), who obtained Lemma 5.5.8 as well as a slightly weaker version of Theorem 5.5.9 (with $\log\log N$ in the bound for the remainder replaced by $(\log N)^3$); Arcones and Giné (1995) obtained Theorem 5.5.9 by combining Stute's approach with the law of the iterated logarithm for empirical processes and for degenerate U-processes.

6

General Decoupling Inequalities for Tangent Sequences

This chapter provides an introduction to the theory of decoupling of tangent sequences, which consists mainly of inequalities comparing functionals of dependent variables to functionals of conditionally independent (decoupled) random variables having equivalent conditional distributions given the immediate past. This tangency property of "shared conditional distributions given the immediate past" is the prevailing concept that has driven the development of the theory. It is surprising that this (minimal) assumption gives rise to such a wealth of interesting and useful results.

Both, decoupling of U-statistics and decoupling of randomly stopped sums, fall within the framework of decoupling of tangent sequences, however, although the general results presented in this chapter are quite powerful, they fall short of implying the full strength of the results presented in Chapters 2 and 3.

Section 1 presents several definitions and illustrates their meaning by providing examples dealing with sampling without replacement from finite populations, U-statistics, quadratic forms, martingale transforms of independent random variables and randomly stopped sums. In this section a proposition is included that, broadly speaking, states that given an arbitrary sequence of random variables adapted to an increasing sequence of sigma fields, one can always find another sequence that has the same conditional distributions as the original one but, in addition, has a conditional independence property. Because of this result, the decoupling inequalities presented in the following sections are broadly applicable.

Section 2 contains inequalities for the product of dependent random variables and for the moment generating function of sums of dependent random variables. A key feature of these results is that they provide for decoupling with constrained random variables. As applications of these results we present sharp inequalities comparing the tail probabilities of the maximum (minimum) of a sequence of dependent random variables to the maximum (minimum) of decoupled (conditionally independent) random variables. Another key application included in this section is a symmetrization inequality for moment generating functions of martingales.

Sections 3 and 4 deal with the problem of approximating the tail probabilities and moments of the maximum of the absolute value of a sum of dependent random variables, in Section 3, in the context of Lenglart-type inequalities, and, in Section 4, in the context of good-lambda inequalities.

Section 5 introduces the concept of differential subordination for two sequences of variables adapted to a common increasing filtration. Subordination of the sequence $\{X_i\}$ by the sequence $\{Y_i\}$ (almost surely $|X_i| \leq |Y_i|$ for all i) leads to the domination of the pth moment of the first sum by a constant times pth moment of the second sum, whenever the variables are conditionally symmetric given the immediate past. These results are closely related to the decoupling inequalities presented in the previous section. As an application of differential subordination we present a square function inequality for the sum of conditionally symmetric random variables.

Section 6 explores the connection between decoupling inequalities and two classical results in probability theory, the Burkholder–Davis–Gundy and Rosenthal's inequalities. In particular, it is shown that decoupling inequalities are sometimes to be preferred to square function inequalities and that decoupling inequalities imply a quite general decoupling inequality for martingales. As an application, an easy proof of the right-hand side of the Burkholder–Davis–Gundy inequality for martingales is given.

Section 7 contains a historical account of the results obtained throughout the chapter.

6.1 Some Definitions and Examples

We begin by recalling the two special cases we have developed extensively in Chapter 2 through Chapter 6. Let $\{X_i\}$ be a sequence of independent random variables and let $\{\tilde{X}_i\}$ be an independent copy of $\{X_i\}$. In Chapter 2 we provided an in-depth study of the properties of randomly stopped sums $S_T = \sum_{j=1}^{T} X_j$ where T is a stopping time adapted to $\sigma(X_1, \ldots, X_n, \ldots)$ by means of decoupling inequalities comparing S_T to $\tilde{S}_T = \sum_{j=1}^{T} \tilde{X}_j$. For simplicity of exposition we assume that $T \leq n$. In the latter chapters we studied U-statistics and extensions, $U_n = \sum_{j=1}^{n} \sum_{i=1, i \neq j}^{n} f(X_i, X_j)$, for a bivariate symmetric function $f(x, y) = f(y, x)$ by comparing it to $\tilde{U}_n = \sum_{j=1}^{n} \sum_{i=1, i \neq j}^{n} f(X_i, \tilde{X}_j)$. There is a common feature in both S_T and U_n that becomes apparent upon an easy re-writing and that motivates seeking for extensions of the decoupling inequalities introduced in earlier chapters. More precisely, upon rewriting $S_T = \sum_{i=1}^{n} X_i I_{T \geq i}$, $\tilde{S}_T = \sum_{i=1}^{n} \tilde{X}_i I_{T \geq i}$ (since $T \leq n$), and $U_n' = U_n/2 = \sum_{j=2}^{n} \sum_{i=1}^{j-1} f(X_i, X_j)$ and $\tilde{U}_n' = \tilde{U}_n/2 =$

$\sum_{j=2}^{n}\sum_{i=1}^{j-1} f(X_i, \tilde{X}_j)$ and letting $d_j = X_j I_{T\geq j}$, $e_j = \tilde{X}_j I_{T\geq j}$ in the case of randomly stopped sums and $d_j = \sum_{i=1}^{j-1} f(X_i, X_j)$, $e_j = \sum_{i=1}^{j-1} f(X_i, \tilde{X}_j)$, in the case of U-statistics with $\mathcal{F}_j = \sigma(X_1, \ldots, X_j; \tilde{X}_1, \ldots, \tilde{X}_j)$ then, the conditional laws of d_j and e_j given $\mathcal{F}_{j-1}$ coincide, that is $\mathcal{L}(d_j|\mathcal{F}_{j-1}) = \mathcal{L}(e_j|\mathcal{F}_{j-1})$. The reason for this is that, fixing the first $j-1$ Xs and $\tilde{X}$s, the only additional randomness comes through $X_j, \tilde{X}_j$ which are identically distributed. Observe that since we assume that $T \leq n$, $S_T = \sum_{j=1}^{n} d_j$ and $\tilde{S}_T = \sum_{j=1} e_j$, and similarly for U-statistics since $U_n' = \sum_{j=2}^{n} d_j$ with $\tilde{U}_n' = \sum_{j=2}^{n} e_j$ for appropriately defined d_js and e_js. Therefore, one hopes that a single result would produce inequalities for both U-statistics and randomly stopped sums. This is indeed the case and more surprisingly one can develop an umbrella theory that includes all sequences of (adapted) dependent variables. In this section we lay out the theoretical foundations that permit this extension by means of introducing the main definitions and examples of the theory of decoupling inequalities.

Suppose $(\Omega, \mathcal{F}, P; \{\mathcal{F}_i\})$ is a filtered probability space with $\{\mathcal{F}_i\}_{i=1}^{\infty}$ a nondecreasing sequence of σ-fields and $\mathcal{F}_i \subset \mathcal{F}$ for all i.

DEFINITION 6.1.1. *Let $\{e_i\}$ and $\{d_i\}$ be two sequences of random variables adapted to the σ-fields $\{\mathcal{F}_i\}$. Then $\{e_i\}$ and $\{d_i\}$ are tangent with respect to $\{\mathcal{F}_i\}$ if, for all i,*

$$\mathcal{L}(d_i|\mathcal{F}_{i-1}) = \mathcal{L}(e_i|\mathcal{F}_{i-1}), \tag{6.1.1}$$

where $\mathcal{L}(d_i|\mathcal{F}_{i-1})$ denotes the conditional probability law of d_i given $\mathcal{F}_{i-1}$.

The inequality in (6.1.1) will be written alternatively as $d_i \sim_{\mathcal{F}_{i-1}} e_i$.

DEFINITION 6.1.2. *A sequence of random variables $\{d_i\}$ adapted to $\{\mathcal{F}_i\}$ is said to be a sequence of conditionally symmetric random variables whenever $d_i \sim_{\mathcal{F}_{i-1}} -d_i$, for all i.*

DEFINITION 6.1.3. *A sequence $\{e_i\}$ of random variables adapted to an increasing sequence of σ-fields $\mathcal{F}_i$ contained in $\mathcal{F}$ is said to satisfy the CI condition* (conditional independence) *if there exists a σ-algebra $\mathcal{G}$ contained in $\mathcal{F}$ such that $\{e_i\}$ is conditionally independent given $\mathcal{G}$ and $\mathcal{L}(e_i|\mathcal{F}_{i-1}) = \mathcal{L}(e_i|\mathcal{G})$.*

DEFINITION 6.1.4. *A sequence $\{e_i\}$ which satisfies the CI condition and which is also* tangent *to $\{d_i\}$ is said to be a* decoupled *tangent sequence to $\{d_i\}$.*

A key observation, to be extensively used below, is that decoupled sequences always exist. We state this crucial fact as a proposition.

PROPOSITION 6.1.5. *For any sequence of random variables $\{d_i\}$ adapted to an increasing sequence $\mathcal{F}_i$ of σ-algebras, there always exists a* decoupled sequence *$\{e_i\}$ (on a possibly enlarged probability space) which is* tangent *to the original sequence and in addition* conditionally independent *given a* master *σ-field $\mathcal{G}$. Frequently $\mathcal{G} = \sigma(\{d_i\})$.*

The *decoupled* sequence of Proposition 6.1.5 can be recursively defined as follows: Let e_1 be an independent copy of d_1. At the ith stage, given $\{d_1, \ldots, d_{i-1}\}$, d_i and e_i are independent and sampled according to the probability law $\mathcal{L}(d_i|\mathcal{F}_{i-1})$. In this case, if we let $\tilde{\mathcal{F}}_i = \sigma(\mathcal{F}_i, e_1, \ldots, e_i)$ then $\{e_i\}$ is $\{\tilde{\mathcal{F}}_i\}$-tangent to $\{d_i\}$ and, moreover, $\{e_i\}$ satisfies the CI condition with respect to $\mathcal{G} = \sigma(\{d_i\})$. Therefore, any sequence of random variables $\{d_i\}$ has a *decoupled version* $\{e_i\}$. It is easy to see that one can always take $\mathcal{F}_i = \sigma(d_1, \ldots, d_i)$. The following diagram illustrates the idea:

$$\begin{array}{ccccccccccccc} d_1 & \to & d_2 & \to & d_3 & \to & d_4 & \to & \ldots. & \to & d_{i-1} & \to & d_i \\ & \searrow & & \searrow & & \searrow & & \searrow & & & & \searrow & \\ e_1 & & e_2 & & e_3 & & e_4 & & \ldots & & e_{i-1} & & e_i. \end{array}$$

The above construction can be made precise as follows. Let $(S, \mathcal{S})$ be a measurable space consisting of a complete separable metric space S with its Borel σ-algebra $\mathcal{S}$. Let $\{d_i\}$ be a sequence of S-valued random variables adapted to an increasing sequence $\mathcal{F}_i$ of σ-algebras, $\mathcal{F}_i$ contained in $\mathcal{S}$. If the random variables d_i are defined on the probability space (Ω, Σ, P), we consider the enlarged measurable space $(\tilde{\Omega}, \tilde{\Sigma}) = (S^{\mathbb{N}} \times \Omega, \mathcal{S}^{\mathbb{N}} \otimes \Sigma)$, and note that if $\tilde{P}$ is any probability measure on $(\tilde{\Omega}, \tilde{\Sigma})$ whose projection onto Ω $(\tilde{\Omega} \to \Omega, (\{s_k\}, \omega) \to \omega)$ coincides with P, then the sequence $\tilde{d}_i(\{s_k\}, \omega) := d_i(\omega)$, $i = 1, \ldots$, is stochastically equivalent to the sequence $\{d_i\}$, that is, for any finite subset $\{i_1, \ldots, i_k\}$ of $\mathbb{N}$, $\mathcal{L}(d_{i_1}, \ldots, d_{i_k}) = \mathcal{L}(\tilde{d}_{i_1}, \ldots, \tilde{d}_{i_k})$. Since S is a Polish space, the (regular) conditional probability laws $\mathcal{L}(d_i|\mathcal{F}_{i-1})(\omega)$ exist (e.g. Dudley, 1989, Theorem 10.2.2); then, the conditional law given ω of the sequence $\{e_i\}$ must be $\otimes_{i=1}^{\infty}(\mathcal{L}(d_i|\mathcal{F}_{i-1})(\omega))$ and, if we want e_i to be the coordinates of $S^{\mathbb{N}}$, this and the above mentioned projection property force $\tilde{P}$ to be

$$\tilde{P}\{A \times B\} = \int_B (\otimes_{i=1}^{\infty} \mathcal{L}(d_i|\mathcal{F}_{i-1})(\omega))(A) dP(\omega).$$

Formally, $\tilde{d}_i(\{s_k\}, \omega) := d_i(\omega)$, and $e_i(\{s_k\}, \omega) = s_i$, $i \in \mathbb{N}$, as random variables defined on the probability space $(\tilde{\Omega}, \tilde{\Sigma}, \tilde{P})$, satisfy the following: if we set $\tilde{\mathcal{F}}_i = \mathcal{S}^i \otimes \mathcal{F}_i$, and $\tilde{\mathcal{G}} = \{S^{\mathbb{N}}, \phi\} \otimes \Sigma$, then the sequences $\{d_i\}$ and $\{\tilde{d}_i\}$ are stochastically equivalent, both sequences $\{\tilde{d}_i\}$ and $\{e_i\}$ are adapted to $\{\tilde{\mathcal{F}}_i\}$, and $\{e_i\}$ is CI with respect to $\tilde{\mathcal{G}}$ contained in $\tilde{\Sigma}$. That is, $\{e_i\}$ is a decoupled version of $\{\tilde{d}_i\}$, and $\{\tilde{d}_i\}$ is stochastically indistinguishable from $\{d_i\}$. We then have Proposition 6.1.5 for S valued d_i.

REMARK 6.1.6. A fact to be used in future sections is that Proposition 6.1.5 and our constructive approach allow for the possibility of obtaining two or more tangent decoupled sequences to any sequence $\{d_i\}$. If we desire two tangent decoupled sequences, in the constructive approach, we simply take two $\mathcal{F}_{i-1}$ independent repetitions of d_i, one of them will be e_i and the other $\tilde{e}_i$. This approach will make the sequence $\{e_i\}$ $\mathcal{G}$-independent from $\{\tilde{e}_i\}$ and in addition each sequence will be a decoupled tangent version of $\{d_i\}$. Because of this, $\mathcal{L}(\tilde{e}_i|\mathcal{F}_{i-1}) = \mathcal{L}(e_i|\mathcal{F}_{i-1}) =$

$\mathcal{L}(e_i|\mathcal{G})$ and $\{d_i - e_i\}$ and $\{e_i - \tilde{e}_i\}$ are $\mathcal{F}_i'$ tangent and conditionally symmetric, with $\mathcal{F}_i' = \sigma(\mathcal{F}_i, e_1, \ldots, e_i, \tilde{e}_1, \ldots, \tilde{e}_i)$.

In what follows we present examples obtaining a *decoupled* version for discrete time processes.

Our first example shows the importance of *decoupling* in the study of simple random sampling without replacement from finite populations. To do this, *conditionally independent* sampling is introduced and then compared to simple random sampling without replacement. For convenience, the notation for random variables and their associated sample values is the same.

EXAMPLE 6.1.7. Consider drawing a sample of size n from a box with N balls $\{b_1, \ldots, b_N\}$, $n \leq N < \infty$. The sequence $\{d_i\}_{i=1}^n$ will represent a sample without replacement, $\{e_i\}_{i=1}^n$ a *conditionally independent* sample and $\{z_i\}_{i=1}^n$ a sample with replacement. In obtaining a *conditionally independent* sequence proceed as follows. At the ith stage of a simple random sample without replacement both d_i and e_i are obtained by sampling uniformly from $\{b_1, \ldots, b_N\} \backslash \{d_1, \ldots, d_{i-1}\}$. This may be attained by randomization if N is known or by selectively returning balls to the box after sampling if N is unknown. More precisely, we can obtain both sequences in the following way: at the ith stage at first we draw e_i, return the ball, then draw d_i and put the ball aside. It is easy to see that the above procedure will make $\{e_i\}_{i=1}^n$ *tangent* to $\{d_i\}_{i=1}^n$ with $\mathcal{F}_n = \sigma(d_1, \ldots, d_n; e_1, \ldots, e_n)$ (see Definition 6.1.1). Moreover, $\{e_i\}_{i=1}^n$ is *conditionally independent* given $\mathcal{G} = \sigma(d_1, \ldots, d_N)$.

The results of this chapter (e.g. Corollary 6.2.5 with $g = 1$) could be compared with Hoeffding's inequality for sampling without replacement which states that if $\{d_i\}$ is obtained from simple random sampling without replacement and $\{z_i\}$ from simple random sampling with replacement from the same finite population, then for any continuous and convex function Ψ, $\mathbb{E}\Psi(\sum_{i=1}^n d_i) \leq \mathbb{E}\Psi(\sum_{i=1}^n z_i)$. This result has been extensively applied in making inferences about simple random sampling without replacement by using results known for simple random sampling with replacement.

EXAMPLE 6.1.8. In this example we provide a *conditionally independent* sequence to an autoregressive model. Let $d_0 = 0$ and, for all $i \geq 1$, $d_i = \theta d_{i-1} + \varepsilon_i$ where $|\theta| < 1$ and ε_i is a sequence of i.i.d. mean zero random variables. Then, a *conditionally independent* sequence *tangent* to $\{d_i\}$ is $\{e_i\}$ where for each i, $e_i = \theta d_{i-1} + \tilde{\varepsilon}_i$ with $\{\tilde{\varepsilon}_i\}$ an independent copy of $\{\varepsilon_i\}$.

EXAMPLE 6.1.9. Let $\{X_i\}$ be a sequence of independent random variables, and let $\{\tilde{X}_i\}$ be an independent copy of $\{X_i\}$. Let N be a stopping time adapted to $\sigma(X_1, \ldots, X_i)$, in particular, independent of $\{\tilde{X}_i\}$. Then $\{X_i I_{N \geq i}\}$ and $\{\tilde{X}_i I_{N \geq i}\}$ are tangent with respect to $\mathcal{F}_i = \sigma(X_1, \ldots, X_i, \tilde{X}_1, \ldots, \tilde{X}_i)$.

Here is a detailed proof of the claim in Example 6.1.9:

PROOF. Since the $\{X_i\}$ are independent, X_i is independent of $\mathcal{F}_{i-1}$. Furthermore, $I_{N\geq i} = 1 - I_{N<i}$, is in $\mathcal{F}_{i-1}$. Therefore, for all $A \in \mathcal{F}_{i-1}$

$$\begin{aligned}
&\int_A I_{X_i I_{N\geq i} > y} dP \\
&= \int_{A\cap\{N\geq i\}} I_{X_i I_{N\geq i} > y} dP + \int_{A\cap\{N<i\}} I_{0>y} dP \\
&= \int_{A\cap\{N\geq i\}} I_{X_i > y} dP + I_{y<0} \Pr\{A \cap \{N < i\}\} \\
&= \int I_{X_i > y} I_{A\cap\{N\geq i\}} dP + I_{y<0} \Pr\{A \cap \{N < i\}\} \\
&= \int I_{\tilde{X}_i > y} dP \int I_{A\cap\{N\geq i\}} dP + I_{y<0} \Pr\{A \cap \{N < i\}\} \quad \text{(by indep.)} \\
&= \int_A I_{\tilde{X}_i I_{N\geq i} > y} dP. \qquad \square
\end{aligned}$$

Extending Example 6.1.9, our next example provides a more general framework to study randomly stopped sums. More precisely, Example 6.1.10 shows how to obtain *conditionally independent* processes *tangent* to randomly stopped sums, U-statistics, quadratic forms and martingale transforms of independent variables.

EXAMPLE 6.1.10. Let $\{X_i\}$ be a sequence of independent random variables, $\{\tilde{X}_i\}$ an independent copy of $\{X_i\}$ and N a stopping time adapted to $\{\sigma(X_1, \ldots, X_i)\}$. Let $f_j : \mathbb{R}^j \to \mathbb{R}$, $\quad j = 1, 2, \ldots$, be a sequence of measurable functions. Then the sequence

$$\{f_j(X_1, X_2, \ldots, X_{j-1}; X_j) I_{N\geq j}\}$$

is tangent to

$$\{f_j(X_1, \ldots, X_{j-1}; \tilde{X}_j) I_{N\geq j}\}$$

with respect to $\mathcal{F}_n = \sigma(X_1, \ldots, X_n; \tilde{X}_1, \ldots, \tilde{X}_n)$.

Using Lemma 2.3.3 and (6.2.12) in combination with Example 6.1.10, one can compare the tail probabilities of the random variables $\max_{j\leq N} X_j$ to those of $\max_{j\leq N} \tilde{X}_j$. More generally one can compare the tail probabilities of $\max_{j\leq N} f(X_1, \ldots, X_j)$ to those of $\max_{j\leq N} f(X_1, \ldots, X_{j-1}, \tilde{X}_j)$. Moreover, the use of Corollary 6.2.4 permits the comparison of the moment generating function of the sum $\sum_{j=1}^{\infty} f_j(X_1, \ldots, X_{j-1}; X_j) I_{N\geq j}$ to the moment generating function (MGF) of $\sum_{j=1}^{\infty} f_j(X_1, X_2, \ldots, X_{j-1}; \tilde{X}_j) I_{N\geq j}$. That is, we can compare the MGF's of $\sum_{j=1}^{N} f_j(X_1, \ldots, X_{j-1}; X_j)$ and $\sum_{j=1}^{N} f_j(X_1, \ldots, X_{j-1}; \tilde{X}_j)$, and the latter is a sum of conditionally independent random variables given $\mathcal{Z} = \sigma(X_1, X_2, \ldots)$.

In particular, the case of U-statistics with kernel f can be obtained by taking

$$f_j(X_1, \ldots, X_{j-1}; X_j) = \sum_{i=1}^{j-1} f(X_i, X_j)$$

to compare $\sum_{j=2}^{N} \sum_{i=1}^{j-1} f(X_i, X_j)$ to $\sum_{j=2}^{N} \sum_{i=1}^{j-1} f(X_i, \tilde{X}_j)$. If we take

$$f_j(X_1, \ldots, X_{j-1}; X_j) = \sum_{i=1}^{j-1} a_{ij} X_i X_j$$

(with $\{a_{ij}\}$ constants), then we get decoupling of quadratic forms. Finally, taking $f_j(X_1, \ldots, X_{j-1}, X_j) = h_{j-1}(X_1, \ldots, X_{j-1})X_j$ for functions $h_j : \mathbb{R}^j \to \mathbb{R}$ and assuming the Xs to be independent mean zero, we get results for martingale transforms of independent random variables comparing $\sum_{j=2}^{n} h_{j-1}(X_1, \ldots, X_{j-1})X_j$ to $\sum_{j=2}^{n} h_{j-1}(X_1, \ldots, X_{j-1})\tilde{X}_j$.

In the following section we will introduce a simple but powerful decoupling inequality comparing the moment generating function of a sequence of dependent random variables to that of a conditionally independent sequence.

6.2 Exponential Decoupling Inequalities for Sums

In this section we provide decoupling inequalities for the moment generating functions of pairs of tangent sequences when one of them satisfies the CI condition. In view of Proposition 6.1.5, our comparisons serve as bounds for functionals of arbitrary real random variables.

The key to all developments is the following result. For the sake of clarity, we state Theorem 6.2.1 in two parts.

THEOREM 6.2.1.A. *Let $\{d_i\}$ be a sequence of positive variables adapted to an increasing sequence of σ-fields $\{\mathcal{F}_i\}$ contained in $\mathcal{F}$. Then, on a possibly enlarged probability space, there exists an increasing sequence of σ-fields $\{\tilde{\mathcal{F}}_i\}$ contained in $\{\tilde{\mathcal{F}}\}$, a σ-field $\mathcal{G}$ contained in $\tilde{\mathcal{F}}$ and sequence $\{e_i\}$ satisfying the CI (conditional independence) condition given $\mathcal{G}$, with $\{e_i\}$ $\{\tilde{\mathcal{F}}_i\}$-tangent to $\{d_i\}$ and such that for all $\mathcal{G}$-measurable random variables $g \geq 0$,*

$$\mathbb{E}(g\Pi_{i=1}^{n} d_i)^{1/2} \leq (\mathbb{E}g\Pi_{i=1}^{n} e_i)^{1/2}. \tag{6.2.1}$$

THEOREM 6.2.1.B. *Let $\{d_i\}$ be a sequence of positive variables adapted to an increasing sequence of σ-fields $\{\mathcal{F}_i\}$ contained in $\mathcal{F}$. Let $\{e_i\}$ be any $\{\mathcal{F}_i\}$-tangent sequence to $\{d_i\}$ with $\{e_i\}$ satisfying the CI (conditional independence) condition given $\mathcal{G} \subseteq \mathcal{F}$. Then, for all $\mathcal{G}$-measurable functions $g \geq 0$ the following inequality holds,*

$$\mathbb{E}(g\Pi_{i=1}^{n} d_i)^{1/2} \leq (\mathbb{E}g\Pi_{i=1}^{n} e_i)^{1/2}. \tag{6.2.2}$$

(Recall that $\mathcal{G}$ may be taken to equal $\sigma(\{d_i\})$).

The proof of the above result is based on the following lemma which (in the case $K = 1$) basically consists of a variation on Jensen's inequality allowing for a change of measure.

LEMMA 6.2.2. *Let X, Y be two non-negative random variables with $X = 0$ when $Y = 0$, and $\mathbb{E}\frac{X}{Y} \le K$ for some constant K. Then,*

$$\mathbb{E}\sqrt{X} \le \sqrt{K\mathbb{E}Y}. \tag{6.2.3}$$

PROOF. $\mathbb{E}\sqrt{X} = \mathbb{E}(\sqrt{\frac{X}{Y}} \times \sqrt{Y}) \le \sqrt{\mathbb{E}\frac{X}{Y}} \times \sqrt{\mathbb{E}Y} \le \sqrt{K\mathbb{E}Y}$, by Hölder's inequality. □

PROOF OF THEOREM 6.2.1B. We will need the following identity

$$\mathbb{E}\left(\frac{\Pi_{i=1}^n d_i}{\Pi_{i=1}^n \mathbb{E}(d_i|\mathcal{F}_{i-1})}\right) = 1, \tag{6.2.4}$$

which will be derived using an induction argument along with the properties of conditional expectation. More formally, if we assume (6.2.4) is valid for $n - 1$ then,

$$\begin{aligned}\mathbb{E}\left(\frac{\Pi_{i=1}^n d_i}{\Pi_{i=1}^n \mathbb{E}(d_i|\mathcal{F}_{i-1})}\right) &= \mathbb{E}\left[\frac{\Pi_{i=1}^{n-1} d_i}{\Pi_{i=1}^{n-1} \mathbb{E}(d_i|\mathcal{F}_{i-1})} \times \frac{\mathbb{E}(d_n|\mathcal{F}_{n-1})}{\mathbb{E}(d_n|\mathcal{F}_{n-1})}\right]\\ &= \mathbb{E}\left(\frac{\Pi_{i=1}^{n-1} d_i}{\Pi_{i=1}^{n-1} \mathbb{E}(d_i|\mathcal{F}_{i-1})}\right) = 1.\end{aligned}$$

Also, since $\{e_i\}$ is tangent to $\{d_i\}$ and *conditionally independent* given $\mathcal{G}$,

$$\begin{aligned}g\Pi_{i=1}^n \mathbb{E}(d_i|\mathcal{F}_{i-1}) &= g\Pi_{i=1}^n \mathbb{E}(e_i|\mathcal{F}_{i-1}) = \Pi_{i=1}^n g\mathbb{E}(e_i|\mathcal{G})\\ &= g\mathbb{E}(\Pi_{i=1}^n e_i|\mathcal{G}) = \mathbb{E}(g\Pi_{i=1}^n e_i|\mathcal{G}),\end{aligned} \tag{6.2.5}$$

since g is $\mathcal{G}$-measurable. Next, we replace X by $g\Pi_{i=1}^n d_i$ and Y by $g\Pi_{i=1}^n \mathbb{E}(d_i|\mathcal{F}_{i-1})$ in Lemma 6.2.2 to get

$$\mathbb{E}\sqrt{g\Pi_{i=1}^n d_i} \le \sqrt{1 \times \mathbb{E}[g\mathbb{E}(\Pi_{i=1}^n e_i|\mathcal{G})]} = \sqrt{\mathbb{E}g\Pi_{i=1}^n e_i}.$$ □

This theorem has far-reaching implications in the study of the properties of dependent variables as can be seen by the following corollaries.

COROLLARY 6.2.3. *Let $\{D_i\}$, $\{E_i\}$ be $\{\mathcal{F}_i\}$-tangent events (their indicator variables are tangent). Assume that $\{E_i\}$ is decoupled (CI). Then, for any set G measurable with respect to $\sigma(\{D_i\}_{i=1}^\infty)$,*

$$\Pr\{\cap_{i=1}^n D_i\} \le \Pr\{\cap_{i=1}^n E_i | \cap_{i=1}^n D_i\}, \tag{6.2.6}$$

more generally,

$$\Pr\{\cap_{i=1}^n D_i \cap G\} \le \Pr\{\cap_{i=1}^n E_i \cap G | \cap_{i=1}^n D_i \cap G\}, \tag{6.2.7}$$

and finally,

$$\Pr\{\cap_{i=1}^n D_i\} \le \Pr\{\cap_{i=1}^n E_i\}^{1/2}. \tag{6.2.8}$$

PROOF. We will first prove (6.2.7). Let $d_i = I_{D_i}$, $e_i = I_{E_i}$ and $g = I_G \Pi_{i=1}^n d_i$. Then, from Theorem 6.2.1,

$$\Pr\{\cap_{i=1}^n D_i \cap G\} = \mathbb{E}(\Pi_{i=1}^n d_i I_G) \le \sqrt{\mathbb{E}(\Pi_{i=1}^n d_i I_G \Pi_{i=1}^n e_i)}$$
$$= \sqrt{\Pr\{\cap_{i=1}^n D_i \cap \cap_{i=1}^n E_i \cap G\}}.$$

The proof of (6.2.7) is now completed by dividing over by $\sqrt{\Pr\{\cap_{i=1}^n D_i \cap G\}}$. Equations (6.2.6) and (6.2.8) follow by letting $G = \Omega$ in (6.2.7). □

Corollary 6.2.3 is of particular importance in the study of the first and last order statistics of a sequence of dependent random variables. If we let $\{d_i\}$ and $\{e_i\}$ be two tangent sequences with $\{e_i\}$ satisfying the CI condition then, letting $D_i = \{d_i > t\}$ and $E_i = \{e_i > t\}$ above, we get

$$\Pr\Big\{\min_{1\le i\le n} d_i > t\Big\} \le \Pr\Big\{\min_{1\le i\le n} e_i > t \Big| \min_{i\le n} d_i > t\Big\} \tag{6.2.9}$$

and

$$\Pr\Big\{\min_{1\le i\le n} d_i > t\Big\} \le \Pr\Big\{\min_{1\le i\le n} e_i > t\Big\}^{1/2}, \tag{6.2.10}$$

for all t. Similarly, letting $D_i = \{d_i \le t\}$ and $E_i = \{e_i \le t\}$, we get

$$\Pr\Big\{\max_{1\le i\le n} d_i \le t\Big\} \le \Pr\Big\{\max_{1\le i\le n} e_i \le t \Big| \max_{i\le n} d_i \le t\Big\} \tag{6.2.11}$$

and

$$\Pr\Big\{\max_{1\le i\le n} d_i \le t\Big\} \le \Pr\Big\{\max_{1\le i\le n} e_i \le t\Big\}^{1/2}, \tag{6.2.12}$$

for all t. Recall the related inequality of Lemma 2.3.3 which was used to develop the decoupling inequalities for randomly stopped maxima,

$$\Pr\Big\{\max_{1\le i\le n} d_i > t\Big\} \le 2\Pr\Big\{\max_{1\le i\le n} e_i > t\Big\}. \tag{6.2.13}$$

In fact, one can use (6.2.12) to improve one side of (6.2.13) when the sequence $\{e_i\}$ is *conditionally independent* by using the inequality $1-(1-x)^{1/2} \ge x/2$, for $0 \le x \le 1$. More precisely,

$$\Pr\Big\{\max_{1\le i\le n} d_i > t\Big\} \ge 1 - \sqrt{\Pr\Big\{\max_{1\le i\le n} e_i \le t\Big\}}$$
$$= 1 - \sqrt{1 - \Pr\Big\{\max_{1\le i\le n} e_i > t\Big\}}$$
$$\ge \frac{1}{2}\Pr\Big\{\max_{1\le i\le n} e_i > t\Big\}.$$

The following example shows that (6.2.6) of Corollary 6.2.3 is sharp for all n.

EXAMPLE 6.2.4. Let $\{D_i\}$ be a sequence of independent events. Let $\{E_i\}$ be an independent copy of $\{D_i\}$. Then, these sequences are $\mathcal{F}_i$-tangent with respect to $\mathcal{F}_n = \sigma(\{D_i\}, \{E_i\}, i = 1, \ldots, n)$: in this case

$$\Pr\{\cap_{i=1}^n D_i\} = \Pr\{\cap_{i=1}^n E_i\} = \Pr\{\cap_{i=1}^n E_i | \cap_{i=1}^n D_i\}, \tag{6.2.14}$$

since the two sequences are independent.

In the sequel, most of the results could be stated in two parts as is the case with Theorem 6.2.1 A and Theorem 6.2.1 B. For conciseness we state them only in one of the two forms.

Among the corollaries to Theorem 6.2.1, we start with the following exponential inequality comparing the moment generating function of an arbitrary sequence to that of a *conditionally independent* sequence.

COROLLARY 6.2.5. *Let $\{d_i\}$ be a sequence of random variables adapted to an increasing sequence of σ-fields $\{\mathcal{F}_i\}$ contained in $\mathcal{F}$. Let $\{e_i\}$ be any $\{\mathcal{F}_i\}$-tangent sequence to $\{d_i\}$ with $\{e_i\}$ satisfying the CI (conditional independence) condition given $\mathcal{G}$. Then, for all $\mathcal{G}$-measurable random variables $g \geq 0$ and all finite λ,*

$$\mathbb{E}\Big[g \exp\Big(\lambda \sum_{i=1}^n d_i\Big)\Big] \leq \sqrt{\mathbb{E}\Big[g^2 \exp\Big(2\lambda \sum_{i=1}^n e_i\Big)\Big]}. \tag{6.2.15}$$

Note that if $g \geq 1$ and d_i is mean zero for all i then, using Jensen's inequality, one can see that the second expectation above is greater than or equal to one and hence the square root symbol may be removed.

COROLLARY 6.2.6. *Under the assumptions of Corollary 6.2.5, for $\lambda > 0$,*

$$\mathbb{E}\Big[g \exp\Big(|\lambda \sum_{i=1}^n d_i|\Big)\Big] \leq 2\sqrt{\mathbb{E}\Big[g^2 \exp\Big(2\lambda |\sum_{i=1}^n e_i|\Big)\Big]}. \tag{6.2.16}$$

PROOF. To see this just use Corollary 6.2.5 twice. □

The following corollary acts as a symmetrization result for martingales.

COROLLARY 6.2.7. *Let $\{d_i\}$ be a mean-zero martingale difference sequence adapted to an increasing sequence of σ-fields $\{\mathcal{F}_i\}$ contained in $\mathcal{F}$. Then, on a possibly enlarged probability space, there exists a sequence of σ-fields $\{\tilde{\mathcal{F}}_i\}$ contained in $\tilde{\mathcal{F}}$, a sigma field $\mathcal{G}$ and a sequence $\{e_i\}$ satisfying the CI (conditional independence) condition given $\mathcal{G}$ with $\{e_i\}$ $\{\tilde{\mathcal{F}}_i\}$-tangent to $\{d_i\}$ such that, for all $\mathcal{G}$-measurable random variables $g \geq 0$ and all finite λ,*

$$\mathbb{E}\Big[g \exp\Big(\lambda \sum_{i=1}^n d_i\Big)\Big] \leq \sqrt{\mathbb{E}\Big[g^2 \exp\Big(4\lambda \sum_{i=1}^n \varepsilon_i e_i\Big)\Big]}, \tag{6.2.17}$$

for a Rademacher sequence $\{\varepsilon_i\}$ *(that is, independent variables with* $\Pr\{\varepsilon_i = 1\} = \Pr\{\varepsilon_i = -1\} = 1/2$*) independent of* $\{d_i\}$ *and* $\{e_i\}$.

PROOF. In Corollary 6.2.5, we obtained a decoupling of the left-hand side. Next, observe that since $\{d_i\}$ is a mean-zero martingale difference, the same is true for $\{e_i\}$. Let $\{\tilde{e}_i\}$ be a second sequence tangent to $\{d_i\}$ satisfying the CI condition and satisfying the property that given $\mathcal{G}$, $\{e_i\}$ and $\{\tilde{e}_i\}$ are independent and for each fixed i, e_i, $\tilde{e}_i$ are i.i.d. given $\mathcal{G}$. Then,

$$\left(\mathbb{E}\left[g \exp\left(\lambda \sum_{i=1}^{n} d_i\right)\right]\right)^2 \le \mathbb{E}\left[g^2 \exp\left(2\lambda \sum_{i=1}^{n} e_i\right)\right]$$

$$= \mathbb{E}\left[g^2 \exp\left(2\lambda \sum_{i=1}^{n} e_i - \mathbb{E}\left(\sum_{i=1}^{n} \tilde{e}_i | \mathcal{G}\right)\right)\right] \quad \text{(mean zero property)}$$

$$= \mathbb{E}\left[g^2 \exp\left(2\mathbb{E}\left(\lambda \sum_{i=1}^{n} e_i - \sum_{i=1}^{n} \tilde{e}_i | \mathcal{G}, \{e_i\}\right)\right)\right] \quad \text{(independence)}$$

$$\le \mathbb{E}\left[g^2 \exp\left(2\lambda\left(\sum_{i=1}^{n} e_i - \sum_{i=1}^{n} \tilde{e}_i\right)\right)\right] \quad \text{(conditional Jensen's)}$$

$$= \mathbb{E}\left[g^2 \exp\left(2\lambda \sum_{i=1}^{n} (e_i - \tilde{e}_i)\varepsilon_i\right)\right] \quad \text{(conditional i.i.d. and symmetry)}$$

$$\le \frac{1}{2}\mathbb{E}\left[g^2 \exp\left(4\lambda \sum_{i=1}^{n} e_i \varepsilon_i\right)\right] + \frac{1}{2}\mathbb{E}\left[g^2 \exp\left(4\lambda \sum_{i=1}^{n} \tilde{e}_i \varepsilon_i\right)\right] \quad \text{(convexity)}$$

$$= \mathbb{E}\left[g^2 \exp\left(4\lambda \sum_{i=1}^{n} e_i \varepsilon_i\right)\right].$$

□

The following corollary provides a decoupling inequality for U-statistics and multilinear forms (see Example 6.1.10). It is interesting to see that in this special case, one can obtain highly improved constants to the related special case of Theorem 3.1.1 in several instances. In particular, if $f_{i_1 \dots i_k}(z_{i_1}, \dots, z_{i_k}) \ge 0$ the exponent $1/2^{k-1}$ can be dropped since the right-hand expectation is greater than or equal to 1.

COROLLARY 6.2.8. *Let* $\{Z_i^1\}, \{Z_i^2\}, \dots, \{Z_i^k\}$ *be independent copies of a sequence of independent random variables* $\{Z_i\}$. *For* $k \ge 2$ *let* $f_{i_1 i_2 \dots i_k}$ *be an array of functions from* $\mathbb{R}^k$ *into* $\mathbb{R}$. *Then for every finite t,*

$$\mathbb{E} \exp\left(t \sum f_{i_1 \dots i_k}(Z_{i_1}, \dots, Z_{i_k})\right) \le \left[\mathbb{E} \exp\left(2^{k-1} t \sum f_{i_1 \dots i_k}(Z_{i_1}^1, \dots, Z_{i_k}^k)\right)\right]^{1/2^{k-1}}, \tag{6.2.18}$$

where $\sum_{1 \le i_1 < i_2 < \dots < i_k \le n} = \sum$.

PROOF. The proof follows by iteration observing that from Example 6.1.10 the Z_{i_k} can be decoupled. Next, one can see by changing the order of summation that the variables in the other entries can be decoupled as well (one by one). □

6.3 Tail Probability and L_p Inequalities for Tangent Sequences I

We continue by developing inequalities comparing the tail probabilities of sums of tangent sequences but first we introduce some notation that will simplify our proofs considerably.

NOTATION. Let $\{e_n\}$, $\{d_n\}$ be two sequences of random variables in a Hilbert space $\mathbf{H}$ and let

$$\begin{aligned} F_n^* &= \max_{1\le j\le n} \|\sum_{i=1}^{j} d_i\|, \qquad & G_n^* &= \max_{1\le j\le n} \|\sum_{i=1}^{j} e_i\|, \\ d_n^* &= \max_{1\le j\le n} \|d_j\|, \qquad & e_n^* &= \max_{1\le j\le n} \|e_j\|. \end{aligned} \tag{6.3.1}$$

The inequality in the following theorem, which is the main result of this section, provides tail probability comparisons between the sums of two tangent sequences.

THEOREM 6.3.1. (LENGLART-TYPE INEQUALITY) *Suppose $\{d_i\}$ and $\{e_i\}$ are two $\{\mathcal{F}_i\}$-tangent sequences of random variables, either $\mathbf{H}$-valued and conditionally symmetric or real and non-negative. Then for any stopping time N adapted to $\{\mathcal{F}_n\}$ and all $p > 0$,*

$$\Pr\{F_N^{*p} > x\} \le \frac{3C_p}{x}\mathbb{E}(G_N^{*p} \wedge y) + 4\Pr\{2^p G_N^{*p} > y\} \tag{6.3.2}$$

for all $x, y > 0$, where $0 < C_p < \infty$ is a constant depending on p only.

In particular, if the d_is are conditionally symmetric then, for $p = 2$,

$$\Pr\{F_N^{*2} > x\} \le \frac{24}{x}\mathbb{E}(G_N^{*2} \wedge y) + 4\Pr\{2G_N^{*2} > y\}. \tag{6.3.3}$$

If the d_is are non-negative then for $p = 1$,

$$\Pr\{F_N > x\} \le \frac{2}{x}\mathbb{E}(G_N \wedge y) + 4\Pr\{G_N > y\}. \tag{6.3.4}$$

REMARK 6.3.2. In this section and subsequent sections, several decoupling inequalities will be stated with N being a stopping time. In general, one can recover this case by the seemingly less general case of N fixed. This can be done by incorporating the information contained in N by means of indicator variables as is done in Example 6.3.3 below. We have stated the results in terms of stopping times to stress the fact that stopping times are important and to avoid, in some cases, the need to pass to a limit to obtain the results for general stopping times.

Before proving Theorem 6.3.1 we will present some applications.

EXAMPLE 6.3.3. Consider a sequence $\{X_i\}$ of independent random variables, $\{\tilde{X}_i\}$ an independent copy of $\{X_i\}$ and a stopping time N adapted to $\{\sigma(X_1, \ldots, X_i)\}_{i=1}^{\infty}$ (and independent of $\{\tilde{X}_i\}$). Let

$$f_j(X_1, \ldots, X_{j-1}; X_j) = \sum_{i=1}^{j-1} f(X_i, X_j) \quad \text{with} \quad f \geq 0.$$

Then by Example 6.1.10, we know that

$$\left\{\sum_{i=1}^{j-1} f(X_i, X_j) I_{N \geq j}\right\} \quad \text{and} \quad \left\{\sum_{i=1}^{j-1} f(X_i, \tilde{X}_j) I_{N \geq j}\right\}$$

are tangent with respect to $\{\mathcal{F}_j\}$ with $\mathcal{F}_j = \sigma(X_1, \ldots, X_j; \tilde{X}_1, \ldots, \tilde{X}_j)$. Therefore these sequences satisfy the conditions in Theorem 6.3.1 and

$$\begin{aligned} F_N^* &= \sum_{j=2}^{N} f_j(X_1, \ldots, X_{j-1}; X_j) \\ &= \sum_{j=2}^{N} \sum_{i=1}^{j-1} f(X_i, X_j), \\ G_N^* &= \sum_{j=2}^{N} \sum_{i=1}^{j-1} f(X_i, \tilde{X}_j), \end{aligned}$$

which implies that for all $z, k > 0$, (with $x = kz$, $y = z$),

$$\begin{aligned} \Pr\left\{\sum_{j=1}^{N} \sum_{i=1}^{j-1} f(X_i, X_j) > kz\right\} &= \Pr\{F_N^* > kz\} \quad (\text{since } f \geq 0) \\ &\leq \frac{2}{kz} \mathbb{E}(G_N^* \wedge z) + 4 \Pr\{G_N^* > z\} \\ &\leq \frac{2}{k} + 4 \Pr\{G_N > z\} \\ &= \frac{2}{k} + 4 \Pr\left\{\sum_{j=1}^{N} \sum_{i=1}^{j-1} f(X_i, \tilde{X}_j) > z\right\}. \end{aligned}$$

This result should be compared with Theorem 3.4.1. One major advantage of the inequality in Example 6.3.3 over the one presented in Theorem 3.4.1 is that it permits the introduction of the stopping time N. However, in Theorem 3.4.1, the restrictions on f are drastically relaxed and the additive factor $\frac{2}{k}$ is removed (other constants are involved).

EXAMPLE 6.3.4. Let $\{X_i\}$ be a sequence of i.i.d. standard normal random variables. Let $\{\tilde{X}_i\}$ be an independent copy of $\{X_i\}$ and N a stopping time adapted to $\{\sigma(X_1, \ldots, X_i)\}$ and independent of $\{\tilde{X}_i\}$. Then

$$\{X_i I_{N \geq i}\} \quad \text{and} \quad \{\tilde{X}_i I_{N \geq i}\}$$

are tangent with respect to $\{\sigma(X_1, \dots, X_i, \tilde{X}_1, \dots, \tilde{X}_i)\}$, and conditionally symmetric. Therefore, by Theorem 6.3.1, letting $x = kz$ and $y = 2z$ we get that for all $z > 0$ and all positive constants $k > 0$,

$$\begin{aligned}
\Pr\Big\{\Big(\sum_{i=1}^{N} X_i\Big)^2 > kz\Big\} &\le \Pr\{F_N^{*2} > kz\} \qquad (F_N^* = \max_{1\le j\le N}\Big|\sum_{i=1}^{j} X_i\Big|) \\
&\le \frac{24}{kz}\mathbb{E}\big(G_N^{*2} \wedge 2z\big) + 4\Pr\{2G_N^{*2} > 2z\} \quad (\text{since } 3C_2 \le 24) \\
&\le \frac{48}{k} + 4\Pr\{G_N^{*2} > z\} \qquad (6.3.5) \\
&\le \frac{48}{k} + 8\Pr\Big\{\Big(\sum_{i=1}^{N} \tilde{X}_i\Big)^2 > z\Big\} \quad (\text{by conditioning and Lévy's inequality}) \\
&= \frac{48}{k} + 8\Pr\{N\tilde{X}_1^2 > z\}.
\end{aligned}$$

This result provides a total decomposition of the dependence present on $\sum_{i=1}^{N} X_i$ at the price of adding the constants.

The proof of Theorem 6.3.1 will use the following straightforward observation on Doob's inequality.

LEMMA 6.3.5. *If $\{M_n, \mathcal{F}_n\}$ is a martingale with values in a Hilbert space* **H**, *or a non-negative submartingale, and $p > 1$, then*

$$\mathbb{E}\big(\sup_k \|M_k\|^p\big) \le \Big(\frac{p}{p-1}\Big)^p \mathbb{E}\|M_\infty\|^p, \qquad (6.3.6)$$

PROOF. If M_n is a martingale in **H**, then $\|M_n\|$ is a non-negative sub-martingale since $\|M_n\| = \mathbb{E}\big(\frac{M_{n+1}\cdot M_n}{\|M_n\|}\big|\mathcal{F}_n\big) \le \mathbb{E}\big(\frac{\|M_{n+1}\|}{\|M_n\|}\|M_n\|\big|\mathcal{F}_n\big)$. Therefore we can use Doob's inequality for real martingales to obtain (6.3.6). □

PROOF OF THEOREM 6.3.1. Without loss of generality, we may assume that the stopping time N is bounded by some finite constant k. The general result follows by applying the special case to the new stopping time $N \wedge k$ and using the Monotone Convergence Theorem as k tends to infinity. For simplicity of exposition, we will first prove the theorem for the case of $p = 2$, and d_i conditionally symmetric for all i. We will finish the proof by pointing out how to extend this result to the case of non-negative d_is and all $p > 0$.

The main idea of the proof is to use a new stopping time S_y (defined below) and Markov's inequality to transform the problem to one involving expectations. The stopping time S_y will include a truncation on $\{d_i\}_{i=1}^{\infty}$. This truncation is designed to allow the use of expectation results while avoiding integrability problems. Two key tools to be used to solve this problem are Lemma 2.3.3 and the following equation,

$$\mathbb{E}\Big\|\sum_{i=1}^{N\wedge S_y} d_i I_{\|d_i\|^2\le y}\Big\|^2 = \mathbb{E}\Big\|\sum_{i=1}^{N\wedge S_y} e_i I_{\|e_i\|^2\le y}\Big\|^2, \tag{6.3.7}$$

which follows from the orthogonality property of martingales since the $\{d_i\}$ are conditionally symmetric (and hence symmetric) and $N \le k < \infty$. A detailed proof of this is given in the lines leading to (6.3.9) below. The above step is the one where the decoupling of the original sequence happens.

Let $S_y = \inf\{n < \infty : G_n^{*2} \vee e_n^{*2} \vee d_n^{*^2} > y\}$, with $S_y = N$ if none of the variables within the inf sign exceeds y. This stopping time will be used to control the growth rate of F_N^* by that of $G_N^* \vee e_N^* \vee d_N^*$.

Beginning the formal part of the proof we have that,

$$\begin{aligned}\Pr\{F_N^{*2} > x\} &= \Pr\{F_N^{*2} > x, S_y > N\} + \Pr\{F_N^{*2} > x, S_y \le N\}\\ &\le \Pr\{F_N^{*2} > x, S_y > N\} + \Pr\{S_y \le N\}\\ &\le \Pr\{F_N^{*2} > x, S_y > N\} + \Pr\{G_N^{*2} > y\} + \Pr\{e_N^{*2} > y\} + \Pr\{d_N^{*2} > y\}.\end{aligned}$$

We obtain a preliminary bound on the last term of the above expression. Since $\{e_i\}$ and $\{d_i\}$ are $\{\mathcal{F}_n\}$ tangent, it is easy to see that for all $p > 0$, $\{\|e_i\|^p I_{N\ge i}\}$ and $\{\|d_i\|^p I_{N\ge i}\}$ are also tangent with respect to $\{\mathcal{F}_n\}$. Therefore, by Lemma 2.3.3,

$$\begin{aligned}\Pr\{d_N^{*p} > y\} &= \Pr\{\sup_i \|d_i I_{N\ge i}\|^p > y\}\\ &\le 2\Pr\{\sup_i \|e_i I_{N\ge i}\|^p > y\} = 2\Pr\{e_N^{*p} > y\}.\end{aligned}$$

Next, we use Markov's inequality to obtain,

$$\Pr\{F_N^{*^2} > x, S_y > N\} \le \frac{1}{x}\mathbb{E}(F_N^{*^2} I_{S_y>N})$$

and

$$\begin{aligned}\mathbb{E}F_N^{*2} I_{S_y>N} &= \mathbb{E}\max_{1\le j\le N}\Big\|\sum_{i=1}^{j} d_i\Big\|^2 I_{S_y>N}\\ &= \mathbb{E}\max_{1\le j\le N}\Big\|\sum_{i=1}^{j} d_i I_{\|d_i\|^2\le y}\Big\|^2 I_{S_y>N} \quad \text{(by def. of } S_y)\\ &\le \mathbb{E}\max_{1\le j\le N\wedge S_y}\Big\|\sum_{i=1}^{j} d_i I_{\|d_i\|^2\le y}\Big\|^2 \qquad (6.3.8)\\ &= \mathbb{E}\sup_j\Big\|\sum_{i=1}^{j} d_i I_{\|d_i\|^2\le y} I_{N\wedge S_y\ge i}\Big\|^2\\ &\le 4\mathbb{E}\Big\|\sum_{i=1}^{\infty} d_i I_{\|d_i\|^2\le y} I_{N\wedge S_y\ge i}\Big\|^2 \quad \text{(by Lemma 6.3.5)}\end{aligned}$$

$$= 4\mathbb{E}\Big\| \sum_{i=1}^{N\wedge S_y} d_i I_{\|d_i\|^2\le y}\Big\|^2.$$

The last inequality follows since $\{d_i\}$ is a sequence of conditionally symmetric random variables, which gives that $M_n = \sum_{i=1}^n d_i I_{\|d_i\|^2<y} I_{N\wedge S_y \ge i}$ is a martingale. Continuing, from our last bound we have

$$\begin{aligned}
&4\mathbb{E}\Big\| \sum_{i=1}^{\infty} d_i I_{\|d_i\|^2\le y} I_{N\wedge S_y\ge i}\Big\|^2 \\
&= 4\Big(\sum_{i=1}^{\infty} \mathbb{E}\|d_i\|^2 I_{\|d_i\|^2\le y} I_{N\wedge S_y\ge i}\Big) \quad \text{(since we have a martingale and } N\le k) \\
&= 4\mathbb{E}\big(\sum_{i=1}^{\infty} \mathbb{E}(\|d_i\|^2 I_{\|d_i\|^2\le y}|\mathcal{F}_{i-1}) I_{N\wedge S_y\ge i}\big) \\
&= 4\mathbb{E}\Big(\sum_{i=1}^{\infty} \mathbb{E}\big(\|e_i\|^2 I_{\|e_i\|^2\le y}\big|\mathcal{F}_{i-1}\big) I_{N\wedge S_y\ge i}\Big) \quad \text{(since } \{d_i\} \text{ and } \{e_i\} \text{ are tangent)} \\
&= 4\mathbb{E}\Big\| \sum_{i=1}^{N\wedge S_y} e_i I_{\|e_i\|^2\le y}\Big\|^2 \\
&\le 4\mathbb{E} \max_{1\le j\le N\wedge S_y} \Big\| \sum_{i=1}^{j} e_i I_{\|e_i\|^2\le y}\Big\|^2 \qquad (6.3.9) \\
&\le 8\mathbb{E} \max_{1\le j\le N\wedge S_y} \Big\| \sum_{i=1}^{j-1} e_i I_{\|e_i\|^2\le y}\Big\|^2 \\
&\quad + 8\mathbb{E} \max_{j\le N\wedge S_y} \|e_j\|^2 I_{\|e_j\|^2\le y} \quad ((x+y)^2 \le 2(x^2+y^2)) \\
&\le 8\mathbb{E} \max_{1\le j\le (N\wedge S_y)-1} \Big\| \sum_{i=1}^{j} e_i\Big\|^2 + 8\mathbb{E} \max_{j\le N\wedge S_y} \big(\|e_j\|^2\wedge y\big) \quad \text{(using the def. of } S_y) \\
&\le 8\mathbb{E}\Big(\max_{j\le N} \Big\| \sum_{j=1}^{j} e_i\Big\|^2 \wedge y\Big) + 8\mathbb{E} \max_{j\le N}(\|e_j\|^2\wedge y).
\end{aligned}$$

Hence

$$\begin{aligned}
&\Pr\big\{F_N^{*2} > x\big\} \\
&\quad \le \frac{8}{x}\big[\mathbb{E}(G_N^{*2}\wedge y) + \mathbb{E}(\max_{j\le N} \|e_j\|^2\wedge y)\big] + \Pr\big\{G_N^{*2} > y\big\} + 3\Pr\big\{e_N^{*2} > y\big\}.
\end{aligned}$$

Recall that, since $(x+y)^2 \le 2(x^2+y^2)$,

$$\max_{i\le n} \|e_i\|^2 = \max_{i\le n} \|G_i - G_{i-1}\|^2 \le 2G_n^{*2}$$

and, therefore,

$$\Pr\{F_N^{*2} > x\} \le \frac{24}{x}\mathbb{E}(G_N^{*2} \wedge y) + 4\Pr\{2G_N^{*2} > y\}.$$

Finally, the case when N is assumed to be finite follows by using the dominated convergence theorem, observing that, for the stopping times $N \wedge k$, the sequences $F_{N\wedge k}^*$, and $G_{N\wedge k}^*$ are increasing.

Next, we study how to transform the above result into one for L_p norms, $0 < p < 2$.

LEMMA 6.3.6. *Let U, V random variables satisfying the inequality*

$$\Pr\{|U| > y\} \le k_1 \frac{\mathbb{E}(|V| \wedge y)}{y} + k_2 \Pr\{k_3|V| > y\}. \tag{6.3.10}$$

Then,

$$\mathbb{E}|U|^p \le C_{k_1,k_2,k_3 p}\mathbb{E}|V|^p,$$

where $C_{k_1,k_2,p}$ depends on k_1, k_2 and p only, $0 < p < 1$.

The proof of this lemma can be obtained by integration and a change of variables. Letting $y = 2x$ in (6.3.3), and applying Lemma 6.3.6, we get

$$\mathbb{E}F_N^{*p} \le C_p\mathbb{E}G_N^{*p}, \tag{6.3.11}$$

for all $0 < p < 2$. In the next section we extend (6.3.11) to cover all $p > 0$.

In what follows, we outline how the validity of (6.3.11) for all $p > 0$ completes the proof of Theorem 6.3.1. The proof is exactly the same from the beginning, replacing $\|d_i\|^2$, $\|e_i\|^2$, F_N^{*2}, G_N^{*2}, d_N^{*2}, e_N^{*2} by $\|d_i\|^p$, $\|e_i\|^p$, F_N^{*p}, G_N^{*p}, d_N^{*p}, e_N^{*p}. Once one reaches the end of line (6.3.8) a use of (6.3.11) gives the bound,

$$\mathbb{E}\max_{j\le N\wedge S_y}\left\|\sum_{i=1}^{j} d_i I_{\|d_i\|^p\le y}\right\|^p \le C_p\mathbb{E}\max_{j\le N\wedge S_y}\left\|\sum_{i=1}^{j} e_i I_{\|e_i\|^p\le y}\right\|^p, \tag{6.3.12}$$

which corresponds to (6.3.9) in the case $p = 2$. From this point on, the proof is basically the same except for the difference in constants related to the use of the inequality $(x + y)^p \le (2(x \vee y))^p = 2^p(x^p + y^p)$ for all $x, y, p > 0$. In the case the d_is are non-negative, when $p = 1$, one can use the same approach as the one used in the case $p = 2$ by replacing $\|d_i\|^2$, $\|e_i\|^2$, F_N^{*2}, G_N^{*2} by d_i, e_i, F_N and G_N, respectively. The key step consists on observing that

$$\mathbb{E}\sum_{i=1}^{N\wedge S_y} d_i I_{d_i\le y} = \mathbb{E}\sum_{i=1}^{N\wedge S_y} e_i I_{e_i\le y}, \tag{6.3.13}$$

which can be seen to be valid by using the dominated convergence theorem since the variables are non-negative, and tangent. The case of non-negative variables and all $p > 0$ follows from (6.4.3) below which contains an extension of (6.3.11) to the case of all $p > 0$ and all non-negative variables. □

6.4 Tail Probability and Moment Inequalities for Tangent Sequences II: Good-Lambda Inequalities

In this section we introduce further inequalities comparing the tail probabilities and the L_Φ norms of two tangent sequences, where the variables involved are either conditionally symmetric or non-negative, for functions $\Phi \in \mathcal{A}_\alpha$.

THEOREM 6.4.1. (GOOD-LAMBDA INEQUALITY) *Fix $\alpha > 0$. Let $\{d_i\}$, $\{e_i\}$ be two $\{\mathcal{F}_n\}$-tangent sequences of random variables, either $\mathbf{H}$-valued and conditionally symmetric or real and non-negative. Then for any stopping time N adapted to $\{\mathcal{F}_n\}$, for all functions $\Phi \in \mathcal{A}_\alpha$ (see 2.2.1) and all $\delta > 0$, there is a $\beta > 1$ and a function $\phi(\beta, \delta, c_\alpha)$ satisfying that for β large and δ small enough, $0 < \beta\phi(\beta, \delta, \alpha) < 1$, such that*

$$\Pr\{\Phi(F_N^*) > \beta y, \Phi(d_N^*) \vee \Phi(e_N^*) \vee \Phi(G_N^*) \le \delta y\} \le \phi(\beta, \delta, \alpha) \Pr\{\Phi(F_N^*) > y\} \tag{6.4.1}$$

for all $y > 0$. In addition, if the d_is are assumed to be conditionally symmetric,

$$\Pr\{F_N^{*2} > \beta y, d_N^{*2} \vee e_N^{*2} \vee G_N^{*2} \le \delta y\} \le \frac{216\delta}{(\beta - 9(1+\delta))} \Pr\{F_N^{*2} > y\}, \tag{6.4.2}$$

whenever $\beta > 9(1+\delta)$.

A closely related result consists in the following comparison of moments.

THEOREM 6.4.2. *Let $\{d_i\}$ and $\{e_i\}$ be two $\{\mathcal{F}_n\}$-tangent sequences of random variables, either $\mathbf{H}$-valued and conditionally symmetric or real and non-negative. Then for any stopping time N adapted to $\{\mathcal{F}_n\}$ and for all $\Phi \in \mathcal{A}_\alpha$, there is a c_α, $0 < c_\alpha < \infty$ such that,*

$$\mathbb{E}\Phi(F_N^*) \le c_\alpha \mathbb{E}\Phi(G_N^*). \tag{6.4.3}$$

We will prove Theorem 6.4.1 first.

PROOF OF THEOREM 6.4.1. As in the proof of Theorem 6.3.1, we will present the proof of the case $\Phi(x) = x^2$ and conditionally symmetric variables first, and will only later outline its extensions to the case of all symmetric or non-negative variables and all $\Phi \in \mathcal{A}_\alpha$. Recall that $F_n = \sum_{i=1}^n d_i$ and $G_n = \sum_{i=1}^n e_i$. For $1 \le m < n < \infty$, let $F_{(m,n]}^* = \max_{m<j\le n} \|\sum_{i=m+1}^j d_i\|$ and $G_{(m,n]}^* = \sup_{m<j\le n} \|\sum_{i=m+1}^j e_i\|$. We also let, $d_i' = d_i I_{\|d_i\|^2 \le y}$, $e_i' = e_i I_{\|e_i\|^2 \le y}$, $F_n' = \sum_{i=1}^n d_i'$, while defining $d_i'^*$, $e_i'^*$, $F_i'^*$, G_i', $G_i'^*$, $F_{(m,n]}'^*$, $G_{(m,n]}'^*$ in a similar way by replacing d_i by d_i' and e_i by e_i'. Note that $F_{(m,n]}^* = \max_{m<j\le n} \|F_j - F_m\| \le \max_{m<j\le n} \|F_j\| + \|F_m\| \le 2\max_{j\le n} \|F_j\| = 2F_n^*$. Likewise, $F_{(m,n]}'^* \le 2F_n'^*$, and similarly $G_{(m,n]}'^* \le 2G_n'^*$, two inequalities to be used later.

Let

$$T_y = \begin{cases} \inf\{n < \infty : F_n^{*2} > y\} & \text{if the set is not empty} \\ \infty & \text{otherwise,} \end{cases}$$

and

$$S_{\delta y} = \begin{cases} \inf\{n < \infty : d_n^{*2} \vee e_n^{*2} \vee G_n^{*2} > \delta y\} & \text{if the set is not empty} \\ \infty & \text{otherwise.} \end{cases}$$

We observe first that on the set where $N = T_y < S_y$,

$$F_N^{*2} = F_{T_y}^* \leq 2(F_{T_y-1}^{*2} + d_{T_y}^{*2}) \leq 2(y + \delta y) = y2(1+\delta).$$

Since $\beta > 9(1+\delta)$, this implies that $\Pr\{F_N^{*2} > \beta y, N = T_y < S_{\delta y}\} = 0$. Using this observation in the third line below, we have

$$\begin{aligned}
&\Pr\{F_N^{*2} > \beta y, d_N^{*2} \vee e_N^{*2} \vee G_N^{*2} \leq \delta y\} \\
&\quad = \Pr\{F_N^{*2} > \beta y, N < S_{\delta y}\} \\
&\quad = \Pr\{F_N^{*2} > \beta y, T_y < N < S_{\delta y}\} + \Pr\{F_N^{*2} > \beta y, T_y = N < S_{\delta y}\} \\
&\quad = \Pr\{F_N'^{*2} > \beta y, T_y < N < S_{\delta y}\} \quad (\text{since } d_i' = d_i \text{ on } N < S_{\delta y}) \\
&\quad \leq \Pr\{3^2[(F_{T_{y-1}}'^*)^2 + (d_{T_y}'^*)^2 + (F_{(T_y,N]}'^*)^2] > \beta y, T_y < N < S_{\delta y}\} \\
&\quad \leq \Pr\{9(F_{(T_y,N]}'^*)^2 > (\beta - 9(1+\delta))y, T_y < N < S_{\delta y}\} \\
&\quad \leq \frac{9}{(\beta - 9(1+\delta))y} \mathbb{E}\big(F_{(T_y,N]}'^*\big)^2 I_{T_y < N < S_{\delta y}} \\
&\quad \leq \frac{9}{(\beta - 9(1+\delta))y} \mathbb{E}\big(F_{(T_y,N\wedge S_{\delta y}]}'^*\big)^2 \\
&\quad \leq \frac{9}{(\beta - 9(1+\delta))y} \mathbb{E} \max_{j \leq N \wedge S_{\delta y}} \Big\| \sum_{i=T_y+1}^{j} d_i' \Big\|^2 \qquad (6.4.4) \\
&\quad \leq \frac{9}{(\beta - 9(1+\delta))y} \left(\frac{2}{2-1}\right)^2 \mathbb{E} \Big\| \sum_{i=T_y+1}^{N \wedge S_{\delta y}} d_i' \Big\|^2 \quad (\text{by Doob's inequality}) \\
&\quad = \frac{36}{(\beta - 9(1+\delta))y} \sum_{i=1}^{\infty} \mathbb{E}\|d_i'\|^2 I_{N \wedge S_{\delta y} \geq i \geq T_y+1} \\
&\quad = \frac{36}{(\beta - 9(1+\delta))y} \sum_{i=1}^{\infty} \mathbb{E}\Big[\mathbb{E}\big(\|d_i'\|^2 [I_{N \wedge S_{\delta y} \geq i} - I_{i-1 < T_y+1}] \big| \mathcal{F}_{i-1}\big)\Big] \\
&\quad = \frac{36}{(\beta - 9(1+\delta))y} \sum_{i=1}^{\infty} \mathbb{E}\Big[\mathbb{E}\big(\|e_i'\|^2 [I_{N \wedge S_{\delta y} \geq i} - I_{i-1 < T_y+1}] \big| \mathcal{F}_{i-1}\big)\Big] \\
&\quad = \frac{36}{(\beta - 9(1+\delta))y} \sum_{i=1}^{\infty} \mathbb{E}\|e_i'\|^2 [I_{N \wedge S_{\delta y} \geq i} - I_{T_y > i-2}] \\
&\quad = \frac{36}{(\beta - 9(1+\delta))y} \mathbb{E}\Big(\sum_{i=T_y+1}^{N \wedge S_{\delta y}} \|e_i'\|^2\Big)
\end{aligned}$$

$$
\begin{aligned}
&= \frac{36}{(\beta - 9(1+\delta))y} \mathbb{E} \Big\| \sum_{i=T_y+1}^{N \wedge S_{\delta y}} e_i' \Big\|^2 \\
&= \frac{36}{(\beta - 9(1+\delta))y} \mathbb{E} \max_{j \le N \wedge S_{\delta y}} \Big\| \sum_{i=T_y+1}^{j} e_i' \Big\|^2 \qquad (6.4.5) \\
&= \frac{36}{(\beta - 9(1+\delta))y} \mathbb{E}\big(G'^*_{(T_y, N \wedge S_{\delta y}]}\big)^2.
\end{aligned}
$$

We remark that the equalities connecting (6.4.4) to (6.4.5) are valid (trivially) if N is bounded since then we have a martingale. An argument using the monotone convergence theorem similar to the one used to prove Lenglart's inequality holds here to extend them to the general case. Next we will bound the last term in the above string of inequalities:

$$
\begin{aligned}
\mathbb{E}\big(G^{1*}_{(T_y, N \wedge S_{\delta y}]}\big)^2 &\le 2\mathbb{E}\big((G'^*_{(T_y, N \wedge S_{\delta y}-1]})^2 + (e'^*_{N \wedge S_{\delta y}})^2\big) I_{T_y < N \wedge S_{\delta y}} \qquad (6.4.6) \\
&\le 2\mathbb{E}\big((G'^*_{(T_y, N \wedge S_{\delta y}-1]})^2 + \delta y\big) I_{T_y < N \wedge S_{\delta y}} \\
&\text{(using } e_n' = e_n \text{ on } n < S_{\delta y} \text{ and } e_i'^* \le \delta y \text{ one gets)} \\
&\le 2\mathbb{E}\big(2G^{*2}_{S_{\delta y} \wedge N-1} + \delta y\big) I_{T_y < N \wedge S_{\delta y}} \\
&\le 2(2\delta y + \delta y) \Pr\{T_y < N \wedge S_{\delta y}\} \\
&\le 6\delta y \Pr\{T_y < N\} = 6\delta y \Pr\{F_N^{*2} > y\}.
\end{aligned}
$$

Finally, putting terms together one gets

$$
\Pr\{F_N^{*2} > \beta y, d_N^{*2} \vee e_n^{*2} \vee G_N^{*2} \le \delta y\} \le \frac{216\delta}{(\beta - 9(1+\delta))} \Pr\{F_N^{*2} > y\},
$$

hence finishing the proof of Theorem 6.4.1 for $\Phi(x) = |x|^2$. □

PROOF OF THEOREM 6.4.2 AND COMPLETION OF THE PROOF OF THEOREM 6.4.1. Using Theorem 6.4.1 as a starting point, observe that if $\Phi \in \mathcal{A}_\alpha$, then $\Psi(x) = \Phi(\sqrt{x}) \in \mathcal{A}_\alpha$. Therefore, taking β large and *delta* small so that $\beta\phi(\beta, \delta, \alpha) < 1$ in Theorem 6.4.1 together with Lemma 2.2.2, we conclude that, for all functions $\Phi \in \mathcal{A}_\alpha$,

$$
\mathbb{E}\Phi(F_N^*) \le c_\alpha \mathbb{E}\Phi(G_N^* \vee d_N^* \vee e_N^*) \le c_\alpha\big[\mathbb{E}\Phi(G_N^*) + \mathbb{E}\Phi(d_N^*) + \mathbb{E}\Phi(e_N^*)\big].
$$

Using Lemma 2.3.3 we get for all $y > 0$, $\Pr\{\Phi(d_N^*) > y\} \le 2\Pr\{\Phi(e_N^*) > y\}$ which gives, upon integrating ,

$$
\mathbb{E}\Phi(d_N^*) \le 2\mathbb{E}\Phi(e_N^*) \le 2\mathbb{E}\Phi(2G_N^*) \le 2^{\alpha+1}\mathbb{E}\Phi(G_N^*),
$$

and we obtain

$$
\mathbb{E}\Phi(F_N^*) \le c_\alpha \mathbb{E}\Phi(G_N^*), \qquad (6.4.7)
$$

completing the proof of Theorem 6.4.1, modulo Theorem 6.4.1 for all $\Phi \in \mathcal{A}_\alpha$.

The proof of (6.4.1) for all $\Phi \in \mathcal{A}_\alpha$ and conditionally symmetric random variables follows by analogy with the proof of Theorem 6.3.1, case $\Phi(x) = |x|^2$, and using (6.4.7) as follows. Adapt the method of proof of the case $p = 2$ by replacing $F_N^{*2}, d_N^{*2}, e_N^{*2}, G_N^{*2}, d_i'e_i'$ and related quantities by the corresponding ones involving Φ such as, $\Phi(F_N^*)$, $\Phi(d_N^*)$, $\Phi(e_N^*)$, $\Phi(G_N^*)$, $d_i' = d_i I_{\Phi(\|d_i\|)\le y}$, $e_i' = e_i I_{\Phi(\|e_i\|)\le y}$, and re-define the stopping times as follows:

$$T_y = \begin{cases} \inf\{n < \infty : \Phi(F_n^*) > y\} & \text{if the set is not empty} \\ \infty & \text{otherwise.} \end{cases}$$

and

$$S_{\delta y} = \begin{cases} \inf\{n < \infty; \Phi(d_n^*) \vee \Phi(e_n^*) \vee \Phi(G_n^*) > \delta y\} & \text{if the set is not empty} \\ \infty & \text{otherwise.} \end{cases}$$

Apply when needed the growth property of the functions $\Phi \in \mathcal{A}_\alpha$ that we now recall: for all $x, y, z > 0$, $\Phi(x+y+z) \le \Phi(3(x \vee y \vee z)) \le 3^\alpha(\Phi(x) + \Phi(y) + \Phi(z))$.

Using the above re-definitions, proceed with the proof as in the case of $p = 2$ until line (6.4.4), and replace the inequalities from (6.4.4) to (6.4.5) by the bound

$$\mathbb{E}\Phi\Big(\max_{j \le N \wedge S_{\delta y}} \Big\| \sum_{i=T_y+1}^{j} d_i' \Big\|\Big) \le c_\alpha \mathbb{E}\Phi\Big(\max_{j \le N \wedge S_{\delta y}} \Big\| \sum_{i=T_y+1}^{j} e_i' \Big\|\Big), \tag{6.4.8}$$

which is easily obtained from (6.4.7) by observing that the truncated variables are $\mathcal{F}_n$-tangent and conditionally symmetric. The remaining steps of the proof, from (6.4.5) on, follow essentially the same pattern as the one outlined in the case p=2.

Concerning non-negative variables, inequalities (6.4.1) and (6.4.3) are treated similarly. One first obtains (6.4.1) for $\Phi(x) = |x|$, re-defining d_i', e_i' and the stopping times as $d_i' = d_i I_{\|d_i\|\le y}$, $e_i' = e_i I_{\|e_i\|\le y}$, and

$$T_y = \begin{cases} \inf\{n < \infty; F_n > y\} & \text{if the set is not empty} \\ \infty & \text{otherwise.} \end{cases}$$

and

$$S_{\delta y} = \begin{cases} \inf\{n < \infty : d_n^* \vee e_n^* \vee G_n > \delta y\} & \text{if the set is not empty} \\ \infty & \text{otherwise.} \end{cases}$$

Then we replace lines (6.4.4) to (6.4.5) by the equality

$$\mathbb{E} \sum_{T_y+1}^{N \wedge S_{\delta y}} d_i' = \mathbb{E} \sum_{T_y+1}^{N \wedge S_{\delta y}} e_i', \tag{6.4.9}$$

which is valid since the d_i, e_is are non-negative and tangent. Proceeding as before one obtains easily the good lambda inequality for $\Phi(x) = |x|$. This result in turn provides (6.4.3) for non-negative variables which in turn extends the good lambda inequality to the case of functions in $\mathcal{A}_\alpha$ by the method outlined for the case of symmetric variables, therefore finishing the proofs of theorems 6.4.1 and 6.4.2. □

The following corollary is important in cases when the variables involved do not satisfy the conditions of non-negativity or conditional symmetry. These conditions are replaced by the CI (conditional independence) condition of definitions 6.1.3 and 6.1.4.

COROLLARY 6.4.3. *Let $\{d_i\}$ and $\{e_i\}$ be two $\{\mathcal{F}_i\}$-tangent sequences of $\mathbf{H}$-valued random variables. Assume that $\{e_i\}$ satisfies the CI (conditional independence) condition. Then, for all $\Phi \in \mathcal{A}_\alpha$, there is $c_\alpha \in (0, \infty)$ such that,*

$$\mathbb{E}\Phi(F^*_\infty) \leq c_\alpha \mathbb{E}\Phi(G^*_\infty). \tag{6.4.10}$$

PROOF. Using Proposition 6.1.5 (see Remark 6.1.6), we can obtain an additional sequence $\{\tilde{e}_i\}$, tangent to $\{d_i\}$ and $\mathcal{G}$ conditionally independent. Moreover, $\{d_i - \tilde{e}_i\}$ and $\{e_i - \tilde{e}_i\}$ are $\{\mathcal{F}_i\}$-tangent and conditionally symmetric. Finally, the sequences $\{e_i\}$ and $\{\tilde{e}_i\}$ have the same joint distributions. Then, letting $N = \infty$ in Theorem 6.4.2 and using the triangle inequality as well as the growth property of Φ we get,

$$\begin{aligned}
&\mathbb{E}\Phi\big(\sup_n \|\sum_{i=1}^n d_i\|\big) \\
&\leq 2^\alpha\Big[\mathbb{E}\Phi\big(\sup_n \|\sum_{i=1}^n d_i - e_i\|\big) + \mathbb{E}\Phi\big(\sup_n \|\sum_{i=1}^n e_i\|\big)\Big] \qquad (6.4.11)\\
&\leq c_\alpha\Big[\mathbb{E}\Phi\big(\sup_n \|\sum_{i=1}^n e_i - \tilde{e}_i\|\big) + \mathbb{E}\Phi\big(\sup_n \|\sum_{i=1}^n e_i\|\big)\Big] \\
&\leq c_\alpha\Big[\mathbb{E}\Phi\big(\sup_n \|\sum_{i=1}^n e_i\|\big) + \mathbb{E}\Phi\big(\sup_n \|\sum_{i=1}^n \tilde{e}_i\|\big) + \mathbb{E}\Phi\big(\sup_n \|\sum_{i=1}^n e_i\|\big)\Big] \\
&= c_\alpha \mathbb{E}\Phi\big(\sup_n \|\sum_{i=1}^n e_i\|\big).
\end{aligned}$$

□

When comparing Corollary 6.4.3 to Corollary 6.2.6, which are inequalities relating dependent variables to conditionally independent variables, a natural line of research consists in trying to extend (6.4.10) to include more general functions like exponentials. We develop this line of research in Chapter 7. The basic idea is that one can prove variations of (6.4.10) where the constant c_α has the form $c_\alpha = C^\alpha$ for a universal constant C, with $0 < C < \infty$. Having this inequality allows, by adding over integer powers, the development of exponential inequalities of the type presented in (6.2.16) when $g = 1$.

As a second application to Theorem 6.4.1, one obtains the following asymptotic result comparing the tail probabilities of two tangent sequences.

COROLLARY 6.4.4. *Fix $\alpha > 0$. Let $\{d_i\}$ and $\{e_i\}$ be two $\{\mathcal{F}_i\}$-tangent sequences of random variables, either $\mathbf{H}$-valued and conditionally symmetric or real and* *non-negative. Then, for any stopping time N adapted to $\{\mathcal{F}_n\}$ and for all functions $\Phi \in \mathcal{A}_\alpha$, there is $c_\alpha \in (0, \infty)$ such that*

$$\limsup_{y\to\infty} \Phi(y)\Pr\{F_N^* > y\} \le c_\alpha \limsup_{y\to\infty} \Phi(y)\Pr\{G_N^* > y\}. \tag{6.4.12}$$

PROOF. Use Lemma 2.2.3. □

6.5 Differential Subordination and Applications

A concept closely linked to the concept of tangent sequences is differential subordination. We begin by providing a definition.

DEFINITION 6.5.1. *Let $\{d_i\}$ and $\{e_i\}$ be two $\mathcal{F}_n$-adapted sequences of variables. Then, $\{d_i\}$ is said to differentially subordinate to $\{e_i\}$ (or to be differentially subordinated to $\{e_i\}$) if $\|d_i\| \le \|e_i\|$ a.e. for all $i \ge 1$.*

Several of the results proved in the previous sections for conditionally symmetric tangent sequences turn out to be also valid for conditionally symmetric differentially subordinated sequences. Lemma 2.3.3 extends directly (with the same proof), while the extensions of Theorem 6.3.1, Theorem 6.4.1 and Theorem 6.4.2 only require minor modifications.

LEMMA 6.5.2. *Suppose the sequence $\{d_i\}$ is differentially subordinated to the sequence $\{e_i\}$. Then, for all $t > 0$,*

$$\Pr\Big\{\sup_{j\le\infty} \|d_j\| > t\Big\} \le \Pr\Big\{\sup_{j\le\infty} \|e_j\| > t\Big\}. \tag{6.5.1}$$

THEOREM 6.5.3. (LENGLART-TYPE INEQUALITY) *Suppose $\{d_i\}$ and $\{e_i\}$ are two $\{F_i\}$ adapted sequences of $\mathbf{H}$-valued conditionally symmetric variables. Assume that $\{d_i\}$ is differentially subordinated to $\{e_i\}$. Then for any stopping time N adapted to $\{\mathcal{F}_n\}$ and all $p > 0$,*

$$\Pr\big\{F_N^{*p} > x\big\} \le \frac{3C_p}{x}\mathbb{E}\big(G_N^{*p} \wedge y\big) + 4\Pr\big\{2^p G_N^{*p} > y\big\} \tag{6.5.2}$$

for all $x, y > 0$, where $0 < C_p < \infty$ is a constant depending on p only.

The reader is encouraged to find improvements on the constants above.

PROOF. Basically the same as that of Theorem 6.3.1, replacing the bound on (6.3.9) by

$$\begin{aligned}\mathbb{E}\Big\|\sum_{i=1}^{\infty} d_i' I_{N\wedge S_y \ge i}\Big\|^2 &= \mathbb{E}\sum_{i=1}^{N\wedge S_y} \|d_i'\|^2 \\ &= \mathbb{E}\sum_{i=1}^{N\wedge S_y - 1} \|d_i'\|^2 + \mathbb{E}\|d_{N\wedge S_y}'\|^2\end{aligned}$$

$$
\begin{aligned}
&\le \mathbb{E} \sum_{i=1}^{N\wedge S_y - 1} \|d_i\|^2 + \mathbb{E}\big(\|d_{N\wedge S_y}\|^2 \wedge y\big) \\
&\le \mathbb{E} \sum_{i=1}^{N\wedge S_y - 1} \|e_i\|^2 + \mathbb{E}\big(\|e_{N\wedge S_y}\|^2 \wedge y\big) \quad \text{(by diff. subord.)} \\
&= \mathbb{E} \sum_{i=1}^{N\wedge S_y - 1} \|e_i'\|^2 + \mathbb{E}\big(\|e_{N\wedge S_y}\|^2 \wedge y\big) \\
&= \mathbb{E} \sum_{i=1}^{N\wedge S_y} \|e_i'\|^2 + \mathbb{E}\big(\|e_{N\wedge S_y}\|^2 \wedge y\big) - \mathbb{E}\|e'_{N\wedge S_y}\|^2 \\
&= \mathbb{E}\Big\| \sum_{i=1}^{N\wedge S_y} e_i' \Big\|^2 + \mathbb{E}\big(\|e_{N\wedge S_y}\|^2 \wedge y\big) - \mathbb{E}\|e'_{N\wedge S_y}\|^2 \\
&\le 2\Big[\mathbb{E}\Big\| \sum_{i=1}^{N\wedge S_y - 1} e_i' \Big\|^2 + \mathbb{E}\|e'_{N\wedge S_y}\|^2\Big] + \mathbb{E}\big(\|e_{N\wedge S_y}\|^2 \wedge y\big) - \mathbb{E}\|e'_{N\wedge S_y}\|^2 \\
&\le 2\Big[\mathbb{E}\Big(\Big\| \sum_{i=1}^{N\wedge S_y - 1} e_i \Big\|^2 \wedge y\Big) + \mathbb{E}\big(\|e_{N\wedge S_y}\|^2 \wedge y\big)\Big] \\
&\le 6\mathbb{E}\Big(\max_{j\le N} \Big\| \sum_{i=1}^{j} e_j \Big\|^2 \wedge y\Big). \qquad \square
\end{aligned}
$$

THEOREM 6.5.4. (GOOD-LAMBDA INEQUALITY) *Fix $\alpha > 0$. Let $\{d_i\}$, $\{e_i\}$ be two $\{\mathcal{F}_n\}$-adapted sequences of $\mathbf{H}$-valued conditionally symmetric variables. Assume that $\{d_i\}$ is differentially subordinated to $\{e_i\}$. Then for any stopping time N adapted to $\{\mathcal{F}_n\}$ and for all functions $\Phi \in \mathcal{A}_\alpha$ and all $\delta > 0$, there is a $\beta > 1$ and a function $\phi(\beta, \delta, c_\alpha)$, satisfying that for β large and δ small enough, $0 < \beta\phi(\beta, \delta, \alpha) < 1$, such that*

$$\Pr\big\{\Phi(F_N^*) > \beta y,\ \Phi(d_N^*) \vee \Phi(e_N^*) \vee \Phi(G_N^*) \le \delta y\big\} \le \phi(\beta, \delta, \alpha) \Pr\big\{\Phi(F_N^*) > y\big\} \tag{6.5.3}$$

for all $y > 0$,

PROOF. The proof of this is equivalent to the one given for tangent sequences up to line (6.4.4) where a device as the one used in the proof of Theorem 6.5.3 is applied to get the result. □

THEOREM 6.5.5. *Let $\{d_i\}$ and $\{e_i\}$ be two $\{\mathcal{F}_i\}$-adapted sequences of $\mathbf{H}$-valued conditionally symmetric variables. Assume that $\{d_i\}$ is differentially subordinated to $\{e_i\}$. Then, for any stopping time N adapted to $\{\mathcal{F}_n\}$ and for all $\Phi \in \mathcal{A}_\alpha$ there is $c_\alpha \in (0, \infty)$ such that*

$$\mathbb{E}\Phi(F_N^*) \le c_\alpha \mathbb{E}\Phi(G_N^*). \tag{6.5.4}$$

PROOF. It follows using Theorem 6.5.4, Lemma 2.2.1 and Lemma 6.5.2. □

An application of these results gives the Burkholder–Davis–Gundy inequality for conditionally symmetric random variables as follows. Let $\{d_i\}$ be a sequence of conditionally symmetric variables. Then $M_n = \sum_{i=1}^n d_i$ is a martingale. Using the above results we will obtain a comparison between the square function and the maximal function of this martingale. Consider the sequence of vectors $u_1 = (d_1, 0, 0, \dots, 0, \dots)$, $u_2 = (0, d_2, 0, \dots, 0, \dots)$ and in general $u_i = (0, 0, \dots, d_i, 0, \dots, 0, \dots)$ with d_i at the ith position and zero everywhere else. Likewise, set $v_1 = (d_1, 0, \dots, 0, \dots)$, $v_2 = (d_2, 0, \dots, 0, \dots)$ and in general $v_n = (d_n, 0, \dots, 0, \dots)$. Then, $\|u_i\| = \|v_i\|$ and therefore $\{u_i\}$ is a differential subordinate of $\{v_i\}$ and vice-versa. Moreover, the two sequences consist of conditionally symmetric variables. Setting $U_n = \sum_{i=1}^n u_i$ and $V_n = \sum_{i=1}^n v_i$ we have that $\|U_n\| = \sqrt{\sum_{i=1}^n d_i^2}$, while $\|V_n\| = |\sum_{i=1}^n d_i|$, giving us the following result.

COROLLARY 6.5.6. *Let $\{d_i\}$ be a sequence of conditionally symmetric real random variables. Then, for all functions $\Phi \in A_\alpha$, there exist universal constants $0 < c_\alpha, C_\alpha < \infty$ for which*

$$c_\alpha \mathbb{E}\Phi\left(\sqrt{\sum_{i=1}^{\infty} d_i^2}\right) \le \mathbb{E}\Phi\left(\sup_n \left|\sum_{i=1}^{n} d_i\right|\right) \le C_\alpha \mathbb{E}\Phi\left(\sqrt{\sum_{i=1}^{\infty} d_i^2}\right). \tag{6.5.5}$$

PROOF. Using Theorem 6.5.5, we get

$$\mathbb{E}\Phi\left(\sup_n \left|\sum_{i=1}^{n} d_i\right|\right) = \mathbb{E}\Phi\left(\sup_n \|V_n\|\right) \le C_\alpha \mathbb{E}\Phi\left(\sup_n \|U_n\|\right) = C_\alpha \mathbb{E}\Phi\left(\sqrt{\sum_{i=1}^{\infty} d_i^2}\right).$$

The reverse bounds follows using the same approach. □

6.6 Decoupling Inequalities Compared to Martingale Inequalities

In this section we explore the connections between decoupling inequalities and the typical martingale inequalities.

We begin by extending two results which are important in the development of square function inequalities for martingales.

THEOREM 6.6.1. *Let $\{Z_i\}$ be a sequence of non-negative variables adapted to $\{\mathcal{F}_n\}$. Fix $\alpha > 0$. If Φ is a convex function with $\Phi \in \mathcal{A}_\alpha$ for some $\alpha > 0$, there exists a constant $0 < c_\alpha < \infty$ such that,*

$$\mathbb{E}\Phi\Big(\sum_{i=1}^{n} \mathbb{E}(Z_i|\mathcal{F}_{i-1})\Big) \le C_\alpha \mathbb{E}\Phi\Big(\sum_{i=1}^{n} Z_i\Big). \tag{6.6.1}$$

If Φ is a concave increasing function with $\Phi(0) = 0$ then there exists a constant $0 < C < \infty$ such that,

$$\mathbb{E}\Phi\Big(\sum_{i=1}^{n} Z_i\Big) \le C\mathbb{E}\Phi\Big(\sum_{i=1}^{n} \mathbb{E}\big(Z_i|\mathcal{F}_{i-1}\big)\Big). \tag{6.6.2}$$

PROOF. It follows from Proposition 6.1.5 that one can find a sequence $\{\tilde{Z}_i\}$ on a possibly enlarged probability space such that $\{\tilde{Z}_i\}$ is $\{\tilde{\mathcal{F}}_i\}$-tangent to $\{Z_i\}$ with $\tilde{\mathcal{F}}_i = \sigma(Z_1, \ldots, Z_i; \tilde{Z}_1, \ldots, \tilde{Z}_i)$ and such that $\{\tilde{Z}_i\}$ satisfies the CI condition. Moreover, $\{\tilde{Z}_i\}$ is conditionally independent given $\mathcal{G} = \sigma(\{Z_i\})$. From these observations it follows that,

$$\mathcal{L}(\tilde{Z}_i|\mathcal{G}) = \mathcal{L}(\tilde{Z}_i|\tilde{\mathcal{F}}_{i-1}) = \mathcal{L}(Z_i|\tilde{\mathcal{F}}_{i-1}) = \mathcal{L}(Z_i|\mathcal{F}_{i-1}). \tag{6.6.3}$$

Therefore, from Theorem 6.4.2, we get that, for all $\Phi \in \mathcal{A}_\alpha$,

$$\begin{aligned}
c_\alpha \mathbb{E}\Phi\Big(\sum_{i=1}^{n} Z_i\Big) &\ge \mathbb{E}\Phi\Big(\sum_{i=1}^{n} \tilde{Z}_i\Big) \\
&= \mathbb{E}\Big(\mathbb{E}\big[\Phi\big(\sum_{i=1}^{n} \tilde{Z}_i\big)|\mathcal{G}\big]\Big) \\
&\ge \mathbb{E}\Big(\Phi\big(\mathbb{E}\big[\sum_{i=1}^{n} \tilde{Z}_i|\mathcal{G}\big]\big)\Big) \\
&= \mathbb{E}\Big(\Phi\big(\sum_{i=1}^{n} \mathbb{E}[\tilde{Z}_i|\mathcal{G}]\big)\Big) \\
&= \mathbb{E}\Big(\Phi\big(\sum_{i=1}^{n} \mathbb{E}[Z_i|\mathcal{F}_{i-1}]\big)\Big),
\end{aligned}$$

where the last line follows from (6.6.3). For the proof of (6.6.2) observe that if Φ is concave increasing with $\Phi(0) = 0$ then $\Phi(cx) \le c\Phi(x)$ and therefore, Φ is in $\mathcal{A}_1$. Therefore, another use of Theorem 6.4.2 changing the $\ge$ signs by $\le$ in the chain of inequalities above (to account for the use of concavity instead of convexity) gives the result. □

The following is the Burkholder–Davis–Gundy inequality. It extends Corollary 6.5.6 to the case of martingales at the price of requiring convexity of the function Φ.

THEOREM 6.6.2. *Fix $\alpha > 0$. Let Φ be a convex function with $\Phi \in \mathcal{A}_\alpha$. Let $M_n = \sum_{i=1}^n d_i$ with $\{M_n, \mathcal{F}_n\}$ a real-valued martingale. Then, there are constants $0 < c_\alpha, C_\alpha < \infty$ for which,*

$$c_\alpha \mathbb{E}\Phi\left(\sqrt{\sum_{i=1}^{\infty} d_i^2}\right) \le \mathbb{E}\Phi\Big(\sup_n \Big\|\sum_{i=1}^{n} d_i\Big\|\Big) \le C_\alpha \mathbb{E}\Phi\left(\sqrt{\sum_{i=1}^{\infty} d_i^2}\right). \tag{6.6.4}$$

We divide the proof into two parts. The proof of the upper bound follows easily by combining the result in the conditionally symmetric case (Corollary 6.5.6) with decoupling inequalities, and will be given first. The proof of the lower bound is more complicated and will require the use of several lemmas.

PROOF OF RIGHT-HAND SIDE. By an easy centering argument we may assume that $\mathbb{E}(d_1|\mathcal{F}_0) = 0$. On a possibly enlarged probability space we can take $\{e_i\}$ to be a decoupled $\{\mathcal{F}_i\}$-tangent sequence to $\{d_i\}$. Letting $\mathcal{G} = \sigma(\{d_i\})$ then, $\mathbb{E}(\sum_{i=1}^n e_i|\mathcal{G}) = 0$. Furthermore, $\{d_i - e_i\}$ is $\{\mathcal{F}_i\}$ conditionally symmetric and $\{d_i^2\}$ is $\{\mathcal{F}_i\}$-tangent to $\{e_i^2\}$. Then, since Φ is a convex function,

$$
\begin{aligned}
&\mathbb{E}\Phi\Big(\sup_n \Big|\sum_{i=1}^n d_i\Big|\Big)\\
&= \mathbb{E}\Phi\Big(\sup_n \Big|\sum_{i=1}^n d_i - \mathbb{E}\Big(\sum_{i=1}^n e_i|\mathcal{G}\Big)\Big|\Big)\\
&= \mathbb{E}\Phi\Big(\sup_n \Big|\mathbb{E}\Big(\sum_{i=1}^n d_i - \sum_{i=1}^n e_i|\mathcal{G}\Big)\Big|\Big)\\
&\le \mathbb{E}\Phi\Big(\sup_n \Big|\sum_{i=1}^n (d_i - e_i)\Big|\Big) \qquad \text{(Jensen's inequality)}\\
&\le C_\alpha \mathbb{E}\Phi\left(\sqrt{\sum_{i=1}^n (d_i - e_i)^2}\right) \qquad \text{(Corollary 6.5.6)}\\
&\le C_\alpha \mathbb{E}\Phi\left(\sqrt{2\Big(\sum_{i=1}^n d_i^2 + \sum_{i=1}^h e_i^2\Big)}\right) \qquad [(a-b)^2 \le 2(a^2+b^2)]\\
&\le \frac{C_\alpha}{2}\left[\mathbb{E}\Phi\left(2\sqrt{2\sum_{i=1}^n d_i^2}\right) + \mathbb{E}\Phi\left(2\sqrt{2\sum_{i=1}^n e_i^2}\right)\right] \qquad \text{(by convexity)}\\
&\le \frac{C_\alpha}{2}\left[\mathbb{E}\Phi\left(2\sqrt{2\sum_{i=1}^n d_i^2}\right) + \mathbb{E}\Phi\left(2\sqrt{2\sum_{i=1}^n d_i^2}\right)\right] \qquad \text{(by Theorem 6.4.2)}\\
&\le C_\alpha \mathbb{E}\Phi\left(\sqrt{\sum_{i=1}^n d_i^2}\right).
\end{aligned}
$$

by factoring out and adjusting constants. □

We continue by presenting the lemmas needed in the proof of the left-hand side of Theorem 6.6.2. The following notation will be used extensively.

$$M_n = \sum_{i=1}^n d_i, \qquad M^* = \sup_j |M_j|, \tag{6.6.5}$$

$$d_n^* = \sup_{j\le n} |d_j|, \qquad d^* = \sup_j |d_j|, \qquad V^* = \sup_j V_j,$$

$$S_n(M) = \sqrt{\sum_{i=1}^{n} d_i^2}, \qquad S(M) = \sqrt{\sum_{i=1}^{\infty} d_i^2}. \tag{6.6.6}$$

LEMMA 6.6.3. *Let $\{M_n, \mathcal{F}_n\}$ be an L_1-martingale. Assume that $|d_j| \le V_j$ for all j, for a sequence of random variables $\{V_n\}$ satisfying that $V_n \in \mathcal{F}_{n-1}$ for all n. Then, for all $\beta > 1, 0 < \delta^2 < \beta^2 - 1$ and all $y > 0$,*

$$\Pr\{S(M) > \beta y, M^* \vee V^* \le \delta y\} \le \frac{9\delta^2}{\beta^2 - \delta^2 - 1}\Pr\{S(M) > y\}. \tag{6.6.7}$$

PROOF. Let

$$\begin{aligned}
\tau &= \inf\{n \ge 1 : S_n(M) > y\}, \qquad \nu = \inf\{n \ge 1 : S_n(M) > \beta y\},\\
\sigma &= \inf\{n \ge 0 : M_n^* \vee V_{n+1} > \delta y\},
\end{aligned} \tag{6.6.8}$$

$$N_n = \sum_{j=1}^{n} d_j I_{\{\tau < j \le \nu\wedge\sigma\}} = \sum_{j=\tau+1}^{\nu\wedge\sigma\wedge n} d_j, \quad N^* = \sup_j |N_j| \quad \text{and} \quad N = N_\infty.$$

Then, $\{N_n, \mathcal{F}_n\}$ is a martingale. It is easy to see that $N_n^* = 0$ on the set where $\{\tau \ge \sigma\}$. Therefore,

$$\begin{aligned}
N^* &\le (M_\sigma^* + M_\tau^*) I_{\tau<\sigma<\infty} + (M_\infty^* + M_\infty^*) I_{\tau<\sigma=\infty}\\
&\le \big((M_{\sigma-1}^* + V_\sigma^*) + M_\tau\big) I_{\tau<\sigma<\infty} + 2\delta y I_{\tau<\sigma=\infty}\\
&= 3\delta y I_{\tau<\sigma<\infty} + 2\delta y I_{\tau<\sigma=\infty}\\
&\le 3\delta y.
\end{aligned} \tag{6.6.9}$$

The definitions of τ and σ give,

$$\begin{aligned}
\mathbb{E}(N^*)^2 &= \mathbb{E}(N^*)^2 I_{S(M)\le y} + \mathbb{E}(N^*)^2 I_{S(M)>y}\\
&\le \mathbb{E}(N^*)^2 I_{\tau=\infty} + \mathbb{E}(N^*)^2 I_{S(M)>y}\\
&\le (3\delta y)^2 \Pr\{S(M) > y\},
\end{aligned}$$

using (6.6.9) and the fact that $N^* = 0$ whenever $\tau \ge \sigma$. Observe that on the set $\{\nu < \infty, \sigma = \infty\}$, $|d_\tau| \le V_\tau \le \delta y$ and hence,

$$S(N)^2 = S_\nu^2(N) - S_\tau^2(N) = S_\nu^2(N) - S_{\tau-1}^2(N) - d_\tau^2 \ge (\beta^2 - 1 - \delta^2)y^2. \tag{6.6.10}$$

It is also important to remark that since N_n is a martingale,

$$\mathbb{E}S^2(N) = \mathbb{E}\Big(\sum_{\tau+1}^{\nu\wedge\sigma} d_i\Big)^2 \le \mathbb{E}(N^*)^2. \tag{6.6.11}$$

Therefore,

$$
\begin{aligned}
&\Pr\{S(M) > \beta y, M^* \vee V^* \le \delta y\} \\
&\quad = \Pr\{\nu < \infty, \sigma = \infty\} \\
&\quad = \Pr\{\nu < \infty, \sigma = \infty, S^2(N) \ge (\beta^2 - 1 - \delta^2)y^2\} \\
&\quad \le \frac{1}{(\beta^2 - 1 - \delta^2)y^2} \mathbb{E}S^2(N) \\
&\quad \le \frac{1}{(\beta^2 - 1 - \delta^2)y^2} \mathbb{E}(N^*)^2 \\
&\quad \le \frac{9\delta^2}{\beta^2 - 1 - \delta^2} \Pr\{S(M) > y\},
\end{aligned}
$$

concluding the proof of the lemma. □

Using the above and Lemma 2.2.2 we obtain the following,

COROLLARY 6.6.4. *Fix $\alpha > 0$. Under the assumptions of Lemma 6.6.3 and for all $\Phi \in \mathcal{A}_\alpha$,*

$$\mathbb{E}\Phi(S(M)) \le C_\alpha\big[\mathbb{E}\Phi(M^*) + \mathbb{E}\Phi(V^*)\big], \tag{6.6.12}$$

for a constant C_α, $0 < C_\alpha < \infty$ depending on α only.

PROOF OF THEOREM 6.6.2, LEFT-HAND SIDE. We begin by introducing *Davis' decomposition* of a martingale into a sum of two martingales which are more manageable. Let $g_n = \sum_{i=1}^n a_i$, $h_n = \sum_{i=1}^n b_i$ with,

$$a_i = d_i I_{|d_i| \le 2d^*_{i-1}} - \mathbb{E}\big(d_i I_{|d_i| \le 2d^*_{i-1}} \big| \mathcal{F}_{i-1}\big),$$

and

$$b_i = d_i I_{|d_i| > 2d^*_{i-1}} - \mathbb{E}\big(d_i I_{|d_i| > 2d^*_{i-1}} \big| \mathcal{F}_{i-1}\big). \tag{6.6.13}$$

Then since $\mathbb{E}(d_i | \mathcal{F}_{i-1}) = 0$ implies that

$$\mathbb{E}\big(d_i I_{|d_i| \le 2d^*_{i-1}} \big| \mathcal{F}_{i-1}\big) = -\mathbb{E}\big(d_i I_{|d_i| > 2d^*_{i-1}} \big| \mathcal{F}_{i-1}\big), \tag{6.6.14}$$

we have the decomposition $M_n = g_n + h_n$. It is easy to see that $|a_n| \le 4d^*_{n-1}$. We will need this and the following bounds:

$$A = \sum_{i=1}^\infty |d_i| I_{|d_i| > 2d^*_{i-1}} \le 2d^*, \tag{6.6.15}$$

and

$$B = \sum_{i=1}^\infty \big|\mathbb{E}\big(d_i I_{|d_i| \le d^*_{i-1}} \big| \mathcal{F}_{i-1}\big)\big| \le \sum_{i=1}^\infty \mathbb{E}\big(|d_i| I_{|d_i| > d^*_{i-1}} \big| \mathcal{F}_{i-1}\big). \tag{6.6.16}$$

To see (6.6.15) observe that on the set $\{|d_i| > 2d^*_{i-1}\}$, $|d_i| + 2d^*_{i-1} \le 2|d_i| \le 2d^*_i$. Therefore,

$$\sum_{i=1}^{\infty} |d_i| I_{|d_i| > 2d^*_{i-1}} \le 2\sum_{i=1}^{\infty}(d^*_i - d^*_{i-1}) = 2d^*.$$

The proof of (6.6.16) follows using (6.6.14) and Jensen's inequality. These observations and the fact that $d^* \le 2M^*$ give immediately that,

$$\mathbb{E}\Phi(A) \le \mathbb{E}\Phi(2d^*) \le \mathbb{E}\Phi(4M^*) \le 4^{\alpha}\mathbb{E}\Phi(M^*), \tag{6.6.17}$$

and likewise, by Theorem 6.6.1 and (6.6.17),

$$\mathbb{E}\Phi(B) \le C_\alpha \mathbb{E}\Phi(A) \le 4^{\alpha} C_\alpha \mathbb{E}\Phi(M^*). \tag{6.6.18}$$

It is easy to see that

$$S(M) \le S(g) + S(h) \le S(g) + A + B, \tag{6.6.19}$$

and

$$g^* \le M^* + h^* \le M^* + A + B. \tag{6.6.20}$$

Therefore, using $\Phi(x+y+z) \le \Phi(3\max\{x, y, z\}) \le 3^{\alpha}\{\Phi(x) + \Phi(y) + \Phi(z)\}$, for $x, y, z \ge 0$ and re-adjusting constants we get,

$$\begin{aligned}
\mathbb{E}\Phi(S(M)) &\le C_\alpha\big[\mathbb{E}\Phi(S(g)) + \mathbb{E}\Phi(A) + \mathbb{E}\Phi(B)\big] \\
&\le C_\alpha\big[\mathbb{E}\Phi(g^*) + \mathbb{E}\Phi(d^*)\big] \\
&\qquad\qquad \text{[by Lemma 6.6.3, (6.6.17), (6.6.18) and (2.2.4)]} \\
&\le C_\alpha\big[\mathbb{E}\Phi(M^*) + \mathbb{E}\Phi(A) + \mathbb{E}\Phi(B) + \mathbb{E}\Phi(d^*)\big] \quad \text{[by (6.6.20)]} \\
&\le C_\alpha\big[\mathbb{E}\Phi(M^*) + \mathbb{E}\Phi(d^*)\big] \quad \text{[by (6.6.17) and (6.6.18)]} \\
&\le C_\alpha \mathbb{E}\Phi(M^*), \quad (\text{since } d^* \le 2M^*)
\end{aligned}$$

We have therefore completed the proof of the left-hand side of Theorem 6.6.2. □

We continue with a result which highlights the importance of decoupling in the study of the properties of martingales.

COROLLARY 6.6.5. *Fix $\alpha > 0$. Let Φ be a convex function with $\Phi \in \mathcal{A}_\alpha$. Let $M_n = \sum_{i=1}^n d_i$ where $\{M_n, \mathcal{F}_n\}$ is a real-valued martingale with martingale difference sequence $\{d_i\}$. Let $\{e_i\}$ be any sequence which is $\{\mathcal{F}_i\}$-tangent to $\{d_i\}$. Then, there are constants $0 < c_\alpha, C_\alpha < \infty$ for which*

$$c_\alpha \mathbb{E}\Phi\Big(\sup_n \Big\| \sum_{i=1}^n e_i \Big\|\Big) \le \mathbb{E}\Phi\Big(\sup_n \Big\| \sum_{i=1}^n d_i \Big\|\Big) \le C_\alpha \mathbb{E}\Phi\Big(\sup_n \Big\| \sum_{i=1}^n e_i \Big\|\Big). \tag{6.6.21}$$

PROOF. Observe that $\{e_i^2\}$ is $\{\mathcal{F}_i\}$-tangent to $\{d_i^2\}$. Also, since $\Phi(x) \in \mathcal{A}_\alpha$ then, $\Psi(x) = \Phi(\sqrt{x}) \in \mathcal{A}_\alpha$. Then, using Theorem 6.4.2 with $N = \infty$, on the variables $d_i^2, e_i^2 \geq 0$ one gets,

$$\mathbb{E}\Phi\left(\sqrt{\sum_{i=1}^{\infty} d_i^2}\right) = \mathbb{E}\Psi\Big(\sum_{i=1}^{\infty} d_i^2\Big)$$
$$\leq C_\alpha \mathbb{E}\Psi\Big(\sum_{i=1}^{\infty} e_i^2\Big)$$
$$= C_\alpha \mathbb{E}\Phi\left(\sqrt{\sum_{i=1}^{\infty} e_i^2}\right),$$

which, by the same argument, is less than or equal to $C_\alpha \mathbb{E}\Phi\big(\sqrt{\sum_{i=1}^{\infty} d_i^2}\big)$. Observe that, $M_n = \sum_{i=1}^{n} d_i$ being a martingale, $\sum_{i=1}^{n} e_i$ is also a martingale since, by tangency, the e_is have the same conditional distributions as the d_is. Therefore, using the above inequalities and (6.6.4) (the Burkholder–Davis–Gundy inequality) on both martingales we get,

$$\mathbb{E}\Phi\left(\sup_n \Big\|\sum_{i=1}^{n} e_i\Big\|\right) \leq C_\alpha \mathbb{E}\Phi\left(\sqrt{\sum_{i=1}^{\infty} e_i^2}\right) \leq C_\alpha \mathbb{E}\Phi\left(\sqrt{\sum_{i=1}^{\infty} d_i^2}\right)$$
$$\leq \mathbb{E}\Phi\Big(\sup_n \Big\|\sum_{i=1}^{n} d_i\Big\|\Big) \leq C_\alpha \mathbb{E}\Phi\left(\sqrt{\sum_{i=1}^{\infty} d_i^2}\right)$$
$$\leq C_\alpha \mathbb{E}\Phi\left(\sqrt{\sum_{i=1}^{\infty} e_i^2}\right) \leq C_\alpha \mathbb{E}\Phi\Big(\sup_n \Big\|\sum_{i=1}^{n} e_i\Big\|\Big). \qquad \square$$

Next, we will illustrate the interest of this result.

EXAMPLE 6.6.6. Let $\{X_i\}$ be a sequence of independent mean zero random variables. Let T be a stopping time adapted to $\{X_i\}$. Let $\{\tilde{X}_i\}$ be an independent copy of $\{X_i\}$ (then $\{\tilde{X}_i\}$ is independent of T as well). Then, for all $p > 0$, there exist constants $0 < c_p, C_p < \infty$ for which

$$c_p \mathbb{E}\Big(\sum_{i=1}^{T} X_i^2\Big)^{p/2} \leq \mathbb{E}\sup_{n\leq T}\Big|\sum_{i=1}^{n} X_i\Big|^p \leq C_p \mathbb{E}\Big(\sum_{i=1}^{T} X_i^2\Big)^{p/2}, \tag{6.6.22}$$

the square function inequality, and

$$c_p \mathbb{E}\sup_{n\leq T}\Big|\sum_{i=1}^{n} \tilde{X}_i\Big|^p \leq \mathbb{E}\sup_{n\leq T}\Big|\sum_{i=1}^{n} X_i\Big|^p \leq C_p \mathbb{E}\sup_{n\leq T}\Big|\sum_{i=1}^{n} \tilde{X}_i\Big|^p, \tag{6.6.23}$$

the decoupling inequality. To see this, let $d_i = X_i I_{T\geq i}$ and $e_i = \tilde{X}_i I_{T\geq i}$ in Theorem 6.6.2 and Corollary 6.6.5. We remark that (6.6.23) is a special case of Theorem 2.3.1. If we assume that $\{X_i\}$ is a sequence of independent symmetric α-stable random variables with $1 < p < \alpha \leq 2$, then a use of Levy's inequality conditioning on T gives,

$$\mathbb{E}T^{\frac{p}{2}}\mathbb{E}|X_1|^p \leq \mathbb{E}\sup_{n\leq T}\left|\sum_{i=1}^{n}\tilde{X}_i\right|^p \leq 2\mathbb{E}T^{\frac{p}{2}}\mathbb{E}|X_1|^p, \tag{6.6.24}$$

which in turn (using (6.6.23)) implies that

$$c_p\mathbb{E}T^{\frac{p}{2}}\mathbb{E}|X_1|^p \leq \mathbb{E}\sup_{n\leq T}\left|\sum_{i=1}^{n}X_i\right|^p \leq C_p\mathbb{E}T^{\frac{p}{2}}\mathbb{E}|X_1|^p, \tag{6.6.25}$$

Therefore, the decoupling bounds in (6.6.23), (6.6.25) are more effective than the square function inequalities of (6.6.22) which are less informative.

In the case of $\Phi(x) = |x|^p$, $p \geq 2$, and with $\{d_i\}$ a martingale difference sequence, there is a simple proof of the decoupling inequalities found in Theorem 6.4.2 and Corollary 6.6.5 by an iterated use of the following classical martingale inequality. For a proof of this result see Hall and Heyde (1980); the right-hand side of this inequality, with best asymptotic constants, is given in Theorem 7.3.2.

THEOREM 6.6.7. *Let $p \geq 2$. Let $M_n = \sum_{i=1}^n d_i$ where $\{M_n, \mathcal{F}_n\}$ is a martingale with martingale difference sequence $\{d_i\}$. Set $\|X\|_p = (\mathbb{E}|X|^p)^{1/p}$ for any random variable X, and $U_{n,p} = \left(\sum_{j=1}^n \|d_j\|_p^p\right)^{\frac{1}{p}}$, $s_n(M) = \left(\sum_{i=1}^n \mathbb{E}(|d_i|^2|\mathcal{F}_{i-1})\right)^{1/2}$. Then, there are constants $0 < c_p, C_p < \infty$ for which*

$$c_p\left[\|s_n(M)\|_p^p + U_{n,p}^p\right] \leq \|M_n\|_p^p \leq C_p\left[\|s_n(M)\|_p^p + U_{n,p}^p\right]. \tag{6.6.26}$$

For the sake of clarity we state the decoupling result that emanates from Theorem 6.6.7:

COROLLARY 6.6.8. *Let $p \geq 2$. Let $\{d_i\}$ be a martingale difference sequence adapted to $\{\mathcal{F}_i\}$. Let $\{e_i\}$ be any sequence which is $\{\mathcal{F}_i\}$-tangent to $\{d_i\}$. Then, using the constants of Theorem 6.6.7 we have that*

$$\frac{c_p}{C_p}\mathbb{E}\left|\sum_{i=1}^n e_i\right|^p \leq \mathbb{E}\left|\sum_{i=1}^n d_i\right|^p \leq \frac{C_p}{c_p}\mathbb{E}\left|\sum_{i=1}^n e_i\right|^p. \tag{6.6.27}$$

PROOF. With the notation of Theorem 6.6.7, and setting $N_n = \sum_{i=1}^n e_i$, $V_{n,p} = \left(\sum_{j=1}^n \|e_j\|_p^p\right)^{1/p}$, it is clear that $s_n(M) = s_n(N)$ since both quantities are obtained using only the conditional distributions of the $\{d_i\}$s and $\{e_i\}$s respectively, and these two sequences are tangent. Tangency also implies that d_i and e_i have the same distributions. Therefore, we also have that $U_{n,p} = V_{n,p}$. The proof is completed by using these observations and applying (6.6.26) twice, the first time on N_n and the second on M_n, adjusting the constants involved. □

REMARK 6.6.9. McConnell (1989) extended Corollary 6.6.8 to the case when the $\{d_i\}$, $\{e_i\}$ are assumed to have values in any UMD (uniform martingale difference) Banach space E for p with $1 < p < \infty$. Indeed, if (6.6.27) holds for any $1 < p < \infty$ then E must be UMD. Examples of UMD spaces include l_p and $L_p[0, 1]$. The spaces l_1, $L_1[0, 1]$ are not UMD.

Closing a theoretical circle, in Chapter 7 we will show how an improvement on the constant of (6.4.10) permits the use of decoupling and of sharp results for sums of independent random variables to obtain the best rate for the constant in the upper bound of (6.6.26). Therefore, martingale inequalities imply decoupling inequalities which then improve them.

6.7 References for Chapter 6

The concept of tangent processes can be traced back to Jacod (1984). The main body of definitions, examples and Proposition 6.1.5 are due to Kwapień and Woyczynski (1991). See also Kwapień and Woyczynsk (1986;1989) and Hitczenko (1990a). An account of this can be found in Kwapień and Woyczynski (1992). Parallel to their work, Jakubowski (1986) developed some of the ideas under the name of "The Principle of Conditioning". See also Hill (1982) for related results. Hoeffding (1963) contains the result presented in Example 6.1.7 for comparing sampling without replacement to sampling with replacement. Example 6.1.7 and the constructive approach for obtaining CI sequences found after Proposition 6.1.5 come from de la Peña (1994) (see also Pinelis (1995)). Section 6.2 basically comes from de la Peña (1994) and de la Peña (1997). Theorems 6.3.1 and 6.4.1 come from de la Peña (1993). Independently Kwapień and Woyczynski (1992) obtained related results (with improved constants but otherwise weaker). Theorem 6.4.2 was obtained in special cases by Zinn (1985) and in general by Hitczenko (1988) (see also Hitczenko (1990) for sharp constants in the case of non-negative variables). Corollary 6.4.3 can be found in Kwapień and Woyczynski (1992) (Proposition 5.7.1 of that work) and Corollary 6.4.4 comes from de la Peña (1993) while McConnell (1989) contains a related result. The definition of differential subordination can be found in Burkholder (1988). Theorem 6.5.3 and Theorem 6.5.4 come from de la Peña (1993) done independently and concurrently of related (weaker) results in Kwapień and Woyczynski (1992). Theorem 6.5.5 in the case of martingales and $\Phi(x) = |x|^p$ for $p > 1$ was studied by Burkholder (1988) (with sharp constants) Kwapień and Woyczynski (1992). The technique used to obtain Corollary 6.5.6 from Theorem 6.5.5 can be found in Burkholder (1988) and Burkholder (1989). Equation (6.6.1) of Theorem 6.6.1 comes from Burkholder, Davis and Gundy (1972) and (6.6.2) can be found in Burkholder (1973). The proof we present of this theorem is due to Kwapień and Woyczynski (1992). Theorem 6.6.2 is the well known Burkholder–Davis–Gundy (1972) inequality for martingales. Our proof of the upper bound of this result is new and uses decoupling. Corollary 6.6.5 was proved for the case

$\Phi(x) = |x|^p$, $p > 1$ in Kwapień and Woyczynski (1992) and was announced in Remark 5.4.2 of that work. See also Wang (1991) for sharp constants for p in different ranges. Example 6.6.6 comes from de la Peña (1998). Theorem 6.6.7 is a martingale version of Rosenthal's inequality found in Burkholder (1973). Corollary 6.6.8 is due to Zinn (1985). For the results summarized in Remmark 6.6.9 see McConnell (1989) (see also Hitczenko (1988a)). See Burkholder (1981) for information on UMD spaces.

7

Conditionally Independent Sequences

Chapter 6 was devoted to the study of the general properties of tangent sequences. In particular, in Sections 6.3 and 6.4 results were presented for comparing the tail probabilities and moments of tangent sequences when the variables involved are non-negative, conditionally symmetric, or form a martingale difference sequence. Two results from that chapter (Corollary 6.2.5 and Corollary 6.4.3) dealt with comparisons between the moment generating function and the L_p-norm of an arbitrary sum and its decoupled (conditionally independent) counterpart. In this chapter we continue to study this type of comparison. The main advantage realized is in the increased accuracy of the results obtained which in several problems translates into the almost direct transfer of results for sums of independent random variables to the case of sums of dependent random variables.

In Section 1 we present the principle of conditioning and some of its applications. This principle shows that the convergence properties of decoupled sums are shared by the original sums. The applications given include a proof of the three series theorem for sums of dependent variables, and of the martingale central limit theorem based on the corresponding results for sums of independent random variables. Section 2 presents an application of the principle of conditioning to obtain a central limit theorem for a sequence of (dependent) two by two tables. This result has several important consequences. It has practical implications in biostatistical applications including clinical follow-up studies, where it can be used to obtain information necessary for testing hypotheses about the odds ratios of a sequence of tables. Also, it is noted that other available limit theorems do not seem to apply in this case. In particular, the central limit theorem for martingales falls short in that the sizes of the increments cannot be controlled directly by the Lindeberg condition. A reduction to sums of independent random variables through decoupling permits the application of Lindeberg's central limit theorem directly. Section 3 presents a decoupling inequality for the L_p-norm ($p \geq 1$) of a sum of arbitrary random variables and shows that this norm is bounded by a universal constant times the L_p-norm of its decoupled counterpart. This result sharpens Corollary 6.4.3, which provided the result with the constant dependent on p. The extension is very powerful in that it permits the almost direct transfer (with constants having the right order of magnitude) of several inequalities for sums of independent random variables to the case of dependent random variables including martingales. As an application, we present a proof of Burkholder's extension of Rosenthal's inequality to martingales (with constants having an optimal growth rate) by applying decoupling and Rosenthal's inequality for sums of independent mean zero random variables with sharp constants. The chapter ends with an example showing that the highly desired strict decoupling inequality for tail probabilities can not hold in general.

7.1 The Principle of Conditioning and Related Results

In this section we provide the principle of conditioning. The principle states roughly that results that hold for all sequences of independent variables extend directly to the case of dependent variables. Two different versions of this principle will be covered. The first one concerns almost sure results, and the second pertains to weak limit theorems. The importance of this approach is underlined by the fact that, as special cases, one obtains both the conditional version of the three-series theorem and a form of the martingale central limit theorem based on the corresponding results for sums of independent random variables.

We begin with a statement of the classical three series theorem.

THEOREM 7.1.1. *Let $\{y_i\}$ be a sequence of independent random variables. Then $\sum_{i=1}^{\infty} y_i$ converges almost surely if and only if,*

i) $\sum_{i=1}^{\infty} \Pr\{|y_i| > 1\} < \infty$,

ii) $\sum_{i=1}^{\infty} \mathbb{E} y_i I_{|y_i| \leq 1}$ *converges,*

iii) $\sum_{i=1}^{\infty} \mathrm{Var}(y_i I_{|y_i| \leq 1}) < \infty$.

We will use this result to obtain its extension to the case of dependent variables stated next.

THEOREM 7.1.2. *Let $\{d_i\}$ be a sequence of variables adapted to an increasing sequence of σ-fields $\{\mathcal{F}_i\}$. Then $\sum_{i=1}^{\infty} d_i$ converges almost surely if*

i) $\sum_{i=1}^{\infty} \Pr\{|d_i| > 1|\mathcal{F}_{i-1}\} < \infty$, *almost surely,*

ii) $\sum_{i=1}^{\infty} \mathbb{E}(d_i I_{|d_i| \leq 1}|\mathcal{F}_{i-1})$ *converges almost surely,*

iii) $\sum_{i=1}^{\infty} \mathrm{Var}(d_i I_{|d_i| \leq 1}|\mathcal{F}_{i-1}) < \infty$, *almost surely.*

A key step in the proof of Theorem 7.1.2 is the following result that follows from Theorem 6.3.1.

COROLLARY 7.1.3. *Let $\{e_i\}$ be a decoupled sequence $\{\mathcal{F}_i\}$–tangent to $\{d_i\}$. Then the almost sure convergence of $\sum_{i=1}^{\infty} e_i$ implies that of $\sum_{i=1}^{\infty} d_i$.*

PROOF. We will use the following elementary facts valid for any sequence of random variables $\{X_j\}$:

a) $X_j \to X$ almost surely for some random variable X if and only if

$$\sup_{n:n>j} |X_n - X_j| \to 0 \quad \text{in probability.} \tag{7.1.1}$$

b) $X_j \to 0$ in probability if and only if for every $\varepsilon > 0$,

$$\Pr\{|X_n| > \varepsilon\} < \varepsilon, \tag{7.1.2}$$

for sufficiently large n.

Let $\{e_i\}$, $\{\tilde{e}_i\}$ and $\{\tilde{d}_i\}$ be three decoupled sequences each one $\{\mathcal{F}_i\}$-tangent to $\{d_i\}$. For each fixed $j \geq 1$, it is easy to see that $\{e_i\}_{i=j+1}^{\infty}$ $\{\tilde{e}_i\}_{i=j+1}^{\infty}$ $\{\tilde{d}_i\}_{i=j+1}^{\infty}$ are decoupled sequences $\{\mathcal{F}_i\}_{i=j+1}^{\infty}$-tangent to $\{d_i\}_{i=j+1}^{\infty}$. Moreover we can pick these sequences so that $\{\tilde{d}_i - \tilde{e}_i\}_{i=j+1}^{n}$ is a decoupled conditionally symmetric sequence $\{\mathcal{F}_i\}_{i=j+1}^{\infty}$-tangent to $\{d_i - e_i\}_{i=j+1}^{n}$. Therefore, from (6.3.3) of Theorem 6.3.1 with $x = \frac{\varepsilon}{4}$ and $y = 2\varepsilon^2$ we have that for $\varepsilon < 1/192$,

$$\begin{aligned}
&\Pr\Big\{\sup_{n\geq m>j}\Big(\sum_{i=j+1}^{m} d_i\Big)^2 > \varepsilon\Big\} \\
&\leq \Pr\Big\{\sup_{n\geq m>j} 2\Big[\Big(\sum_{i=j+1}^{m}(d_i - e_i)\Big)^2 + \Big(\sum_{i=j+1}^{m} e_i\Big)^2\Big] > \varepsilon\Big\} \\
&\leq \Pr\Big\{\sup_{n\geq m>j}\Big(\sum_{i=j+1}^{m}(d_i - e_i)\Big)^2 > \frac{\varepsilon}{4}\Big\} + \Pr\Big\{\sup_{n\geq m>j}\Big(\sum_{i=j+1}^{m} e_i\Big)^2 > \frac{\varepsilon}{4}\Big\} \\
&\leq 24\cdot 2\varepsilon^2\cdot\frac{4}{\varepsilon} + 4\Pr\Big\{2\sup_{n\geq m>j}\Big(\sum_{i=j+1}^{m}(\tilde{d}_i - \tilde{e}_i)\Big)^2 > 2\varepsilon^2\Big\} \\
&\quad + \Pr\Big\{\sup_{n\geq m>j}\Big(\sum_{i=j+1}^{m} e_i\Big)^2 > \frac{\varepsilon}{4}\Big\} \\
&\leq 192\varepsilon + 8\Pr\Big\{\sup_{n\geq m>j}\Big(\sum_{i=j+1}^{m} \tilde{d}_i\Big)^2 > \frac{\varepsilon^2}{4}\Big\} + \Pr\Big\{\sup_{n\geq m>j}\Big(\sum_{i=j+1}^{m} e_i\Big)^2 > \frac{\varepsilon}{4}\Big\} \\
&\leq 192\varepsilon + 9\Pr\Big\{\sup_{n\geq m>j}\Big(\sum_{i=j+1}^{m} e_i\Big)^2 > \frac{\varepsilon^2}{4}\Big\} \\
&\leq 192\Big[\varepsilon + \Pr\Big\{\sup_{n\geq m>j}\Big(\sum_{i=j+1}^{m} e_i\Big)^2 > \frac{\varepsilon^2}{4}\Big\}\Big].
\end{aligned} \tag{7.1.3}$$

The monotone convergence theorem allows taking $n \to \infty$. Setting $X_n = \sum_{i=1}^{n} e_i$, we have that $X_n - X_j = \sum_{i=j+1}^{n} e_i$. Using (7.1.1) it is easy to see that convergence of $\sum_{i=1}^{\infty} e_i$ implies $\Pr\{\sup_{\infty>m>j} (\sum_{i=j+1}^{m} e_i)^2 > \varepsilon\} \to 0$ as $j \to \infty$. This and (7.1.3) give that, for a numerical constant $c_1 > 0$,

$$\Pr\Big\{\sup_{\infty>m>j} c_1(\sum_{i=j+1}^{m} d_i)^2 > c_1\varepsilon\Big\} \leq c_1\varepsilon$$

for $j \to \infty$ large enough. Hence, using (7.1.2) and (7.1.1), $\sum_{i=1}^{\infty} d_i$ converges almost surely. □

PROOF OF THEOREM 7.1.2. Using Corollary 7.1.3, the proof requires showing that i)–iii) implies almost sure convergence of $\sum_{i=1}^{\infty} e_i$. Since $\{e_i\}$ satisfies the CI condition of Definition 6.1.3, $\mathcal{L}(e_i|\mathcal{F}_{i-1}) = \mathcal{L}(e_i|\mathcal{G})$, and $\{e_i\}$ is conditionally independent given $\mathcal{G}$. Therefore, i)–iii) (using regular conditional probabilities) above are equivalent to the almost sure convergence of

i′) $\sum_{i=1}^{n} \Pr\{|e_i| > 1|\mathcal{G}\} < \infty$,

ii′) $\sum_{i=1}^{n} \mathbb{E}(e_i I_{|e_i|\leq 1}|\mathcal{G})$ converges,

iii′) $\sum_{i=1}^{n} \mathrm{Var}(e_i I_{|e_i|\leq 1}|\mathcal{G}) < \infty$. An application of Theorem 7.1.1 conditionally on $\mathcal{G}$ gives that $\sum_{i=1}^{\infty} e_i$ converges conditionally on $\mathcal{G}$ and hence (by (7.1.1)) it also converges without conditioning. □

In what follows we describe the mode of convergence that we will be using in presenting the principle of conditioning for weak convergence. Let μ_n, μ be measurable mappings from $(\Omega, \mathcal{F}, P)$ into $\mathcal{P}(\mathbb{R})$, the space of probability distributions on $\mathbb{R}$ equipped with the topology of weak convergence. Then convergence in probability in $\mathcal{P}(\mathbb{R})$ of μ_n to μ (weak convergence in probability) will be denoted by $\mu_n \Rightarrow_P \mu$.

THEOREM 7.1.4. *Let $\{d_{ni}\}$ be an array of real random variables adapted to $\{\mathcal{F}_{ni}\}$. For each n, let $\{e_{ni}\}$ be $\mathcal{F}_{ni}$-tangent to $\{d_{ni}\}$ and satisfy the CI property with respect to $\mathcal{G}_n$. Set $U_n = \sum_{i=1}^{\infty} d_{ni}$ and $V_n = \sum_{i=1}^{\infty} e_{ni}$. If $\mathcal{L}(V_n|\mathcal{G}_n) \Rightarrow_P \mu$, where μ is a nonrandom probability distribution with characteristic function $\hat{\mu}(t) \neq 0$ for all t, then $\mathcal{L}(U_n)$ also converges to μ.*

PROOF. It suffices to show that, for all t, $\mathbb{E}e^{itU_n} \to \hat{\mu}(t)$ whenever $\mathcal{L}(V_n|\mathcal{G}_n) \Rightarrow_P \mu$. By the argument of subsequences, this assumption implies that, almost surely, $\lim_n \mathbb{E}(e^{itV_n}|\mathcal{G}_n) = \hat{\mu}(t)$. We will begin by assuming that $1 \geq |\mathbb{E}(e^{itV_n}|\mathcal{G}_n)| \geq \beta > 0$ for all $n \geq 1$. This assumption implies that $|\mathbb{E}(e^{ite_{nj}}|\mathcal{F}_{n,j-1})| \geq \beta$, for all n, j as will be shown by contradiction. Suppose that this is not the case, i.e. there exists n, j for which $|\mathbb{E}(e^{ite_{nj}}|\mathcal{F}_{n,j-1})| < \beta \leq 1$. Recalling the properties of $\{e_{nj}\}$,

$$\mathbb{E}\big(e^{itV_n}|\mathcal{G}_n\big) = \prod_{i=1}^{\infty} \mathbb{E}\big(e^{ite_{nj}}|\mathcal{F}_{n,j-1}\big) = \prod_{i=1}^{\infty} \mathbb{E}\big(e^{itd_{nj}}|\mathcal{F}_{n,j-1}\big).$$

Using this identity and the fact that $|\mathbb{E}(e^{ite_{nj}}|\mathcal{F}_{n,j-1})| \le 1$ for all j, n, we have that $|\mathbb{E}(e^{itV_n}|\mathcal{G}_n)| = |\Pi_{j=1}^{\infty}\mathbb{E}(e^{ite_{nj}}|\mathcal{F}_{n,j-1})| = \Pi_{j=1}^{\infty}|\mathbb{E}(e^{ite_{nj}}|\mathcal{F}_{n,j-1})| < \beta \le 1$, contradicting our original assumption. The above fact (to avoid dividing by 0) and an argument similar to the one used to obtain (6.2.4) gives

$$\mathbb{E}\Big(\frac{e^{itU_n}}{\mathbb{E}(e^{itV_n}|\mathcal{G}_n)}\Big) = 1.$$

Therefore,

$$\lim_{n\to\infty}\big|\mathbb{E}e^{itU_n} - \hat{\mu}(t)\big| = \lim_n \Big|\mathbb{E}e^{itU_n} - \hat{\mu}(t)\mathbb{E}\Big(\frac{e^{itU_n}}{\mathbb{E}(e^{itV_n}|\mathcal{G}_n)}\Big)\Big|$$
$$\le \lim_n \mathbb{E}\Big|1 - \frac{\hat{\mu}(t)}{\mathbb{E}(e^{itV_n}|\mathcal{G}_n)}\Big| = 0,$$

where the last line follows by the identity $|e^{itx}| = 1$ and the dominated convergence theorem.

The result in its full generality is obtained by the following argument. Let $A_{nk} = \big\{|\Pi_{l=1}^{k}\mathbb{E}(e^{ite_{nl}}|\mathcal{F}_{n,l-1})| \ge \frac{1}{2}|\hat{\mu}(t)|\big\}$. Then, $A_{nk} \in \mathcal{F}_{n,k-1}$ and $A_{nk} \in \mathcal{G}_n$. Hence $d'_{nk} = d_{nk}I_{A_{nk}}$ is $\mathcal{F}_{nk}$-tangent to $e'_{nk} = e_{nk}I_{A_{nk}}$, with the last sequence satisfying the CI condition with respect to $\mathcal{G}_n$. Set $U'_n = \sum_{i=1}^{\infty} d'_{nk}$ and $V'_n = \sum_{i=1}^{\infty} e'_{nk}$. Observe that $\mathbb{E}(e^{ite'_{nk}}|\mathcal{F}_{n,k-1}) = I_{A_{nk}}\mathbb{E}(e^{ite_{nk}}|\mathcal{F}_{n,k-1}) + I_{A^c_{nk}}$. We will use this to show that $|\mathbb{E}(e^{itV'_n}|\mathcal{G}_n)| \ge \frac{1}{2}|\hat{\mu}(t)|$. To see this note that

$$\Big|\mathbb{E}\big(e^{itV'_n}|\mathcal{G}_n\big)\Big| = \prod_{k=1}^{\infty}\Big(I_{A_{nk}}\big|\mathbb{E}\big(e^{ite_{nk}}|\mathcal{F}_{n,k-1}\big)\big| + I_{A^c_{nk}}\Big)$$
$$= \sum_{l=1}^{\infty}\Big[\prod_{k=1}^{l} I_{A_{nk}}\big|\mathbb{E}\big(e^{itY'_{nk}}|\mathcal{F}_{n,k-1}\big)\big|\Big]\Big[\prod_{k=l+1}^{\infty} I_{A^c_{nk}}\Big]$$
$$= \sum_{l=1}^{\infty} I_{A_{nl}}\prod_{k=1}^{l}\big|\mathbb{E}\big(e^{ite_{nl}}|\mathcal{F}_{n,k-1}\big)\big| I_{A^c_{n,l+1}} \ge \frac{1}{2}|\hat{\mu}(t)|.$$

The above follows by using the definition of A_{nk}, and the following properties of the sequence $\{A_{nk}\}$:

1) $\{A_{nk}\}$ is decreasing in k,

2) $\{A^c_{nk}\}$ is increasing in k,

3) $\sum_{l=1}^{\infty} I_{A_{n,l}} I_{A^c_{n,l+1}} = 1$.

The proof is completed by observing that since $\Pr\{\cap_{k=1}^{\infty} A_{n,k}\} \to 1$ we have that $U_n - U'_n \to 0$ and $V_n - V'_n \to 0$ in probability. This gives $\lim_{n\to\infty}\mathbb{E}e^{itV'_n} = \hat{\mu}(t)$ and hence $\lim_{n\to\infty}\mathbb{E}e^{itU'_n} = \hat{\mu}(t)$ which in turn gives $\lim_{n\to\infty}\mathbb{E}e^{itU_n} = \hat{\mu}(t)$. □

EXAMPLE 7.1.5. Let $\{z_i\}$ be a sequence of independent normal random variables with $\mathbb{E}z_i = 0$ and $\mathbb{E}z_i^2 = \sigma_i^2$, with $\sigma = 1$. Let $\{g_{n,i}\}$ be a sequence of functions mapping R^i into R^1. We will show that $\sum_{i=1}^{n} g_{n,i}(z_1, \ldots, z_{i-1})z_i$ converges in distribution to cz_0, where c is a constant, whenever $\sum_{i=1}^{n} g^2_{n,i}(z_1, \ldots, z_{i-1})\sigma_i^2$ converges in probability to c^2. To this end, let $\{\tilde{z}_i\}$ be an independent copy of $\{z_i\}$.

Set $d_{ni} = g_{n,i}(z_1, \ldots, z_{i-1})z_i$ and $e_{ni} = g_{n,i}(z_1, \ldots, z_{i-1})\tilde{z}_i$. Then $\{d_{ni}\}$ is $\mathcal{F}_{n,i}$-tangent to $\{e_{ni}\}$ for $\mathcal{F}_{n,i} = \sigma(z_1, \ldots, z_i; \tilde{z}_1, \ldots, \tilde{z}_i)$. Furthermore, $\{e_{ni}\}$ satisfies the CI condition with respect to $\mathcal{G}_n = \sigma(z_i : i = 1, \ldots, n)$. Set $U_n = \sum_{i=1}^n d_{ni}$ and $V_n = \sum_{i=1}^n e_{ni}$. It is easy to see that $V_n = \sum_{i=1}^n g_{n,i}(z_1, \ldots, z_{i-1})\tilde{z}_i$ is equal in conditional distribution to $\left(\sqrt{\sum_{i=1}^n g_{n,i}^2(z_1, \ldots, z_{i-1})\sigma_i^2}\right)\tilde{z}_0$. From this it is immediate that $\mathcal{L}(V_n|\mathcal{G})$ converges in probability to $c\hat{z}_0$. A use of Theorem 7.1.4 gives the result.

Our second application of the principle of conditioning consists of an extension of the Lindeberg-Feller central limit theorem for sums of independent random variables to the case of martingales. We begin by stating this important result (e.g., Araujo and Giné, 1980, or Dudley, 1989).

THEOREM 7.1.6. *For each $n \geq 1$, let $\{Y_{ni}\}$, $i = 1, \ldots,$ be an array of independent random variables with (1) $\mathbb{E}Y_{ni} = 0$, (2) $\lim_{n\to\infty} \sum_{i=1}^\infty \mathbb{E}Y_{ni}^2 = 1$ and (3) $\lim_{n\to\infty} \sum_{i=1}^\infty \mathbb{E}Y_{ni}^2 I_{|Y_{ni}|>\varepsilon} = 0$ for all $\varepsilon > 0$. Then, $\mathcal{L}(\sum_{i=1}^\infty Y_{ni}) \to_w N(0, 1)$.*

An important extension of this result is the Brown–Eagleson central limit theorem for martingales.

THEOREM 7.1.7. *For each $n \geq 1$, let $\{x_{ni}\}$, $\mathcal{F}_{n,i}$, $i = 1, \ldots,$ be a martingale difference array such that, in probability,*

1) *for every $n, i \geq 1$, $\mathbb{E}|x_{ni}| < \infty$ and $\mathbb{E}(x_{ni}|\mathcal{F}_{n,i-1}) = 0$,*

2) *$\lim_{n\to\infty} \sum_{i=1}^\infty \mathbb{E}(x_{ni}^2|\mathcal{F}_{n,i-1}) = 1$ and*

3) *$\lim_{n\to\infty} \sum_{i=1}^\infty \mathbb{E}(x_{ni}^2 I_{|x_{ni}|>\varepsilon}|\mathcal{F}_{n,i-1}) = 0$ for all $\varepsilon > 0$. Then, $\mathcal{L}(\sum_{i=1}^\infty x_{ni}) \to_w N(0, 1)$.*

PROOF. Using Proposition 6.1.5 (on a possibly enlarged probability space) for each n, we can find sigma fields $\tilde{\mathcal{F}}_{n,i}$, $\tilde{\mathcal{G}}_n$ and a $\tilde{\mathcal{F}}_{n,i}$-adapted sequence $\{y_{ni}\}$, $\tilde{\mathcal{F}}_{n,i}$-tangent to $\{x_{ni}\}$, such that given $\tilde{\mathcal{G}}_n$, $\{y_{ni}\}$ satisfies the CI condition and, in particular, is a sequence of $\tilde{\mathcal{G}}_n$-independent random variables. Furthermore, the CI condition guarantees that almost surely (a) $\mathbb{E}(y_{ni}|\tilde{\mathcal{G}}_n) = \mathbb{E}(x_{ni}|\tilde{\mathcal{F}}_{n,i-1}) = 0$, (b) $\mathbb{E}(y_{ni}^2|\tilde{\mathcal{G}}_n) = \mathbb{E}(x_{ni}^2|\tilde{\mathcal{F}}_{n,i-1})$ and (c) $\mathbb{E}(y_{ni}^2 I_{|y_{ni}|>\varepsilon}|\tilde{\mathcal{G}}_n) = \mathbb{E}(x_{ni}^2 I_{\|x_{ni}|>\varepsilon}|\tilde{\mathcal{F}}_{n,i-1})$ for all $\varepsilon > 0$. Using (a), (b) and (c) it is easy to translate conditions (1), (2) and (3) for the martingale (conditionally on $\tilde{\mathcal{G}}_n$) into the conditions for the $\tilde{\mathcal{G}}_n$-independent random sequence $\{y_{ni}\}$. Moreover, the convergence in probability of (2) and (3) translates into almost sure convergence along a subsequence $\{y_{n'i}\}$. Applying the Lindeberg-Feller Theorem along this subsequence guarantees that $\mathcal{L}(\sum_{i=1}^\infty y_{n'i}|\tilde{\mathcal{G}}_n) \to_w N(0, 1)$ almost surely. This in turn implies that that $\mathcal{L}(\sum_{i=1}^\infty y_{ni}|\tilde{\mathcal{G}}_n) \Rightarrow_P N(0, 1)$, as n tends to ∞. Therefore, by the principle of conditioning (Theorem 7.1.4), $\mathcal{L}(\sum_{i=1}^\infty x_{ni}) \to_w N(0, 1)$. □

As an example of a practical application of Theorem 7.1.4, we will present below the central limit theorem for a sequence of two by two tables.

7.2 Analysis of a Sequence of Two-by-Two Tables

In several statistical applications results are summarized by two by two tables. An example of this involves the study of the relationship between the occurrence of a disease and exposure to a possible risk factor for a group of subjects. A two by two table that can be used to study such problems is one in which N_1 and N_2 correspond to the number of persons in the exposed and unexposed groups, and M_1 and M_2 represent the number of diseased and disease-free individuals. The variable X corresponds to the number of diseased individuals who belong to the group exposed to the risk factor. The data can be summarized in a 2×2 table as follows:

	Diseased	Disease-free	Total
Exposed	X	$N_1 - X$	N_1
Unexposed	$M_1 - X$	$X + N_2 - M_1$	N_2
Total	M_1	M_2	N

when the values on the margins are fixed, a distribution that can be used in studying the properties of X is the noncentral hypergeometric. Consider tossing two coins respectively N_1 and N_2 times and with success probability p_1 and p_2. Let M_1 denote the number of heads and M_2 the number of tails out of the total number of tosses $N = N_1 + N_2$. Let X be the number of heads obtained when flipping the first coin. Then, the conditional probability of X given (M_i, N_i), $i = 1, 2$, is given by,

$$P\{X = x\} = \binom{N_1}{x}\binom{N_2}{M_1 - x}\theta^x \Big/ \sum_{u=L}^{S} \binom{N_1}{u}\binom{N_2}{M_1 - u}\theta^u, \tag{7.2.1}$$

for $L \leq x \leq S$ with $L = \max(0, M_1 - N_2)$ and $S = \min(N_1, M_1)$. The parameter θ is called the odds ratio, and is given by the formula $\theta = \{p_1/(1 - p_1)\}/\{p_2/(1 - p_2)\}$. Observe that the case in which $p_1 = p_2$ or equivalently $\theta = 1$ gives rise to the hypergeometric distribution,

$$P\{X = x\} = \binom{N_1}{x}\binom{N_2}{M_1 - x} \Big/ \binom{N}{M_1}, \tag{7.2.2}$$

for $L \leq x \leq S$.

In several situations it is convenient to summarize results in a sequence of two by two tables. One example of this involves the study of the relationship between the occurrence of a disease and exposure to a possible risk factor for a group of people over an extended period of time. A sequence of two by two tables that can be used to study such processes is one in which, at time k, $N_1^{(k)}$ and $N_2^{(k)}$ correspond to the number of persons in the exposed and unexposed groups, and $M_1^{(k)}$ and $M_2^{(k)}$ represent the number of diseased and disease-free individuals. The variable $X^{(k)}$ corresponds to the number of diseased individuals who belong to the group exposed to the risk factor at time k.

The study of the limiting properties of a general sequence of 2×2 tables can be used among other things to obtain information necessary to test the null hypothesis that across tables all the odds ratios are equal to θ. In particular, the hypothesis $\theta = 1$ signifies that the disease rates for both the exposed and unexposed groups are the same.

The sequence $\{X^{(k)}, N_i^{(k)}, M_i^{(k)} : i = 1, 2, k \geq 1\}$, is adapted to a filtration $\{\mathcal{F}_k, k \geq 1\}$ with $X^{(k)} \in \mathcal{F}_k$ and $N_i^{(k)}, M_i^{(k)} \in \mathcal{F}_{k-1}$, $i = 1, 2$. The nature of follow-up studies typically makes the tables dependent. Nevertheless, this dependence is often such that the conditional distribution of the kth table given its margins is noncentral hypergeometric with individual odds ratios. That is, the conditional distribution of $X^{(k)}$ given $\mathcal{F}_{k-1}$ is noncentral hypergeometric (θ_k) with,

$$P(X^{(k)} = x) = \binom{N_1^{(k)}}{x}\binom{N_2^{(k)}}{M_1^{(k)} - x}\theta_k^x \Big/ \sum_{u=L^{(k)}}^{S^{(k)}} \binom{N_1^{(k)}}{u}\binom{N_2^{(k)}}{M_1^{(k)} - u}\theta_k^u, \quad (7.2.3)$$

for $L^{(k)} \leq x \leq S^{(k)}$ with $L^{(k)} = \max\{0, M_1^{(k)} - N_2^{(k)}\}$ and $S^{(k)} = \min(N_1^{(k)}, M_1^{(k)})$. Finally the odds ratio for the kth table is given by $\theta_k = \{p_1^{(k)}/(1 - p_1^{(k)})\}/\{p_2^{(k)}/(1 - p_2^{(k)})\}$ with $\theta_k \varepsilon (0, \infty)$. In order to facilitate the study of sequential procedures, an additional variable is introduced. Let K be a stopping time adapted to $\sigma(\{\mathcal{F}_k\})$. Under the assumption that all the odds ratios are the same ($\theta^{(k)} = \theta$ for all $k \geq 1$), the $X^{(k)}$ are hypergeometric random variables and it is easy to see that in this case $\mathbb{E}^{(k-1)}(X^{(k)}) = M_1^{(k)} N_1^{(k)}/N^{(k)}$, $\text{Var}^{(k-1)}(X^{(k)}) = M_1^{(k)} M_2^{(k)} N_1^{(k)} N_2^{(k)}/\{(N^{(k)} - 1)(N^{(k)})^2\}$ where $\mathbb{E}_\theta^{(k-1)}$, $\text{Var}_\theta^{(k-1)}$ denote the conditional expectations and variances given $\mathcal{F}_{k-1}$.

The random variable U given by the formula

$$U = \sum_{k=1}^{K_n} w_k \big(X^{(k)} - E_\theta^{(k-1)}(X^{(k)})\big) \big/ \big(\sum_{k=1}^{K} w_k^2 \,\text{Var}_\theta^{(k-1)}(X^{(k)})\big)^{1/2}, \quad (7.2.4)$$

where $w_k \in \mathcal{F}_{k-1}$ are weights, and U is the normalized weighted log-rank test statistic for testing the null hypothesis that all the odds ratios are the same.

Owing to the difficulty in obtaining the exact distribution of U, an asymptotic approximation is essential in carrying any inference procedure. Such an approximation is introduced in Theorem 7.2.1. The approximation we will present can be better understood in the sense of double arrays as all the variables in (7.2.4) are supposed to depend on n.

THEOREM 7.2.1. *Let $\{X_n^{(k)}, N_{i,n}^{(k)}, M_{i,n}^{(k)} : i = 1, 2, k \geq 1\}$ be a sequence of two by two tables with weights $w_{n,k}$. Assume that there exists a deterministic sequence $q_n \to \infty$ such that*

$$\frac{\sum_{k=1}^{K_n} w_{n,k}^2 \,\text{Var}_\theta^{(k-1)}(X_n^{(k)})}{q_n} \to 1 \quad (7.2.5)$$

in probability and that

$$\frac{\max_{1 \leq k \leq K_n} w_{n,k}^2}{q_n} \to 0, \quad (7.2.6)$$

in probability. Then $U_n \to_d z$, where U_n is defined by (7.2.4) with $X_n^{(k)}$ instead of $X^{(k)}$, and where z is a standard normal variable.

The use of Theorem 7.1.4 ("the principle of conditioning") in obtaining the proof of Theorem 7.2.1 appears to be necessary since the usual CLT for martingales does not seem to apply directly. Moreover, by using the principle of conditioning, the proof becomes very transparent. One first constructs a sequence $\{Y_n^{(k)}\}$ tangent to $\{X_n^{(k)}\}$ and uses the fact that the sequence $\{Y_n^{(k)}\}$ satisfies the CI condition which, along with Theorem 7.1.4, reduces the proof to the case of sums of independent variables. In order to obtain the necessary tangent sequence we need to introduce a lemma that provides a representation of the noncentral hypergeometric random variable as a sum of independent random variables (see Kou and Ying, 1996, for its proof).

LEMMA 7.2.2. *Let X be a noncentral hypergeometric random variable. Let $\eta_1, \eta_2, \ldots$ be independent uniform $(0, 1)$ random variables. Then, there exist non-negative constants $\lambda_1 \geq 0, \lambda_2 \geq 0, \ldots, \lambda_s \geq 0$, such that X has the same distribution as*

$$\sum_{i=1}^{S} I_{\eta_i \leq (1+\theta^{-1}\lambda_i)^{-1}}. \tag{7.2.7}$$

Moreover, $-\lambda_1, \ldots, -\lambda_S$ are the roots of the polynomial

$$\phi(z) = \sum_{u=L}^{S} \binom{N_1}{u}\binom{N_2}{M_1 - u} z^u. \tag{7.2.8}$$

Using this lemma one can derive the following useful expressions for the mean and variance of a noncentral hypergeometric variable with parameter θ.

$$\mathbb{E}_\theta X = \sum_{i=1}^{S} \frac{1}{1+\theta^{-1}\lambda_i}, \qquad \mathrm{Var}_\theta\, X = \sum_{i=1}^{S} \frac{\theta^{-1}\lambda_i}{(1+\theta^{-1}\lambda_i)^2}. \tag{7.2.9}$$

It is important to underline that (7.2.7) not only represents the noncentral hypergeometric as a sum of independent Bernoulli random variables but also that the λ_i do not depend on the value of the odds ratio θ. This can be verified by observing that $\phi(z)$ does not depend on θ. We refer the reader to the original papers (see the comments at the end of the chapter) for the proof of Lemma 7.2.2 and its consequences. In this book we will only concentrate on the decoupling aspects of the problem.

PROOF OF THEOREM 7.2.1. By (7.2.5) and Slutzky's theorem, it suffices to prove that

$$U_n' = \frac{\sum_{k=1}^{K_n} w_{n.k}^2 (X_n^{(k)} - \mathbb{E}_\theta^{(k-1)}(X_n^{(k)}))}{\sqrt{q_n}} \to_d z, \tag{7.2.10}$$

where z is a standard normal random variable. A further reduction is obtained by working along appropriate subsequences since then the convergence in probability in (7.2.5) and (7.2.6) can be taken to be almost sure convergence.

In order to use Theorem 7.1.4, we need to find a tangent sequence which also satisfies the CI condition.

Let $\{\eta_j\}_{j=1}^{\infty}$ be a sequence of independent random variables with uniform distribution on (0, 1). The sequence $\{\eta_j\}$ is assumed to be independent of the filtration $\{\mathcal{F}_{n,k} : k \geq 0, n \geq 1\}$ and, therefore, also independent of the original sequence of two by two tables. Let $S_n^{(0)} = 0$ and

$$\eta_{n,i}^{(k)} = \eta_{S_n^{(1)}+\cdots+S_n^{(k-1)}+i} \text{for all } i = 1, \ldots, S_n^{(k)}, k \geq 1, \text{ and } n \geq 1. \tag{7.2.11}$$

Set

$$\begin{aligned} \mathcal{H}_{n,k} &= \sigma\left(\mathcal{F}_{n,k}, \{\eta_{n,j}^{(l)}, j = 1, \ldots, S_n^{(l)}, l = 1, \ldots, k\}\right) \\ &= \sigma\left(\mathcal{F}_{n,k}, \{\eta_i\}_{i=1}^{S_n^{(1)}+\cdots+S_n^{(k)}}\right), \qquad n = 1, 2, \ldots. \end{aligned} \tag{7.2.12}$$

Using the above lemma, we conclude that for all fixed $k \geq 1$ one can find $\lambda_{1,n}^{(k)}, \ldots, \lambda_{S_n^{(k)},n}^{(k)}$, depending on $N_{1,n}^{(k)}, N_{2,n}^{(k)}, M_{1,n}^{(k)}, M_{2,n}^{(k)}$, such that

$$\sum_{i=1}^{S_n^{(k)}} I_{\eta_{n,i}^{(k)} \leq (1+\theta^{-1}\lambda_{n,i}^{(k)})^{-1}} \tag{7.2.13}$$

has the same distribution as $X_n^{(k)}$. We will use this to obtain a tangent sequence that will allow us to use the principle of conditioning in showing (7.2.10).

Let

$$\tilde{X}_{n,k} = \frac{w_{n,k}}{\sqrt{q_n}}\left(X_n^{(k)} - E_\theta^{(k-1)}(X_n^{(k)})\right) I_{K_n \geq k}, \tag{7.2.14}$$

and

$$\tilde{Y}_{n,k} = \frac{w_{n,k}}{\sqrt{q_n}}\left(\sum_{i=1}^{S_n^{(k)}}\left[I_{\eta_{n,i}^{(k)} \leq (1+\theta^{-1}\lambda_{n,i}^{(k)})^{-1}} - (1+\theta^{-1}\lambda_{n,i}^{(k)})^{-1}\right]\right) I_{K_n \geq k}. \tag{7.2.15}$$

Once we show that $\{\tilde{X}_{n,k}\}$ is tangent to $\{\tilde{Y}_{n,k}\}$ and that the second sequence satisfies the CI condition with respect to $\mathcal{G}_n$ (to be defined later), (7.2.10) and Theorem 7.1.4 imply that in order to show that $U_n' = \sum_{k=1}^{\infty} \tilde{X}_{n,k} \to_d z$ it is enough to show that

$$\mathcal{L}\left(\sum_{k=1}^{\infty} \tilde{Y}_{n,k} \middle| \mathcal{G}_n\right) \Rightarrow_P z. \tag{7.2.16}$$

To see that $\{\tilde{X}_{n,k}\}$ is tangent to $\{\tilde{Y}_{n,k}\}$ with respect to $\{\mathcal{H}_{n,k}\}$ we observe that, since $N_{i,n}^{(k)}, M_{i,n}^{(k)} \in \mathcal{F}_{n,k-1}$, $S_n^{(k)} \in \mathcal{F}_{n,k-1}$ and hence also $S_n^{(k)} \in \mathcal{H}_{n,k-1}$. Since K_n is an $\mathcal{F}_{n,k}$ stopping time, $I_{K_n \geq k} \in \mathcal{F}_{n,k-1}$. We also recall that $w_{n,k} \in \mathcal{F}_{n,k-1}$. Therefore we also have that both $w_{n,k}$ and $I_{K_n \geq k}$ are in $\mathcal{H}_{n,k-1}$. Using these facts, the formula for $E_\theta^{(k-1)} X_n^{(k)}$ in (7.2.9) and (7.2.11) it is easy to see that $\tilde{Y}_{n,k}$ and $\tilde{X}_{n,k}$ have the same conditional distribution given $\mathcal{H}_{n,k-1}$ and therefore the two sequences in question are are tangent with respect to $\{\mathcal{H}_{n,k}\}$.

We continue by showing that $\{\tilde{Y}_{n,k}\}$ satisfies the CI condition with respect to $\mathcal{G}_n = \vee_{k\geq 1}\mathcal{F}_{n,k}$. To this end, observe that $\mathcal{G}_n$ is contained in $\vee_{k\geq 1}\mathcal{H}_{n,k}$, and that, by (7.2.7) and (7.2.15), $\mathcal{L}(\tilde{Y}_{n,k}|\mathcal{H}_{n,k-1}) = \mathcal{L}(\tilde{Y}_{n,k}|\mathcal{F}_{n,k-1}) = \mathcal{L}(\tilde{Y}_{n,k}|\mathcal{G}_n)$. It should also be observed that conditional on $\mathcal{G}_n$ the $\lambda_{n,i}^{(k)}$ become constants and the sequence $\{\eta_{n,j}^{(k)}\}$ is a sequence of independent random variables and therefore, $\tilde{Y}_{n,k}$ is a sequence of independent random variables given $\mathcal{G}_n$, completing the claim that the CI condition holds.

Recall that from (7.2.16), in order to show (7.2.10) and hence prove the theorem, it is enough to show that $\mathcal{L}(\sum_{k=1}^\infty \tilde{Y}_{n,k}|\mathcal{G}_n) \Rightarrow_P Z$.

Let

$$y_{n,k,i} = \frac{w_{n,k}}{\sqrt{q_n}}\Big[I_{\eta_{n,i}^{(k)}\leq(1+\theta^{-1}\lambda_{n,i}^{(k)})^{-1}} - (1+\theta^{-1}\lambda_{n,i}^{(k)})^{-1}\Big] I_{K_n\geq k} I_{S_n^{(k)}\geq i}. \tag{7.2.17}$$

Then, given $\mathcal{G}_n$ $y_{n,k,i}$ is a sequence of independent mean zero variables (since $\{\eta_j\}$ is independent of $\{\mathcal{F}_{n,k}\}$). In the spirit of (7.2.11) the sequence $\{y_{n,j}\}$ is defined as follows. Set $S_n^{(0)} = 0$ then

$$y_{n,S_n^{(0)}+\cdots+S_n^{(k-1)}+i} = y_{n,k,i}. \tag{7.2.18}$$

Therefore,

$$\tilde{U}_n' = \sum_{k=1}^\infty \tilde{Y}_{n,k} = \sum_{k=1}^{K_n}\sum_{i=1}^{S_n^{(k)}} y_{n,k,i} = \sum_{i=1}^{\sum_{k=1}^{K_n} S_n^{(k)}} y_{n,i} = \sum_{i=1}^\infty y_{n,i}, \tag{7.2.19}$$

and, given $\mathcal{G}_n$, $\{y_{n,i}\}$ is a sequence of independent random variables with mean zero.

It appears to be easier at this time to again invoke (7.2.5) and replace q_n by $\sum_{k=1}^{K_n} w_{n,k}^2 \operatorname{Var}_\theta^{(k-1)}(X_n^{(k)})$ in (7.2.17) to obtain a sequence $\{e_{n,k,i}\}$,

$$e_{n,k,i} = \frac{w_{n,k}\big[I_{\eta_{n,i}^{(k)}\leq(1+\theta^{-1}\lambda_{n,i}^{(k)})^{-1}} - (1+\theta^{-1}\lambda_{n,i}^{(k)})^{-1}\big] I_{K_n\geq k} I_{S_n^{(k)}\geq i}}{\sqrt{\sum_{k=1}^{K_n} w_{n,k}^2 \operatorname{Var}_\theta^{(k-1)}(X^{(k)})}}. \tag{7.2.20}$$

Then, as before,

$$e_{n,S_n^{(0)}+\cdots+S_n^{(k-1)}+i} = e_{n,k,i}. \tag{7.2.21}$$

Again by Slutzky's Theorem and (7.2.5), conditionally on $\mathcal{G}_n$ the two sums $\sum_{i=1}^\infty y_{n,i}$ and $\sum_{i=1}^\infty e_{n,i}$ converge to the same limit as $n\to\infty$. The advantage of using this new sum is that in view of (7.2.9), conditional on $\mathcal{G}_n$ we have a sum of independent mean zero random variables and in addition

$$\sum_{i=1}^\infty \mathbb{E}(e_{n,i}^2|\mathcal{G}_n) = 1. \tag{7.2.22}$$

In order to use the Lindeberg-Feller central limit theorem (Theorem 7.1.6) we need to show that the Lindeberg condition holds. Let $\delta > 0$ be given. Equations (7.2.20) and (7.2.22) together give,

$$\begin{aligned}\sum_{i=1}^{\infty} \mathbb{E}\big(|e_{n,i}|^{2+\delta}|\mathcal{G}_n\big) &\le \sup_{l \le K_n}\left(\frac{2w_{n,l}}{\sqrt{\sum_{k=1}^{K_n} w_{n,k}^2 Var_\theta^{(k-1)}(X^{(k)})}}\right)^{\delta} \sum_{i=1}^{\infty}\mathbb{E}\big(e_{n,i}^2|\mathcal{G}_n\big)\\ &= \sup_{l \le K_n}\left(\frac{2w_{n,l}}{\sqrt{\sum_{k=1}^{K_n} w_{n,k}^2 Var_\theta^{(k-1)}(X^{(k)})}}\right)^{\delta} \to_P 0,\end{aligned} \tag{7.2.23}$$

where the last term above goes to zero by (7.2.5) and (7.2.6). Regarding Lindeberg's condition for all $\varepsilon > 0$, as discussed in the beginning of the proof, we can assume, by going along a subsequence if necessary, that the mode convergence in (7.2.3) is almost sure convergence; therefore, along such a subsequence $\{n'\}$,

$$\sum_{i=1}^{\infty} \mathbb{E}\big(e_{n',i}^2 I_{|e_{n',i}|>\varepsilon}|\mathcal{G}_{n'}\big) \le \frac{1}{\varepsilon^{\delta}}\sum_{i=1}^{\infty}\mathbb{E}\big(|e_{n',i}|^{2+\delta}|\mathcal{G}_{n'}\big), \tag{7.2.24}$$

converges to zero almost surely by (7.2.23). Therefore, $\mathcal{L}\big(\sum_{i=1}^{\infty} e_{n',i}\big|\mathcal{G}_{n'}\big) \to_w N(0,1)$ almost surely and hence, going back to (7.2.16), $\mathcal{L}(\sum_{i=1}^{\infty} \tilde{Y}_{n',k}|\mathcal{G}_{n'}) \to_w N(0,1)$ almost surely, which implies that, along the original sequence, $\mathcal{L}(\sum_{i=1}^{\infty} \tilde{Y}_{n,k}|\mathcal{G}_n) \Rightarrow_P N(0,1)$. By the principle of conditioning, this implies (7.2.10), thus completing the proof of the theorem. □

7.3 Sharp L_p Comparison of Sums of Arbitrarily Dependent Variables to Sums of CI Variables

Let $\{d_i\}$ be a sequence adapted to an increasing filtration. From Proposition 6.1.5 we know that, on a possibly enlarged probability space, one can find a filtration $\{\mathcal{F}_i\}$, a σ algebra $\mathcal{G}$ contained in $\mathcal{F}_\infty$, and a sequence $\{e_i\}$ which is $\{\mathcal{F}_i\}$–tangent to $\{d_i\}$, with $\{e_i\}$ satisfying the CI condition [$\{e_i\}$ is conditionally independent given $\mathcal{G}$ and $\mathcal{L}(e_i|\mathcal{F}_{i-1}) = \mathcal{L}(e_i|\mathcal{G})$]. Recall that $\{e_i\}$ is said to be a decoupled version of $\{d_i\}$. In this section we prove the following inequality comparing any two tangent sequences when one of them satisfies the CI condition.

THEOREM 7.3.1. *For all $p \ge 1$, and for all adapted sequences $\{d_i\}$ of random variables,*

$$\mathbb{E}\Big|\sum_{i=1}^{n} d_i\Big|^p \le K_0^p\,\mathbb{E}\Big|\sum_{i=1}^{n} e_i\Big|^p, \tag{7.3.1}$$

where $\{e_i\}$ is any decoupled sequence tangent to $\{d_i\}$, and K_0, $0 < K_0 < \infty$ is a universal constant not depending on n, p or the distribution of the variables involved.

In the case of non-negative variables the constant K_0^p can be replaced by 2^{p-1} (Hitczenko, 1994b). In here, we will only present the proof of the general result. A key advantage of inequality (7.3.1) is that the constant K_0 does not depend on n, p or the distribution of the variables: this fact makes it widely applicable and powerful, as can be seen from the following Burkholder's martingale version of Rosenthal's inequality with constants having the optimal growth rate.

THEOREM 7.3.2. *There exists a universal constant K_1, with $0 < K_1 < \infty$, such that if $\{d_i\}$ is a mean zero martingale difference sequence, then, for all $p \geq 2$,*

$$\mathbb{E}\Big|\sum_{i=1}^n d_i\Big|^p \leq K_1^p\Big(\frac{p}{\log p}\Big)^p\Big[\mathbb{E}\Big(\sum_{i=1}^n \mathbb{E}(d_i^2|\mathcal{F}_{i-1})\Big)^{p/2} + \sum_{i=1}^n \mathbb{E}|d_i|^p\Big]. \tag{7.3.2}$$

PROOF OF THEOREM 7.3.2. If the variables the d_is are independent and have mean zero then inequality (7.3.2) reduces to the right-hand side of Theorem 1.5.11 (see the paragraph after Theorem 1.5.11 for the mean zero case). In the context of Theorem 7.3.1, since in this case $\{d_i\}$ is a mean zero martingale difference sequence, $\{e_i\}$ satisfies the CI condition and hence is a sequence of independent mean zero variables given $\mathcal{G}$.

From Theorem 7.3.1 (which will be proved later), Proposition 6.1.5 and Theorem 1.5.11 conditionally on $\mathcal{G}$, we have

$$\begin{aligned}
\mathbb{E}\Big|\sum_{i=1}^n d_i\Big|^p &\leq K_0^p\mathbb{E}\Big|\sum_{i=1}^n e_i\Big|^p = K_0^p\mathbb{E}\Big(\mathbb{E}\big[|\sum_{i=1}^n e_i|^p\,|\mathcal{G}\big]\Big)\\
&\leq K_0^p(2K)^p\Big(\frac{p}{\log p}\Big)^p\mathbb{E}\Big[\Big(\sum_{i=1}^n \mathbb{E}(e_i^2|\mathcal{G})\Big)^{p/2} + \sum_{i=1}^n \mathbb{E}(|e_i|^p|\mathcal{G})\Big]\\
&= K_0^p(2K)^p\Big(\frac{p}{\log p}\Big)^p\Big[\mathbb{E}\Big(\sum_{i=1}^n \mathbb{E}(e_i^2|\mathcal{G})\Big)^{p/2} + \sum_{i=1}^n \mathbb{E}|e_i|^p\Big]\\
&= K_1^p\Big(\frac{p}{\log p}\Big)^p\Big[\mathbb{E}\Big(\sum_{i=1}^n \mathbb{E}(d_i^2|\mathcal{F}_{i-1})\Big)^{p/2} + \sum_{i=1}^n \mathbb{E}|d_i|^p\Big],
\end{aligned}$$

completing the proof of Theorem 7.3.2. □

In what follows we show how to reduce the problem of proving Theorem 7.3.1 to the special case when the $\{d_i\}$ are assumed to be conditionally symmetric. Suppose that (7.3.1) holds for conditionally symmetric variables. Note that,

$$\mathbb{E}\Big|\sum_{i=1}^n d_i\Big|^p \leq 2^{p-1}\Big[\mathbb{E}\Big|\sum_{i=1}^n (d_i - e_i)\Big|^p + \mathbb{E}\Big|\sum_{i=1}^n e_i\Big|^p\Big]. \tag{7.3.3}$$

We remark that $\{d_i - e_i\}$ is a sequence of conditionally symmetric variables since d_i, e_i are i.i.d. given $\mathcal{F}_{i-1}$. Now, using Proposition 6.1.5, we take $\{\tilde{d}_n - \tilde{e}_n\}$ to be a decoupled sequence tangent to $\{d_n - e_n\}$. It is easy to see that this can be done even with the additional condition that $\{\tilde{d}_i\}$ and $\{\tilde{e}_i\}$ are decoupled versions of $\{d_i\}$ having the same joint distribution as $\{e_i\}$. Using (7.3.3) followed by (7.3.1) on the sequence $\{d_i - e_i\}$ gives,

$$\begin{aligned}
\mathbb{E}\Big|\sum_{i=1}^n d_i\Big|^p &\le 2^{p-1}K_0^p\mathbb{E}\Big|\sum_{i=1}^n(\tilde{d}_i - \tilde{e}_i)\Big|^p + 2^{p-1}\mathbb{E}\Big|\sum_{i=1}^n e_i\Big|^p \\
&\le 2^{p-1}\cdot 2^{p-1}K_0^p\Big[\mathbb{E}\Big|\sum_{i=1}^n \tilde{d}_i\Big|^p + \mathbb{E}\Big|\sum_{i=1}^n \tilde{e}_i\Big|^p\Big] + 2^{p-1}\mathbb{E}\Big|\sum_{i=1}^n e_i\Big|^p \\
&\le \frac{1}{2}(2(4K_0)^p + 2^p)\mathbb{E}\Big|\sum_{i=1}^n e_i\Big|^p \\
&\le \frac{1}{2}(8K_0 + 2)^p\mathbb{E}\Big|\sum_{i=1}^n e_i\Big|^p. \qquad (7.3.4)
\end{aligned}$$

Therefore, if Theorem 7.3.1 is valid for conditionally symmetric variables $\{d_i\}$, with the constant K_0 then it is also valid for general variables with the constant $8K_0 + 2$.

Two important tools in the proof of Theorem 7.3.1 are the following decoupling-based operators, one of which extends the conditional variance operator for martingales. In the context of Theorem 7.3.1, let $f_n = \sum_{i=1}^n d_i$, $f = \sum_{i=1}^\infty d_i$, $g_n = \sum_{i=1}^n e_i$, $g = \sum_{i=1}^\infty e_i$. For $\mathcal{G} = \sigma(d_1, \ldots, d_n, \ldots)$, define

$$T_p(g_m^n) = \Big(\mathbb{E}\Big[|\sum_{i=m+1}^n e_i|^p|\mathcal{G}\Big]\Big)^{1/p} \quad \text{and} \quad T_p^*(g_m^n) = \Big(\mathbb{E}\Big[\max_{j\le n}|\sum_{i=m+1}^j e_i|^p|\mathcal{G}\Big]\Big)^{1/p}, \qquad (7.3.5)$$

with $T_p(g_0^n) = T_p(g^n)$ and $T_p^*(g_0^n) = T_p^*(g^n)$. The reason for the nonstandard notation in (7.3.5) will become apparent later. In the case $\{d_i\}$ is a mean zero martingale difference sequence its decoupled version $\{e_i\}$ is also a mean zero martingale difference sequence. Moreover, $T_2(g^n)^2$ is the conditional variance of f_n as we show next.

$$\begin{aligned}
T_2(g^n)^2 = \mathbb{E}\Big[|\sum_{i=1}^n e_i|^2|\mathcal{G}\Big] &= \text{Var}\,\Big(\sum_{i=1}^n e_i|\mathcal{G}\Big) \\
= \sum_{i=1}^n \text{Var}\,(e_i|\mathcal{G}) &= \sum_{i=1}^n \mathbb{E}(e_i^2|\mathcal{G}) \\
= \sum_{i=1}^n \mathbb{E}(e_i^2|\mathcal{F}_{i-1}) &= \sum_{i=1}^n \mathbb{E}(d_i^2|\mathcal{F}_{i-1}),
\end{aligned}$$

where we used the conditional independence of $\{e_i\}$ in the third equality. A key feature of these operators is that they can be used to obtain bounds on the L_p norms of $\sum_{i=1}^n e_i$ and $\max_{j\le n}|\sum_{i=1}^j e_i|$ as we show next.

$$\|T_p(g^n)\|_p = (\mathbb{E}[T_p(g^n)]^p)^{1/p} = \Big(\mathbb{E}\Big[\mathbb{E}(|\sum_{i=1}^n e_i|^p|\mathcal{G})\Big]\Big)^{1/p}$$

$$= \left(\mathbb{E}\Big|\sum_{i=1}^{n} e_i|^p\right)^{1/p} = \Big\|\sum_{i=1}^{n} e_i\Big\|_p,$$

and similarly $\|T_p^*(g^n)\|_p = \|\max_{j\le n}|\sum_{i=1}^{j} e_i|\,\|_p$.

Therefore, the proof of inequality (7.3.1) can be reduced to showing that

$$\|\max_{j\le n}|f_j|\|_p \le K_0\|T_p^*(g^n)\|_p = K_0\Big\|\max_{j\le n}|\sum_{i=1}^{j} e_i|\Big\|_p \le 2K_0\Big\|\sum_{i=1}^{n} e_i\Big\|_p, \quad (7.3.6)$$

for a universal constant K_0 not depending on p or the distribution of the variables involved, and for all sequences of conditionally symmetric random variables $\{d_i\}$. In (7.3.6), the last inequality follows by conditional Levy's inequality and the fact that the e_is are conditionally symmetric. The proof of (7.3.6), and hence of Theorem 7.3.1, will be obtained by means of a sequence of lemmas.

LEMMA 7.3.3. *There exist a constant $\kappa > 0$ and $A(k) > 0$, $0 < \varepsilon(k) < 1$ (depending on k only) with $\varepsilon(k) \to 0$ such that for all $k \ge \kappa$, every $p \ge 1$ and all conditionally symmetric martingale difference sequences $\{d_i\}$ the following inequality is true:*

$$\Pr\{|f_n| \ge A(k)\|T_p(g^n)\|_\infty\} \le \varepsilon(k)^p, \quad n = 1, 2, \ldots, \quad (7.3.7)$$

where $\{e_i\}$ is a decoupled version of $\{d_i\}$ and $f_n = \sum_{i=1}^{n} d_i$, $g_n = \sum_{i=1}^{n} e_i$.

PROOF. Let $\{e_i\}$ be a decoupled sequence tangent to $\{d_i\}$. For a function $c(k)$ to be defined later let

$$d_i' = d_i I_{|d_i|\le c(k)\|T_p(g^n)\|_\infty}, \quad d_n^* = \max_{i\le n}|d_i|, \quad d_n'^* = \max_{i\le n}|d_i'|,$$

$$e_i' = e_i I_{|e_i|\le c(k)\|T_p(g^n)\|_\infty}, \quad e_n^* = \max_{i\le n}|e_i|, \quad e_n'^* = \max_{i\le n}|e_i'|.$$

Then

$$\Pr\Big\{\sum_{i=1}^{n} d_i \ge A(k)\|T_p(g^n)\|_\infty\Big\}$$
$$\le \Pr\Big\{\sum_{i=1}^{n} d_i \ge A(k)\|T_p(g^n)\|_\infty,\ d_n^* \le c(k)\|T_p(g^n)\|_\infty\Big\} \quad (7.3.8)$$
$$+ \Pr\Big\{d_n^* \ge c(k)\|T_p(g^n)\|_\infty\Big\}.$$

The second summand can be bounded as follows. First one uses Lemma 2.3.3 (decoupling for maxima) followed by an application of the triangle inequality to transform the problem into one on maxima of sums. The next step consists of conditional use of Lévy's inequality, to get

$$\Pr\Big\{d_n^* \ge c(k)\|T_p(g^n)\|_\infty\Big\}$$
$$\le 2\Pr\Big\{e_n^* \ge c(k)\|T_p(g^n)\|_\infty\Big\}$$

$$
\begin{aligned}
&= 2\Pr\Big\{\max_{j\le n} \Big|\sum_{i=1}^{j} e_i - \sum_{i=1}^{j-1} e_i\Big| \ge c(k)\|T_p(g^n)\|_\infty\Big\} \\
&\le 2\Pr\Big\{2\max_{j\le n} \Big|\sum_{i=1}^{j} e_i\Big| \ge c(k)\|T_p(g^n)\|_\infty\Big\} \\
&= 2\mathbb{E}\Big[\Pr\Big\{2\max_{j\le n} \Big|\sum_{i=1}^{j} e_i\Big| \ge c(k)\|T_p(g^n)\|_\infty \Big| \mathcal{G}\Big\}\Big] \\
&\le 2\mathbb{E}\Big[2\Pr\Big\{2\Big|\sum_{i=1}^{n} e_i\Big| \ge c(k)\|T_p(g^n)\|_\infty \Big| \mathcal{G}\Big\}\Big] \qquad \text{(by Levy's inequality)} \\
&= 4\Pr\Big\{2\Big|\sum_{i=1}^{n} e_i\Big| \ge c(k)\|T_p(g^n)\|_\infty\Big\} \\
&\le 4\Big(\frac{2}{c(k)}\Big)^p \frac{\mathbb{E}\big|\sum_{i=1}^n e_i\big|^p}{\|T_p(g^n)\|_\infty^p} \le 4\Big(\frac{2}{c(k)}\Big)^p, \qquad (7.3.9)
\end{aligned}
$$

whenever $p \ge 1$. We will pick $c(k) \ge 8$ and k large to obtain the result.

The first term can be handled by using decoupling for exponentials (Corollary 6.2.5 with $g = 1$) in combination with results for sums of independent random variables. In particular, in line 9 of the proof we use the exponential inequality given in Theorem 1.2.8. (eq. (1.2.10)). In the context of that theorem, taking $\delta' = \delta$ and $\lambda = \frac{\delta' p}{c(k)\|T_p(g^n)\|_\infty}$, with $c(k) \ge 1$. we have,

$$
\begin{aligned}
&\Pr\Big\{\sum_{i=1}^{n} d_i \ge A(k)\|T_p(g^n)\|_\infty, d_n^* \le c(k)\|T_p(g^n)\|_\infty\Big\} \\
&\le \Pr\Big\{\sum_{i=1}^{n} d_i' \ge A(k)\|T_p(g^n)\|_\infty\Big\} \\
&\le \Pr\Big\{\exp\Big(\frac{\lambda}{2}\sum_{i=1}^{n} d_i'\Big) \ge \exp\Big(\frac{\lambda}{2}A(k)\|T_p(g^n)\|_\infty\Big)\Big\} \\
&\le \exp\Big(-A(k)\frac{\lambda}{2}\|T_p(g^n)\|_\infty\Big)\mathbb{E}\exp\Big(\frac{\lambda}{2}\sum_{i=1}^{n} d_i'\Big) \qquad \text{(by Markov's inequality)} \\
&\le \sqrt{\exp(-A(k)\lambda\|T_p(g^n)\|_\infty)\mathbb{E}\exp(\lambda\sum_{i=1}^{n} e_i')} \qquad \text{(from Corollary 6.2.5)} \\
&= \sqrt{\exp(-A(k)\lambda\|T_p(g^n)\|_\infty)\mathbb{E}\Big(\mathbb{E}[\exp(\lambda\sum_{i=1}^{n} e_i')|\mathcal{G}]\Big)} \\
&\le \sqrt{\exp(-A(k)\lambda\|T_p(g^n)\|_\infty)\Big\|\mathbb{E}[\exp(\lambda\sum_{i=1}^{n} e_i')|\mathcal{G}]\Big\|_\infty}
\end{aligned}
$$

$$\le \sqrt{\exp(-A(k)\lambda\|T_p(g^n)\|_\infty)\Big\|3\exp\Big(C\lambda\Big[\Big(\mathbb{E}[|\sum_{i=1}^n e_i'|^p|\mathcal{G}]\Big)^{\frac{1}{p}} + \|e_n'^*\|_\infty\Big]\Big)\Big\|_\infty}$$

[(1.2.10)]

$$\le \sqrt{\exp(-A(k)\lambda\|T_p(g^n)\|_\infty)\Big\|3\exp\Big(C\lambda\Big[(2\mathbb{E}[|\sum_{i=1}^n e_i|^p|\mathcal{G}]\Big)^{1/p} + \|e_n'^*\|_\infty\Big]\Big)\Big\|_\infty}$$

[(1.4.6)]

$$\le \sqrt{3\exp(-A(k)\lambda\|T_p(g^n)\|_\infty)\exp\Big(C\lambda\big[2\|T_p(g^n)\|_\infty + c(k)\|T_p(g^n)\|_\infty\big]\Big)}$$

$$\le \sqrt{3\exp\Big(-\lambda\big(A(k) - 3Cc(k)\big)\|T_p(g^n)\|_\infty\Big)}, \tag{7.3.10}$$

where we used equation (1.4.7) of Lemma 1.4.3, conditionally on $\mathcal{G}$, on the conditionally independent and symmetric variables $\{e_i\}$, as well as the fact that $c(k) \ge 1$ in the last line. We observe that the λ we have chosen guarantees that the conditions of Theorem 1.2.8 are satisfied (conditionally on $\mathcal{G}$). To see this, note that for $c(k) \ge 1$,

$$\lambda \le \frac{\delta' p}{c(k)\|T_p(g^n)\|_\infty} \le \frac{\delta' p}{T_p(g^n) \vee \|e_n'^*\|_\infty}. \tag{7.3.11}$$

We stress the fact that C and $\delta' = \delta$ are universal constants coming from Theorem 1.2.8. Taking $A(k) = 2(k + 1.5)Cc(k)$, the above gives

$$\Pr\Big\{\sum_{i=1}^n d_i \ge A(k)\|T_p(g^n)\|_\infty, d_n^* \le c(k)\|T_p(g^n)\|_\infty\Big\} \le \sqrt{3}\exp\big(-\delta' Ckp\big).$$

Applying the bounds to f_n and $-f_n$ we get

$$\begin{aligned}
\Pr\Big\{\Big|\sum_{i=1}^n d_i\Big| \ge A(k)\|T_p(g^n)\|_\infty\Big\} &= \Pr\Big\{\sum_{i=1}^n d_i \ge A(k)\|T_p(g^n)\|_\infty\Big\} \\
&\quad + \Pr\Big\{-\sum_{i=1}^n d_i \ge A(k)\|T_p(g^n)\|_\infty\Big\} \\
&\le 2\Big[4\cdot\Big(\frac{2}{c(k)}\Big)^p + \sqrt{3}\exp(-\delta' Ckp)\Big] \\
&\le 2(1+\sqrt{3})\exp(-\delta' Ckp) \\
&\le \big[2(1+\sqrt{3})\exp\{-\delta' Ck\}\big]^p \\
&= \varepsilon(k)^p,
\end{aligned}$$

with

$$c(k) = 8e^{\delta' Ck} \quad \text{and} \quad \varepsilon(k) = 2(1+\sqrt{3})\exp(-\delta' Ck) \tag{7.3.12}$$

and κ large enough so that for $k \ge \kappa$, $c(k) \ge 8$ and $\varepsilon(k) < 1$. □

Next we introduce a maximal operator related to $T_p^*(g^n)$. Let

$$W_p(e_n^*) = \big(\sup_{j\le n} \mathbb{E}[|e_j|^p|\mathcal{G}]\big)^{\frac{1}{p}}. \tag{7.3.13}$$

The following lemma shows how the growth rate of $\max_{j\le n}|f_j| = \max_{j\le n}\big|\sum_{i=1}^j d_i\big|$ can be controlled by using $T_p^*(g^n)$, $W_p(e_n^*)$ and d_n^*.

LEMMA 7.3.4. *In the context of Lemma 7.3.3 for all $\lambda > 0$, $p \ge 1$, there exist $\beta(k)$, $\delta(k) > 0$ and a universal constant κ such that for all $k \ge \kappa$,*

$$\Pr\big\{\max_{j\le n}|f_j| \ge \beta(k)\lambda,\, T_p^*(g^n)\vee W_p(e_n^*)\vee d_n^* < \delta(k)\lambda\big\} \le \varepsilon(k)^p \Pr\big\{\max_{j\le n}|f_j| \ge \lambda\big\},$$

with $\varepsilon(k) < 1$ given by (7.3.12) and $T_p^*(g^n)$ by (7.3.5).

PROOF. Let

$\nu = \inf\{j \le n : |f_j| \ge \beta(k)\lambda\}$, with $\nu = \infty$ otherwise,

$\mu = \inf\{n : |f_n| \ge \lambda\}$, with $\mu = \infty$ otherwise,

$\tau = \inf\{n : T_p^*(g^n) \vee W_p(e_n^*) \vee d_n^* \ge \delta(k)\lambda\}$, with $\tau = \infty$ otherwise.

Then, whenever $\nu < \infty$ and $\tau = \infty$,

$$\begin{aligned}\Big|\sum_{i=1}^{\nu\wedge\tau} d_i\Big| &= \Big|\sum_{i=1}^{\mu-1} d_i + d_\mu + \sum_{i=\mu+1}^{\nu\wedge\tau} d_i\Big| \\ &\le \Big|\sum_{i=1}^{\mu-1} d_i\Big| + \big|d_\mu\big| + \Big|\sum_{i=\mu+1}^{\nu\wedge\tau} d_i\Big| \\ &\le \lambda + \delta(k)\lambda + \Big|\sum_{i=\mu+1}^{\nu\wedge\tau} d_i\Big|.\end{aligned}$$

Therefore, on the set

$$\big\{\max_{j\le n}|f_j| \ge \beta(k)\lambda,\ T_p^*(g^n)\vee W_p(e_n^*)\vee d_n^* < \delta(k)\lambda\big\} = (\nu \le n, \tau = \infty),$$

$$\Big|\sum_{i=\mu+1}^{\nu\wedge\tau} d_i\Big| \ge \big(\beta(k) - 1 - \delta(k)\big)\lambda.$$

Let $f_\mu^{\nu\wedge\tau} = \sum_{i=\mu+1}^{\nu\wedge\tau} d_i$ and $g_\mu^{\nu\wedge\tau} = \sum_{i=\mu+1}^{\nu\wedge\tau} e_i$. Then,

$$T_p^*(g_\mu^{\nu\wedge\tau}) = \Big(\mathbb{E}\big[|\max_{j\le\nu\wedge\tau}|\sum_{i=\mu+1}^{j} e_i|\,|^p|\mathcal{G}\big]\Big)^{1/p},$$

and on the same set

$$\begin{aligned}\|T_p^*(g_\mu^{\nu\wedge\tau})\|_\infty &= \|T_p^*(g_\mu^{\nu\wedge\tau})\|_\infty I_{\mu<\nu\le n} \\ &\le \big(\|T_p^*(g^{\nu\wedge\tau})\|_\infty + \|T_p^*(g^\mu)\|_\infty\big) I_{\mu<\nu\le n} \\ &\le 2\delta(k)\lambda \le 4\delta(k)\lambda.\end{aligned}$$

Therefore, as long as $\frac{(\beta(k)-1-\delta(k))}{4\delta(k)} \geq A(k)$,

$$\begin{aligned}
\Pr\Big\{\max_{j\le n}|f_j| \geq \beta(k)\lambda,\, T_p^*(g^n)\vee W_p(e_n^*)\vee d_n^* < \delta(k)\lambda\Big\} \\
&= \Pr\big\{\nu<\infty, \tau=\infty\big\} \\
&\leq \Pr\Big\{|f_\mu^{\nu\wedge\tau}| \geq \big(\beta(k)-1-\delta(k)\big)\lambda,\, \mu<\nu<\infty\Big\} \\
&\leq \Pr\Big\{|f_\mu^{\nu\wedge\tau}| \geq \frac{\lambda\big(\beta(k)-1-\delta(k)\big)}{4\delta(k)\lambda}\|T_p^*(g_\mu^{\nu\wedge\tau})\|_\infty,\, \mu<\nu<\infty\Big\} \\
&\leq \mathbb{E}\Pr\Big\{|f_\mu^{\nu\wedge\tau}| \geq \frac{\beta(k)-1-\delta(k)}{4\delta(k)}\|T_p(g_\mu^{\nu\wedge\tau})\|_\infty,\, \mu<\infty\Big|\mathcal{F}_\mu\Big\} \\
&\leq \varepsilon(k)^p\Pr\{\mu<\infty\} = \varepsilon(k)^p\Pr\{\max_{j\le n}|f_j|\geq\lambda\}.
\end{aligned}$$

The last inequality follows using a conditional version of the inequality preceding (7.3.12), which is available because $f_\mu^{\nu\wedge\tau}$ is a martingale. □

LEMMA 7.3.5. *Let X, Y be random variables with $X, Y \geq 0$. Fix $0 < p < \infty$ and assume that there exist positive numbers $\delta, \beta > 0$ and $0 < \varepsilon < \frac{1}{\beta}$ such that*

$$\Pr\big\{X \geq \beta\lambda, Y < \delta\lambda\big\} \leq \epsilon^p \Pr\{X \geq \lambda\},$$

for all $\lambda > 0$. Then

$$\|X\|_p^p \leq \frac{(\beta/\delta)^p}{1-\beta^p\varepsilon^p}\|Y\|_p^p.$$

PROOF. The proof is similar to the proof of Lemma 2.2.1. It uses the fact that $\mathbb{E}(X/\beta)^p = \int_0^\infty \Pr\{X \geq \beta\lambda^{1/p}\}d\lambda$ and the stated condition. □

LEMMA 7.3.6. *For every $p \geq 1$ and all sequences $\{d_i\}$ of conditionally symmetric random variables, $\|d_n^*\|_p \leq 8\|T_p(g^n)\|_p$.*

PROOF. It follows from Lemma 2.3.3 that

$$\begin{aligned}
\|d_n^*\|_p^p \leq 2\|e_n^*\|_p^p &= 2\mathbb{E}\max_{j\le n}\Big|\sum_{i=1}^j e_i - \sum_{i=1}^{j-1} e_i\Big|^p \\
&\leq 2\,2^{p-1}\Big[\mathbb{E}\max_{j\le n}\Big|\sum_{i=1}^j e_i\Big|^p + \mathbb{E}\max_{j\le n}\Big|\sum e_i\Big|^p\Big] \\
&\leq 4\cdot 2^{p-1}\mathbb{E}\max_{j\le n}\Big|\sum_{i=1}^j e_i\Big|^p = 4\cdot 2^{p-1}\mathbb{E}\Big(\mathbb{E}\big[\max_{j\le n}\big|\sum_{i=1}^j e_i\big|^p\big|\mathcal{G}\big]\Big) \\
&\leq 4\cdot 2^{p-1}2\mathbb{E}\Big(\mathbb{E}\big[\big|\sum_{i=1}^n e_i\big|^p\big|\mathcal{G}\big]\Big) \quad \text{(by Levy's inequality conditionally)} \\
&\leq 8^p\mathbb{E}\Big|\sum_{i=1}^n e_i\Big|^p.
\end{aligned}$$

□

LEMMA 7.3.7. *For every $p \geq 1$ and all sequences $\{d_i\}$, their decoupled counterpart satisfies the inequality $\| W_p(e_n^*)\|_p \leq 4\| T_p(g^n)\|_p$.*

PROOF. Using conditional Jensen's,

$$\| W_p(e_n^*)\|_p^p = \mathbb{E} \max_{j\leq n} \mathbb{E}[|e_i|^p|\mathcal{G}] \leq \mathbb{E}\Big(\mathbb{E}\big[\max_{j\leq n}|e_i|^p|\mathcal{G}\big]\Big) = \|e_n^*\|_p^p.$$

The proof is now completed by following the method of proof given in Lemma 7.3.6. □

PROOF OF THEOREM 7.3.1. Let $\beta(k) = \frac{\exp\{\delta' Ck\}}{4(1+\sqrt{3})}$. Recall that

$$c(k) = 8\exp(\delta' Ck), \qquad A(k) = 2(k+1.5)Cc(k) = 16(k+1.5)C\exp(\delta' Ck).$$

Using the equation $\frac{\beta(k)-1-\delta(k)}{4\delta(k)} = A(k)$ one gets

$$\delta(k) = \frac{\beta(k)-1}{4A(k)+1} = \frac{\frac{\exp\{\delta' Ck\}}{4(1+\sqrt{3})} - 1}{64(k+1.5)C\exp\{\delta' Ck\}+1}.$$

Pick k large enough so that $\beta(k) > 1+\delta(k)$ and observe that, since $p \geq 1$,

$$\beta(k)^p\big(2(1+\sqrt{3})\exp(-\delta' Ck)\big)^p = \beta(k)^p\varepsilon(k)^p \leq \frac{1}{2^p} \leq \frac{1}{2}.$$

From Lemma 7.3.4 and 7.3.5 with $\varepsilon = \varepsilon(k)$, $\beta = \beta(k)$ and $\delta = \delta(k)$, we obtain

$$\big\| \max_{j\leq n}|f_j|\big\|_p^p \leq 2(\beta/\delta)^p\big[\|T_p^*(g^n)\|_p^p + \|W_p(e_n^*)\|_p^p + \|d_n^*\|_p^p\big].$$

Using Lemma 7.3.6 and Lemma 7.3.7, this gives

$$\begin{aligned}
\|f_n\|_p^p &\leq \big\| \max_{j\leq n}|f_j|\big\|_p^p \\
&\leq 2(\beta/\delta)^p\Big[\|T_p^*(g^n)\|_p^p + 4^p\mathbb{E}\Big|\sum_{i=1}^n e_i\Big|^p + 8^p\mathbb{E}\Big|\sum_{i=1}^n e_i\Big|^p\Big] \\
&\leq 2(\beta/\delta)^p\big[2^p+4^p+8^p\big]^p\mathbb{E}\Big|\sum_{i=1}^n e_i\Big|^p \\
&\leq \Big[2(\beta/\delta+14)\Big]^p\mathbb{E}\Big|\sum_{i=1}^n e_i\Big|^p,
\end{aligned}$$

completing the proof of the theorem. □

To end this section we will present an example showing that it is not possible to extend Theorem 7.3.1 to strict tail probability comparisons.

LEMMA 7.3.8. *Let $\{d_i\}$, $\{e_i\}$ be any two tangent sequences with $\{e_i\}$ satisfying the CI condition. Then, there exist no constants C, $0 < C < \infty$, that would make the inequality*

$$\Pr\Big\{\Big|\sum_{i=1}^{\infty} d_i\Big| \geq Ct\Big\} \leq C \Pr\Big\{\Big|\sum_{i=1}^{\infty} e_i\Big| \geq t\Big\}. \tag{7.3.14}$$

valid for all $t > 0$.

PROOF. Let $\{\varepsilon_i\}$, $\{\tilde{\varepsilon}_i\}$ be two independent sequences of independent Bernoulli random variables with

$$\Pr\{\varepsilon_i = 1\} = \Pr\{\varepsilon_i = -1\} = \Pr\{\tilde{\varepsilon}_i = 1\} = \Pr\{\tilde{\varepsilon}_i = -1\} = \frac{1}{2}.$$

Let $\{v_j\}$ be a sequence of predictable variables adapted to $\sigma(\{\varepsilon_i\})$ defined in the following way. For a fixed integer k, let M_1 be an integer to be specified, and for $i = 2, \ldots, k$, let

$$M_i = M_{i-1} + 2^{-(i-1)} M_1. \tag{7.3.15}$$

We will also need the sets Ω_i defined as follows. For $\Omega_1 = \{\varepsilon_1 = \cdots = \varepsilon_{M_1} = 1\}$ set

$$\Omega_i = \Omega_{i-1} \cap \{\varepsilon_{M_{i-1}+1} = \cdots = \varepsilon_{M_i}\}, \quad \text{for } i = 2, \ldots, k. \tag{7.3.16}$$

Then set

$$\begin{aligned} v_1 = \cdots = v_{M_1} &= 1, \\ v_{M_1+1} = \cdots = v_{M_2} &= 2 I_{\Omega_1}, \end{aligned} \tag{7.3.17}$$

and, in general,

$$v_{M_{i-1}+1} = \cdots = v_{M_i} = 2^{i-1} I_{\Omega_{i-1}}, \qquad i = 2, \ldots, k.$$

It is easy to see that, since the v_i are predictable ($v_i \in \mathcal{F}_{i-1}$), the sequences $\{v_i\varepsilon_i\}$ and $\{v_i\tilde{\varepsilon}_i\}$ are tangent with respect to $\mathcal{F}_i = \sigma(\varepsilon_1, \ldots, \varepsilon_i; \tilde{\varepsilon}_1, \ldots, \tilde{\varepsilon}_i)$. Moreover, the sequence $\{v_i\tilde{\varepsilon}_i\}$ satisfies the CI condition with respect to $\mathcal{G} = \sigma(\{\varepsilon_i\}_{i=1}^{\infty})$. Let $v_i = 0$ for $i > M_k$. In order to contradict (7.3.14), it suffices to show that the inequality

$$\Pr\Big\{\Big|\sum_{i=1}^{M_k} v_i\varepsilon_i\Big| \geq Ct\Big\} \leq C \Pr\Big\{\Big|\sum_{i=1}^{M_k} v_i\tilde{\varepsilon}_i\Big| \geq t\Big\}, \tag{7.3.18}$$

does not hold in general. In what follows we will show that, letting $t = M_1$,

$$\Pr\Big\{\Big|\sum_{i=1}^{M_k} v_i\tilde{\varepsilon}_i\Big| \geq 4t\Big\} \leq k2^{-M_1/2^{k-2}} \Pr\Big\{\Big|\sum_{i=1}^{M_k} v_i\varepsilon_i\Big| \geq kt\Big\}, \tag{7.3.19}$$

contradicting (7.3.18) since M_1 can be made arbitrarily large.

It is easy to see that

$$\Pr\Big\{\Big|\sum_{i=1}^{M_k} v_i\varepsilon_i\Big| \ge kt\Big\} = 2\Pr\{\varepsilon_i = 1, i = 1, \ldots, M_k\} = 2\cdot 2^{-M_1(1+2^{-1}+\cdots+2^{-(k-1)})}.$$

It remains to approximate the left-hand side of (7.3.19). Using the fact that the Ωs are nested,

$$\begin{aligned}\Pr\Big\{\Big|\sum_{i=1}^{M_k} v_i\tilde{\varepsilon}_i\Big| \ge 4t\Big\} =& \Pr\Big\{\Big|\sum_{i=1}^{M_k} v_i\tilde{\varepsilon}_i\Big| \ge 4t, \Omega_1^c\Big\}\\ &+ \Pr\Big\{\Big|\sum_{i=1}^{M_k} v_i\tilde{\varepsilon}_i\Big| \ge 4t, \Omega_1\cap\Omega_2^c\Big\} + \cdots\\ &+ \Pr\Big\{\Big|\sum_{i=1}^{M_k} v_i\tilde{\varepsilon}_i\Big| \ge 4t, \Omega_{k-1}\cap\Omega_k^c\Big\}\\ &+ \Pr\Big\{\Big|\sum_{i=1}^{M_k} v_i\tilde{\varepsilon}_i\Big| \ge 4t, \Omega_k\Big\}.\end{aligned}$$

The independence between $\{\tilde{\varepsilon}_i\}$ and $\{v_j, \Omega_j\}$, together with (7.3.16), and (7.3.17), transforms the above bound into

$$\begin{aligned}\Pr\Big\{\Big|\sum_{i=1}^{M_k} v_i\tilde{\varepsilon}_i\Big| \ge 4t\Big\} =& \Pr\Big\{\Big|\sum_{i=1}^{M_1} \tilde{\varepsilon}_i\Big| \ge 4t\Big\}\big(1 - \Pr\{\Omega_1\}\big)\\ &+ \Pr\Big\{\Big|\sum_{i=1}^{M_1} \tilde{\varepsilon}_i + 2\sum_{i=M_1+1}^{M_2}\tilde{\varepsilon}_i\Big| \ge 4t\Big\}\Pr\{\Omega_1\cap\Omega_2^c\} + \cdots\\ &+ \Pr\Big\{\Big|\sum_{i=1}^{M_1} \tilde{\varepsilon}_i + \cdots + 2^{k-1}\sum_{i=M_{k-1}+1}^{M_k}\tilde{\varepsilon}_i\Big| \ge 4t\Big\}\Pr\{\Omega_k\}.\end{aligned}$$

Using the fact that $4t = 4M_1 > M_1 + 2(M_2 - M_1) + 4(M_3 - M_2)$ one can see that the first three terms above are zero. We will show how to bound each of the remaining terms. Recall Hoeffding's exponential inequality (3.2.16) for weighted independent Bernoulli random variables, with $\{a_i\}$ a sequence of constants,

$$\Pr\Big\{\Big|\sum_{i=1}^{n} a_i\varepsilon_i\Big| \ge x\Big\} \le 2\exp\Big(-\frac{x^2}{2\sum_{i=1}^n a_i^2}\Big) \tag{7.3.20}$$

Applying (7.3.20) and (7.3.15) one gets

$$\begin{aligned}\Pr\Big\{\Big|\sum_{i=1}^{M_1} \tilde{\varepsilon}_i + \cdots + 2^{j-1}\sum_{i=M_{j-1}+1}^{M_j}\tilde{\varepsilon}_i\Big| \ge 4t\Big\} &\le 2\exp\Big(\frac{-16t^2}{M_1(1+\cdots+2^{j-1})}\Big)\\ &\le 2\exp(-8M_1/2^j) \le 2\cdot 2^{-8M_1/2^j}.\end{aligned}$$

Moreover,

$$\Pr\{\Omega_{j-1} \cap \Omega_j^c\} \le \Pr\{\Omega_{j-1}\} = 2^{-M_{j-1}} = 2^{-M_1(1+\cdots+2^{-(j-2)})}. \tag{7.3.21}$$

Putting these estimates together one obtains,

$$\begin{aligned}
\Pr\Big\{\Big|\sum_{i=1}^{M_1} \tilde{\varepsilon}_i + \cdots + 2^{j-1} \sum_{i=M_{j-1}+1}^{M_j} \tilde{\varepsilon}_i\Big| \ge 4t\Big\} \Pr\{\Omega_{j-1}\}
&\le 2 \cdot 2^{-8M_1/2^j} 2^{-M_1(1+\cdots+2^{-(j-2)})} \\
&= 2 \cdot 2^{-M_1(1+\cdots+2^{-(k-1)})} \cdot 2^{-8M_1/2^j} \cdot 2^{M_1(2^{-(j-1)}+\cdots+2^{-(k-1)})} \\
&\le \Pr\Big\{\Big|\sum_{i=1}^{M_k} v_i \varepsilon_i\Big| \ge kt\Big\} \cdot 2^{-8M_1/2^j} \cdot 2^{M_1/2^{j-2}} \\
&= 2^{-M_1/2^{j-2}} \Pr\Big\{\Big|\sum_{i=1}^{M_k} v_i \varepsilon_i\Big| \ge kt\Big\}.
\end{aligned}$$

Summing over j gives

$$\begin{aligned}
\Pr\Big\{\Big|\sum_{i=1}^{M_k} v_i \tilde{\varepsilon}_i\Big| \ge 4t\Big\} &\le \sum_{j=4}^{k} 2^{-M_1/2^{j-2}} \Pr\Big\{\Big|\sum_{i=1}^{M_k} v_i \varepsilon_i\Big| \ge kt\Big\} \\
&\le k \cdot 2^{-M_1/2^{k-2}} \Pr\Big\{\Big|\sum_{i=1}^{M_k} v_i \varepsilon_i\Big| \ge kt\Big\},
\end{aligned}$$

completing the proof of (7.3.19) and hence the proof of Lemma 7.3.8. □

A modification of the proof of Theorem 7.3.1 introducing Davis' decomposition for martingales gives the following extension.

THEOREM 7.3.9. *Fix $\alpha > 0$. Let Φ be a function in $\mathcal{A}_\alpha$. Then for all adapted sequences $\{d_i\}$ of random variables,*

$$\mathbb{E}\Phi\Big(\Big|\sum_{i=1}^{n} d_i\Big|\Big) \le C^{\alpha+1} \mathbb{E}\Phi\Big(\Big|\sum_{i=1}^{n} e_i\Big|\Big), \tag{7.3.22}$$

where $\{e_i\}$ is any decoupled sequence tangent to $\{d_i\}$, and C, $0 < C < \infty$ is a universal constant not depending on n, α, Φ or the distribution of the variables involved.

The following lemma shows that Theorem 7.3.9 may not be extended to the case of convex functions.

LEMMA 7.3.10. *For every constant $C > 0$ there exists a convex function $\Phi : \mathbb{R} \to \mathbb{R}$ and a sequence $\{d_i\}$ along with a decoupled tangent sequence $\{e_i\}$ such that,*

$$\mathbb{E}\Phi\Big(\Big|\sum_{i=1}^{\infty} d_i\Big|\Big) \geq C\mathbb{E}\Phi\Big(C\Big|\sum_{i=1}^{\infty} e_i\Big|\Big). \tag{7.3.23}$$

PROOF. To show this, one uses the ideas introduced in Lemma 7.3.8. The proof is omitted. □

The last two lemmas underline the importance of working on inequalities for special structures as in Theorem 3.4.1 that presents a decoupling inequality for the tail probability of U-statistics and multilinear forms.

7.4 References for Chapter 7

Theorem 7.1.1 can be traced back to Khinchin and Kolmogorov (1925). Hall and Heyde (1980) contains several results for martingales as well as a historical account. We refer the reader to this book for the extension of Theorem 7.1.1 found in Theorem 7.1.2. For Corollary 7.1.3 see Kwapień and Woyczynski (1992). Theorem 7.1.4 can be traced back to Jackubowski (1986) and Kwapien and Woyczynski (1992). Theorem 7.1.7 is due to Brown and Eagleson (1971). Theorem 7.2.1 comes from Kou and Ying (1995) and Lemma 7.2.2 from Kou and Ying (1996). Theorem 7.3.1 is from Hitczenko (1994) and with improved constants for sums of non-negative random variables in Hitczenko (1994a). Theorem 7.3.2 is due to Burkholder (1973) and with constants having the correct order of growth from Hitczenko (1990a). Lemma 7.3.3 is from Hitczenko (1994) however, we give a new proof. Lemma 7.3.8 is due to M. Talagrand and comes from Hitczenko (1993). Theorem 7.3.9 and Lemma 7.3.10 are from Hitczenko and Montgomery-Smith (1996), which provides general results extending Theorem 7.3.9 to certain Orlicz and rearrangement spaces.

8

Further Applications of Decoupling

In this chapter we present several additional applications of decoupling. The first section gives several results for randomly stopped U-statistics, including an extension of Wald's equations. This result contains as a special case Wald's second equation for sums and is the basis for extensions of sequential methods to U-statistics. We also develop several bounds (some of which are shown to be sharp) on the L_p-norm of randomly stopped U-statistics and apply these results to prove convergence of moments in Anscombe's theorem for U-statistics. The key tools for the second part of Section 8.1 consist of bounds on the moments of U-statistics including an extension of Lévy's inequality for the L_p-norm of sums of independent mean zero random variables to the case of U-statistics with degenerate kernels (already encountered, albeit implicitly, in Chapter 4). This extension of Lévy's inequality shows that, in certain aspects, U-statistics behave more like sums of independent random variables than like martingales (in other aspects like, e.g., regarding the law of large numbers, converse part—see Section 4.1—, U-statistics do not behave at all as sums of independent variables).

In Section 8.2 we introduce a general class of exponential inequalities. The main results are natural extensions (with similar constants) of several of the classical tail probability exponential inequalities for martingales. These inequalities capture the information on the growth rate of the martingale in relation to its conditional variance in cases when the conditional variance is typically far away from its extreme value, in contrast with the usual bounds, which are in terms of the L_∞ norm of the conditional variance. This comparison is accomplished by an extension that includes tail probability inequalities for the ratio of a martingale over its conditional variance. We include, among others, extensions of Bernstein's and Bennet's inequalities. The key tools used include Corollary 6.2.5 which provides a bound on the moment generating function of sums of dependent variables allowing for constraints.

8.1 Randomly Stopped Canonical U-Statistics

In the first part of this section we use decoupling arguments and martingale inequalities to extend Wald's equation to randomly stopped de-normalized U-statistics. In a second part we apply decoupling to obtain moment bounds for regular as well as for randomly stopped U-statistics. We conclude this section with an extension of Anscombe's central limit theorem for randomly stopped sums of independent random variables to randomly stopped U-statistics, including convergence of moments.

8.1.1 Wald's Equation for Canonical U-Statistics

Let $\{Y_i\}$ be a sequence of i.i.d. random variables and T a stopping time adapted to $\mathcal{F}_n = \sigma(\{Y_i\}_{i=1}^n)$. Let $S_n = \sum_{i=1}^n Y_i$. We recall from Chapter 2 Wald's two equations:

$$\mathbb{E}S_T = 0 \quad \text{if } \mathbb{E}Y_1 = 0 \text{ and } \mathbb{E}T < \infty \text{ (First Equation)}, \tag{8.1.1}$$

and

$$\mathbb{E}S_T^2 = \mathbb{E}Y_1^2\mathbb{E}T \quad \text{if } \mathbb{E}Y_1 = 0.\ \mathbb{E}Y_1^2 < \infty \text{ and } \mathbb{E}T < \infty \text{ (Second Equation)}. \tag{8.1.2}$$

The key observation in developing the proofs of (8.1.1) and (8.1.2) is that when writting $S_T = \sum_{i=1}^{\infty} Y_i I_{T\geq i}$, the components of the sum, that is $Y_i I_{T\geq i}$ have the property that Y_i is independent of $I_{T\geq i}$. This property is primarily responsible for the decoupling that occurs. However, for U-statistics, the increments do not have this property as we show next.

Let $\{X_i\}$ be a sequence of i.i.d. random variables with values in a measurable space $(S, \mathcal{S})$ with T a stopping time adapted to the sequence $\{X_i\}$, that is, such that, for each $j \in \mathbb{N}$, the event $\{T = j\}$ is $\sigma(X_1, \ldots, X_j)$-measurable. Let f be any function mapping $S \times S \to \mathbb{R}$. Then, the T-stopped de-normalized U-statistic with kernel f is defined as

$$\mathcal{U}_T := \sum_{1\leq i<j\leq T} f(X_i, X_j) = \sum_{j=2}^{\infty}\sum_{i=1}^{j-1} f(X_i, X_j) I_{T\geq j} = \sum_{j=2}^{\infty} d_j.$$

Note that the variable $d_j = \sum_{i=1}^{j-1} f(X_i, X_j) I_{T \geq j}$ is formed by the product of

$$\sum_{i=1}^{j-1} f(X_i, X_j) \quad \text{and} \quad I_{T \geq j},$$

which are two highly dependent random variables since the indicator is measurable with respect to $\sigma(X_1, \ldots, X_{j-1})$. This dependence makes the problem much harder and explains why it took so long for the development of the extension of Wald's equation to U-statistics until new techniques were developed. These techniques include decoupling to reduce the level of dependence between the U-statistic and the stopping time and a symmetrization device to freeze the stopping time with the aid of Hölder's inequality. Our extension of Wald's equation to U-statistics is as follows:

THEOREM 8.1.1. *Let $\{X_i\}$ be a sequence of i.i.d. random variables with values in a measure space $(S, \mathcal{S})$ with T a stopping time adapted to $\{X_i\}$. Let f be any measurable function mapping $S \times S \to \mathbb{R}$. Then,*

$$\mathbb{E} \sum_{1 \leq i < j \leq T} f(X_i, X_j) = 0 \tag{8.1.3}$$

whenever the following three conditions hold for some $1 < p \leq 2$:

$$\mathbb{E}(f(X_1, X_2)|X_1) = \mathbb{E}(f(X_1, X_2)|X_2) = 0, \tag{8.1.4}$$

$$\mathbb{E}|f(X_1, X_2)|^p < \infty, \tag{8.1.5}$$

$$\mathbb{E}(T^{\frac{1}{p-1}}) < \infty. \tag{8.1.6}$$

PROOF. Since $\mathcal{U}_n = \sum_{1 \leq i < j \leq n} f(X_i, X_j) = \sum_{j=2}^{n} \sum_{i=1}^{j-1} f(X_i, X_j)$, the degeneracy condition (8.1.4) implies that the sequence $\{\mathcal{U}_n\}_{n=1}^{\infty}$ is a mean zero martingale with respect to the sigma fields $\mathcal{F}_n = \sigma(X_1, \ldots, X_n)$, $n = 1, 2, \ldots$. Therefore, the optional sampling theorem gives that $\mathbb{E}\mathcal{U}_{T_n} = 0$ for the bounded stopping time $T_n = \min(T, n)$. The desired conclusion will follow for general stopping times once we show that

$$\mathbb{E} \sup_n |\mathcal{U}_{T_n}| = \mathbb{E} \max_{n \leq T} |\mathcal{U}_n| < \infty, \tag{8.1.7}$$

since, by the dominated convergence theorem,

$$\mathbb{E}\mathcal{U}_T = \mathbb{E} \lim_{n \to \infty} \mathcal{U}_{T_n} = \lim_{n \to \infty} \mathbb{E}\mathcal{U}_{T_n} = 0. \tag{8.1.8}$$

The key step in showing (8.1.7) involves the use of decoupling arguments. Let $\{\tilde{X}_i\}$ be an independent copy of $\{X_i\}$ and let us set

$$d_j = \sum_{i=1}^{j-1} f(X_i, X_j) I_{T \geq j}, \qquad e_j = \sum_{i=1}^{j-1} f(X_i, \tilde{X}_j) I_{T \geq j}. \tag{8.1.9}$$

Then, $\{e_j\}$ and $\{d_j\}$ are tangent with respect to $\{\mathcal{F}_n'\}$ for all $j \geq 2$, where $\mathcal{F}_n' = \sigma(X_1, \ldots, X_n; \tilde{X}_1, \ldots, \tilde{X}_n)$. A use of Corollary 6.4.3 gives

$$\mathbb{E} \sup_n |\mathcal{U}_{T_n}| = \mathbb{E} \sup_n \Big| \sum_{j=2}^{n} \Big(\sum_{i=1}^{j-1} f(X_i, X_j) \Big) I_{T \geq j} \Big|$$
$$\leq C \mathbb{E} \sup_n \Big| \sum_{j=2}^{n} \Big(\sum_{i=1}^{j-1} f(X_i, \tilde{X}_j) \Big) I_{T \geq j} \Big|. \tag{8.1.10}$$

Observe that conditioning on $(\{X_i\}, T)$, $\sum_{j=2}^{n} \big(\sum_{i=1}^{j-1} f(X_i, \tilde{X}_j) \big) I_{T \geq j}$ is a sum of independent mean-zero random variables (with the sum converging a.s. to a limit as $n \to \infty$). Therefore, by conditional Jensen,

$$\mathbb{E}\Big[\Big| \sum_{j=2}^{n} \Big(\sum_{i=1}^{j-1} f(X_i, \tilde{X}_j) \Big) I_{T \geq j} \Big| \Big| \sigma(\{X_i\}) \Big]$$
$$\leq \mathbb{E}\Big[\Big| \sum_{j=2}^{\infty} \Big(\sum_{i=1}^{j-1} f(X_i, \tilde{X}_j) \Big) I_{T \geq j} \Big| \Big| \sigma(\{X_i\}) \Big]. \tag{8.1.11}$$

Using Proposition 1.1.2 (integrating over t) conditionally on $\{X_i\}$ in the second line below, we can bound the last expectation in (8.1.10) by

$$\mathbb{E} \sup_n \Big| \sum_{j=2}^{n} \Big(\sum_{i=1}^{j-1} f(X_i, \tilde{X}_j) \Big) I_{T \geq j} \Big|$$
$$= \mathbb{E}\Big(\mathbb{E}\Big[\sup_n \Big| \sum_{j=2}^{n} \Big(\sum_{i=1}^{j-1} f(X_i, \tilde{X}_j) \Big) I_{T \geq j} \Big| \Big| \sigma(\{X_i\}) \Big] \Big)$$
$$\leq 9\mathbb{E}\Big(\sup_n \mathbb{E}\Big[\Big| \sum_{j=2}^{n} \Big(\sum_{i=1}^{j-1} f(X_i, \tilde{X}_j) \Big) I_{T \geq j} \Big| \Big| \sigma(\{X_i\}) \Big] \Big)$$
$$\leq 9\mathbb{E}\Big(\sup_n \mathbb{E}\Big[\Big| \sum_{j=2}^{\infty} \Big(\sum_{i=1}^{j-1} f(X_i, \tilde{X}_j) \Big) I_{T \geq j} \Big| \Big| \sigma(\{X_i\}) \Big] \Big)$$
$$= 9\mathbb{E} \Big| \sum_{j=2}^{\infty} \Big(\sum_{i=1}^{j-1} f(X_i, \tilde{X}_j) \Big) I_{T \geq j} \Big| \qquad \text{[by (8.1.11)]}$$
$$= 9\mathbb{E} \Big| \sum_{1 \leq i < j \leq T} f(X_i, \tilde{X}_j) \Big|$$
$$\leq 9\mathbb{E} \Big| \sum_{j=1}^{T} \sum_{i=1}^{T} f(X_i, \tilde{X}_j) \Big|$$
$$+ 9\mathbb{E} \Big| \sum_{1 \leq i < j \leq T} f(X_j, \tilde{X}_i) \Big| + 9\mathbb{E} \Big| \sum_{i=1}^{T} f(X_i, \tilde{X}_i) \Big|$$
$$:= \textbf{(A)} + \textbf{(B)} + \textbf{(C)}.$$

We begin by providing a bound for (**A**). The key observation here is that both $\{X_i\}$ and $\{\tilde{X}_i\}$ appear in $\sum_{j=1}^{T}\sum_{i=1}^{T} f(X_i, \tilde{X}_j)$ in a symmetric fashion. This fact will be used below in combination with an averaging device and Hölder's inequality. We remark that the constant C, $0 < C < \infty$, appearing throughout in the following computations is a universal constant that may change from line to line:

$$\mathbb{E}\Big|\sum_{j=1}^{T}\sum_{i=1}^{T} f(X_i, \tilde{X}_j)\Big| = \mathbb{E}\left(T^{\frac{1}{p}} \frac{|\sum_{j=1}^{T}\sum_{i=1}^{T} f(X_i, \tilde{X}_j)|}{T^{\frac{1}{p}}}\right)$$

$$\leq \big(\mathbb{E}T^{\frac{1}{p-1}}\big)^{\frac{p-1}{p}} \left(\mathbb{E}\frac{|\sum_{j=1}^{T}\sum_{i=1}^{T} f(X_i, \tilde{X}_j)|^p}{T}\right)^{1/p} \quad \text{(by I below)}$$

$$\leq C\big(\mathbb{E}T^{\frac{1}{p-1}}\big)^{\frac{p-1}{p}} \left(\mathbb{E}\frac{\mathbb{E}[(\sum_{j=1}^{T}(\sum_{i=1}^{T} f(X_i, \tilde{X}_j))^2)^{\frac{p}{2}}|\sigma(\{X_r\}, T)]}{T}\right)^{1/p} \quad \text{(by II)}$$

$$\leq C\big(\mathbb{E}T^{\frac{1}{p-1}}\big)^{\frac{p-1}{p}} \left(\mathbb{E}\frac{\sum_{j=1}^{T}\mathbb{E}[|\sum_{i=1}^{T} f(X_i, \tilde{X}_j)|^p|\sigma(\{X_r\}, T)]}{T}\right)^{1/p} \quad \text{(by III)}$$

$$= C\big(\mathbb{E}T^{\frac{1}{p-1}}\big)^{\frac{p-1}{p}} \Big(\mathbb{E}\Big|\sum_{i=1}^{T} f(X_i, \tilde{X}_1)\Big|^p\Big)^{\frac{1}{p}} \quad \text{(by IV)}$$

$$\leq C\big(\mathbb{E}T^{\frac{1}{p-1}}\big)^{\frac{p-1}{p}} \Big(\mathbb{E}\Big[\sum_{i=1}^{T} f^2(X_i, \tilde{X}_1)\Big]^{p/2}\Big)^{1/p} \quad \text{(by V)},$$

$$\leq C\big(\mathbb{E}T^{\frac{1}{p-1}}\big)^{\frac{p-1}{p}} \Big(\mathbb{E}\sum_{i=1}^{T} |f(X_i, \tilde{X}_1)|^p\Big)^{\frac{1}{p}} \quad \text{(by VI)}$$

$$= C\big(\mathbb{E}T^{\frac{1}{p-1}}\big)^{\frac{p-1}{p}} \big(\mathbb{E}T\mathbb{E}|f(X_1, \tilde{X}_1)|^p\big)^{\frac{1}{p}} \quad \text{(by VII)},$$

where I-VII are justified as follows:

I) Hölder's inequality.

II) Lemma 1.4.13 (Marcinkiewicz inequality), conditionally.

III) Since for $x_i \geq 0$ and $0 < \alpha \leq 1$, $\left(\sum_{i=1}^{n} x_i\right)^{\alpha} \leq \sum_{i=1}^{n} x_i^{\alpha}$.

IV) Since $\{\tilde{X}_i\}$ is a sequence of i.i.d. variables.

V) Lemma 1.4.13 conditionally on $\tilde{X}_1$.

VI) Same reason as III.

VII) By Wald's first equation applied conditionally on $\tilde{X}_1$.

The bound for (**B**) uses further decoupling. Building upon (8.1.9), let $\{\tilde{\tilde{X}}_i\}$ be an additional independent copy of $\{X_i\}$, independent of $\{\tilde{X}_i\}$ as well. Set

$$\tilde{d}_j = \sum_{i=1}^{j-1} f(X_j, \tilde{X}_i) I_{T\geq j}, \quad \tilde{e}_j = \sum_{i=1}^{j-1} f(\tilde{\tilde{X}}_j, \tilde{X}_i) I_{T\geq j}. \tag{8.1.12}$$

Then $\{\tilde{e}_j\}$ and $\{\tilde{d}_j\}$ are tangent with respect to the sigma algebras $\tilde{\mathcal{F}}_n = \sigma(X_1, \ldots, X_n; \tilde{X}_1, \ldots, \tilde{X}_n; \tilde{\tilde{X}}_1, \ldots, \tilde{\tilde{X}}_n)$, $n = 1, 2, \ldots$. Therefore,

$$\begin{aligned}
\mathbb{E}\Big|\sum_{1\le i<j\le T} f(X_j, \tilde{X}_i)\Big| &= \mathbb{E}\Big|\sum_{j=2}^{T}\sum_{i=1}^{j-1} f(X_j, \tilde{X}_i)\Big| \\
&\le \mathbb{E}\sup_n\Big|\sum_{j=2}^{T_n}\sum_{i=1}^{j-1} f(X_j, \tilde{X}_i)\Big| \\
&\le C\mathbb{E}\sup_n\Big|\sum_{j=2}^{T_n}\sum_{i=1}^{j-1} f(\tilde{\tilde{X}}_j, \tilde{X}_i)\Big| \quad \text{(by Corollary 6.4.3)} \\
&= C\mathbb{E}\Big[\mathbb{E}\Big(\sup_n\Big|\sum_{j=2}^{n}\sum_{i=1}^{j-1} f(\tilde{\tilde{X}}_j, \tilde{X}_i)I_{T\ge j}\Big|\Big|\{\tilde{X}_i\}, T\Big)\Big] \\
&\le C\mathbb{E}\sup_n\mathbb{E}\Big(\Big|\sum_{j=2}^{n}\sum_{i=1}^{j-1} f(\tilde{\tilde{X}}_j, \tilde{X}_i)I_{T\ge j}\Big|\Big|\{\tilde{X}_i\}, T\Big) \quad \text{(Prop. 1.1.2)} \\
&\le C\mathbb{E}\Big[\mathbb{E}\Big(\Big|\sum_{j=2}^{\infty}\sum_{i=1}^{j-1} f(\tilde{\tilde{X}}_j, \tilde{X}_i)I_{T\ge j}\Big|\Big|\{\tilde{X}_i\}, T\Big)\Big] \quad \text{[as in (8.1.11)]} \\
&= C\mathbb{E}\Big|\sum_{j=2}^{\infty}\sum_{i=1}^{j-1} f(\tilde{\tilde{X}}_j, \tilde{X}_i)I_{T\ge j}\Big| \\
&= C\mathbb{E}\Big|\sum_{j=2}^{T}\sum_{i=1}^{j-1} f(\tilde{\tilde{X}}_j, \tilde{X}_i)\Big|.
\end{aligned}$$

At this point it is important to observe that we have now completely decoupled the stopping time from the variables. Using Hölder's inequality followed by Lemma 1.4.13 (applied conditionally) and the concavity of $f(x) = x^{p/2}$, $x \ge 0$, we get,

$$\begin{aligned}
\mathbb{E}\Big|\sum_{j=2}^{n}\sum_{i=1}^{j-1} f(\tilde{\tilde{X}}_j, \tilde{X}_i)\Big| &\le \Big(\mathbb{E}\Big|\sum_{j=2}^{n}\sum_{i=1}^{j-1} f(\tilde{\tilde{X}}_j, \tilde{X}_i)\Big|^p\Big)^{\frac{1}{p}} \\
&\le C\Big(\mathbb{E}\Big[\sum_{j=2}^{n}\Big(\sum_{i=1}^{j-1} f(\tilde{\tilde{X}}_j, \tilde{X}_i)\Big)^2\Big]^{\frac{p}{2}}\Big)^{\frac{1}{p}} \quad \text{(by Lemma 1.4.13)} \\
&\le C\Big[\sum_{j=2}^{n}\mathbb{E}\Big|\sum_{i=1}^{j-1} f(\tilde{\tilde{X}}_j, \tilde{X}_i)\Big|^p\Big]^{\frac{1}{p}} \quad \text{(by concavity of } |x|^{\frac{p}{2}}) \\
&\le C\Big[\sum_{j=2}^{n}\mathbb{E}\Big(\sum_{i=1}^{j-1} f^2(\tilde{\tilde{X}}_j, \tilde{X}_i)\Big)^{\frac{p}{2}}\Big]^{\frac{1}{p}} \quad \text{(by Lemma 1.4.13)}
\end{aligned}$$

$$\le C\Big[\sum_{j=2}^{n}(j-1)\mathbb{E}\big|f(\tilde{\tilde{X}}_j,\tilde{X}_1)\big|^p\Big]^{\frac{1}{p}}$$
$$\le Cn^{\frac{2}{p}}\big(\mathbb{E}|f(X_1,X_2)|^p\big)^{\frac{1}{p}}.$$

Gathering the results obtained above one gets

$$\begin{aligned}
(\mathbf{B}) &= 9\mathbb{E}\Big|\sum_{1\le i<j\le T} f(X_j,\tilde{X}_i)\Big| \\
&\le C\mathbb{E}\Big|\sum_{j=2}^{T}\sum_{i=1}^{j-1} f(\tilde{\tilde{X}}_j,\tilde{X}_i)\Big| \\
&= C\sum_{n=1}^{\infty}\mathbb{E}\Big|\sum_{j=2}^{n}\sum_{i=1}^{j-1} f(\tilde{\tilde{X}}_j,\tilde{X}_i)\Big|\Pr\{T=n\} \\
&\le C\big(\mathbb{E}|f(X_1,X_2)|^p\big)^{\frac{1}{p}}\sum_{n=1}^{\infty} n^{\frac{2}{p}}\Pr\{T=n\} \\
&= C\big(\mathbb{E}|f(X_1,X_2)|^p\big)^{\frac{1}{p}}\mathbb{E}T^{\frac{2}{p}} \\
&\le C\big(\mathbb{E}|f(X_1,X_2)|^p\big)^{\frac{1}{p}}\mathbb{E}T^{\frac{1}{p-1}}.
\end{aligned}$$

We complete the proof by providing a bound on (**C**). A use of Hölder's inequality followed by application of Lemma 1.4.13 conditionally on $\{X_i, T\}$, gives

$$\begin{aligned}
\mathbb{E}\Big|\sum_{i=1}^{T} f(X_i,\tilde{X}_i)\Big| &\le \Big(\mathbb{E}\Big|\sum_{i=1}^{T} f(X_i,\tilde{X}_i)\Big|^p\Big)^{\frac{1}{p}} \le C\Big[\mathbb{E}\Big(\sum_{i=1}^{T} f^2(X_i,\tilde{X}_i)\Big)^{\frac{p}{2}}\Big]^{\frac{1}{p}} \\
&\le C\Big(\mathbb{E}\sum_{i=1}^{T}|f(X_i,\tilde{X}_i)|^p\Big)^{\frac{1}{p}} = C\big(\mathbb{E}T\,\mathbb{E}|f(X_1,\tilde{X}_1)|^p\big)^{\frac{1}{p}},
\end{aligned}$$

where the last line follows by a conditional use of Wald's first equation. □

It is convenient to point out that Theorem 8.1.1 contains as a special case Wald's second equation. More precisely, in the case $p = 2$, the conditions on T and on $f(X_1, X_2)$ in Theorem 8.1.1 are

$$\mathbb{E}T < \infty \quad \text{and} \quad \mathbb{E}f^2(X_1,X_2) < \infty. \tag{8.1.13}$$

Taking $f(x, y) = xy$ gives,

$$0 = \mathbb{E}\sum_{1\le i<j\le T} X_iX_j = \frac{1}{2}\mathbb{E}\Big[\Big(\sum_{i=1}^{T} X_i\Big)^2 - \sum_{i=1}^{T} X_i^2\Big], \tag{8.1.14}$$

whenever

$$\mathbb{E}X_1 = 0 \quad \text{and} \quad \mathbb{E}X_1^2 < \infty. \tag{8.1.15}$$

It is easy to see that this is exactly Wald's second equation. That is (8.1.15) implies

$$\mathbb{E}\Big(\sum_{i=1}^{T} X_i\Big)^2 = \mathbb{E}\sum_{i=1}^{T} X_i^2 = \mathbb{E}T\mathbb{E}X_1^2, \tag{8.1.16}$$

where the last identity is obtained by using Wald's first equation (8.1.1) on $Y_i = X_i^2 - EX_i^2$.

Theorem 8.1.1 admits a multivariate extension given in Theorem 8.1.2 below. In obtaining it we had to develop a new decoupling inequality to take into account the increased degree of dependence found in the multivariate case. We give this decoupling inequality in Theorem 8.1.3.

THEOREM 8.1.2. *Let $\{X_i\}$ be a sequence of i.i.d. random variables with values in a measure space $(S, \mathcal{S})$ and common probability law P. Let $\mathcal{F}_n = \sigma(X_1, \ldots, X_n)$. Let $f : S^k \to \mathbb{R}$ be a p-integrable P-canonical kernel, that is, f is a measurable function that satisfies the following two properties: f is symmetric, i.e.,*

$$f(X_1, \ldots, X_k) = f(X_{s(1)}, \ldots, X_{s(k)}), \tag{8.1.17}$$

for any permutation s of $(1, \ldots, k)$ and f is centered and completely degenerate, i.e.

$$\mathbb{E}\big(f(X_1, \ldots, X_k)|X_h\big) = 0, \tag{8.1.18}$$

for all $1 \le h \le k$, and

$$\mathbb{E}\big|f(X_1, \ldots, X_k)\big|^p < \infty \quad \text{for some } 1 < p \le 2. \tag{8.1.19}$$

Then, if T is an $\{\mathcal{F}_n\}$ adapted stopping time,

$$\mathbb{E}\sum_{1\le i_1<\cdots<i_k\le T} f(X_{i_1}, \ldots, X_{i_k}) = 0, \tag{8.1.20}$$

whenever $\mathbb{E}T^{\rho(k,p)} < \infty$, where $\rho(k, p) = (p^{1/(k-1)} - 1)^{-1}$.

Since $(p^{1/(k-1)} - 1)^{-1} < (k-1)/\ln p$, the rate of growth of $\rho(k, p)$ is at most linear in k. Observe that the case $k = 2$ gives Theorem 8.1.1. The key to obtaining Theorem 8.1.2 is the following decoupling inequality for randomly stopped U-statistics.

THEOREM 8.1.3. *Fix α, $1 \le \alpha \le 2$. Let $\{X_i\}$ and $\mathcal{F}_n$ be as in Theorem 8.1.2. Let $\{X_i^1\}, \ldots, \{X_i^k\}$ be $k \ge 2$ independent copies of $\{X_i\}$. Let $f : S^k \to \mathbb{R}$ be an α-integrable P-canonical kernel. Then, if T is an $\{\mathcal{F}_n\}$ adapted stopping time,*

$$\mathbb{E}\max_{k\le n\le T}\Big|\sum_{1\le i_1<i_2<\cdots<i_k\le n} f(X_{i_1}, X_{i_2}, \ldots, X_{i_k})\Big|^\alpha \tag{8.1.21}$$

$$\le C_{k,\alpha}\sum_{h=0}^{k-1}\sum_{t=1}^{k-h}\mathbb{E}\Big|\sum_{i_{h+1}=1}^{T}\cdots\sum_{i_{h+t}=1}^{T} f_{T,k,h,t}(X_{i_{h+1}}^{h+1}, \ldots, X_{i_{h+t}}^{h+t}, \ldots, X_{i_{h+t}}^{k})\Big|^\alpha.$$

where $f_{T,k,h,t}$ *is given by the formula*

$$f_{T,k,h,t}(X_{i_{h+1}}^{h+1}, \dots, X_{i_{h+t}}^{h+t}, \dots, X_{i_{h+t}}^{k}) \qquad (8.1.22)$$
$$= \sum_{1 \le i_1 < \dots < i_h \le T} f(X_{i_1}, \dots, X_{i_h}, X_{i_{h+1}}^{h+1}, \dots, X_{i_{h+t}}^{h+t}, \dots, X_{i_{h+t}}^{k}),$$

and $C_{k,\alpha}$ *is a constant depending on* α, k *only.*

The bound provided by Theorem 8.1.3 is sharp. This fact can be verified by noting that if the stopping time is assumed to be independent of all the variables involved, then the upper and lower bounds are of the same order of magnitude. We refer the reader to de la Peña and Lai (1997a) for the proofs of both, Theorem 8.1.2 and theorem 8.1.3. It shares some aspects similar to the proof of Theorem 2.1.4. However, the high dependence present in U-statistics (as opposed to sums of independent random variables) makes it rather involved. The techniques used include the K-function from Chapter 1 and an argument of subsequences designed to trap the stopping time and gradually decouple the U-statistic from the stopping time.

8.1.2 Moment Bounds for Regular and Randomly Stopped U-Statistics

We begin with a well known moment bound for sums of centered independent random variables. It is a particular case of Rosenthal's inequality (Theorem 1.5.11 and the paragraph following it), but it has a direct, more elementary proof via the Marcinkiewicz and the generalized Minkowski inequalities.

LEMMA 8.1.4. *Let* $\{X_i\}$ *be a sequence of i.i.d. real random variables such that* $\mathbb{E}|X_i|^p < \infty$ *for some* $p \ge 2$ *and* $\mathbb{E}X_i = 0$. *Then,*

$$\mathbb{E} \max_{1 \le k \le n} \Big|\sum_{i=1}^{k} X_i\Big|^p \le C_p n^{p/2} \mathbb{E}|X_1|^p, \qquad (8.1.23)$$

for a universal constant $0 < C_p < \infty$ *depending on* p *only.*

PROOF. Using Lemma 1.4.13 (Marcinkiewicz inequalities) in the first inequality that follows and Minkowski's inequality for the $L_{p/2}$-norm of $\sum X_i^2$ in the second, we obtain

$$\mathbb{E} \max_{1 \le k \le n} \Big|\sum_{i=1}^{k} X_i\Big|^p \le C_p \mathbb{E}\Big(\sum_{i=1}^{n} X_i^2\Big)^{p/2}$$
$$\le C_p \Big[\sum_{i=1}^{n} \big(\mathbb{E}|X_i|^p\big)^{2/p}\Big]^{p/2} = C_p n^{p/2} \mathbb{E}|X_1|^p. \qquad \square$$

As a first step towards extending this moment inequality, first to U-statistics and then to randomly stopped U-statistics, we begin with Lévy's inequality for canonical U-statistics, already encountered, without making it explicit, in the proof of Lemma 4.3.2. We recall that the index set I_n^k is the set of k-tuples $\mathbf{i} = (i_1, \ldots, i_k)$ of natural numbers such that $1 \le i_j \le n$ and $i_j \ne i_\ell$ whenever $j \ne \ell$.

LEMMA 8.1.5. *Let $\{X_i\}$ be a sequence of independent random variables with values in $(S, \mathcal{S})$ and let $f : S^k \to \mathbb{R}$, $k \ge 2$, be a measurable function such that, for some $p \ge 1$ and any set of different subindices $i_1, \ldots, i_k$, $\mathbb{E}|f(X_{i_1}, \ldots, X_{i_k})|^p < \infty$ and $\mathbb{E}\big(f(X_{i_1}, \ldots, X_{i_k})\big|X_{i_1}, \ldots, X_{i_{\ell-1}}, X_{i_{\ell+1}}, \ldots, X_{i_k}\big) = 0$, $1 \le \ell \le k$. Let $\{X_i^1\}, \ldots, \{X_i^k\}$ be k independent copies of $\{X_i\}$ and let $\{\varepsilon_i^j : i \in \mathbb{N}, 1 \le j \le k\}$ be a collection of independent Rademacher variables independent from all the X^j variables. Then, there exists a constant depending only on k and p such that*

$$\mathbb{E} \max_{k \le n \le m} \Big|\sum_{I_n^k} f(X_{i_1}, \ldots, X_{i_k})\Big|^p \le C_{k,p}\mathbb{E}\Big|\sum_{I_m^k} \varepsilon_{i_1}^1 \cdots \varepsilon_{i_k}^k f(X_{i_1}^1, \ldots, X_{i_k}^k)\Big|^p \tag{8.1.24}$$

for all $m \ge k$.

PROOF. We use the device of incorporating the maximum into the norm, as in (3.1.17) and in the proofs of Lemma 4.3.2 and Theorem 4.3.3: setting

$$f_r := (0, \overset{r-1)}{\ldots}, 0, f, \ldots, f) \in \ell_m^\infty, \quad r = k, \ldots, m,$$

we have

$$\max_{k \le n \le m} \Big|\sum_{\mathbf{i} \in I_n^k} f(X_{i_1}, \ldots, X_{i_k})\Big| = \Big\|\sum_{\mathbf{i} \in I_m^k} f_{|\mathbf{i}|}(X_{i_1}, \ldots, X_{i_k})\Big\|,$$

where $\|(x_1, \ldots, x_m)\| := \max_{i \le m} |x_i|$ is the norm of ℓ_m^∞. Then, the first part of Theorem 3.1.1 on decoupling of U-statistics applies to the ℓ_m^∞-valued kernels f_r, and so does the argument in the proof of Theorem 3.5.3 on randomization (canonical case), thus concluding

$$\mathbb{E} \max_{k \le n \le m} \Big|\sum_{I_n^k} f(X_{i_1}, \ldots, X_{i_k})\Big|^p \le C_{k,p}\mathbb{E} \max_{k \le n \le m} \Big|\sum_{I_n^k} \varepsilon_{i_1}^1 \cdots \varepsilon_{i_k}^k f(X_{i_1}^1, \ldots, X_{i_k}^k)\Big|^p.$$

Now, the maximum at the right side of this inequality is removed by repeated conditional application of the integral form of Lévy's inequality, very much as in the second part of the proof of Lemma 4.3.2. □

Now we can extend Lemma 8.1.4 to canonical U-statistics.

THEOREM 8.1.6. *Let $\{X_i\}$ be a sequence of i.i.d. random variables with values in a measurable space $(S, \mathcal{S})$. Let $p \ge 2$ and let $f : S^k \to \mathbb{R}$ be a measurable function of k variables for some $k \ge 2$ such that $\mathbb{E}|f(X_1, \ldots, X_k)|^p < \infty$ and $\mathbb{E}(f(X_1, \ldots, X_k)|X_1, \ldots, X_{j-1}, X_{j+1}, \ldots, X_k) = 0$ for all $1 \le j \le k$. Then,*

$$\mathbb{E} \max_{k \le n \le m} \Big|\sum_{I_n^k} f(X_{i_1}, \ldots, X_{i_k})\Big|^p \le C_{k,p} m^{\frac{kp}{2}} \mathbb{E}\Big|f(X_1, \ldots, X_k)\Big|^p \tag{8.1.25}$$

for all $m \ge k$.

PROOF. By repeated application of Khinchin's inequality (1.3.1) (or, alternatively, by hypercontractivity of Rademacher chaos, Theorem 3.2.2) and by Minkowski's inequality for the $L_{p/2}$-norm applied to the sum of the variables $f^2(X_{i_1}^1, \ldots, X_{i_k}^k)$, we have

$$\begin{aligned}
&\mathbb{E}\Big|\sum_{I_m^k} \varepsilon_{i_1}^1 \cdots \varepsilon_{i_k}^k f(X_{i_1}^1, \ldots, X_{i_k}^k)\Big|^p \\
&= \mathbb{E}\Big[\mathbb{E}\Big(\Big|\sum_{I_m^k} \varepsilon_{i_1}^1 \cdots \varepsilon_{i_k}^k f(X_{i_1}^1, \ldots, X_{i_k}^k)\Big|^p \Big| \{X_i^j\}\Big)\Big] \\
&\leq (p-1)^{kp/2} \mathbb{E}\Big(\sum_{I_m^k} f^2(X_{i_1}^1, \ldots, X_{i_k}^k)\Big)^{p/2} \\
&\leq (p-1)^{kp/2} \Big[\sum_{I_m^k} \big(\mathbb{E}|f(X_{i_1}^1, \ldots, X_{i_k}^k)|^p\big)^{2/p}\Big]^{p/2} \\
&\leq (p-1)^{kp/2} m^{kp/2} \mathbb{E}|f(X_1, \ldots, X_k)|^p.
\end{aligned}$$

This estimate together with Lemma 8.1.5 gives the theorem. □

The bound given by Theorem 8.1.6 is sharp as can be seen by considering the U-statistic $\sum_{I_k^m} \varepsilon_{i_1} \cdots \varepsilon_{i_k}$. By hypercontractivity, the L_p-norm of this statistic is of the order of its L_2-norm, which is of the order of $m^{k/2}$.

Theorem 8.1.6 extends to randomly stopped U-statistics, at the price of some extra integrability on the kernel. This is the content of the next result.

THEOREM 8.1.7. *Let $\{X_i\}$ be a sequence of i.i.d. random variables with values in a measurable space $(S, \mathcal{S})$ and common probability law P. Let $p \geq 2$ and let $f : S^k \to \mathbb{R}$ be a P-canonical kernel of k variables for some $k \geq 2$ (that is, f is symmetric in its entries, centered and completely degenerate), such that $\mathbb{E}|f(X_1, \ldots, X_k)|^{kp} < \infty$ Let T be a stopping time adapted to $\{X_i\}$ such that $\mathbb{E}T^{\frac{kp}{2}} < \infty$. Then,*

$$\mathbb{E}\max_{n \leq T}\Big|\sum_{1 \leq i_1 < \cdots < i_k \leq n} f(X_{i_1}, \ldots, X_{i_k})\Big|^p \leq C_{k,p} \mathbb{E}T^{\frac{kp}{2}} \big(\mathbb{E}|f(X_1, \ldots, X_k)|^{kp}\big)^{\frac{1}{k}}, \tag{8.1.26}$$

for some universal constant $C_{k,p}$ depending on (k, p) only.

PROOF. Let $\{X_i^1\}, \ldots, \{X_i^k\}$ be k independent copies of $\{X_i\}$. Set

$$\mathcal{G}_n = \sigma\big(\{X_i\}_{i=1}^\infty, \{X_i^l\}_{i=1}^n, \quad l = 1, \ldots, k\big) \qquad n \in \mathbb{N}.$$

Then, it is easy to see that, due to the degeneracy of f and the independence of T and $\{X_i^l\}$, $l = 1, \ldots, k$.

$$\mathbb{E}\Big[\sum_{i_{k-1}=i_k}^{T} \cdots \sum_{i_1=1}^{i_2-1} f(X_{i_1}, \ldots, X_{i_{k-1}}, X_{i_k}^k) I_{T \geq i_k} \big| \mathcal{G}_{i_k - 1}\Big] = 0. \tag{8.1.27}$$

It is also clear that the sequence

$$Z_\ell := \Big(\sum_{i_{k-1}=k-1}^{T} \cdots \sum_{i_1=1}^{i_2-1} f(X_{i_1}, \ldots, X_{i_{k-1}}, X_\ell^{(k)}) I_{T \geq \ell} \Big)^2, \; \ell = k, \ldots$$

is adapted to $\{\mathcal{G}_\ell\}_{\ell=k}^{\infty}$. We use these facts in the following chain of inequalities, which constitute the first and main step to verify our claim:

$$\mathbb{E} \max_{n \leq T} \Big| \sum_{1 \leq i_1 < \cdots < i_k \leq n} f(X_{i_1}, \ldots, , X_{i_k}) \Big|^p$$

$$= \mathbb{E} \sup_n \Big| \sum_{i_k=k}^{n} \Big(\sum_{i_{k-1}=k-1}^{i_k} \cdots \sum_{i_1=1}^{i_2} f(X_{i_1}, \ldots, , X_{i_k}) \Big) I_{T \geq i_k} \Big|^p$$

$$\leq C_p \mathbb{E} \Big[\sum_{i_k=k}^{\infty} \Big(\sum_{i_{k-1}=k-1}^{i_k-1} \cdots \sum_{i_1=1}^{i_2-1} f(X_{i_1}, \ldots, X_{i_{k-1}}, X_{i_k}) I_{T \geq i_k} \Big)^2 \Big]^{p/2} \quad \text{(Theorem 6.6.2)}$$

$$\leq C_p \mathbb{E} \Big[\sum_{i_k=k}^{\infty} \Big(\sum_{i_{k-1}=k-1}^{i_k-1} \cdots \sum_{i_1=1}^{i_2-1} f(X_{i_1}, \ldots, X_{i_{k-1}}, X_{i_k}^k) I_{T \geq i_k} \Big)^2 \Big]^{p/2} \quad \text{(Theorem 6.4.2)}$$

$$= C_p \mathbb{E} \Big[\sum_{i_k=k}^{\infty} \Big(\mathbb{E}\Big[\sum_{i_{k-1}=k-1}^{T} \cdots \sum_{i_1=1}^{i_2-1} f(X_{i_1}, \ldots, X_{i_{k-1}}, X_{i_k}^k) I_{T \geq i_k} \big| \mathcal{G}_{i_k-1} \Big] \Big)^2 \Big]^{p/2} \quad [(8.1.27)]$$

$$\leq C_p \mathbb{E} \Big[\sum_{i_k=k}^{\infty} \mathbb{E}\Big(\Big[\sum_{i_{k-1}=k-1}^{T} \cdots \sum_{i_1=1}^{i_2-1} f(X_{i_1}, \ldots, X_{i_{k-1}}, X_{i_k}^k) I_{T \geq i_k} \Big]^2 \big| \mathcal{G}_{i_k-1} \Big) \Big]^{p/2} \quad \text{(Jensen)}$$

$$\leq C_p \mathbb{E} \Big[\sum_{i_k=k}^{\infty} \Big(\sum_{i_{k-1}=k-1}^{T} \cdots \sum_{i_1=1}^{i_2-1} f(X_{i_1}, \ldots, X_{i_{k-1}}, X_{i_k}^k) I_{T \geq i_k} \Big)^2 \Big]^{p/2} \quad \text{(Theorem 6.6.1)}$$

$$\leq C_p \mathbb{E} \max_{m \leq T} \Big| \sum_{i_k=1}^{m} \sum_{i_{k-1}=k-1}^{T} \sum_{i_{k-2}=k-2}^{i_{k-1}} \cdots \sum_{i_1=1}^{i_2-1} f(X_{i_1}, \ldots, X_{i_{k-1}}, X_{i_k}^k) \Big|^p$$

(Theorem 6.6.2)

$$\leq C_p \mathbb{E} \Big| \sum_{i_k=1}^{T} \sum_{i_{k-1}=k-1}^{T} \sum_{i_{k-2}=k-2}^{i_{k-1}} \cdots \sum_{i_1=1}^{i_2-1} f(X_{i_1}, \ldots, X_{i_{k-1}}, X_{i_k}^k) \Big|^p \quad [(1.2.9) \text{ and } (1.2.8)]$$

$$\leq C_p \mathbb{E} \Big[T^{\frac{p}{2}} \Big| \sum_{i_{k-1}=k-1}^{T} \sum_{i_{k-2}=k-2}^{i_{k-1}} \cdots \sum_{i_1=1}^{i_2-1} f(X_{i_1}, \ldots, X_{i_{k-1}}, X_1^k) \Big|^p \Big], \tag{8.1.28}$$

where the last inequality follows by conditioning on $(\{X_i\}, T)$ and (8.1.23). Continuing from the above bound, and using of Hölder's inequality, we obtain

$$\leq C_p (\mathbb{E} T^{\frac{kp}{2}})^{\frac{1}{k}}$$
$$\times \Big(\mathbb{E} \Big| \sum_{i_{k-1}=k-1}^{T} \sum_{i_{k-2}=k-2}^{i_{k-1}-1} \cdots \sum_{i_1=1}^{i_2-1} f(X_{i_1}, \ldots, X_{i_{k-2}}, X_{i_{k-1}}, X_1^{(k)}) \Big|^{\frac{kp}{k-1}} \Big)^{\frac{k-1}{k}}$$

$$\leq C_p(\mathbb{E}T^{\frac{kp}{2}})^{\frac{1}{k}}$$

$$\times \left(\mathbb{E}\max_{m\leq T}\left|\sum_{i_{k-1}=k-1}^{m}\sum_{i_{k-2}=k-2}^{i_{k-1}-1}\cdots\sum_{i_1=1}^{i_2-1} f(X_{i_1},\ldots,X_{i_{k-2}},X_{i_{k-1}},X_1^{(k)})\right|^{\frac{kp}{k-1}}\right)^{\frac{k-1}{k}}$$

$$\vdots$$

$$\leq C_p(\mathbb{E}T^{\frac{kp}{2}})^{\frac{1}{k}} \tag{8.1.29}$$

$$\times \left(\mathbb{E}T^{\frac{kp}{2(k-1)}}\left|\sum_{i_{k-2}=k-2}^{T}\sum_{i_{k-3}=k-3}^{i_{k-2}-2}\cdots\sum_{i_1=1}^{i_2-1} f(X_{i_1},\ldots,X_{i_{k-2}},X_{i_{k-1}}^{(k-2)},X_1^{(k)})\right|^{\frac{kp}{k-1}}\right)^{\frac{k-1}{k}},$$

where the last line follows by (8.1.23) (conditionally on $X_1^{(k)}$) with p replaced by $kp/(k-1)$ and using the original chain of inequalities. Repeating the argument leading to (8.1.29) $k-2$ times, with an adequate re-definition of $\mathcal{G}_n$, gives that the previous quantities are dominated as follows:

$$\leq C_{p,k}(\mathbb{E}T^{\frac{kp}{2}})^{\frac{k-1}{k}}\left(\mathbb{E}\left|\sum_{i_1=1}^{T} f(X_{i_1},X_1^{(2)},\ldots,X_1^{(k)})\right|^{kp}\right)^{\frac{1}{k}}$$

$$\leq C_{p,k}\mathbb{E}T^{\frac{kp}{2}}\left(\mathbb{E}\left|f(X_1^{(1)},X_1^{(2)},\ldots,X_1^{(k)})\right|^{kp}\right)^{\frac{1}{k}}. \qquad \square$$

8.1.3 Moment Convergence in Anscombe's Theorem for U-Statistics

A modification of Theorem 8.1.7 will allow us to obtain moment convergence in Anscombe's theorem for U-statistics. Before presenting the modification we introduce several versions of Anscombe's theorem. We recall that the notation $\rightarrow_{\mathcal{L}}$ is used to denote convergence in law or in distribution of the variables involved. The notation $\rightarrow_w$ is used for weak convergence of probability measures.

THEOREM 8.1.8. *Let $\{X_i\}$ be a sequence of i.i.d. random variables with $\mathbb{E}X_i = 0$ and $\mathbb{E}X_i^2 = \sigma^2$, $0 < \sigma < \infty$. Set $S_n = \sum_{i=1}^n X_i$. Let $N(t)$ be an integer valued random process satisfying*

$$\frac{N(t)}{t} \rightarrow \eta, \tag{8.1.30}$$

in probability as $t \rightarrow \infty$ for $0 < \eta < \infty$. Then,

$$\frac{S_{N(t)}}{\sigma\sqrt{N(t)}} \rightarrow_{\mathcal{L}} z, \tag{8.1.31}$$

and,

$$\frac{S_{N(t)}}{\sigma\sqrt{\eta t}} \rightarrow_{\mathcal{L}} z, \tag{8.1.32}$$

as $t \rightarrow \infty$, where z is a standard normal random variable.

PROOF. By re-scaling, we may assume that $\sigma^2 = 1$. Let $n_{0t} = [\eta t]$. Then,

$$\frac{S_{N(t)}}{\sqrt{N(t)}} = \left(\frac{S_{n_{0t}}}{\sqrt{n_{0t}}} + \frac{S_{N(t)} - S_{n_{0t}}}{\sqrt{n_{0t}}}\right)\sqrt{\frac{n_{0t}}{N(t)}}.$$

The central limit theorem guarantees that the variables $S_{n_{0t}}/\sqrt{n_{0t}}$ converge in distribution to z. Moreover, since (8.1.30) implies that $\sqrt{n_{0t}/N(t)} \to 1$ in probability, we only need to show that, as $t \to \infty$,

$$\frac{S_{N(t)} - S_{n_{0t}}}{\sqrt{n_{0t}}} \to 0$$

in probability. To do this, for $0 < \epsilon < \frac{1}{3}$ let

$$n_{1t} = [n_{0t}(1 - \varepsilon^3)] + 1 \qquad \text{and} \qquad n_{2t} = [n_{0t}(1 + \varepsilon^3)].$$

Then,

$$\begin{aligned}
&\Pr\{|S_{N(t)} - S_{n_{0t}}| > \varepsilon\sqrt{n_{0t}}\} \\
&\quad = \Pr\{|S_{N(t)} - S_{n_{0t}}| > \varepsilon\sqrt{n_{0t}}, N(t) \in [n_{1t}, n_{2t}]\} \\
&\qquad + \Pr\{|S_{N(t)} - S_{n_{0t}}| > \varepsilon\sqrt{n_{0t}}, N(t) \in [n_{1t}, n_{2t}]^c\} \\
&\quad \le \Pr\Big\{\max_{n_{1t} \le n \le n_{2t}} |S_n - S_{n_{0t}}| > \varepsilon\sqrt{n_{0t}}\Big\} + \Pr\{N(t) \in [n_{1t}, n_{2t}]^c\} \\
&\quad \le \Pr\Big\{\max_{n_{1t} \le n \le n_{0t}} |S_n - S_{n_{0t}}| > \varepsilon\sqrt{n_{0t}}\Big\} \\
&\qquad + \Pr\Big\{\max_{n_{0t} \le n \le n_{2t}} |S_n - S_{n_{0t}}| > \varepsilon\sqrt{n_{0t}}\Big\} + \Pr\{N(t) \in [n_{1t}, n_{2t}]^c\} \\
&\quad \le \frac{\mathrm{Var}(S_{n_{0t}} - S_{n_{1t}})}{\varepsilon^2 n_{0t}} + \frac{\mathrm{Var}(S_{n_{2t}} - S_{n_{0t}})}{\varepsilon^2 n_{0t}} + \Pr\{N(t) \in [n_{1t}, n_{2t}]^c\} \qquad (8.1.33) \\
&\quad \le \frac{[\eta t] - [[\eta t](1 - \varepsilon^3)]}{\varepsilon^2 [\eta t]} + \frac{[[\eta t](1 + \varepsilon^3)] - [\eta t]}{\varepsilon^2 [\eta t]} + \Pr\Big\{\frac{N(t)}{t} \in \Big[\frac{n_{1t}}{t}, \frac{n_{2t}}{t}\Big]^c\Big\} \\
&\quad \le \varepsilon + \varepsilon + \varepsilon = 3\varepsilon,
\end{aligned}$$

where (8.1.33) follows from Kolmogorov's inequality and the last line is valid for all $t \ge t_\varepsilon$ large enough since $N(t)/t \to \eta$ in probability. This proves (8.1.31), and (8.1.32) follows from (8.1.30), (8.1.31) and Slutzky's theorem. □

Anscombe's result for general variables uses a regularity condition which we introduce next. A sequence of random variables $\{Y_i\}$ is said to satisfy the condition of *uniform continuity in probability* if for all $\varepsilon > 0$ there exists $\delta > 0$ for which

$$\Pr\Big\{\max_{0 \le k \le n\delta} |Y_{n+k} - Y_n| \ge \varepsilon\Big\} \le \varepsilon \qquad \text{for all } n \ge 1. \tag{8.1.34}$$

THEOREM 8.1.9. *Let $\{Y_i\}$ be a sequence of variables satisfying condition (8.1.34). Assume that $\{N(t)\}$ is a sequence of integer valued random variables such that*

$$\frac{N(t)}{t} \to \eta, \tag{8.1.35}$$

in probability for some constant $\eta > 0$. *Then,*

$$Y_{N(t)} - Y_{[t\eta]} \to 0 \quad \text{in probability as } t \to \infty. \tag{8.1.36}$$

If in addition $Y_n \to_{\mathcal{L}} Y$ *as* $n \to \infty$ *for some random variable* Y, *then* $Y_{N(t)} \to_{\mathcal{L}} Y$ *as* $t \to \infty$.

PROOF. Without loss of generality we may assume that $\eta = 1$. Let $\epsilon > 0$ be given, and $\delta > 0$ to be chosen later. Then,

$$\begin{aligned}\Pr\{|Y_{N(t)} - Y_{[t]}| > \varepsilon\} &\leq \Pr\left\{|Y_{N(t)} - Y_{[t]}| > \varepsilon, \left|\frac{N(t)}{[t]} - 1\right| \leq \delta\right\} \\ &\quad + \Pr\left\{|Y_{N(t)} - Y_{[t]}| > \varepsilon, \left|\frac{N(t)}{[t]} - 1\right| > \delta\right\} \\ &\leq \Pr\left\{\max_{|k-[t]|\leq\delta[t]} |Y_k - Y_{[t]}| > \varepsilon\right\} + \Pr\left\{\left|\frac{N(t)}{[t]} - 1\right| > \delta\right\} \\ &\leq 2\varepsilon.\end{aligned}$$

To see this, note that, by condition (8.1.34), the first term is bounded by ε and, by condition (8.1.35), one can pick $\delta = \delta(\varepsilon)$ so that the second term is bounded by ε as well. Convergence in distribution of $Y_{N(t)}$ follows easily since $Y_{[t]}$ converges in distribution to Y and $Y_{N(t)} = Y_{[t]} + (Y_{N(t)} - Y_{[t]})$, where the second term converges to zero in probability, □

Here is Anscombe's theorem for degenerate U-statistics:

THEOREM 8.1.10. *Let* $\{X_i\}$ *be a sequence of i.i.d. random variables with values in a measurable space* $(S, \mathcal{S})$ *and with common probability law* P. *Let* $f : S^k \to \mathbb{R}$ *be a* P*-canonical kernel of* k *variables for some* $k \geq 2$ *such that* $\mathbb{E}(f(X_1, \ldots, X_k))^2 < \infty$. *Set*

$$\mathcal{U}_n := \sum_{1\leq i_1<\cdots<i_k\leq n} f(X_{i_1}, \ldots, X_{i_k}). \tag{8.1.37}$$

Let $\{N(t)\}$ *be a family of integer valued random processes such that*

$$\frac{N(t)}{t} \to \eta \tag{8.1.38}$$

in probability for some constant $\eta > 0$. *Then*

$$N(t)^{-\frac{k}{2}}\mathcal{U}_{N(t)} \to_{\mathcal{L}} \eta^{\frac{k}{2}}(k!)^{-1/2}K_P(f), \tag{8.1.39}$$

and

$$t^{-\frac{k}{2}}\mathcal{U}_{N(t)} \to \eta^{\frac{k}{2}}(k!)^{-1/2}K_P(f), \tag{8.1.40}$$

where $K_P(f)$ *is the* P*-isonormal Gaussian chaos process evaluated at* f *(see (4.2.16)) or, what is the same, a multiple Wiener integral.*

PROOF. By re-scaling, we may assume that $Ef^2(X_1, \ldots, X_k) = 1$. Let $n_{0t} = [\eta t]$. Then,

$$\frac{\mathcal{U}_{N(t)}}{N(t)^{\frac{k}{2}}} = \Big(\frac{\mathcal{U}_{n_{0t}}}{n_{0t}^{k/2}} + \frac{\mathcal{U}_{N(t)} - \mathcal{U}_{n_{0t}}}{n_{0t}^{k/2}}\Big)\Big(\frac{n_{0t}}{N(t)}\Big)^{\frac{k}{2}}. \tag{8.1.41}$$

Convergence in distribution of $\mathcal{U}_{n_{0t}}/\sqrt{n_{0t}}$ to $\eta^{\frac{k}{2}} K_P(f)/\sqrt{n_{0t}}$ is guaranteed by the central limit theorem for canonical U-statistics given in Theorem 4.2.4. Moreover, (8.1.38) implies that $(n_{0t}/N(t))^{\frac{k}{2}} \to 1$ in probability. Therefore, we only need to show that

$$\frac{\mathcal{U}_{N(t)} - \mathcal{U}_{n_{0t}}}{n_{0t}^{k/2}} \to 0 \tag{8.1.42}$$

in probability as $t \to \infty$. To do this, we set, in analogy with the proof of Theorem 8.1.8,

$$n_{1t} = [n_{0t}(1-\varepsilon^3)^{\frac{1}{k}}] + 1 \qquad \text{and} \qquad n_{2t} = [n_{0t}(1+\varepsilon^3)^{\frac{1}{k}}]. \tag{8.1.43}$$

for $0 < \epsilon < \frac{1}{10}$. The function f being canonical, $\mathcal{U}_n$ is a martingale with respect to $\sigma(X_1, \ldots, X_n)$, and therefore, so is $\mathcal{U}_n - \mathcal{U}_r$, $n \geq r$, for fixed r. Then, again in analogy with the previous proof, but using Doob's maximal inequality for $\mathcal{U}_n - \mathcal{U}_r$ instead of Kolmogorov's maximal inequality, we have

$$\begin{aligned}
&\Pr\big\{|\mathcal{U}_{N(t)} - \mathcal{U}_{n_{0t}}| > \varepsilon n_{0t}^{\frac{k}{2}}\big\}\\
&= \Pr\big\{|\mathcal{U}_{N(t)} - \mathcal{U}_{n_{0t}}| > \varepsilon n_{0t}^{\frac{k}{2}}, N(t) \in [n_{1t}, n_{2t}]\big\}\\
&\quad + \Pr\big\{|\mathcal{U}_{N(t)} - \mathcal{U}_{n_{0t}}| > \varepsilon n_{0t}^{\frac{k}{2}}, N(t) \in [n_{1t}, n_{2t}]^c\big\}\\
&\leq \Pr\big\{\max_{n_{1t}\leq n\leq n_{2t}} |\mathcal{U}_n - \mathcal{U}_{n_{0t}}| > \varepsilon n_{0t}^{\frac{k}{2}}\big\} + \Pr\big\{N(t) \in [n_{1t}, n_{2t}]^c\big\}\\
&\leq \Pr\big\{\max_{n_{1t}\leq n\leq n_{0t}} |\mathcal{U}_n - \mathcal{U}_{n_{0t}}| > \varepsilon n_{0t}^{\frac{k}{2}}\big\}\\
&\quad + \Pr\big\{\max_{n_{0t}\leq n\leq n_{2t}} |\mathcal{U}_n - \mathcal{U}_{n_{0t}}| > \varepsilon n_{0t}^{\frac{k}{2}}\big\} + \Pr\big\{N(t) \in [n_{1t}, n_{2t}]^c\big\}\\
&\leq \Pr\big\{\max_{n_{1t}\leq n\leq n_{0t}} |\mathcal{U}_n - \mathcal{U}_{n_{1t}}| > \varepsilon n_{0t}^{\frac{k}{2}}/2\big\} + \Pr\big\{|\mathcal{U}_{n_{0t}} - \mathcal{U}_{n_{1t}}| > \varepsilon n_{0t}^{\frac{k}{2}}/2\big\}\\
&\quad + \Pr\big\{\max_{n_{0t}\leq n\leq n_{2t}} |\mathcal{U}_n - \mathcal{U}_{n_{0t}}| > \varepsilon n_{0t}^{\frac{k}{2}}\big\} + \Pr\big\{N(t) \in [n_{1t}, n_{2t}]^c\big\}\\
&\leq 8\frac{\operatorname{Var}(\mathcal{U}_{n_{0t}} - \mathcal{U}_{n_{1t}})}{\varepsilon^2 n_{0t}^k} + \frac{\operatorname{Var}(\mathcal{U}_{n_{2t}} - \mathcal{U}_{n_{0t}})}{\varepsilon^2 n_{0t}^k} + \Pr\big\{N(t) \in [n_{1t}, n_{2t}]^c\big\}\\
&\leq \frac{8\Big[\binom{[\eta t]}{k} - \binom{[[\eta t](1-\varepsilon^3)^{\frac{1}{k}}]}{k}\Big]}{\varepsilon^2[\eta t]^k} + \frac{\binom{[[\eta t](1+\varepsilon^3)^{\frac{1}{k}}]}{k} - \binom{[\eta t]}{k}}{\varepsilon^2[\eta t]^k} + \Pr\Big\{\frac{N(t)}{t} \in [\frac{n_{1t}}{t}, \frac{n_{2t}}{t}]^c\Big\}\\
&\leq 8\varepsilon + \varepsilon + \varepsilon = 10\varepsilon,
\end{aligned}$$

for all $t \geq t_\varepsilon$ large enough since $N(t)/t \to \eta$ in probability. This proves (8.1.39) while (8.1.40) follows easily from (8.1.38), (8.1.39) and Slutzky's theorem. □

The modification of Theorem 8.1.8 needed to prove moment convergence in Anscombe's theorem for U-statistics is as follows.

COROLLARY 8.1.11. *Let $m \geq k$. Under the assumptions of Theorem 8.1.7 equation (8.1.26) can be extended to*

$$\begin{aligned} \mathbb{E} \sup_{n \geq m} \Big| \sum_{i_k = m}^{n} I_{T \geq i_k} \sum_{1 \leq i_1 < \cdots < i_{k-1} < i_k} f(X_{i_1}, \ldots, X_{i_k}) \Big|^p \\ \leq C_{k,p} \big(\mathbb{E} T^{\frac{kp}{2}} I_{T \geq m}\big)^{\frac{1}{k}} \big(\mathbb{E} T^{\frac{kp}{2}}\big)^{\frac{(k-1)}{k}} \big(\mathbb{E}|f(X_1, \ldots, X_k)|^{kp}\big)^{\frac{1}{k}}. \end{aligned} \tag{8.1.44}$$

PROOF. Replace $i_k = k$ by $i_k = m$ in the expression following (8.1.27), as well as $T^{\frac{p}{2}}$ by $T^{\frac{p}{2}} I_{T \geq m}$ in (8.1.28), and, in line (8.1.29), $(\mathbb{E} T^{\frac{kp}{2}})^{\frac{1}{k}}$ by $(\mathbb{E} T^{\frac{kp}{2}} I_{T \geq m})^{\frac{1}{k}}$. □

Next we give the key uniform integrability result required for moment convergence in Anscombe's theorem for U-statistics. It is well known [see Chow and Yu (1981, Lemma 5)] that, for a subset A of $(0, \infty)$ and a sequence of i.i.d. random variables $\{Z_i\}$, with $\mathbb{E} Z_1 = 0$ and $\mathbb{E}|Z_1|^p < \infty$ for some $p \geq 2$, if $\{T(a), a \in A\}$ is a family of stopping times for which $\{(a^{-1} T(a))^{\frac{p}{2}}, a \in A\}$ is uniformly integrable, then

$$\left\{ \left| \frac{\sum_{i=1}^{T(a)} Z_i}{\sqrt{a}} \right|^p, a \in A \right\} \quad \text{is uniformly integrable.} \tag{8.1.45}$$

This uniform integrability result extends to U-statistics as follows:

COROLLARY 8.1.12. *Under the assumptions of Theorem 8.1.7, let $p \geq 1$ and let A be a subset of $(0, \infty)$. Assume that $\mathbb{E}|f(X_1, \ldots, X_k)|^{kp} < \infty$ and let $\{T(a), a \in A\}$, be a family of stopping times adapted to the sequence $\{X_i\}$ such that $\{(a^{-1} T(a))^{\frac{kp}{2}}, a \in A\}$ is uniformly integrable. Then $\{|a^{-\frac{k}{2}} \mathcal{U}_{T(a)}|^p, a \in A\}$ is uniformly integrable, where $\mathcal{U}_n = \sum_{1 \leq i_1 < \cdots < i_k \leq n} f(X_{i_1}, \ldots, X_{i_k})$.*

PROOF. Set $m_{a,\lambda} = [\lambda a]$ and observe that

$$\begin{aligned} |\mathcal{U}_{T(a)}| \leq \max_{n \leq m_{a,\lambda}} \Big| \sum_{1 \leq i_1 < \cdots < i_k \leq n} f(X_{i_1}, \ldots, X_{i_k}) \Big| \\ + \Big| \sum_{i_k = m_{a,\lambda}}^{\infty} I_{T(a) \geq i_k} \sum_{1 \leq i_1 < \cdots < i_{k-1} < i_k} f(X_{i_1}, \ldots, X_{i_k}) \Big|. \end{aligned} \tag{8.1.46}$$

For $\varepsilon > 0$ we can pick λ large enough so that

$$\big[\mathbb{E}(a^{-1} T(a))^{\frac{kp}{2}} I_{T(a) \geq m_{a,\lambda}}\big]^{\frac{1}{k}} \leq \varepsilon \qquad \text{for all } a \in A.$$

Furthermore, $\sup_{a\varepsilon A} \mathbb{E}(a^{-1}T(a))^{\frac{kp}{2}} = K < \infty$. Therefore, (8.1.44) gives that, for all $a \in A$,

$$\mathbb{E}\Big| \sum_{i_k=m_{a.\lambda}}^{\infty} I_{T(a)\geq i_k} \sum_{1\leq i_1<\cdots<i_{k-1}<i_k} f(X_{i_1},\ldots,X_{i_k})\Big|^p \leq C_{k,p} K^{\frac{(k-1)}{k}} \big(\mathbb{E}|f(X_1,\ldots,X_k)|^{kp}\big)^{\frac{1}{k}} \varepsilon a^{\frac{kp}{2}}.$$

Using this, the uniform integrability of $\{|a^{-\frac{k}{2}} U_{T(a)}|\}^p$ will be established if one can prove uniform integrability of

$$\left\{\Big(\max_{n\leq m_{a.\lambda}}\Big| \sum_{1\leq i_1<\cdots<i_k\leq n} f(X_{i_1},\ldots,X_{i_k})\Big|/m_{a,\lambda}^{\frac{k}{2}}\Big)^p, a \in A\right\} \tag{8.1.47}$$

for all $\lambda > 0$. Now, the bound given in (8.1.25) with kp replacing p shows that the kth moments of the variables in (8.1.47) are uniformly bounded. Since $k > 1$, these variables are uniformly integrable, proving the corollary. □

Theorem 8.1.10 along with Corollary 8.1.12 give moment convergence in Anscombe's theorem for U-statistics as shown next.

COROLLARY 8.1.13. *Let $p \geq 2$. Under the assumptions of Theorem 8.1.7 and, in addition, the integrability condition*

$$\mathbb{E}|f(X_1,\ldots,X_k)|^{kp} < \infty, \tag{8.1.48}$$

if $T(a)$ is a stopping time adapted to the sequence $\{X_i\}$ such that, for some constant $c > 0$,

$$\lim_{a\to\infty} \mathbb{E}|a^{-1}T(a) - c|^{\frac{kp}{2}} = 0, \tag{8.1.49}$$

then

$$\lim_{a\to\infty} \mathbb{E}|a^{-\frac{k}{2}} \mathcal{U}_{T(a)}|^p = \mathbb{E}\big|c^{\frac{k}{2}} K_P(f)/\sqrt{k!}\big|^p, \tag{8.1.50}$$

where $K_P(f)$ is the Gaussian chaos variable given in Theorem 8.1.10.

PROOF. Condition 8.1.49 implies the uniform integrability of

$$\{|T(a)/a|^{kp/2} : 0 < a < \infty\}$$

which, by Corollary 8.1.12, in turn implies the uniform integrability of $\{|a^{-\frac{k}{2}} U_{T(a)}|^p\}$. This and Anscombe's Theorem for U-statistics (Theorem 8.1.10) give the result. □

8.2 A General Class of Exponential Inequalities for Martingales and Ratios

In this section we apply decoupling techniques to obtain exponential inequalities for the tail probability of the ratio of a martingale over its conditional variance. The results presented strictly extend several of the classical exponential inequalities for sums of independent random variables and martingales. In summary, what we prove is that, when extending results for sums of independent random variables to martingales and to the ratio of a martingale over its conditional variance, one should replace the variance by the conditional variance and the exponential of a function of the variance by the expectation of the exponential of the same function of the conditional variance (see Lemma 4.1.9). Let $\{d_i\}$ be a martingale difference sequence. The typical approach for obtaining exponential inequalities for $M_n = \sum_{i=1}^n d_i$ begins as follows. For $\lambda > 0$,

$$\Pr\{\sum_{i=1}^n d_i \geq x\} = \Pr\Big\{\exp(\lambda \sum_{i=1}^n d_i) \geq \exp(\lambda x)\Big\}$$
$$\leq \inf_{\lambda>0} \exp(-\lambda x)\mathbb{E}\exp\Big(\lambda \sum_{i=1}^n d_i\Big).$$

Corollary 6.2.5 (with $g = 1$) can be used here to bound the above expression by

$$\sqrt{\inf_{\lambda>0} \exp(-2\lambda x)\mathbb{E}\exp\Big(2\lambda \sum_{i=1}^n e_i\Big)}, \qquad (8.2.1)$$

where $\{e_i\}$ is a decoupled version of $\{d_i\}$ and hence conditionally independent given $\mathcal{G} = \sigma(\{d_i\})$. It is natural to conjecture that, by conditioning on $\mathcal{G}$, one could use the above to extend results for sums of independent random variables to martingales as was the case in extending Rosenthal's inequality for sums of independent random variables to martingales in Chapter 7. As it stands, without modification the extension is not possible since (8.2.1) does not carry enough information on the d_is to allow for this to happen. However, use of the full strength of Corollary 6.2.5, including appropriate constraints on $\{d_i\}$ and, in particular, information on the conditional variance of $\sum_{i=1}^n d_i$, through the random variable g of that corollary, allows one to develop a unifying theory of exponential inequalities for martingales and the ratio of a martingale over its conditional variance.

REMARK 8.2.1. In this section we will use the following notation. If X is a positive random variable on the probability space $(\Omega, \mathcal{F}, P)$ and A is an $\mathcal{F}$-measurable set, then $\mathbb{E}(X|A) = \int_A X dP / P\{A\}$.

We begin with extensions of Bernstein's and Bennet's inequalities (Lemma 4.1.9 and Remark 4.1.11).

THEOREM 8.2.2. *Let $\{d_i, \mathcal{F}_i\}$ be a martingale difference sequence with $M_n = \sum_{i=1}^n d_i$. Assume that $\mathbb{E}(d_j|\mathcal{F}_{j-1}) = 0$ and $\mathbb{E}(d_j^2|\mathcal{F}_{j-1}) = \sigma_j^2 < \infty$, and set $V_n^2 = \sum_{j=1}^n \sigma_j^2$. Furthermore, assume that there exist a constant $c \in (0, \infty)$ such that, almost surely, $\mathbb{E}(|d_j|^k|\mathcal{F}_{j-1}) \leq (k!/2)\sigma_j^2 c^{k-2}$ or $\Pr\{|d_j| \leq c|\mathcal{F}_{j-1}\} = 1$ for all $k > 2$. Then, for all $x, y > 0$,*

$$\Pr\Big\{\sum_{i=1}^n d_i > x,\, V_n^2 \leq y \text{ for some } n\Big\} \leq \exp\Bigg(-\frac{x^2}{y(1+\sqrt{1+\frac{2cx}{y}})+cx}\Bigg)$$

$$\leq \exp\Big(-\frac{x^2}{2(y+cx)}\Big). \tag{8.2.2}$$

PROOF. We will use the exponential inequality for sums of independent random variables given in Lemma 4.1.9, which gives that if ξ_i are i.i.d., mean zero, bounded by c and with sum of variances equal to v_n^2, then, for all $0 < r < 1/c$,

$$\mathbb{E}\exp\Big(r\sum_{i=1}^n \xi_i\Big) \leq \exp\Big(\frac{v_n^2 r^2}{2(1-cr)}\Big). \tag{8.2.3}$$

Let $\mathcal{G} = \sigma(\{d_i\})$ and let τ be the stopping time defined as $\tau = \min\{n : M_n > x\}$ if the minimum exists and $\tau = \infty$ otherwise. Taking

$$A = \{M_n \geq x \text{ and } V_n^2 \leq y \text{ for some } n\},$$

we have that $\tau < \infty$, $M_\tau > x$ and $V_\tau^2 \leq y$ on A. Moreover, observe that $I_A \leq I_{\tau<\infty} I_A \leq I_A I_{\tau<\infty} I_{M_\tau > x}$ implies $\Pr\{A\} \leq \Pr\{M_\tau > x, A\}$. Applying Markov's inequality, Fatou's lemma (valid since $\tau < \infty$ on A) and Corollary 6.2.5 with $g = \exp(-\frac{\lambda}{2}x) I_{M_\tau > x, A}$ we get,

$$P\{A\} \leq \Pr\Big\{\sum_{i=1}^\tau d_i > x,\, A\Big\} \leq \inf_{\lambda>0} \mathbb{E}\exp\Big(\frac{\lambda}{2}\Big(\sum_{i=1}^\tau d_i - x\Big)\Big) I_{M_\tau > x, A}$$

$$= \inf_{\lambda>0} \mathbb{E} \liminf_n \exp\Big(\frac{\lambda}{2}\Big(\sum_{i=1}^{\tau\wedge n} d_i - x\Big)\Big) I_{M_\tau > x, A}$$

$$\leq \inf_{\lambda>0} \lim_{n\to\infty}\inf \mathbb{E}\exp\Big(\frac{\lambda}{2}\Big(\sum_{i=1}^{\tau\wedge n} d_i - x\Big)\Big) I_{M_\tau > x, A}$$

$$\leq \inf_{\lambda>0} \lim_{n\to\infty} \sqrt{\mathbb{E}\exp\Big(\lambda\Big(\sum_{i=1}^{\tau\wedge n} e_i - x\Big)\Big) I_{M_\tau > x, A}}$$

$$= \inf_{\lambda>0} \lim_{n\to\infty} \sqrt{\mathbb{E}\Big[I_{M_\tau > x, A} \exp(-\lambda x)\mathbb{E}\Big(\exp\{\lambda\sum_{i=1}^{\tau\wedge n} e_i\}|\mathcal{G}\Big)\Big]},$$

where the last equality follows since $I_{M_\tau > x, A}$ is $\mathcal{G}$ measurable. Observe that, since $\{d_i\}$ and $\{e_i\}$ are tangent and $\{e_i\}$ is conditionally independent given $\mathcal{G}$, the moment assumptions on the distribution of d_i transfer to conditions on the e_is and therefore we can apply (8.2.3) to obtain,

$$\mathbb{E}\Big(\exp(\lambda\sum_{i=1}^{\tau\wedge n} e_i)\big|\mathcal{G}\Big) \le \exp\big(h(\lambda)V_{\tau\wedge n}^2\big), \tag{8.2.4}$$

where $h(\lambda) = \frac{\lambda^2}{2(1-\lambda c)}$. Replacing this in the above bound, one obtains

$$\Pr\Big\{\sum_{i=1}^{\tau} d_i > x, A\Big\} \le \inf_{\lambda>0}\lim_{n\to\infty}\sqrt{\mathbb{E}\exp\big(-(\lambda x - h(\lambda)V_{\tau\wedge n}^2)\big)I_{M_\tau>x,A}}.$$

Since the variable inside the expectation is dominated by

$$\exp\big(-(\lambda x - h(\lambda)V_\tau^2)\big)I_{M_\tau>x,A},$$

and $V_\tau \le y$ on A, the dominated convergence theorem yields

$$\Pr\Big\{\sum_{i=1}^{\tau} d_i > x, A\Big\} \le \inf_{\lambda>0}\sqrt{\mathbb{E}\exp\big(-(\lambda x - h(\lambda)V_\tau^2)\big)I_{M_\tau>x,A}}.$$

Dividing both sides by $\sqrt{\Pr\big\{M_\tau > x, A\}}$ gives

$$\Pr\{\sum_{i=1}^{\tau} d_i > x, A\} \le \inf_{\lambda>0}\mathbb{E}\Big[\exp\big(-(\lambda x - h(\lambda)V_\tau^2)\big)\big|(M_\tau > x, A)\Big].$$

Then, since $\tau < \infty$, $M_\tau > x$ and $V_\tau^2 \le y$ on A, we have

$$\Pr\Big\{\sum_{i=1}^{n} d_i > x, V_n^2 \le y \text{ for some } n\Big\} \le \inf_{\lambda>0}\exp\big(-(\lambda x - h(\lambda)y)\big). \tag{8.2.5}$$

The proof is completed by minimizing the above expression. □

In the special case of sums of independent random variables setting $y = V_n^2$, one gets exactly Lemma 4.1.9. Letting $y = \|V_n\|_\infty^2$ gives the usual result for martingales. In cases where typically V_n^2 is much smaller than $\|V_n\|_\infty^2$ the replacement of the conditional variance by its L_∞-norm might be providing a skewed assessment of the size of $\sum_{i=1}^n d_i$. In these instances, a closer measure of the size of $\sum_{i=1}^n d_i$ in relation to V_n^2 is given by the following result.

COROLLARY 8.2.3. *Let $\{d_i, \mathcal{F}_i\}$ satisfy the same hypotheses as in Theorem 8.2.2, but now let V_n^2 denote either of $\sum_{j=1}^n \sigma_j^2$ or $\|\sum_{i=1}^n \sigma_j^2\|_\infty$. Then, for all $\mathcal{F}_\infty$ measurable sets A, $x > 0$,*

$$\Pr\Big\{\frac{\sum_{i=1}^n d_i}{V_n^2} > x, A\Big\} \le \mathbb{E}\Big[\exp\Big(-\Big(\frac{x^2}{1+cx+\sqrt{1+2cx}}\Big)V_n^2\Big)\Big|\frac{M_n}{V_n^2} > x, A\Big], \tag{8.2.6}$$

$$\Pr\left\{\frac{\sum_{i=1}^n d_i}{V_n^2} > x,\, A\right\} \le \sqrt{\mathbb{E}\exp\Big(-\Big(\frac{x^2}{1+cx+\sqrt{1+2cx}}\Big)V_n^2\Big)1_A}, \tag{8.2.7}$$

and

$$\Pr\left\{\frac{\sum_{i=1}^n d_i}{V_n^2} > x,\, \frac{1}{V_n^2} \le y \text{ for some } n\right\} \le \exp\Big(-\frac{1}{y}\Big(\frac{x^2}{1+cx+\sqrt{1+2cx}}\Big)\Big)$$
$$\le \exp\Big(-\frac{x^2}{2y(1+cx)}\Big). \tag{8.2.8}$$

PROOF. Applying Markov's inequality first followed by a use of Corollary 6.2.5 with $g = \exp(-\frac{\lambda}{2}V_n^2 x)I_{M_n > V_n^2 x, A}$, $\mathcal{G} = \sigma(\{d_i\})$, we obtain

$$\Pr\Big\{\sum_{i=1}^n d_i > V_n^2 x,\, A\Big\} \le \inf_{\lambda>0} \mathbb{E}\exp\Big(\frac{\lambda}{2}\big(\sum_{i=1}^n d_i - V_n^2 x\big)\Big) I_{M_n > V_n^2 x, A}$$
$$\le \inf_{\lambda>0}\sqrt{\mathbb{E}\exp\Big(\lambda\big(\sum_{i=1}^n e_i - V_n^2 x\big)\Big)I_{M_n > V_n^2 x, A}}$$
$$= \inf_{\lambda>0}\sqrt{\mathbb{E}\Big[I_{M_n > V_n^2 x, A}\exp(-\lambda V_n^2 x)\mathbb{E}\Big(\exp\big(\lambda\sum_{i=1}^n e_i\big)\big|\mathcal{G}\Big)\Big]},$$

where the last equality follows since the variables outside the conditional expectation are $\mathcal{G}$ measurable. Using (8.2.4) above one obtains

$$\Pr\Big\{\sum_{i=1}^n d_i > V_n^2 x,\, A\Big\} \le \inf_{\lambda>0}\sqrt{\mathbb{E}\exp\Big(-\Big(\lambda x - \frac{\lambda^2}{2(1-\lambda c)}\Big)V_n^2\Big)I_{M_n > V_n^2 x, A}}$$
$$\le \sqrt{\mathbb{E}\exp\Big(-\Big(\frac{x^2}{1+cx+\sqrt{1+2cx}}\Big)V_n^2\Big)I_{M_n > V_n^2 x, A}}. \tag{8.2.9}$$

Dividing both sides by $\sqrt{\Pr\{M_n > V_n^2 x, A\}}$ gives (8.2.6), (8.2.7) is an easy consequence of (8.2.9).

Equation (8.2.8) is obtained by adapting the stopping time argument already used in the previous theorem. Let $A = \big\{\frac{M_n}{V_n} > x,\, \frac{1}{V_n^2} \le y \text{ for some } n\big\}$. This is possible since Corollary 6.2.5 works when we replace n by $\tau \wedge n$, where

$$\tau = \inf\Big\{n \ge 1 : \frac{\sum_{i=1}^n d_i}{V_n^2} > x\Big\}. \tag{8.2.10}$$

We have that $\tau < \infty$, $\frac{M_\tau}{V_\tau^2} > x$ and $\frac{1}{V_\tau^2} \le y$ on A. Moreover, observe that $I_A \le I_{\tau<\infty} I_A \le I_A I_{\tau<\infty} I_{\frac{M_\tau}{V_\tau^2} > x}$ implies $\Pr\{A\} \le \Pr\{\frac{M_\tau}{V_\tau^2} > x, A\}$. Applying Markov's inequality, Fatou's lemma (valid since $\tau < \infty$ on A) and Corollary 6.2.5 with $g = \exp\big(-\frac{\lambda}{2} x V_{\tau\wedge n}^2\big) I_{\frac{M_\tau}{V_\tau^2} > x, A}$ we get,

$$
\begin{aligned}
P\{A\} &\le \Pr\Big\{\frac{M_\tau}{V_\tau^2} > x,\, A\Big\} \le \inf_{\lambda>0} \mathbb{E}\exp\Big(\frac{\lambda}{2}(M_\tau - xV_\tau^2)\Big) I_{\frac{M_\tau}{V_\tau^2} > x, A} \\
&= \inf_{\lambda>0} \mathbb{E} \liminf_n \exp\Big(\frac{\lambda}{2}(M_{\tau\wedge n} - xV_{\tau\wedge n}^2)\Big) I_{\frac{M_\tau}{V_\tau^2} > x, A} \\
&\le \inf_{\lambda>0} \lim \inf_{n\to\infty} \mathbb{E}\exp\Big(\frac{\lambda}{2}(M_{\tau\wedge n} - xV_{\tau\wedge n}^2)\Big) I_{\frac{M_\tau}{V_\tau^2} > x, A} \\
&\le \inf_{\lambda>0} \lim_{n\to\infty} \sqrt{\mathbb{E}\exp\Big(\lambda\Big(\sum_{i=1}^{\tau\wedge n} e_i - xV_{\tau\wedge n}^2\Big)\Big) I_{\frac{M_\tau}{V_\tau^2} > x, A}} \\
&= \inf_{\lambda>0} \lim_{n\to\infty} \sqrt{\mathbb{E}\Big[I_{\frac{M_\tau}{V_\tau^2} > x, A} \exp\big(-\lambda x V_{\tau\wedge n}^2\big)\mathbb{E}\big(\exp\{\lambda \sum_{i=1}^{\tau\wedge n} e_i\} | \mathcal{G}\big)\Big]},
\end{aligned}
$$

where the last equality follows since $I_{\frac{M_\tau}{V_\tau^2} > x, A}$ is $\mathcal{G}$ measurable. Replacing (8.2.4) above one obtains

$$
\Pr\Big\{\frac{M_\tau}{V_\tau^2} > x,\, A\Big\} \le \inf_{\lambda>0} \lim_{n\to\infty} \sqrt{\mathbb{E}\exp\big(-(\lambda x - h(\lambda))V_{\tau\wedge n}^2\big) I_{\frac{M_\tau}{V_\tau^2} > x, A}}.
$$

The minimizing λ is such that $(\lambda x - h(\lambda)) > 0$. For this λ, the variable inside the expectation is dominated by $I_{\frac{M_\tau}{V_\tau^2} > x, A}$, and $\frac{1}{V_\tau^2} \le y$ on A, the dominated convergence theorem yields

$$
\Pr\Big\{\frac{M_\tau}{V_\tau^2} > x,\, A\Big\} \le \inf_{\lambda>0} \sqrt{\mathbb{E}\exp\big(-(\lambda x - h(\lambda))V_\tau^2\big) I_{\frac{M_\tau}{V_\tau^2} > x, A}}.
$$

Dividing both sides by $\sqrt{\Pr\big\{\frac{M_\tau}{V_\tau^2} > x,\, A\}}$ gives

$$
\Pr\{\frac{M_\tau}{V_\tau^2} > x,\, A\} \le \inf_{\lambda>0} \mathbb{E}\Big[\exp\big(-(\lambda x - h(\lambda))V_\tau^2\big)\Big| \frac{M_\tau}{V_\tau^2} > x,\, A\Big].
$$

Then, since $\tau < \infty$, $\frac{M_\tau}{V_\tau^2} > x$ and $\frac{1}{V_\tau^2} \le y$ on A, we have

$$
\Pr\Big\{\frac{\sum_{i=1}^n d_i}{V_n^2} > x,\, \frac{1}{V_n^2} \le y \text{ for some } n\Big\} \le \inf_{\lambda>0} \exp\big(-\frac{1}{y}(\lambda x - h(\lambda))\big).
$$

The proof is completed by minimizing the above expression. □

If one has a mean zero martingale, letting $V_n^2 = \| \sum_{i=1}^{\infty} \sigma_i^2 \|_{\infty}$ (a finite constant) in (8.2.6) or (8.2.8), a change of variables gives,

$$\Pr\Big\{\sum_{i=1}^{n} d_i > x\Big\} \leq \exp\Big(-\frac{x^2}{V_n^2(1+\sqrt{1+\frac{2cx}{V_n^2}})+cx}\Big) \\ \leq \exp\Big(\frac{-x^2}{2(V_n^2+cx)}\Big), \tag{8.2.11}$$

which gives Bennet's inequality [equation (4.1.19)] and Bernstein's inequality [equation (4.1.18)] in the case of sums of independent mean zero random variables. Therefore, the decoupling approach provides extensions of the typical exponential inequalities for sums of independent random variables and martingales to the case of a martingale over its conditional variance.

In the case of sums of *uniformly bounded* independent random variables, Prohorov's inequality (see, e.g., Stout, 1974, page 262) is sharper than Bernstein's, and is best possible. It also extends to martingales whenever the martingale differences are uniformly bounded:

THEOREM 8.2.4. *Let $\{d_i, \mathcal{F}_i\}$, σ_j^2 and V_n^2 be as in Corollary 8.2.3 and let us further assume $|d_j| \leq c$, for some $0 < c < \infty$. Then, for all $x, y > 0$,*

$$\Pr\Big\{\sum_{i=1}^{n} d_i > x,\ V_n^2 \leq y \text{ for some } n\Big\} \leq \exp\Big(-\frac{x}{2c}\operatorname{arc\,sinh}\frac{cx}{2y}\Big). \tag{8.2.12}$$

PROOF. The proof is very similar to the one given in Theorem 8.2.2. We start with some preliminary facts used to get a bound analogous to (8.2.3) which we will use in the proof.

Let $\{X_i\}$ be a sequence of independent random variables with $\mathbb{E}X_i = 0$, $\mathbb{E}X_i^2 = \sigma_i^2 < \infty$. Let $S_n = \sum_{i=1}^{n} X_i$, and $v_n^2 = \sum_{i=1}^{n} \sigma_i^2$, and assume $|X_i| \leq c$ for all i. Then

$$\mathbb{E}\exp\Big(\lambda\sum_{i=1}^{n} X_i\Big) \leq \exp\Big(\frac{\lambda}{c}\sinh(\lambda c)v_n^2\Big). \tag{8.2.13}$$

To see this, by independence, we only need to check that

$$\mathbb{E}\exp(\lambda X_n) \leq \exp\left(\sigma_n^2 \cdot \frac{\lambda}{c}\sinh(\lambda c)\right).$$

Using the inequalities

$$x \leq \exp(x-1) \quad \text{and} \quad e^x - x - 1 \leq e^x + e^{-x} - 2 \leq x\sinh x,$$

which are valid for all real x, gives,

$$\begin{aligned}\mathbb{E}\exp(\lambda X_n) &\le \exp\big(\mathbb{E}\exp(\lambda X_n) - 1\big) \\ &= \exp\big(\mathbb{E}[\exp(\lambda X_n) - \lambda X_n - 1]\big) \quad (\text{since } \mathbb{E}X_i = 0) \\ &\le \exp\big(\mathbb{E}\lambda X_n \sinh \lambda X_n\big) \\ &\le \exp\big(\mathbb{E}(\lambda|X_n| \sinh \lambda|X_n|)\big) \\ &= \exp\Big(\mathbb{E}\lambda^2 X_n^2 \frac{\sinh \lambda|X_n|}{\lambda|X_n|}\Big) \\ &\le \exp\Big(\mathbb{E}X_n^2 \cdot \frac{\lambda}{c} \sinh \lambda c\Big),\end{aligned}$$

completing the proof of (8.2.13).

Let now $\mathcal{G} = \sigma(\{d_i\})$ and let τ be the stopping $\tau = \min\{n : M_n > x\}$ if the minimum exists and $\tau = \infty$ otherwise. Then, using (8.2.13) and the conditional independence of $\{e_i\}$ it is easy to see that the following bound holds:

$$\mathbb{E}\Big(\exp(\lambda \sum_{i=1}^{\tau\wedge n} e_i)\big|\mathcal{G}\Big) \le \exp\big(\frac{\lambda}{c}\sinh(\lambda c) V_{\tau\wedge n}^2\big). \tag{8.2.14}$$

The proof of Theorem 8.2.4 proceeds exactly as the proof of Theorem 8.2.2, replacing (8.2.4) by (8.2.14) and taking $h(\lambda) = \frac{\lambda}{c}\sinh(\lambda c)$ in (8.2.5) to obtain,

$$\Pr\Big\{\sum_{i=1}^n d_i > x,\ V_n^2 \le y \text{ for some n}\Big\} \le \inf_{\lambda>0} \exp\Big(-\big(\lambda x - \frac{\lambda}{c}\sinh(\lambda c) y\big)\Big).$$

The minimizing λ is $\lambda_0 = \frac{1}{c}\operatorname{arc\,sinh}\frac{cx}{2y}$ which gives $\frac{x}{2} = \frac{y}{c}\sinh\lambda_0 c$, completing the poof of the theorem. □

For normalized sums we can add a twist to the inequality by replacing V_n^2 by the linear combination $\alpha + \beta V_n^2$:

THEOREM 8.2.5. *Under the same hypotheses as Theorem 8.2.4, we have that, for all $\mathcal{F}_\infty$-measurable sets A and all $\alpha,\ \beta,\ x \ge 0$,*

$$\begin{aligned}&\Pr\Big\{\frac{\sum_{i=1}^n d_i}{\alpha + \beta V_n^2} > x,\ A\Big\} \\ &\quad\le \exp\Big(-\frac{\alpha x}{c}\operatorname{arc\,sinh}\frac{\beta c x}{2}\Big) \\ &\qquad\times \mathbb{E}\Big[\exp\Big(-\big(\frac{\beta x}{2c}\operatorname{arc\,sinh}\frac{\beta c x}{2}\big)V_n^2\Big)\big|M_n > (\alpha + \beta V_n^2)x,\ A\Big],\end{aligned} \tag{8.2.15}$$

and

$$\begin{aligned}&\Pr\Big\{\frac{\sum_{i=1}^n d_i}{\alpha + \beta V_n^2} > x,\ A\Big\} \\ &\quad\le \sqrt{\exp\Big(-\frac{\alpha x}{c}\operatorname{arc\,sinh}\frac{\beta c x}{2}\Big)\mathbb{E}\exp\Big(-\big(\frac{\beta x}{2c}\operatorname{arc\,sinh}\frac{\beta c x}{2}\big)V_n^2\Big)1_A}.\end{aligned} \tag{8.2.16}$$

Moreover,

$$\Pr\Big\{\frac{\sum_{i=1}^n d_i}{\alpha+\beta V_n^2} > x,\ \frac{1}{V_n^2}\le y \text{ for some } n\Big\} \le \exp\Big(-\frac{\alpha x}{c}\operatorname{arc\,sinh}\frac{\beta c x}{2}\Big)\exp\Big(-\frac{\beta c x}{2y}\operatorname{arc\,sinh}\frac{\beta c x}{2}\Big). \tag{8.2.17}$$

PROOF. Using Markov's inequality and Corollary 6.2.5 we obtain,

$$\Pr\Big\{\sum_{i=1}^n d_i > (\alpha+\beta V_n^2)x,\ A\Big\}$$

$$\le \inf_{\lambda>0}\exp\Big(\frac{\lambda}{2}\alpha x\Big)\mathbb{E}\exp\Big(\frac{\lambda}{2}\big(\sum_{i=1}^n d_i - V_n^2\beta x\big)\Big)I_{M_n>(\alpha+\beta V_n^2)x,A}$$

$$\le \inf_{\lambda>0}\sqrt{\exp(-\lambda\alpha x)\mathbb{E}\exp\Big(\lambda\big(\sum_{i=1}^n e_i - V_n^2\beta x\big)\Big)I_{M_n>(\alpha+\beta V_n^2)x,A}}$$

$$\le \inf_{\lambda>0}\sqrt{\exp(-\lambda\alpha x)\mathbb{E}\Big(\mathbb{E}\big[\exp([\frac{\lambda}{c}\sinh\lambda c-\lambda\beta x]V_n^2)I_{M_n>(\alpha+\beta V_n^2)x,A}|\mathcal{G}\big]\Big)}$$

[conditioning on $\mathcal{G}$ and using (8.2.13)]

$$\le \sqrt{\exp(-\frac{\alpha x}{c}\operatorname{arc\,sinh}\frac{\beta c x}{2})\mathbb{E}\Big(\mathbb{E}\big[\exp(-\frac{V_n^2\beta x}{2c}\operatorname{arc\,sinh}\frac{\beta c x}{2})I_{M_n>(\alpha+\beta V_n^2)x,A}|\mathcal{G}\big]\Big)},$$

In the above, $\lambda_0=\frac{1}{c}\operatorname{arc\,sinh}\frac{\beta c x}{2}$ gives $\frac{\beta x}{2}=\frac{\sinh(\lambda_0\cdot c)}{c}$. The proof of (8.2.15) is completed by removing the conditional expectation and dividing both sides above by

$$\sqrt{\Pr\{M_n>(\alpha+\beta V_n^2)x,\ A\}}.$$

Equation (8.2.16) is an easy consequence of (8.2.15). As for (8.2.17), the proof is obtained as in Theorem 8.2.2, letting

$$A=\Big\{\frac{\sum_{i=1}^n d_i}{\alpha+\beta V_n^2} > x,\ \frac{1}{V_n^2}\le y \text{ for some } n\Big\},$$

and

$$\tau=\min\Big\{n:\frac{M_n}{\alpha+\beta V_n^2}>x\Big\}$$

if the minimum exists and $\tau=\infty$ otherwise. □

In the special case $\alpha=0$ and $\beta=1$, and $A=\frac{1}{V_n^2}\le y$ (8.2.15) gives,

$$\Pr\Big\{\frac{\sum_{i=1}^n d_i}{V_n^2} > x,\ \frac{1}{V_n^2}\le y\Big\} \le \exp\Big(-\Big(\frac{x}{2yc}\operatorname{arc\,sinh}\frac{cx}{2}\Big)\Big). \tag{8.2.18}$$

A further reduction to the case $V_n^2 = \| \sum_{i=1}^n \sigma_i^2 \|_\infty^2 = \frac{1}{y}$, with x replaced by $\frac{x}{V_n^2}$ gives,

$$\Pr\Big\{\sum_{i=1}^n d_i > x\Big\} \le \exp\Big(-\Big(\frac{x}{2c} \operatorname{arc\,sinh} \frac{cx}{2\| \sum_{i=1}^n \sigma_i^2 \|_\infty}\Big)\Big), \tag{8.2.19}$$

which is a special case of (8.2.12). We finish this section by stressing the fact that the technique introduced in this section can be used to extend a wide class of exponential inequalities going beyond problems involving martingales. We refer the reader to the references for these extensions.

8.3 References for Chapter 8

For references to Wald's equations see Chapter 2. Theorem 8.1.1 comes from de la Peña and Lai (1997). Theorem 8.1.2 and Theorem 8.1.3 come from de la Peña and Lai (1997a). They can be traced back to Chow, de la Peña and Teicher (1993) and de la Peña (1992b) which treated the case of multilinear forms. Lemma 8.1.4 dates back to Brillinger (1962). Lemma 8.1.5 can be traced back to de la Peña (1992). In the case of bi-linear forms it is found in de la Peña and Klass (1994). Theorem 8.1.6 and its proof based on decoupling come from Giné and Zinn (1992a) (see also Lai and Wang, 1993, for another proof). Theorem 8.1.7 is from de la Peña and Lai (1997a). Theorem 8.1.8 is a version of Anscombe's theorem due to Rényi (1957) (see also Chung (1974) and Gut (1988)). Theorem 8.1.9 is the general version of Ancombe's (1952) theorem. See also Ghosh, Mukhopadhyay and Sen (1997) for several applications. Theorem 8.1.10 is Anscombe's theorem for degenerate U-statistics and we do not have an exact reference for this result. Corollary 8.1.11, Corollary 8.1.12 aned Corollary 8.1.13 come from de la Peña and Lai (1997a).

Theorem 8.2.2 in the case of independent random variables is basically due to Bernstein [see Bennet (1962)] and Bennet (1962) and in the martingale case by Pinelis and Utev (1989) and Pinelis (1994) (in these two papers the exponents need to be fixed). See also Freedman (1975). Theorem 8.2.2 (which is probably available elsewhere) and Corollary 8.2.3 come from de la Peña (1997). Theorem 8.2.4 is a slight extension of the exponential inequality of Hitczenko (1990a) and Levental (1989) (equation (8.2.19)) that can be found in de la Peña (1997). Theorem 8.2.5 comes from de la Peña (1997) which develops the technique for extending exponential inequalities used in here. Moreover, that paper contains several other exponential inequalities including results for continuous time martingales and for the sum of conditionally symmetric random variables over its sum of squares without any integrability assumptions.

References

Alexander, K. (1987). The central limit theorem for empirical processes over Vapnik–Červonenkis classes. *Ann. Probab. 15*, 178–203.

Andersen, N. T. and Dobrić, V. (1987). The central limit theorem for stochastic processes. *Ann. Probab. 15*, 164–177.

Anscombe, F. J. (1952). Large-sample theory of sequential estimation. *Proc. Cambridge Philos. Soc. 48*, 600–607.

Araujo, A. and Giné, E. (1980). *The Central Limit Theorem for Real and Banach Valued Random Variables*. Wiley, New York.

Arcones, M. A. (1993). The law of the iterated logarithm for U-processes. *J. Mult. Analysis 47*, 139–151.

Arcones, M. A. (1995). A Bernstein type inequality for U-statistics and U-processes. *Statist. Probab. Letters 22*, 223–230.

Arcones, M. A.; Chen, Z. and Giné, E. (1994). Estimators related to U-processes with applications to multivariate medians: asymptotic normality. *Ann. Statist. 22*, 1460–1477.

Arcones, M. A. and Giné, E. (1992). On the bootstrap of U and V-statistics. *Ann. Statist. 20*, 655–674.

Arcones, M. A. and Giné, E. (1993). Limit theorems for U-processes. *Ann. Probab. 21*, 1494–1542.

Arcones, M. A. and Giné, E. (1993a). On decoupling, series expansions, and tail behavior of chaos processes. *J. Theoret. Probab. 6*, 101–122.

Arcones, M. A. and Giné, E. (1994). U-processes indexed by Vapnik–Červonenkis classes of functions with applications to asymptotics and bootstrap of U-statistics with estimated parameters. *Stoch. Proc. Appl. 52*, 17–38.

Arcones, M. A. and Giné, E. (1995). On the law of the iterated logarithm for canonical U-statistics and processes. *Stoch. Proc. Appl. 58*, 217–245.

Azéma, J.; Gundy, R. F. and Yor, M. (1979). Sur l'intégrabilité uniforme des martingales continues, Sém. de Probab. XIV, 1978–79. *Lecture Notes in Math. 784*, 53–61. Springer-Verlag, Berlin.

Beckner, W. (1975). Inequalities in Fourier Analysis. *Ann. Math. 102*, 159–182.

Bennett, G. (1962). Probability inequalities for sums of independent random variables. *J. Amer. Stat. Assoc. 57*, 33–45.

Berk, R. H. (1966). Limiting behavior of posterior distributions when the model is incorrect. *Ann. Math. Statist. 37*, 51–58.

Blackwell, D. (1946). On an equation of Wald. *Ann. Math. Stat. 17*, 84–87.

Bonami, A. (1970). Étude des coefficients de Fourier des fonctions de $L^p(G)$. *Ann. Inst. Fourier 20*, 335–402.

Borell, C. (1979). On the integrability of Banach space valued Walsh polynomials. *Séminaire de Probabilités XIII, Lecture Notes in Math. 721*, 1–3. Springer, Berlin.

Borell, C. (1984). On polynomial chaos and integrability. *Probab. Math. Statist. 3*, 191–203.

Borovskikh, Yu. V. (1996). *U-statistics in Banach Spaces*. VSP, Utrecht, The Netherlands.

Borovskikh, Yu. V. and Korolyuk, V. S. (1994). *Theory of U-statistics*. Kluwer, Dordrecht, The Netherlands.

Bourgain, J. and Tzafriri, L. (1987). Invertibility of "large" submatrices with applications to the geometry of Banach spaces and harmonic analysis. *Israel J. Math. 57*, 137–224.

Bretagnolle, J. (1983). Lois limites du bootstrap de certaines fonctionelles. *Ann. Inst. H. Poincaré, Sect. B., 3*, 256–261.

Brillinger, D. R. (1962). A note on the rate of convergence of a mean. *Biometrika 49*, 574–576.

Brown, B. M. (1971). Martingale central limit theorems. *Ann. Math. Statist. 42*, 59–66.

Brown, B. M. and Eagleson, G. K. (1971). Martingale convergence to infinitely divisible laws with finite variances. *Trans. Amer. Math. Soc. 162*, 449–453.

Burkholder, D. L. (1973). Distribution function inequalities for martingales. *Ann. Probab. 1*, 19–42.

Burkholder, D. L. (1979). Weak inequalities for exit times and analytic functions. *Probability Theory Banach Center Publications 5*. PWN-Polish Scientific Publishers, Warsaw.

Burkholder, D. L. (1981). A geometric characterization of Banach spaces in which martingale difference sequences are unconditional. *Ann. Probab. 9*, 997–1011.

Burkholder, D. L. (1983). A geometric condition that implies the existence of certain singular integrals of Banach-space-valued functions. *Conference on harmonic analysis in honor of Antoni Zygmund, Vol. I, II (Chicago, Ill., 1981)* 270–286. Wadsworth Math. Ser., Wadsworth, Belmont, Calif.

Burkholder, D. L. (1988). Sharp inequalities for martingales and stochastic integrals. *Astérisque 157–158*, 75–94.

Burkholder, D. L. (1989). Differential subordination of harmonic functions and martingales. Proceedings of the seminar on harmonic analysis and partial differential equations (El Escorial, Spain). *Lecture Notes in Math. 1384*, 1–23.

Burkholder, D. L.; Davis, B. and Gundy, R. F. (1972). Inequalities for convex functions of quasi-linear operators on martingales. *Proc. of the Sixth Berkeley Symp. on Mathematical Statistics and Probability* 2, 223–240.

Burkholder, D. L. and Gundy, R. F. (1970). Extrapolation and interpolation of quasi-linear operators on martingales. *Acta Math. 124*, 249–304.

Carl, A. and Pajor, A. (1988). Gelfand numbers of operators in a Hilbert space. *Inventiones Math. 94*, 479–504.

Chen, Z. (1995). Bounds for the breakdown point of the simplicial median. *J. Multivariate Anal. 55*, 1–13.

Chow, Y. S. and Teicher, H. (1988). *Probability Theory: Independence, Interchangeability, Martingales*. Springer-Verlag, New York.

Chow, Y. S., de la Peña, V. H. and Teicher, H. (1993). Wald's lemma for a class of de-normalized U-statistics. *Ann. of Probab. 21*, 1151–1158.

Chow, Y. S.; Robbins, H. and Siegmund, D. (1971). *Great Expectations: The Theory of Optimal Stopping*. Houghton-Mifflin, Boston.

Chung, K. L. (1974). *A Course in Probability Theory*. 2nd ed. Academic Press, New York.

Cohn, D. L. (1980). *Measure Theory*. Birkhäuser, Boston.

Cuzick, J., Giné, E., Zinn, J. (1995). Laws of large numbers for quadratic forms, maxima of products and truncated sums of i.i.d. random variables. *Ann. Probab. 23*, 292–333.

de Acosta, A. (1987). A decoupling inequality for multilinear forms of stable vectors. *Probab. Math. Statist. 8*, 71–76.

Dehling, H. (1989). Complete convergence of triangular arrays and the law of the iterated logarithm for degenerate U-statistics. *Stat. Probab. Letters 7*, 319–321.

Dehling, H. (1989a). The functional law of the iterated logarithm for von Mises functionals and multiple Wiener integrals. *J. Multivariate Analysis 28*, 177–189.

Dehling, H.; Denker, M. and Philipp, W. (1986). A bounded law of the iterated logarithm for Hilbert space valued martingales and its application to U-statistics. *Prob. Th. Rel. Fields 72*, 111–131.

Dehling, H. and Utev, S. A. (1996). The law of the iterated logarithm for degenerate U-statistics. In *Probability theory and Mathematical Statistics*, I. A. Ibragimov and A. Yu. Zaitsev eds., pp. 19–28. Gordon and Breach, New York.

de la Peña, V. H. (1988). L-Bounds of best possible type for martingales, degenerate U-statistics, and certain linear forms. Ph. D. dissertation, Dept. Statistics, Univ. California, Berkeley.

de la Peña, V. H. (1990). Bounds for the expectation of functions of martingales and sums of positive rv's in terms of norms of sums of independent random variables. *Proc. Amer. Math. Soc. 108*, 233–239.

de la Peña, V. H. (1992). Decoupling and Khintchine inequalities for U-statistics. *Ann. Probab. 20*, 1877–1892.

de la Peña, V. H. (1992a). Nuevas desigualdades para U-estadísticas y gráficas aleatorias. *Proceedings of IV Congreso Latinoamericano en Probabilidad y Estadística Matemática*, in *Contribuciones en Probabilidad y Estadística.* Edited by E. Cabaña, L. Gorostiza, P. Morettin, R. Rebolledo, V. Yohai and J. Ortega. Publisher, Instituto Nacional de Estadística, Geografía e Informática, Aguascalientes, Ags., México, (in Spanish). 290–296.

de la Peña, V. H. (1992b). Sharp bounds on the L_p norm of a randomly stopped multilinear form with an application to Wald's equation. *"Probability in Banach Spaces, VIII"*. Edited by R. M. Dudley, M. G. Hahn, and J. Kuelbs. Birkhauser, New York, New York. pp. 69–79.

de la Peña, V. H. (1993). Inequalities for tails of adapted processes with an application to Wald's lemma. *Journal of Theoretical Probability 6*, 285–302.

de la Peña, V. H. (1994). A Bound on the moment generating function of a sum of dependent variables with an application to simple random sampling without replacement. *Annales de L'Institute Henry Poincare. Probabilités et Statistiques 30*, 197–211.

de la Peña, V. H. (1995). Correction to: A Bound on the moment generating function of a sum of dependent variables with an application to simple random sampling without replacement. *Annales de L'Institute Henry Poincare. Probabilités et Statistiques 31*, 703–704.

de la Peña, V. H. (1997). A general class of exponential inequalities for martingales and ratios. *University of Aarhus Research Report Series*. No. 378. To appear in Ann. Probab.

de la Peña, V. H. (1997a). From boundary crossing of non-random functions to first passage times of processes with independent increments. Submitted.

de la Peña, V. H. (1998). Decoupling inequalities: A second generation of martingale inequalities in *Probability towards 2000*, L. Accardi and C. C. Heyde Eds. *Springer Lecture Notes in Statistics 128*, 151–164. Springer, New York.

de la Peña, V. H. (1998a). On Wald's equation and first exit times for randomly stopped processes with independent increments. *Proceedings of conference "Probability on Higher Dimensions"*. Oberwolfach, Germany, Summer 1996 in *Progress in Probability 43*, 277–286.

de la Peña, V. H. and Eisenbaum, N. (1994). Decoupling inequalities for the local times of linear Brownian motion. Preprint.

de la Peña, V. H. and Eisenbaum, N. (1997). Exponential Burkholder-Davis-Gundy inequalities. *Bulletin of the London Mathematical Society 29*, 239–242.

de la Peña, V. H., and Govindarajulu, Z. (1992). A note on a second moment of a randomly stopped sum of independent variables. *Statistics and Probability Letters 14*, 275–281.

de la Peña, V. H. and Klass, M. (1994). Order-of-magnitude bounds for expectations involving quadratic forms. *Ann. Probab. 22*, 1044–1077.

de la Peña, V. H. and Lai, T. L. (1997). Wald's equation and asymptotic bias of randomly stopped U-statistics. *Proceedings of the American Mathematical Society 125*, 917–925.

de la Peña, V. H. and Lai, T. L. (1997a). Moments of randomly stopped U-statistics. *Ann. of Probab. 25*, 2055–2081.

de la Peña, V. H. and Montgomery-Smith, S. (1994). Bounds for the tail probabilities of U-statistics and quadratic forms. *Bull. Amer. Math. Soc. 31*, 223–227.

de la Peña, V. H. and Montgomery-Smith, S. (1995). Decoupling inequalities for the tail probabilities of multivariate U-statistics. *Ann. Probab. 23*, 806–816.

de la Peña, V. H., Montgomery-Smith, S. and Szulga, J. (1994). Contraction and decoupling inequalities for multilinear forms in U-statistics. *Ann. Probab. 22*, 1745–1765.

Dilworth, S. J. and Montgomery-Smith, S. J. (1993). The distribution of vector valued Rademacher series. *Ann. Probab. 21*, 2046–2052.

Dudley, R. M. (1966). Weak convergence of probabilities on nonseparable Banach spaces and empirical measures on Euclidean spaces. *Illinois J. Math. 10*, 109–126.

Dudley, R. M. (1967). Measures on nonseparable metric spaces. *Illinois J. Math. 11* 449–453.

Dudley, R. M. (1967a). The size of compacts subsets of Hilbert space and continuity of Gaussian processes. *J. Functional Anal. 1*, 290–330.

Dudley, R. M. (1978). Central limit theorems for empirical measures. *Ann. Probab. 6*, 899–929.

Dudley, R. M. (1984). A course on empirical processes. *Lect. Notes in Math. 1097*, 1–142. Springer, New York.

Dudley, R. M. (1987). Universal Donsker classes and metric entropy. *Ann. Probab. 15*, 1306–1326.

Dudley, R. M. (1989). *Real Analysis and Probability*. Wadsworth & Brooks, Pacific Grove, California.

Dudley, R. M. (1990). Nonlinear functionals of empirical measures and the bootstrap. *Probability in Banach Spaces VII, Progess in Probability 21*, 63–82. Birkhäuser, Boston.

Dudley, R. M. (1998). *Uniform Central Limit Theorems*. Cambridge University Press. Cambridge, U.K. To appear.

Dudley, R. M. and Philipp, W. (1983). Invariance principle for sums of Banach space valued random elements and empirical processes. *Zeits. Wahrsch. v. Geb. 62*, 509–552.

Dümbgen, L. (1990). Limit theorems for the empirical simplicial depth. Inst. für Angewandte Mathematik, Universität Heidelberg. Preprint No. 581.

Dynkin, E. B. and Mandelbaum, A. (1983). Symmetric statistics, Poisson point processes and multiple Wiener integrals. *Ann. Statist. 11*, 739–745.

Etemadi, N. (1985). On some classical results in probability theory. *Sankhyā, Ser. A. 47*, 209–214.

Feller, W. (1966). An Introduction to Probability Theory and Its Applications, Vol. II. Wiley, New York.

Fernique, X. (1975). Régularité des trajectoires des functions aléatoires gaussiennes. *École d'Eté de Probabilités de St-Flour, 1974, Lecture Notes in Math. 480*, 1–96. Springer, Berlin.

Fernique, X. (1983). Régularité des functions aléatoires non gaussiennes. *École d'Eté de Probabilités de St-Flour, 1981, Lecture Notes in Math. 976*, 1–74. Springer, Berlin.

Figiel, T.; Hitczenko, P.; Johnson, W. B.; Schechtman, G.; and Zinn, J. (1997). Extremal properties of Rademacher functions with applications to the Khintchine and Rosenthal inequalities. *Trans. Amer. Math. Soc. 349*, 997–1027.

Frankl, P. (1983). On the trace of finite sets. *J. Comb. Theory 34*, 41–45.

Giné, E. (1997). Decoupling and limit theorems for U-statistics and U-processes. In *Lectures on Probability and Statistics, École d'Eté de Probabilités de Saint-Flour, XXVI-1996* (P. Bernard, ed.), pp. 1–35.

Giné, E. (1997a). Lectures on some aspects of the bootstrap. In *Lectures on Probability and Statistics, École d'Eté de Probabilités de Saint-Flour, XXVI-1996* (P. Bernard, ed.), pp. 37–152.

Giné, E. (1998). A consequence for random polynomials of a result of de la Peña and Montgomery-Smith. In *Probability in Banach Spaces 10, Progress in Probability, Vol. 43* pp. 104–110, Birhäuser Verlag, Basel.

Giné, E., Götze, F. and Mason, D. (1997). When is the student t-statistic asymptotically standard normal? *Ann. Probab. 25*, 1514–1531.

Giné, E. and Zhang, C. H. (1996). On integrability in the LIL for degenerate U-statistics. *J. Theoret. Probab. 9*, 385–412.

Giné, E. and Zinn, J. (1983). Central limit theorems and weak laws of large numbers in certain Banach spaces. *Zeits. Warsch. v. Geb. 62*, 323–354.

Giné, E. and Zinn, J. (1984). Some limit theorems for empirical processes. *Ann. Probab. 12*, 929–989.

Giné, E. and Zinn, J. (1986). Lectures on the central limit theorem for empirical processes. *Probability and Banach Spaces, Zaragoza 1985, Lecture Notes in Math. 1221*, 50–113.

Giné, E. and Zinn, J. (1990). Bootstrapping general empirical measures. *Ann. Probab. 18*, 851–869.

Giné, E. and Zinn, J. (1991). Gaussian characterization of uniform Donsker classes of functions. *Ann. Probab. 19*, 758–782.

Giné, E. and Zinn, J. (1992). On Hoffmann-Jørgensen's inequality for U-processes. *Probability in Banach Spaces 8*, pp. 273–291. Birkhäuser, Boston.

Giné, E. and Zinn, J. (1992a). Marcinkiewicz type laws of large numbers and convergence of moments for U-statistics. *Probability in Banach Spaces 8*, pp. 273–291. Birkhäuser, Boston.

Giné, E. and Zinn, J. (1994). A remark on convergence in distribution of U-statistics. *Ann. Probab. 22*, 117–125.

Goodman, V. (1997). A bounded LIL for second order U-statistics. To appear in *J. Theoret. Probab.*

Ghosh, M.; Mukhopadhyay, N. and Sen, P. K. (1997). *Sequential Estimation*. John Wiley & Sons Inc., New York.

Gregory, G. (1977). Large sample theory for U-statistics and tests of fit. *Ann. Statist. 5*, 110–123.

Gross, L. (1975). Logarithmic Sobolev inequalities. *Amer. J. Math. 97*, 1061–1083.

Gut, A. (1988). *Stopped Random Walks: Limit Theorems and Applications.* Springer-Verlag, New York.

Hall, P. and Heyde, C. C. (1980). *Martingale limit theory and its applications.* Academic Presss, Inc., New York.

Haagerup, U. (1982). The best constants in the Khintchine inequality. *Studia Math. 70*, 231–283.

Halmos, P. R. (1946). The theory of unbiased estimation. *Ann. Math. Statist. 17*, 34–43.

Helmers, J.; Janssen, P. and Serfling, R. (1988). Glivenko–Cantelli properties of some generalized empirical df's and strong convergence of generalized L-statistics. *Probab. Theory Rel. Fields 79*, 75–93.

Heinrich, P. (1996). Zero–one laws for polynomials in Gaussian chaos variables: a simple proof. *J. Theoret. Probab. 9*, 1019–1028.

Hill, T. P. (1982). Conditional generalizations of strong laws which conclude that partial sums converge a.s. *Ann. Probab. 10*, 828–830.

Hitczenko, P. (1988). Comparison of moments for tangent sequences of random variables. *Probab. Theory Rel. Fields 78*, 223–230.

Hitczenko, P. (1988a). On tangent sequences of UMD-space valued random vectors. Manuscript, Warsaw.

Hitczenko, P. (1989). On a conjecture of J. Zinn. *Comment. Math. 28*, 219–222.

Hitczenko, P. (1990). Best constants in the decoupling inequality for non-negative random variables. *Stat. Probab. Letters 9*, 327–329.

Hitczenko, P. (1990a). Best constants in Rosenthal's inequality for martingales. *Ann. Probab. 18*, 1656–1668.

Hitczenko, P. (1991). A remark on the paper "Martingale inequalities in rearrangement invariant function spaces" by W. B. Johnson and G. Schechtman. *Proc. Seminar Func. Anal., Longhorn Notes 1987–89*. Odell, E., Rosenthal, H.P. (eds.). *Lecture Notes in Mathematics 1470*, 177–182, Springer-Verlag.

Hitczenko, P. (1993). Domination inequality for martingale transforms of a Rademacher sequence. *Israel J. Math. 84*, 161–178.

Hitczenko, P. (1994). On the behavior of the constant in a decoupling inequality for martingales. *Proc. Amer. Math. Soc. 121*, 253–258.

Hitczenko, P. (1994a). On a domination of sums of random variables by sums of conditionally independent ones. *Ann. Probab. 22*, 453–468.

Hitczenko, P. (1994b). Sharp inequality for randomly stopped sums of independent non-negative random variables. *Stoch. Proc. Appl. 51*, 63–73.

Hitczenko, P. (1995). Two examples of randomly stopped sums of independent variables. *Interaction between Functional Analysis, Harmonic Analysis, and Probability, Lect. Notes in Pure and applied Mathematics 175*, 213–218, N. Kalton, E. Saab, S.M. Montgomery-Smith, eds, Marcel Dekker, New York.

Hitczenko, P. and Montgomery-Smith, S. J. (1996). Tangent sequences in Orlicz and rearrangement invariant spaces. *Math. Proc. Cambridge Phil. Soc. 119*, 91–101.

Hoeffding, W. (1948). A class of statistics with asymptotically normal distribution. *Ann. Math. Statist. 19*, 293–325.

Hoeffding, W. (1961). The strong law of large numbers for U-statistics. *Inst. Statist. Mimeo Ser. No. 302*, University of North Carolina. Chapel Hill, North Carolina.

Hoeffding, W. (1963). Probability inequalities for sums of bounded random variables. *J. Amer. Statist. Assoc. 58*, 13–30.

Hoffmann-Jørgensen, J. (1974). Sums of independent Banach space valued random variables. *Studia Math. 52*, 159–186.

Hoffmann-Jørgensen, J. (1991). *Stochastic Processes on Polish Spaces.* Aarhus Universitet, Matematisk Inst. Various Publication Series, 39. Aarhus, Denmark.

Hoffmann-Jørgensen, J. (1992). Asymptotic Likelihood Theory. In *Functional Analysis III* Aarhus Universitet, Matematisk Inst. Aarhus, Denmark.

Ibragimov, I. A. and Has'minski, R. Z. (1981). *Statistical Estimation: Asymptotic Theory*. Springer, Berlin.

Jacobson, N. (1951). *Lectures on Abstract Algebra*, Vol. 1. Van Nostrand, Princeton.

Jacod, J. (1984). Une généralisation des semimartingales: les processus admettant un processus a accroissements independants tangent. *Springer Lecture Notes in Math. 1059*, 91–118.

Jakubowski, A. (1986). Principle of conditioning in limit theorems for sums of random variables. *Ann. Probab. 14*, 902–915.

Johnson, W. B. and Schechtman, G. (1989). Sums of independent random variables in rearrangement invariant function spaces. *Ann. Probab. 17*, 789–818.

Johnson, W. B., Schechtman, G., Zinn, J. (1985). Best constants in moment inequalities for linear combinations of independent and exchangeable random variables. *Ann. Probab. 13*, 234–253.

Jones, F. (1993). *Lebesgue Integration on Euclidean Space*. Jones and Bartlett, Boston.

Kahane, J.-P. (1964). Sur les sommes vectorielles $\sum \pm u_n$. *Comptes Rendus Acad. Sci. Paris 259*, 2577–2580.

Kahane, J.-P. (1968). *Some Random Series of Functions*. Heath, Lexington, Massachusetts.

Khinchin, A. (1923). Uber dyadische Brüche. *Math. Zeits. 18*, 109–116.

Khinchin, A. and Kolmogorov, A. (1925). On the convergence of series. *Rec. Math. Soc. Moscow 32*, 668–677.

Klass, M. (1981). A method of approximating expectations of functions of sums of independent random variables. *Ann. Probab. 9*, 413–428.

Klass, M. (1988). A best possible improvement of Wald's equation. functions of sums of independent random variables. *Ann. Probab. 16*, 840–853.

Klass, M. (1990). Uniform lower bounds for randomly stopped Banach space-valued random sums. *Ann. Probab. 18*, 790–809.

Klass, M. and Nowicki, K. (1996). Order of magnitude bounds for expectations of Δ_2-functions of non-negative random bilinear forms and generalized U-statistics. *Ann. Probab. 25*, 1471–1501.

Koltchinskii, V. and Giné, E. (1996). Random matrix approximation of spectra of integral operators. Bernoulli, to appear.

Kou, S. G. and Ying, Z. (1995). Analysis of a sequence of 2×2 two tables. Preprint.

Kou, S. G. and Ying, Z. (1996). Asymptotics for a 2×2 table with fixed margins. *Statistica Sinica 6*, 809–829.

Krakowiak, W. and Szulga, J. (1988). Hypercontraction principle and multilinear forms in Banach spaces. *Probab. Theory Rel. Fields 77*, 325–242.

Krasnoselsky, M. A. and Rutitsky, Y. B. (1961). *Convex functions and Orlicz spaces*. Noordhof, 1961.

Kwapień, S. (1987). Decoupling for polynomial chaos. *Ann. Probab. 15*, 1062–1071.

Kwapień, s., Latała, R. and Oleszkiewicz, K. (1996). Comparison of moments of sums of independent random variables and differential inequalities. *J. Functional Analysis 136*, 258–268.

Kwapień, S. and Szulga, J. (1991). Hypercontraction methods in moment inequalities for series of independent random variables in normed spaces. *Ann. Probab. 19*, 369–379.

Kwapień, S. and Woyczynski, W. (1986). Decoupling of martingale transforms and stochastic integration for processes with independent increments, in *Probability Theory and Harmonic Analysis.* M. Dekker, Inc., 139–148.

Kwapień, S. and Woyczynski, W. (1989). Tangent sequences of random variables: Basic inequalities and their applications, in *Almost Everywhere Convergence,* 237–265. G. A. Edgar and L. Sucheston, Eds., Academic Press.

Kwapień, S. and Woyczynski, W. (1991). Semimartingale integrals via decoupling inequalities and tangent processes. *Probability and Math. Statistics 12*, 165–200.

Kwapień, S. and Woyczynski, W. (1992). *Random Series and Stochastic Integrals: Single and Multiple.* Birkhäuser, Boston.

Lai, T. L. and Wang, J. Q. (1993). Edgeworth expansions for symmetric statistics with applications to bootstrap method. *Statistica Sinica 3*, 517–542.

Latała, R. (1997). Estimation of moments of sums of independent random variables. *Ann. Probab. 25*, 1502–1513.

Latała, R. and Oleszkiewicz, K. (1994). On the best constant in the Khinchin-Kahane inequality. *Studia Math. 109*, 101–104.

Latała, R. and Oleszkiewicz, K. (1995). A note on the constants in the Khinchin–Kahane inequality. Unpublished preprint.

Latała, R. and Zinn, J. (1998). Strong laws of large numbers for U-statistics. Preprint.

Ledoux, M. and Talagrand, M. (1991). *Probability in Banach Spaces: Isoperimetry and Processes.* Springer-Verlag, Berlin.

Lee, A. J. (1990). *U-statistics: Theory and Practice.* Dekker, New York.

Levental, S. A. (1989). A uniform CLT for uniformly bounded families of martingale differences. *J. Theoret. Probab. 2*, 271–287.

Lindenstrauss, J. and Tzafriri, L. (1979). *Classical Banach Spaces, II.* Springer, Berlin.

Littlewood, J. E. (1930). On bounded bilinear forms in an infinite number of variables. *Quart. J. Math., Oxford Ser. 1*, 164–174.

Liu, R. V. (1990). On a notion of data depth based on random simplices. *Ann. Statist. 18*, 405–414.

Lynden-Bell, D. (1971). A method of allowing for known observational selection in small samples applied to 3CR quasars. *Mon. Not. R. Astr. Soc. 155*, 95–118.

Marcus, M. B. (1997). A sufficient condition for continuity of high order Gaussian chaos processes. Preprint.

Marcus, M. B. (1997a). Multiple Wick product chaos processes. Preprint.

Marcus, M. B. and Pisier, G. (1981). *Random Fourier Series with Applications to Harmonic Analysis.* Princeton University Press. Princeton, New Jersey.

Marcus, M. B. and Rosen, J. (1996). Gaussian chaos and sample path properties of additive functionals of symmetric Markov processes. *Ann. Probab. 24*, 1130–1177.

McConnell, T. (1989). Decoupling and stochastic integration in UMD Banach spaces. *Probab. and Math. Stat. 10*, 283–295.

McConnell, T. and Taqqu, M. (1986). Decoupling inequalities for multilinear forms in independent symmetric random variables. *Ann. Probab. 14*, 943–954.

McConnell, T. and Taqqu, M. (1987). Decoupling of Banach valued multilinear forms in independent symmetric Banach valued random variables. *Probab. Theory Rel. Fields 75*, 499–507.

Montgomery-Smith, S. J. (1990). The distribution of Rademacher sums. *Proc. Amer. Math. Soc. 109*, 517–522.

Montgomery-Smith, S. (1993). Comparison of sums of independent identically distributed random variables. *Prob. Math. Statist. 14*, 281–285.

Nelson, E. (1973). The free Markov field. *J. Funct. Anal. 12*, 211–227.

Neveu, J. (1968). *Processus Aléatoires Gaussiens.* Les Presses de l'Univ. de Montréal, Montréal, Canada.

Nolan, D. and Pollard, D. (1987). U-Processes: rates of convergence. *Ann. Statist. 15*, 780–799.

Nolan, D. and Pollard, D. (1988). Functional limit theorems for U-processes. *Ann. Probab. 16*, 1291–1298.

Ossiander, M. (1987). A central limit theorem under metric entropy with L_2-bracketing. *Ann. Probab. 15*, 897–919.

Ottaviani, G. (1939). Sulla teoria astratta del calcolo delle probabilità proposita dal Cantelli. *Giorn. Ist. Ital. Attuari 10*, 10–40.

Paley, R. E. A. C. and Zygmund, A. (1930). On some series of functions, (1). *Proc. Cambridge Philos. Soc. 26*, 337–357.

Paley, R. E. A. C. and Zygmund, A. (1932). A note on analytic functions in the unit circle. Proceedings Cambridge Phil. Soc. 28, 266–272.

Pinelis, I. (1994). Optimum bounds for the distribution of martingales in Banach spaces. *Ann. Probab. 22*, 1679–1706.

Pinelis, I. (1995). Sharp exponential inequalities for the martingales in the 2-smooth Banach spaces and applications to "scalarizing" decoupling. *Proceedings of the Ninth International Conference on Probability in Banach Spaces*, 55–70.

Pinelis, I. and Utev, S. A. (1989). Exact exponential estimates for sums of independent random variables. *Theory Probab. Appl. 34*, 340–346.

Pisier, G. (1975–76). Le théorème limite central et la loi du logarithme itéré dans les espaces de Banach. *Séminaire Maurey–Schwartz, 1975–76*, Exposés III et IV.

Pisier, G. (1976). Sur la loi du logarithme iteré dans les espaces de Banach. In *Probability in Banach Spaces, Lecture Notes in Math. 526*, pp. 203–210. Springer, Berlin.

Pisier, G. (1983). Some applications of the metric entropy condition to harmonic analysis. *Banach Spaces, Analysis and Probability, University of Connecticut, Lecture Notes in Math. 995*, 123–154. Springer, Berlin.

Pisier, G. (1985). Unpublished note on decoupling of multilinear forms in stable variables.

Pollard, D. (1982). A central limit theorem for k-means clustering. *Ann. Probab. 10*, 916–926.

Pollard, D. (1984). *Convergence of Stochastic Processes*. Springer-Verlag, New York.

Pollard, D. (1985). New ways to prove central limit theorems. *Econometric Theory 1*, 295–314.

Pollard, D. (1990). *Empirical Processes: Theory and Methods*. NSF–CBMS Regional Conference Series in Probability and Statistics, Vol. 2. Hayward, California.

Prohorov, Yu. V. (1956). Convergence of random processes and limit theorems in Probability Theory. *Theor. Probab. Appl. 1* 157–214.

Rényi, A. (1957). On the asymptotic distribution of the sum of a random number of independent random variables. *Acta Math. Acad. Sci. Hungar. 8*, 193–199.

Rosenthal, H. P. (1970). On the subspaces of L_p $(p > 2)$ spanned by sequences of independent random variables. *Israel J. Math. 8*, 273–303.

Rubin, M. and Vitale, R. A. (1980). Asymptotic distribution of symmetric statistics. *Ann. Statist. 8*, 165–170.

Sauer, N. (1972). On the density of families of sets. *J. Comb. Theory 13*, 145–147.

Sen, P. K. (1974). On L_p convergence of U-statistics. *Ann. Inst. Statist. Math. 26*, 55–60.

Serfling, R. J. (1971). The law of the iterated logarithm for U-statistics and related von Mises functionals. *Ann. Math. Statist. 42* 1794.

Serfling, R. J. (1980). *Approximation theorems of mathematical statistics*. Wiley, New York.

Shelah, S. (1972). A combinatorial problem: stability and order for models and theories in infinitary languages. *Pacific J. Math. 41*,247–261.

Sherman, R. P. (1994). Maximal inequalities for degenerate U-processes with applications to optimization estimators. *Ann. Statist. 22* 439–459.

Siegmund, D. (1985). *Sequential Analysis. Tests and Confidence Intervals*. Springer Series in Statistics. Springer-Verlag, New York.

Silverman, B. W. (1978). Distances on circles, toruses and spheres. *J. Appl. Probab. 15*, 136–143.

Stout, W. (1974). *Almost Sure Convergence*. Academic Press, New York.

Strobl, F. (1995). On the reversed sub-martingale property of empirical discrepancies in arbitrary sample spaces. *J. Theoret. Probab. 8*, 825–832.

Stute, W. (1993). Almost sure representations of the product limit estimator for truncated data. *Ann. Statist. 21*, 146–156.

Stute, W. (1994). Strong and weak representations of cumulative hazard functions and Kaplan-Meier estimators on increasing sets. *J. Statist. Planning and Inf. 42*, 315–329.

Sudakov, V. N. (1966). Contributed talk at the International Congress of Mathematicians, Moscow. Unpublished.

Sudakov, V. N. (1969). Gaussian measures, Cauchy measures and ε-entropy. *Soviet Math. Dokl. 12*, 310–313.

Szarek, S. J. (1976). On the best constants in the Khintchine inequality. *Studia Math. 58*, 197–208.

Szulga, J. (1993). Decoupling and exchangeability. Preprint.

Talagrand, M. (1987). Regularity of Gaussian processes. *Acta Math. 159*, 99–149.

Talagrand, M. (1987a). The Glivenko–Cantelli problem. *Ann. Probab. 15*, 837–870.

Talagrand, M. (1989). Isoperimetry and integrability of the sum of independent Banach space valued random variables. *Ann. Probab. 17* 1546–1570.

Talagrand, M. (1996). New concentration inequalities in product spaces. *Invent. Math. 126* 505–563.

Teicher, H. (1995). Moments of randomly stopped sums revisited. *J. Theoret. Probab. 8* 779–794.

Teicher, H. (1997). On the Marzinciewicz–Zygmund strong law for U-statistics. *J. Theoret. Probab.*, to appear.

Utev, S. A. (1985). Extremal problems in moment inequalities. In *Limit Theorems in Probability Theory*, Trudy Inst. Math., Novosibirsk. pp. 56–75.

van der Vaart, A. W. and Wellner, J. A. (1989). Prohorov and continuous mapping theorems in the Hoffmann-Jørgensen weak convergence theory with applications to convolution and asymptotic minimax theorems. Preprint, Department of Statistics, University of Waashington.

van der Vaart, A. W. and Wellner, J. A. (1996). *Weak Convergence and Empirical Processes*. Springer-Verlag, Berlin.

Vapnik, V. N. and Červonenkis. A. Ja. (1971). On the uniform convergence of relative frequencies of events to their probabilities. *Theory Prob. Appl. 16*, 164–280.

Vapnik, V. N. and Červonenkis. A. Ja. (1981). Necessary and sufficient conditions for the convergence of means to their expectation. *Theory Prob. Appl. 26*, 532–553.

Wald, A. (1945). Sequential analysis of statistical hypotheses. *Ann. Math. Stat. 16*, 117–186.

Wang, G. (1991). Sharp square-function inequalities for conditionally symmetric martingales. *Trans. Amer. Math. Soc. 328*, 393–419.

Woodroofe, M. (1985). Estimating a distribution function with truncated data. *Ann. Statist. 13*, 163–177.

Zhang, C.-H. (1996). Strong law of large numbers for sums of products. *Ann. Probab. 24* 1589–1615.

Zhang, C.-H. (1997). Some inequalities and strong laws of large numbers for positive and symmetrized U-statistics. To appear.

Zinn, J. (1985). Comparison of martingale difference sequences. In *Probability in Banach Spaces V, Lecture Notes in Math. 1153* 453–457.

Index